ADJUDICATING CLIMATE CHANGE

State, National, and International Approaches

Courts have emerged as a crucial battleground in efforts to regulate climate change. Over the past several years, tribunals at every level of government around the world have seen claims regarding greenhouse gas emissions and impacts. These cases rely on diverse legal theories, but all focus on government regulation of climate change or the actions of major corporate emitters. This book explores climate actions in state and national courts, as well as international tribunals, in order to explain their regulatory significance. It demonstrates the role that these cases play in broader debates over climate policy and argues that they serve as an important force in pressuring governments and emitters to address this crucial problem. As law firms and public interest organizations increasingly develop climate practice areas, this book serves as a crucial resource for practitioners, policymakers, and academics.

William C. G. Burns is the Class of '46 Visiting Professor with the Center for Environmental Studies at Williams College. Most recently, Dr. Burns was a Senior Fellow with the Center for Global Law & Policy at the Santa Clara University School of Law. Additionally, he serves as editor in chief of the *Journal of International Wildlife Law & Policy* and cochair of the International Environmental Law Committee of the American Branch of the International Law Association. He received his B.S. in political science from Bradley University and his Ph.D. in international law from the University of Wales – Cardiff School of Law. Prior to his academic career, he spent more than twenty years in the nongovernmental sector, including as executive director of the GreenLife Society/Pacific Center for International Studies, a think tank that focused on implementation of international wildlife law. He has published more than 70 articles in a range of law, policy, and science journals, including the *Georgetown International Environmental Law Review*, the *Journal of the American Medical Association*, and *Global Change*, and he has served as the coeditor of three books.

Hari M. Osofsky is an associate professor at Washington and Lee University School of Law. She received her B.A. and J.D. from Yale University. She currently is a Ph.D. student in the Department of Geography at the University of Oregon. Her articles have been published in a variety of journals, including the *Washington University Law Quarterly*, *Villanova Law Review*, *Chicago Journal of International Law*, *Stanford Environmental Law Journal*, *Stanford Journal of International Law*, *Virginia Journal of International Law*, and *Yale Journal of International Law*. Her advocacy work has included assisting with Earthjustice's annual submissions to the U.N. Human Rights Commission on environmental rights and with the Inuit Circumpolar Conference's petition on climate change to the Inter-American Commission on Human Rights. She also has taught climate change litigation courses that assisted the Southern Environmental Law Center and Western Environmental Law Center.

Adjudicating Climate Change

STATE, NATIONAL, AND INTERNATIONAL APPROACHES

Edited by
WILLIAM C. G. BURNS
Williams College

HARI M. OSOFSKY
Washington and Lee University

CAMBRIDGE UNIVERSITY PRESS
Cambridge, New York, Melbourne, Madrid, Cape Town, Singapore, São Paulo, Delhi

Cambridge University Press
32 Avenue of the Americas, New York, NY 10013-2473, USA

www.cambridge.org
Information on this title: www.cambridge.org/9780521879705

© Cambridge University Press 2009

This publication is in copyright. Subject to statutory exception
and to the provisions of relevant collective licensing agreements,
no reproduction of any part may take place without the written
permission of Cambridge University Press.

First published 2009

Printed in the United States of America

A catalog record for this publication is available from the British Library.

Library of Congress Cataloging in Publication data

Adjudicating climate change : state, national, and international approaches /
William C. G. Burns & Hari M. Osofsky (eds.).
    p. cm.
Includes bibliographical references and index.
ISBN 978-0-521-87970-5 (hardback)
1. Global warming – Law and legislation.   2. Liability for environmental damages.
3. Global warming – Law and legislation – United States.   4. Liability for
environmental damages – United States.   I. Burns, William C. G.   II. Osofsky,
Hari M., 1972–   III. Title.
K3593.A93 2009
344.04′6342 – dc22   2009000903

ISBN   978-0-521-87970-5 hardback

Cambridge University Press has no responsibility for the persistence or
accuracy of URLs for external or third-party Internet Web sites referred to in
this publication and does not guarantee that any content on such Web sites is,
or will remain, accurate or appropriate. Information regarding prices, travel
timetables, and other factual information given in this work are correct at
the time of first printing, but Cambridge University Press does not guarantee
the accuracy of such information thereafter.

Contents

Foreword		*page* vii
Peter E. Roderick		
Acknowledgments		ix
1	Overview: The Exigencies That Drive Potential Causes of Action for Climate Change William C. G. Burns and Hari M. Osofsky	1

PART I: SUBNATIONAL CASE STUDIES

2	State Action as Political Voice in Climate Change Policy: A Case Study of the Minnesota Environmental Cost Valuation Regulation Stephanie Stern	31
3	Litigating Climate Change at the Coal Mine Lesley K. McAllister	48
4	Cities, Land Use, and the Global Commons: *Genesis* and the Urban Politics of Climate Change Katherine Trisolini and Jonathan Zasloff	72
5	Atmospheric Trust Litigation Mary Christina Wood	99

PART II: NATIONAL CASE STUDIES

6	The Intersection of Scale, Science, and Law in *Massachusetts v. EPA* Hari M. Osofsky	129

7	Biodiversity, Global Warming, and the United States Endangered Species Act: The Role of Domestic Wildlife Law in Addressing Greenhouse Gas Emissions Brendan R. Cummings and Kassie R. Siegel	145
8	An Emerging Human Right to Security from Climate Change: The Case Against Gas Flaring in Nigeria Amy Sinden	173
9	Tort-Based Climate Litigation David A. Grossman	193
10	Insurance and Climate Change Litigation Jeffrey W. Stempel	230

PART III: SUPRANATIONAL CASE STUDIES

11	The World Heritage Convention and Climate Change: The Case for a Climate-Change Mitigation Strategy beyond the Kyoto Protocol Erica J. Thorson	255
12	The Inuit Petition as a Bridge? Beyond Dialectics of Climate Change and Indigenous Peoples' Rights Hari M. Osofsky	272
13	Bringing Climate Change Claims to the Accountability Mechanisms of International Financial Institutions Jennifer Gleason and David B. Hunter	292
14	Potential Causes of Action for Climate Change Impacts under the United Nations Fish Stocks Agreement William C. G. Burns	314
15	Climate Change Litigation: Opening the Door to the International Court of Justice Andrew Strauss	334
16	The Implications of Climate Change Litigation: Litigation for International Environmental Law-Making David B. Hunter	357
17	Conclusion: Adjudicating Climate Change across Scales Hari M. Osofsky	375

Index   387

Foreword

Peter E. Roderick*

The world's political process has been slow to react to the serious, and potentially catastrophic, consequences for life on our planet that flow from the burning of fossil fuel. In one sense, this is understandable: turning around the global energy base is not a simple task. In another sense, it is inexcusable: a myopic failure to act in the face of clear scientific evidence. And among those who have failed to act, until recently, I include the legal profession. But as the pages of this book demonstrate, the long slumber of the lawyers is over.

I was one of those fast asleep. In the late 1980s, long after scientists had been researching the problem, but with global awareness of climate change emerging, I was horrified to realize that as a legal adviser to Shell I was facilitating extraction of the hydrocarbons at the heart of the problem. The obvious answer was to leave the fossil fuel in the ground and to begin the arduous, yet critical, task of "decarbonizing" the world's economy. But I was naive to imagine that hope for such a turnaround would start with the very corporations whose legal structure drives their slavish servicing of the "demands" of the stock exchange.

It took me quite a while though to awaken fully. It was in 2001 that the Intergovernmental Panel on Climate Change published its finding that most of the observed warming at the Earth's surface over the past fifty years was likely to have been due to human activities. If the world's scientists were saying that human activities had led to temperature increases, with the qualitative nature of the effects well understood, then it was time for the courts to have something to say about it. This was the spur for Roda Verheyen and me to begin thinking about enforcement of the law around the world in order to combat climate change. This book tracks much of the development of climate change law in the five years since we scribbled down our thoughts, and spilt our wine, on the tablecloth of a North London restaurant. Alas, the restaurant is no more, but the development of climate change jurisprudence has moved on apace.

What is to be made of this jurisprudence? Its *origin* lies in the inadequate political and corporate response to the planet's biggest threat. Its *content* is a varied, innovative,

---

* Co-Director, Climate Justice Programme, http://www.climatelaw.org/.

barely formed mix across a spectrum of legal theories in domestic and international forums addressing both the causes and effects of climate change. Its *results*, to date, in purely legal and policy terms, include positive outcomes in the United States (for example, under the National Environment Policy Act, the Clean Air Act, and the Endangered Species Act), in Australia (on land use planning decisions in respect of coal mines), in Nigeria (on human rights violations from gas flaring), in Germany (on access to information on export credits), and at the UNESCO World Heritage Committee (adopting a world heritage and climate change strategy in July 2006); while negative judgements in the United States are under appeal with the support of some of the world's top climate scientists. Its *future* can be expected to include more damages cases, such as the one filed by the state of California against the automobile companies in September 2006, and if, as I hope, the May 2006 submission by the Group of 77 and China to the UNFCCC/Kyoto Protocol Compliance Committee concerning the noncompliance by fifteen Annex I countries with Article 3.1 of the Kyoto Protocol is a sign of the increased willingness of developing countries to hold the developed world to account, then perhaps the future will see some of the public international law avenues discussed in this book playing a more important role than they have so far.

The *implications* of the use of petitions and lawsuits to combat climate change should be judged by whether significant reductions in greenhouse gas emissions, and compensation for those who suffer from climate change damage, ensue. Although the effects of climate change are already upon us, I believe it is still possible to avoid the more serious consequences, but only if we make the right choices over the next few years. Until those choices are made, enforcing the law must play its full role.

Of course, going to court isn't the answer. It is rarely better in my experience than the lesser of two evils. The substance and procedure of the law usually favours the rich. Elitist language and a mismatch of resources too often shut out those who could benefit most from justice. And far better that political and corporate leaders would make decisions in the interests of life on the planet without being forced to do so.

But worse would be to allow these pervasive and entrenched imperfections to determine the outcome of the human response to climate change. The individuals, organizations, government entities, and lawyers who have brought these cases, several of whom have written the chapters of this book, are among those who are not prepared to do that, and I salute them. The ultimate justification for law is that it offers the possibility of resolving disputes without us killing each other. And there can be no bigger dispute than over the future of our planet.

Acknowledgments

This book has been vastly improved by the contributions of many people, and we would like to express our gratitude for their insights and efforts. Our editor at Cambridge, John Berger, has patiently shepherded this book through its many iterations as the landscape of climate change litigation evolved. We tremendously appreciate his supportiveness and assistance. Stefanie Herrington, who worked with us throughout her three years as a student at the University of Oregon School of Law, copyedited all of the chapters of the book. Her meticulousness has eliminated countless errors from the book and has made it read much more consistently. We also thank Shana Meyer and Mary Cadette, both project managers at Aptara, for their flexibility and contributions to the book's formatting and uniformity.

The chapter authors, all busy academics or practitioners deeply involved with this litigation, provided the thoughtful analyses of litigation that made this book possible. We particularly appreciate their patience and thoughtfulness as we updated the book over the course of its production for major developments.

We both benefitted from the support of our academic institutions over the course of this project. They provided crucial research support and helpful feedback that made this book possible. In particular, Wil Burns would like to thank Santa Clara University School of Law and the Monterey Institute of International Studies Department of International Environmental Policy, and Hari Osofsky would like to thank Washington and Lee University School of Law, University of Oregon School of Law and Department of Geography, and Whittier Law School.

Last, but certainly not least, we are both deeply appreciative of the loving and patient support of our families – specifically our partners, Tamar Meidav and Josh Gitelson, and our children, Shira Meidav-Burns, Oz Gitelson, and Scarlet Gitelson – over the course of this project. We are grateful for their tolerance of the many hours we spent writing and editing, and for their hugs and smiles to remind us of what really matters.

ADJUDICATING CLIMATE CHANGE

1

Overview: The Exigencies That Drive Potential Causes of Action for Climate Change

William C. G. Burns* and Hari M. Osofsky[†]

The implications of the crystallizing scientific understanding is that the planet is on the verge of dramatic climate change. It is still possible to avoid the most deleterious effects, but only if prompt actions are taken to stabilize global temperature close to its present value.[1]

INTRODUCTION

Over the course of the last few years, climate change litigation has been transformed from a creative lawyering strategy to a major force in transnational regulatory governance of greenhouse gas emissions. This book traces that journey and looks ahead to the future by considering a range of lawsuits and petitions filed in state, national, and international tribunals, as well as some potential causes of action. These actions cover an immense legal terrain but have in common their concern with more effective regulation of greenhouse gas emissions.

This introductory chapter frames the contributions in this book. It first provides an overview of climate change science, including both the current and the projected global impacts of climate change; second, it assesses current institutional responses to climate change and why they have been and likely will continue to be wholly inadequate to confront the looming threat of climate change in this century and beyond; third, it examines current efforts to open a new front to address climate change and climate change litigation; and finally, it provides a synopsis of the chapters that follow.

---

* Class of '46 Visiting Professor, Center for Environmental Studies, Williams College, Williamstown, Massachusetts, wburns@williams.edu, 650-281-9126.
[†] Associate Professor, Washington and Lee University School of Law, Lexington, Virginia, osofskyh@wlu.edu; 540-458-8259.
[1] James E. Hansen, *Dangerous Human-Made Interference with Climate*, Testimony to the Select Committee on Energy Independence and Global Warming, U.S. House of Representatives, Apr. 26, 2007, *available at* http://www.columbia.edu/~jeh1/testimony_26april2007.pdf (last visited May 26, 2008).

1. OVERVIEW OF CLIMATE CHANGE SCIENCE

The most recent assessment by the Intergovernmental Panel on Climate Change (IPCC)[2] concludes that global average surface temperatures have increased by 0.8°C over the last century, with the linear warming trend over the past fifty years twice that of the past century.[3] Moreover, the assessment concluded that "[m]ost of the observed increase in globally averaged temperatures since the mid-20th century is *very likely* due to the observed increase in anthropogenic greenhouse gas concentrations."[4] This section provides an overview of the scientific understanding of the growth of these emissions and its impacts.

The surface of the Earth is heated by solar radiation emanating from the sun at short wavelengths between 0.15 and 5 μm. Each square meter of the Earth receives an average of 342 watts of solar radiation throughout the year.[5] Approximately 26% of this radiation is reflected or scattered back to space by clouds and other atmospheric particles, and another 19% is absorbed by clouds, gases, and atmospheric particles.[6] Fifty-five percent of incoming solar energy passes through the atmosphere. Four percent is reflected from the surface back to space, with the remaining 51% reaching the Earth's surface. The heating of Earth's surfaces results in reradiation of

---

[2] The IPCC was established by the World Meteorological Organization and the United Nations Environment Program in 1988 to review and assess the most recent scientific, technical, and socioeconomic information related to the understanding of climate change, to evaluate proposals for reducing greenhouse gas emissions, and to assess the viability of response mechanisms. G.A. Res. 43/53, U.N. GAOR, 2d Comm., 43rd Sess., Supp. No. 49, at 133, U.N. Doc. A/43/49 (1989). The IPCC provides comprehensive Assessment Reports of the current knowledge and future projections of climate change at regular intervals. The reports are authored by teams of authors from throughout the world from universities, research centers, businesses, and nongovernmental organizations. There were more than 800 contributing authors to the latest report, and more than 2,500 scientific expert reviewers of the report. The First Assessment Report was published in 1990, the Second Assessment Report in 1995, the Third Assessment Report was released in 2001, and the Fourth Assessment Report (designated as "AR4") was released in four volumes throughout 2007. IPCC, *Fact Sheet* (2007), *available at* http://www.ipcc.ch/press/factsheet.htm (last visited May 10, 2007).

[3] *Technical Summary*, in CLIMATE CHANGE 2007: THE PHYSICAL SCIENCE BASIS 5 (S. Solomon et al. eds., 2007), *available at* http://www.ipcc.ch/ipccreports.ar4-wg1.htm (last visited May 25, 2008) [hereinafter THE PHYSICAL SCIENCE BASIS]. Atmospheric temperatures have been rising at a rate of approximately 0.2°C per decade over the past thirty years. James E. Hansen, Green Mountain Chrysler-Plymouth-Dodge-Jeep v. Thomas W. Torti, Nos. 2:05-CV-302 & 2:05-CV-304 (Consolidated), *Declaration of James E. Hansen*, (Vt., 2007), *available at* http://www.columbia.edu/~jeh1/case_for_vermont.pdf (last visited May 25, 2008).

[4] THE PHYSICAL SCIENCE BASIS, *supra* note 3, at 10. *See also* R. Somerville et al., *Historical Overview of Climate Change*, in THE PHYSICAL SCIENCE BASIS, supra note 3, at 105 ("human activities have become a dominant force, and are responsible for most of the warming observed over the past 50 years"). The IPCC defines the term "very likely" as a greater than 90% likelihood of occurrence/outcome. *Id.* at 121.

[5] INTERGOVERNMENTAL PANEL ON CLIMATE CHANGE, CLIMATE CHANGE 2001: THE SCIENTIFIC BASIS, Contribution of Working Group I to the Third Assessment Report of the Intergovernmental Panel on Climate Change 89 (2001) [hereinafter *Climate Change 2001-Scientific*].

[6] Physicalgeography.net, *The Greenhouse Effect*, http://www.physicalgeography.net/fundamentals/7h.html (last visited May 25, 2008).

approximately one-third of this energy, in the form of long-wave band (wavelengths of 3–50 μm) or infrared radiation.[7]

Some of the outgoing infrared radiation is absorbed by naturally occurring atmospheric gases – principally water vapor (H_2O), as well as carbon dioxide (CO_2), ozone (O_3), methane (CH_4), nitrous oxide (N_2O), and clouds.[8] This absorption is termed the "natural greenhouse effect" because these gases, which are termed "greenhouse gases," operate much like a greenhouse: they are "transparent" to incoming shortwave radiation, but "opaque" to outgoing infrared radiation, trapping a substantial portion of such radiation and reradiating much of this energy to the Earth's surface, increasing surface temperatures.[9] While greenhouse gases constitute only 1% of the atmosphere,[10] they are critical to the sustenance of life on Earth, elevating surface temperatures by about 33°C.[11]

Prior to the Industrial Revolution, atmospheric concentrations of naturally occurring greenhouse gases had been relatively stable for 10,000 years.[12] As a consequence, the net incoming solar radiation at the top of the atmosphere was roughly balanced by the net outgoing infrared radiation.[13] However, with the advent of fossil fuel–burning plants to support industry, automobiles, and the energy demands of modern consumers, as well as substantial expansion of other human activities, including agricultural production, "humans began to interfere seriously in the composition of the atmosphere"[14] by emitting large amounts of additional greenhouse gases. The human-driven buildup of greenhouse gases in the atmosphere has resulted in "radiative forcing"; that is, increased levels of these gases result in greater absorption of outgoing infrared radiation, and ultimately an increase in temperatures when a portion of this radiation is reradiated to the Earth's surface.[15]

The most important anthropogenic greenhouse gas over the past two centuries has been carbon dioxide, which is primarily attributable to fossil fuel

---

[7] Somerville et al., *supra* note 3, at 96; Intergovernmental Panel on Climate Change, *Radiative Forcing of Climate Change* 7 (1994) [hereinafter *Radiative Forcing*].
[8] Thomas R. Karl & Kevin E. Trenberth, *Modern Global Climate Change*, 302 SCI. 1719, 1719 (2003).
[9] Stephen H. Schneider, *The Greenhouse Effect: Science and Policy*, 243 SCI. 771, 772 (1989).
[10] UNFCCC Secretariat, *The Greenhouse Effect and the Carbon Cycle*, available at http://unfccc.int/essential_background/feeling_the_heat/items/2903.php (last visited May 25, 2008).
[11] *Id.*
[12] Haroon S. Kheshgi, Steven J. Smith & James A. Edmonds, *Emissions and Atmospheric CO_2 Stabilization*, in 10 MITIGATION & ADAPTATION STRATEGIES FOR GLOBAL CHANGE 213, 214 (2005).
[13] John R. Justus & Susan R. Fletcher, *Global Climate Change*, CRS Issue Brief for Congress 3 (Aug. 13, 2001), *available at* http://www.ncseonline.org/NLE/CRSreports/Climate/clim-2.cfm?&CFID=13638750&CFTOKEN=63020586 (last visited May 25, 2008).
[14] Fred Pearce, *World Lays Odds on Global Catastrophe*, NEW SCI., Apr. 8, 1995, at 4.
[15] UNEP, *Vital Climate Change Graphics* 10 (2005).

> The earth then is radiating less energy to space than it absorbs from the sun. This temporary planetary energy imbalance results in the earth's gradual warming... Because of the large capacity of the oceans to absorb heat, it takes the earth about a century to approach a new balance – that if, for it to once again receive the same amount of energy from the sun it radiates to space. And of course the balance is reset at a higher temperature.

> *See also* James Hansen, *Defusing the Global Warming Time Bomb*, SCI. AM., Mar. 2004, at 71.

combustion,[16] cement production, and land-use change.[17] Carbon dioxide has accounted for 90% or more of the increased greenhouse gas climate forcing in recent years.[18] Since 1751, more than 297 billion metric tons of carbon have been released into the atmosphere from anthropogenic sources, with half of the emissions occurring since 1978.[19] Atmospheric concentrations of carbon dioxide were approximately 280 parts per million (ppm) at the start of the Industrial Revolution in the 1780s. While it took a century and a half to reach atmospheric concentrations of 315 ppm, the trend accelerated in the twentieth century, reaching 360 ppm by the 1990s, and 384 ppm currently,[20] which exceeds atmospheric levels for at least the last 650,000 years,[21] and most likely the last 20 million years.[22]

---

[16] Consumption of crude oil and coal account for almost 77% of fossil fuel carbon dioxide emissions. Climate Change Science Program & Subcommittee on Global Change Research, *Our Changing Planet: The U.S. Climate Change Science Program for Fiscal 2007* 117 (2007). Energy-related carbon dioxide emissions have risen 130-fold since 1850. Pew Center on Global Climate Change, *Climate Change 101: Understanding and Responding to Global Climate Change* 34 (2006), *available at* http://www.pewclimate.org/docUploads/Climate101-FULL_121406_065519.pdf (last visited May 25, 2008). "Worldwide use of coal, oil, and natural gas in 2005 led to the emission of about 7.5 gigatonnes of carbon (GtC) in CO2, an amount that continues to increase year by year." Rosina Bierbaum et al., *Confronting Climate Change: Avoiding the Unmanageable and Managing the Unavoidable*, Scientific Expert Group Report on Climate Change and Sustainable Development 12 (2006), *available at* http://www.unfoundation.org/files/pdf/2007/SEG_Report.pdf (last visited May 25, 2008).

[17] "The additional release in recent years from deforestation and land-use change, mainly in tropical regions, has been estimated variously at between 0.7 GtC/year and 3.0 GtC/year in CO2 . . . a mid-range value of 1.5 GtC/year is often cited." Bierbaum et al., *supra* note 16, at 12–13. This constitutes 20–25% of anthropogenic greenhouse gas emissions. Chatham House/Royal Society for the Protection of Birds, *Workshop on Reducing Emissions from Tropical Deforestation*, Summary Report 1 (2007), *available at* http://www.chathamhouse.org.uk/files/9165_160407workshop.pdf (last visited May 25, 2008); Raymond E. Gullison et al., *Tropical Forests and Climate Change*, 316 SCI. 985, 985 (2007). Deforestation also contributes to warming trends by eliminating possible increased storage of carbon and decreasing evapotranspiration. G. Bala et al., *Combined Climate and Carbon-Cycle Effects of Large-Scale Deforestation*, 104(16) PROC. NAT'L ACAD. SCI. 6550, 6550 (2007). However, deforestation exerts a cooling effect, particularly in seasonally snow-covered high latitudes, by decreasing the albedo (reflectivity) of surfaces. *Id.*

[18] James Hansen & Makiko Sato, *Greenhouse Gas Growth Rates*, 101(46) PROC. NAT'L ACAD. SCI. 16,109, 16,111 (2004).

[19] Climate Change Science Program & Subcommittee on Global Change Research, *Our Changing Planet: The U.S. Climate Change Science Program for Fiscal Year 2007* 117 (2006).

[20] Eric Steig, *The Lag between Temperature and CO2*, RealClimate, Apr. 27, 2007, *available at* http://www.realclimate.org/index.php/archives/2007/04/the-lag-between-temp-and-co2/ (last visited on June 2, 2008). Approximately half of carbon dioxide emissions since 1751 have occurred since 1978. Climate Change Science Program & Subcommittee on Global Change Research, *supra* note 19, at 117. Carbon dioxide emissions grew 80% between 1970 and 2004. IPCC, Working Group III contribution to the Intergovernmental Panel on Climate Change Fourth Assessment Report, CLIMATE CHANGE 2007: MITIGATION OF CLIMATE CHANGE, SUMMARY FOR POLICYMAKERS 3 (2007), *available at* http://www.ipcc.ch/pdf/assessment-report/ar4/wg3/ar4-wg3-spm.pdf (last visited May 25, 2008) [hereinafter MITIGATION OF CLIMATE CHANGE]. Between 2006 and 2007, carbon dioxide emissions rose a startling 20 percent. Malte Meinhaussen et al., *Greenhouse Gas Emissions Targets for Limiting Global Warming to 20°*, 458 NATURE 1158, 1160 (2009).

[21] THE PHYSICAL SCIENCE BASIS, *supra* note 3, at 4.

[22] CNA Corporation, *National Security and the Threat of Climate Change* 56 (2007), *available at* http://www.securityandclimate.cna.org/report/National%20Security%20and%20the%20Threat%20of%20Climate%20Change.pdf (last visited May 25, 2008).

Nitrous oxide emissions, primarily generated through fertilizer production and industrial processes, account for approximately 5% of greenhouse gas forcing in recent years.[23] Atmospheric concentrations of nitrous oxides rose from a preindustrial value of 270 parts per billion (ppb) to 319 ppb in 2005.[24]

Methane emissions, generated primarily through rice cultivation, ruminants, energy production, and landfills, account for approximately 4% of greenhouse gas forcing in recent years.[25] Atmospheric concentrations of methane have increased 153% from preindustrial levels, reaching 1,774 ppb in 2005. This far exceeds the natural range of the last 650,000 years.[26] Overall, the global emissions of the six primary anthropogenic greenhouse gases rose 70% between 1970 and 2004.[27]

The increasing emissions translate into tangible human impacts. The World Health Organization has estimated that warming and precipitation trends over the past thirty years associated with anthropogenic climate change have claimed 150,000 lives annually, primarily attributable to human disease and malnutrition.[28] Recent studies have linked the significant increase in violent weather events over the past several decades to increases in sea surface temperature associated with climate change.[29] Other expressions of climate change include "increasing ground instability of permafrost regions... shifts in ranges and changes in algal, plankton and fish abundance in high-latitude oceans... [and] poleward and upward shifts in ranges in plant and animal species...."[30]

Overall, warming is undoubtedly exerting a substantial and pervasive influence on the globe. As the IPCC recently concluded, "[o]f the more than 29,000 observational data series, from 75 studies, that show significant change in many physical and biological systems, more than 89% are consistent with the direction of change expected as a response to warming."[31] Physical system responses to climate change

---

[23] Hansen & Sato, *supra* note 18, at 16,111.
[24] Intergovernmental Panel on Climate Change, *supra* note 5, at 4.
[25] Hansen & Sato, *supra* note 18, at 16,111.
[26] Intergovernmental Panel on Climate Change, *supra* note 7, at 4. Overall, emissions of the six primary greenhouse gases generated by anthropogenic sources increased 75% between 1970 and 2004. Netherlands Environmental Assessment Agency, *Global Greenhouse Gas Emissions Increased 75% since 1970* (Nov. 13, 2006), *available at* http://www.mnp.nl/en (last visited May 25, 2008).
[27] *Climate Change 2001-Scientific, supra* note 5, at 3.
[28] Jonathan A. Patz et al., *Impact of Regional Climate Change on Human Health*, 438 NATURE 310, 310 (2005).
[29] Greg J. Holland & Peter J. Webster, *Heightened Tropical Cyclone Activity in the North Atlantic: Natural Variability or Climate Trend?*, 365 PHIL. TRANSACTIONS ROYAL SOC'Y A 2695 (2007); K. Emanuel, *Increasing Destructiveness of Tropical Cyclones over the Past 30 Years*, 436 NATURE 686–88 (2005).
[30] Intergovernmental Panel on Climate Change, CLIMATE CHANGE 2007: IMPACTS, ADAPTATION AND VULNERABILITY, Working Group II Contribution to the Intergovernmental Panel on Climate Change Fourth Assessment Report, Summary for Policymakers 2, *available at* http://www.ipcc.ch/pdf/assessment-report/ar4/wg2/ar4-wg2-spm.pdf (last visited on June 2, 2008) [hereinafter IMPACTS, ADAPTATION AND VULNERABILITY].
[31] *Id.* at 2. *See also* Cynthia Rosenzweig et al., *Attributing Physical and Biological Impacts to Anthropogenic Climate Change*, 453 NATURE 353, 353–54 (May 2008) (stating that in a study of 29,500 data series "[n]inety-five per cent of the 829 documented physical changes have been in directions consistent with warming").

over the past three decades include shrinking glaciers on every continent, melting permafrost, shifts in the spring peaks of river discharge, and coastal erosion. Biological effects include phonological changes (such as the timing of blooming of fauna, species' migration and reproduction), and changes in community structure.[32]

However, the greatest trepidation of climate scientists lies in the outlook for this century and beyond, as atmospheric concentrations of greenhouse gases continue to rise. Absent aggressive global efforts to reduce greenhouse gas emissions, atmospheric concentrations of carbon dioxide may reach twice preindustrial levels by as early as 2050,[33] and perhaps triple by the end of the century.[34] The latest assessment by the IPCC projects that a doubling of atmospheric concentrations of carbon dioxide from preindustrial levels is likely to result in temperature increases in the range of 2°–4.5°C, with a best estimate of 3°C.[35] This projection is remarkably consistent with paleoclimatic evidence. "[E]mpirical data climate change over the past 700,000 years yields a climate sensitivity of $\frac{3}{4}$°C for each W/m^2 of forcing, or 3°C for a 4 W/m^2 forcing.[36] However, the time line for these projections may prove to be far too sanguine given a "shocking" rise in global energy demand in the past few years, according to the International Energy Agency (IEA) in its most recent *World Energy Outlook*.[37] The IEA report concludes that world energy demand has accelerated rapidly during this decade, primarily attributable to breakneck economic growth in China and India, and that world energy needs could be 50% higher in 2030 than today.[38] As a consequence, the IEA projects that atmospheric concentrations

---

[32] Rosenzweig et al., *supra* note 31 at 353.
[33] Hansen, *supra* note 1, at 4.
[34] David Talbot, *The Dirty Secret*, TECH. REV. (July/Aug. 2006), *available at* http://www.technologyreview.com/Energy/17054 (last visited May 25, 2008); Stephen F. Lincoln, *Fossil Fuels in the 21st Century*, 34(8) AMBIO 621, 621 (2005).
[35] Intergovernmental Panel on Climate Change, *supra* note 7, at 12. See also Bierbaum et al., *supra* note 16, at x:

> If CO2 emissions and concentrations grow according to mid-range projections, moreover, the global average surface temperature is expected to rise by 0.2°C to 0.4°C per decade throughout the 21st century and would continue to rise thereafter. The cumulative warming by 2100 would be approximately 3°C to 5°C over preindustrial conditions.

[36] Hansen, *supra* note 3, at 7. As Hansen notes, paleoclimatic data is particularly compelling because it also includes any cloud feedbacks that may exist. Cloud feedbacks are recognized by most climatologists as the largest source of uncertainty in climatic modeling. Intergovernmental Panel on Climate Change, *supra* note 7, at 4; Richard A. Kerr, *Three Degrees of Consensus*, 305 SCI. 932, 933 (2004).
[37] *IEA Predicts 'Shocking' Rise in Global Energy Demand*, Environmental Finance Online News, Nov. 8, 2007, *available at* http://www.environmental-finance.com/onlinews/1108iea.html (last visited May 25, 2008). *See also* Josep G. Canadell, *Contributions to Accelerating Atmospheric CO$_2$ Growth from Economic Activity, Carbon Intensity, and Efficiency of Natural Sinks*, PROC. NAT'L ACAD. SCI. EARLY ED., Nov. 13, 2007, http://www.pnas.org/cgi/doi/10.1073/pnas.0702737104 (last visited May 25, 2008) (global carbon dioxide emissions rate increased from 1.3% in the 1990s to 3.3% annually in 2000–2006).
[38] International Energy Agency, World Energy Outlook 2007: China and India Insights 41 (2007), http://www.iea.org/npsum/weo2007sum.pdf (last visited Nov. 15, 2007). The United States, China, and India are slated to construct an additional 850 coal-fired plants by 2012. These plants are projected to produce an additional 2.7 billion tons of carbon dioxide, while the Kyoto Protocol only requires its Parties to reduce their emissions by about 483 million tons. Mark Clayton, *New Coal Plants Bury 'Kyoto,'*

of carbon dioxide could rise to levels that would produce a 6°C increase in global temperatures by 2030.[39]

Moreover, the IPCC's most recent assessment's midrange scenario projects that sea levels will rise between 18 and 59 centimeters (7–23 inches) during the remainder of this century as a consequence of projected warming.[40] However, there is a very real possibility that sea levels will rise much more than this given potential dynamical responses of ice sheets in Greenland and West Antarctica,[41] which may exert substantial positive feedbacks on sea level rise over the next century and beyond.[42]

A recent study that incorporates ice dynamics projects that sea levels will rise between 0.8 and 2.0 meters,[43] "the highest estimates of sea level rise by 2100 that has been published in the literature to date."[44] In the longer term, if annual temperatures increase by more than 3°C in the Antarctic region, which is highly likely by the end of this century, one study projects that globally averaged sea levels could increase by 7 *meters* over a period of 1,000 years or more.[45]

Consistent scientific evidence predicts that climate change will have dire implications for both natural systems and human institutions. In the context of natural systems, the IPCC's most recent assessment concluded that "the resilience of many ecosystems is likely to be exceeded this century by an unprecedented combination of climate change, associated disturbances (e.g., flooding, drought, wildfire, insects, ocean acidification), and other global change drivers (e.g., land use change, pollution, overexploitation of resources)."[46] For example, coral reefs have extremely narrow temperature tolerances of between 25 and 29°C, with some species in Pacific

---

CHRISTIAN SCI. MONITOR, Dec. 23, 2004, *available at* http://www.csmonitor.com/2004/1223/p01s04-sten.html (last visited May 25, 2008).

[39] *IEA Predicts 'Shocking' Rise in Global Energy Demand*, *supra* note 37.

[40] G. A. Meehl, et al., *Global Climate Change Projections*, in IPCC Fourth Assessment Report, THE PHYSICAL SCIENCE BASIS (2007), at 820, http://ipcc-wg1.ucar.edu/wg1/Report/AR4WG1_Ch10.pdf (last visited May 20, 2007). Rising sea levels associated with climate change are attributable primarily to thermal expansion of ocean waters due to warming and glacial melting. Hansen, *supra* note 1, at 16.

[41] A persuasive case is made by Hansen that the IPCC in its Fourth Assessment Report failed to adequately take into account multiple positive feedbacks that could occur in Greenland and West Antarctica should temperatures rise by 2–3°C. These include "reduced surface albedo, loss of buttressing ice shelves, dynamical response of ice streams to increased melt-water and lower sea surface ice altitude," all of which result in massive rises in sea level within a few centuries. James Hansen et al., *Global Temperature Change*, 103(39) PROC. NAT'L ACAD. SCI. 14,288, 14,292 (2006).

[42] J.E. Hansen, *Scientific Reticence and Sea Level Rise*, 3 ENVTL. RES. LETTERS 1, 4 (2007); James Hansen et al., *Climate Change and Trace Gases*, 365 PHIL. TRANSACTIONS ROYAL SOC'Y A 1925, 1936 (2007); Michael Oppenheimer et al., *The Limits of Consensus*, 317 SCI. 1505, 1505 (2007).

[43] W.T. Pfeffer, et al., *Kinematic Constraints on Glacier Contributions to 21st Century Sea-Level Rise*, 321 SCI. 1340, 1342 (2008).

[44] *How Much Will Sea Level Rise?*, RealClimate.org, Sept. 4, 2008, http://www.realclimate.org/index.php?p=598 (last visited Sept. 5, 2008).

[45] Jonathan M. Gregory, Philippe Huybrechts & Sarah C.B. Raper, *Threatened Loss of the Greenland Ice-Sheet*, 428 NATURE 616, 616 (2004). *See also* Julian A. Dowdeswell, *The Greenland Ice Sheet and Global Sea-Level Rise*, 311 SCI. 963, 963 (2006). Hansen also concluded that a 2–3°C increase in temperatures could ultimately result in sea level rise of 25 *meters* over the course of the next few hundred years. *Id.* at 21.

[46] IMPACTS, ADAPTATION AND VULNERABILITY, *supra* note 30, at 5.

island developing countries (PIDCs) currently living near their threshold of thermal tolerance.[47] Projected sea temperature rises in the Pacific region over the next century are likely to result in a "catastrophic decline" in coral cover.[48] Loss of coral reefs could have similar implications in other regions, including the Indian Ocean and Caribbean Sea.[49] Overall, the World Bank has estimated that 50% of the subsistence and artisanal fisheries will be lost in regions where coral reefs die due to coral bleaching attributable to climate change.[50] The massive infusion of carbon dioxide into the world's oceans associated with the growth of anthropogenic emissions also may result in serious declines in coral reef calcification rates, further contributing to their destruction.[51]

In addition, forest ecosystems may be negatively impacted by climate change. Climate change may drive changes in floristic composition in some regions, resulting in changes in forest composition. This could result in the decline of species that sustain assemblages of pollinators, herbivores, symbiotic fungi, and other important species in regions such as the Amazon.[52] In some cases, the loss of key tree species could result in the collapse of entire forest ecosystems.[53]

Climate change may adversely impact a wide array of species through, inter alia, habitat alteration and destruction, changes in phenology (the relationship between climate and periodic biological phenomena, such as hibernation or migration), and direct temperature effects.[54] The IPCC in its Fourth Assessment Report concluded

---

[47] William C.G. Burns, *The Potential Impacts of Climate Change on Pacific Island State Ecosystems*, Occasional Paper of the Pacific Institute for Studies in Development, Environment, and Security, at 4 (Mar. 2000).

[48] Brian C. O'Neill & Michael Oppenheimer, *Climate Change Impacts Are Sensitive to the Concentration Stabilization Path*, 101(47) PROC. NAT'L ACAD. SCI. 16,411, 16,414 (2004) ("Model studies suggest that Earth may enter an era of sustained bleaching and widespread demise of reefs if global mean temperature increases by >1°C from recent levels").

[49] John P. McWilliams et al., *Accelerating Impacts of Temperature-Induced Coral Bleaching in the Caribbean*, 86(8) ECOLOGY 2055, 2059 (2005) (projected warming in the Caribbean could result in "maximum bleaching extent (i.e., 100% of coral-bearing cells) and maximum bleaching intensity (100% of coral colonies)"; Simon D. Donner et al., *Global Assessment of Coral Bleaching and Required Rates of Adaptation Under Climate Change*, 11 GLOBAL CLIMATE CHANGE BIOLOGY 2251, 2256–57 (2005) (severe coral bleaching events could occur every three to five years by 2030 in the majority of the world's coral reefs, and become a biannual event by 2050); Charles R.C. Sheppard, *Coral Decline and Weather Patterns over 20 Years in the Chagos Archipelago, Indian Ocean*, 28(6) AMBIO 472, 475 (1999).

[50] O. Hoegh-Guldberg et al., *Pacific in Peril*, GREENPEACE 54 (Oct. 2000), available at http://www.greenpeace.org/raw/content/australia/resources/reports/climate-change/coral-bleaching-pacific-in-pe.pdf (last visited May 26, 2008).

[51] See William C.G. Burns, *Potential Causes of Action for Climate Change Impacts under the United Nations Fish Stocks Agreement*, in this volume.

[52] William F. Laurance et al., *Pervasive Alteration of Tree Communities in Undisturbed Amazonian Forests*, 428 NATURE 171, 174–75 (2004).

[53] Frank Ackerman & Elizabeth Stanton, *Climate Change: The Costs of Inaction*, REPORT TO FRIENDS OF THE EARTH ENGLAND, WALES AND NORTHERN IRELAND 23 (2006), available at http://www.foe.co.uk/resource/reports/econ_costs_cc.pdf (last visited May 25, 2008).

[54] See James Battin, *Projected Impacts of Climate Change on Salmon Habitat Restoration*, 104 PROC. NAT'L ACAD. SCI. 6720–25 (2007); Mark B. Bush, Miles R. Silman & Dunia H. Urrego, *48,000 Years of Climate and Forest Change in a Biodiversity Hotspot*, 303 SCI. 827, 829 (2004); Andrew R. Blaustein et al., *Amphibian Breeding and Climate Change*, 15(6) CONSERVATION BIOLOGY 1804–09 (2001).

that 20–30% of species would likely face an increased risk of extinction if globally averaged temperatures rise 1.5–2.5°C above 1980–1999 levels, and that 40–70% of species could be rendered extinct should temperature increases exceed 3.5°C.[55] Thus, climate change may pose the greatest global threat to biodiversity in most regions of the world by the middle or latter part of this century.[56]

In terms of human impacts, 100 million people may be imperiled by coastal flooding even under the middle range of projections,[57] with the very future of many small island nations potentially hanging in the balance.[58] Should sea level ultimately rise 4–6 meters, the results would be "globally catastrophic,"[59] resulting in the inundation of large parts of many major cities, including New York, London, Sydney, Vancouver, Mumbai, and Tokyo.[60] "In Florida, Louisiana, the Netherlands, Bangladesh and elsewhere, whole regions and cities may vanish. China's economic powerhouse, Shanghai, has an average elevation of just 4 metres."[61]

There is also likely to be a substantial increase in the incidence of a wide array of deadly diseases. This includes vector-borne infectious diseases such as malaria and dengue fever,[62] as well as water-borne diseases such as cholera and hepatitis A.[63] A 2°C increase in temperature, for example, could lead to 40–60 million additional cases of malaria in Africa and millions of additional deaths.[64]

Global food production potential is anticipated to rise over a range of 1–3°C temperature increases.[65] However, increased temperatures and regional declines in precipitation could exacerbate conditions in arid and semiarid regions,[66] resulting

---

[55] T. FLANNERY, THE WEATHER MAKERS 116 (2005); Craig D. Thomas et al., *Extinction Risk from Climate Change*, 427 NATURE 145, 146–47 (2004).

[56] Thomas et al., *supra* note 55, at 146–47.

[57] Climate Action Network, *Preventing Dangerous Climate Change* 6 (2002).

[58] William C.G. Burns, *Potential Implications of Climate Change for the Coastal Resources of Pacific Island Developing Countries and Potential Legal and Policy Responses*, 8(1) HARV. ASIA-PAC. REV. 1–8 (2005); William C.G. Burns, *The Possible Impacts of Climate Change on Pacific Island State Ecosystems*, Occasional Paper of the Pacific Institute for Studies in Development, Environment & Security 1–19 (Mar. 2000).

[59] Richard A. Kerr, *Global Warming Is Changing the World*, 316 SCI. 188, 190 (2007).

[60] James Hansen, *Climate Catastrophe*, NEW SCI. 33, July 30, 2007, *available at* http://pubs.giss.nasa.gov/docs/2007/2007_Hansen_2.pdf (last visited May 25, 2008).

[61] *Id.* Sea level rise of several meters could compel more than one billion inhabitants to retreat inland "or face exile." *See also* Sujatha Byravan & Sudhir Chella Rajan, *Providing New Homes for Climate Change Exiles*, 6 CLIMATE POL'Y 247, 247 (2006).

[62] IMPACTS, ADAPTATION AND VULNERABILITY, *supra* note 30, ch. 9, Human Health, at sec. 9.7.1.1 (Number of people living in potential transmission zone of malaria may increase by 260–320 million by 2080); John E. Hay et al., *Climate Variability and Change and Sea-Level Rise in the Pacific Islands Region*, SOUTH PACIFIC REGIONAL ENVIRONMENT 69 (2003), *available at* http://www.sprep.org/climate/documents/webi.pdf (last visited May 25, 2008); William C.G. Burns, *Climate Change and Human Health, The Critical Policy Agenda*, 287(17) J. AM. MED. ASS'N 287, 287 (2002).

[63] IPCC, *supra* note 7, at sec. 9.5.1.

[64] Paul Reiter, *Climate Change and Mosquito-Borne Disease*, 109 ENVTL. HEALTH PERSP. 1, 1 (2001).

[65] IMPACTS, ADAPTATION AND VULNERABILITY, *supra* note 30, at 8.

[66] Papua New Guinea & Pacific Island County Unit, The World Bank, *Cities, Seas, and Storms*, in 4 ADAPTING TO CLIMATE CHANGE 13, Nov. 13, 2000; UNEP Information Unit On Climate Change, *Climate Change Scenarios: Why the Poor Are the Most Vulnerable*, Fact Sheet No. 111 (May 1993).

in substantial declines in crop production in many developing nations.[67] This could be especially disastrous in Africa, where close to half of the currently 800 million undernourished people reside.[68] The IPCC in its most recent assessment indicates that yields from rain-fed agriculture could decline by up to 50% by 2020.[69]

The economic implications of climate change could also be extremely serious. A 2005 study for the European Commission projected that the cost of climate change could be more than $100 trillion by the end of this century.[70] Other studies project even potentially direr economic impacts. For example, the German Institute for Economic Research projects that economic damage could reach $20 trillion annually by 2100 under a business-as-usual scenario for greenhouse gas emissions, reducing global economic output by 6–8%.[71] The Stern Review on the Economics of Climate Change for the U.K. government concluded that warming on the higher end of projections could result in a 5–10% loss of GDP, with poorer countries suffering losses in excess of 10%.[72]

2. INTERNATIONAL LEGAL RESPONSES TO CLIMATE CHANGE

The primary international legal response to climate change to date is the United Nations Framework Convention on Climate Change (UNFCCC),[73] which entered into force in 1994 and has been ratified by 189 countries and the European Economic Community.[74] Unfortunately, resistance by several nations, most prominently, the United States and OPEC States, to mandatory reduction targets for greenhouse gas emissions led the drafters to resort instead to "constructive ambiguities" and "guidelines, rather than a legal commitment."[75] Thus, the UNFCCC merely calls

---

[67] Drew T. Shindell et al., *Solar and Anthropogenic Forcing of Tropical Hydrology*, 33 GEOPHYSICAL RES. LETTERS L24706 1, 5 (2006), *available at* http://pubs.giss.nasa.gov/docs/2006/2006_Shindell_etal_4.pdf (last visited May 25, 2008); Mark Spalding, Stephen Grady & Christoph Zöckler, *Changes in Tropical Regions*, in IMPACTS OF CLIMATE CHANGE ON WILDLIFE 28 (Rys E. Green et al. eds., 2002).

[68] CNA CORPORATION, NATIONAL SECURITY AND THE THREAT OF CLIMATE CHANGE 15 (2007), *available at* http://securityandclimate.cna.org (last visited May 25, 2008).

[69] IMPACTS, ADAPTATION AND VULNERABILITY, *supra* note 30, at 10.

[70] PAUL WATKISS ET AL., THE IMPACTS AND COSTS OF CLIMATE CHANGE iv (2005), Commissioned by the European Commission DG Environment, *available at* http://ec.europa.eu/environment/climat/pdf/final_report2.pdf (last visited May 25, 2008).

[71] Ackerman & Stanton, *supra* note 53, at 22.

[72] NICHOLAS STERN, THE ECONOMICS OF CLIMATE CHANGE, Executive Summary ix (2006), *available at* http://www.hm-treasury.gov.uk/media/4/3/Executive_Summary.pdf (last visited May 25, 2008).

[73] United Nations Framework Convention on Climate Change, 31 I.L.M. 849 (May 9, 1992) [hereinafter UNFCCC].

[74] United Nations Framework Convention on Climate Change Secretariat, UNFCCC: *Status of Ratifications*, *available at* http://unfccc.int/kyoto_protocol/status_of_ratification/items/2613.php (last visited May 25, 2008) [hereinafter *Status of Ratifications*].

[75] Ranee Khooshie Lai Panjabi, *Can International Law Improve the Climate? An Analysis of the United Nations Framework Convention on Climate Change Signed at the Rio Summit in 1992*, 18 N.C. J. INT'L L. & COM. REG. 401, 404 (1993).

on the Parties in Annex I (developed countries and economies in transition) to "aim" to return their emissions back to 1990 levels.[76]

By 1995, the greenhouse gas emissions of most developed countries were already well above 1990 levels, and a study by the Organization of Economic Cooperation and Development projected that emissions from industrialized countries would rise between 11 and 24% over the next 15 years.[77] The realization that more substantive measures were necessary ultimately led to the adoption of the Kyoto Protocol to the UNFCCC[78] at the Third Conference of the Parties in 1997. The Protocol entered into force in 2005 and currently has 169 States and the EEC as Parties.[79]

The Protocol calls for industrialized States and States with economies in transition to reduce their aggregate greenhouse gas emissions to at least 5% below 1990 levels in the commitment period of 2008 to 2012.[80] In addition, the Protocol required Parties to begin negotiating commitments for subsequent periods by 2005.[81]

Unfortunately, for several reasons, the Protocol is not the panacea that the popular press sometimes portrays it to be. First, former President George W. Bush announced in 2001 that the United States, responsible for 25% of the world's anthropogenic greenhouse gas emissions, would not become a Party to the Protocol.[82] As an alternative, the former president as part of his 2002 "Clear Skies Initiative" proposed the "Global Climate Initiative (GCI)," which would have sought to reduce the "greenhouse gas intensity" of the U.S. economy by 18% over the following ten years.[83] "Greenhouse gas intensity" is defined as the ratio of greenhouse gases to economic output.[84]

While touted as a bold approach by the Bush administration, in reality, the GCI constituted an extremely tepid response by the world's largest producer of greenhouse gases. While the Kyoto Protocol would have committed the United States to reduce its greenhouse gas emissions by 7% below 1990 levels,[85] under the GCI it was estimated that emissions would rise by 32% above 1990 levels.[86] The GCI ultimately

---

[76] UNFCCC, *supra* note 73, at art. 4(2)(b).
[77] Bas Arts, *New Arrangements in Climate Policy*, 52 CHANGE 1, 2 (2000).
[78] Kyoto Protocol to the United Nations Framework Convention on Climate Change, Dec. 10, 1997, FCCC/CP/1997/L.7/Add. 1, 37 I.L.M. 22.
[79] *Status of Ratifications, supra* note 74.
[80] *Id.* at art. 3(1).
[81] *Id.* at art 3(9); art. 21(7).
[82] Press Release, White House Office of the Press Secretary, President Bush Discusses Global Climate Change (June 11, 2001), *available at* http://www.whitehouse.gov/news/releases/2001/06/20010611-2.html (last visited May 25, 2008).
[83] The White House, *Global Climate Change Policy Book*, Feb. 2002, *available at* http://www.whitehouse.gov/news/releases/2002/02/climatechange.html (last visited May 25, 2008). The proposal also called, inter alia, for increasing funding for climate change research by $700 million in fiscal year 2003. *Id.*
[84] *Id.*
[85] Kyoto Protocol, *supra* note 78, at Annex B.
[86] Detlef van Vuuren et al., *An Evaluation of the Level of Ambition and Implications of the Bush Climate Change Initiative*, 2 CLIMATE POL'Y 293, 295 (2002); A.P.G. de Moor et al., *Evaluating the Bush Climate Change Initiative*, Dutch Ministry of Environment, RIVM Report 278001019/2002 13 (2002).

withered on the vine after failing to clear out of the Senate Environment and Public Works Committee in March 2005.[87] While the Bush administration continued to tout a voluntary, technologically driven approach, the UNFCCC Secretariat recently projected that U.S. greenhouse gas emissions will be more than 32% above 1990 levels by 2010, and more than 50% above 1990 levels by 2020.[88] The steady upward projection of emissions is in no small part attributable to the United States' continued commitment to coal, which produces triple the carbon dioxide per unit of energy as natural gas and double that of oil.[89] Fifty percent of the electricity generated in the United States is currently produced from coal, and there are an estimated 130 new

---

[87] Michael Janofsky, *Bush-Backed Emissions Bill Fails to Reach Senate Floor*, N.Y. TIMES, Mar. 10, 2005, *available at* http://www.nytimes.com/2005/03/10/politics/10enviro.html (last visited May 25, 2008). The United States, China, India, Japan, South Korea, and Australia, responsible for 49% of the world's greenhouse gas emissions, did agree to form the Asia-Pacific Partnership on Clean Development and Climate in 2005. The Partnership's primary objective is to "promote and create an enabling environment for the development, diffusion, deployment and transfer of existing and emerging cost-effective, cleaner technologies and practices...." Potential areas for collaboration include development of energy efficiency programs, clean coal, renewable energy sources, including wind, solar, and geothermal, and carbon sequestration projects. It is contemplated that a nonbinding compact will be established to specify terms of implementation of the Partnership. *Vision Statement of Australia, China, India, Japan, the Republic of Korea and the United States of America for a New Asia-Pacific Partnership on Clean Development and Climate*, July 28, 2005, http://www.pm.gov.au/news/media_releases/media_Release1482.html#statement (last visited Aug. 25, 2005); Anna Matysek et al., *Technology – Its Role in Economic Development and Climate Change* 7, Abare Res. Rep. 06.6 (2006), *available at* http://www.abareconomics.com/publications_html/climate/climate_06/cc_technology.pdf (last visited May 25, 2008). However, the Partnership agreement is not likely to substantially change the terrain, as it does not incorporate legally binding commitments or targets to reduce greenhouse gas emissions, obviating the incentive for the public and private sectors to deploy costly new technologies, and doesn't at this point have a funding mechanism for the programs it outlines, including facilitation of transfers of low-emission technologies to developing countries. Zhong Xiang Zhang, *Energy, Environment and Climate Issues in Asia*, Harvard Project for Asian and International Relations 26 (2006), *available at* http://papers.ssrn.com/sol3/papers.cfm?abstract_id=920756 (last visited May 25, 2008); Richard Black, *Climate Pact: For Good or Bad?*, BBC News, July 28, 2005, *available at* http://news.bbc.co.uk/1/hi/sci/tech/4725681.stm (last visited May 25, 2008). As Anthony Hobley, Chairman of the London Climate Change Services concluded: "This partnership does not provide anything additional to the UNFCCC to which all of the countries involved have already signed up." Liz Bossley, *Asia-Pacific Partnership: Complementing or Competing with Kyoto?*, 48 MIDDLE E. ECON. SURV., No. 32, Aug. 8, 2005, *available at* http://www.mees.com/postedarticles/oped/v48n32-5ODo1.htm (last visited May 25, 2008). Moreover, to date, Australia and the United States have pledged to spend a paltry $455 million over the next five years on clean energy projects. Clair Miller, *New Climate Partnership Makes Little Difference*, 4(2) FRONTIERS IN ECOLOGY & ENV'T 60, 60 (2006).

[88] UNFCCC Secretariat, *Data Appendices to UNFCCC Presentation at the AWG Workshop* 6, Nov. 7, 2006, *available at* http://unfccc.int/files/meetings/cop_12/in-session_workshops/application/pdf/061107_6_ghg_app.pdf (last visited May 25, 2008).

[89] William K. Stevens, *Global Economy Slowly Cuts Use of High-Carbon Energy*, N.Y. TIMES, Oct. 31, 1999, at A12. Coal-burning plants contributed most of the new carbon dioxide emitted by the electric power sector, which in turn has accounted for nearly half of the 18% increase in carbon dioxide emissions in the United States between 1990 and 2004. Megan Tady, *Climate Change Gas Emissions Way Up Nationwide*, AlterNet, Apr. 20, 2007, http://www.alternet.org/story/50624 (last visited May 25, 2008).

coal-fired plants on the drawing boards.[90] As the IPCC recently observed, energy infrastructure decisions during this period of time will exert substantial influence on future greenhouse gas emissions, given the long lifetimes of such facilities.[91]

With the Bush administration's increased openness to international climate negotiations in its last couple of years and with the Obama administration's much greater commitment to the negotiations, the United States appears to be reengaging the world community on this issue. At the G8 Summit in June 2007, the United States joined the other States in adopting "Agenda for Global Growth and Stability," which included a section on addressing climate change. In the Agenda, the G8 States acknowledged the need for "resolute and concerted action" to reduce greenhouse gas emissions, and that "tackling climate change is a shared responsibility for all."[92] However, primarily because of U.S. resistance, the G8 stopped short of agreeing to specific targets and timetables for reducing emissions, rather only pledging to "consider seriously" the decisions made by the European Union, Canada, and Japan to reduce emissions by at least half of 1990 levels by 2050.[93]

Later in 2007, then president Bush invited the EU, the United Nations, and eleven industrial and developing States to work toward a long-term goal for emissions reductions by 2008.[94] However, at the thirteenth Conference of the Parties to the UNFCCC in December, the Bush administration castigated the European Union for proposing that industrialized nations reduce their greenhouse gas emissions by 25–40% by 2020, characterizing the proposal as "totally unrealistic" and "unhelpful."[95] Ultimately, the United States did agree to the Bali Action Plan, which lays out a process for addressing climate change in the long term.[96] However, the Action Plan also reflects the U.S. resistance to binding targets and timetables for reducing emissions. The United States successfully beat back an effort by the European Union to secure an agreement on the part of industrialized nations to

---

[90] Pew Center on Global Climate Change, *Coal and Climate Change Facts*, *available at* http://www.pewclimate.org/global-warming-basics/coalfacts.cfm (last visited May 25, 2008).

[91] MITIGATION OF CLIMATE CHANGE, *supra* note 20, at 18.

[92] G8 Summit, *Growth and Responsibility in the World Economy*, ¶¶40–41, June 7, 2007.

[93] *Id.* at ¶49.

[94] Matt Spetalnick, *Bush Calls Meeting on Global Warming for September*, Planet Ark, Aug. 6, 2007, http://www.planetark.com/dailynewsstory.cfm/newsid/43467/story.htm (last visited May 25, 2008).

[95] David Adam, *U.S. Balks at Bali Carbon Targets*, GUARDIAN UNLIMITED, Dec. 10, 2007, *available at* http://www.guardian.co.uk/environment/2007/dec/10/climatechange.usnews (last visited May 25, 2008). The European Union in late 2008 committed itself to a "triple" initiative to reduce greenhouse gas emissions by 20% below 1990 levels by 2020, to reduce energy consumption by 20% the same date, and to ensure that 20% of EU energy is produced with renewable energy sources by that date. Ian Traynor, *A Mix of Rules and Markets, Held Together by Tradeoffs*, Guardian.co.uk, Dec. 13, 2008, *available at* http://www.guardian.co.uk/environment/2008/dec/13/carbonemissions-emissionstrading (last visited Mar. 10, 2009).

[96] UNFCCC, 13th Conference of the Parties, *Bali Action Plan*, CP.13 (2007), *available at* http://unfccc.int/files/meetings/cop_13/application/pdf/cp_bali_action.pdf (last visited May 25, 2008) [hereinafter *Bali Action Plan*].

reduce their emission by 25–40% by 2012.[97] Thus, the Plan merely establishes a comprehensive process to "enable the full, effective and sustained implementation of the Convention through long-term cooperative action...,"[98] including "[m]easurable, reportable and verifiable nationally appropriate mitigation commitments or actions, including quantified emission limitation and reduction objectives, by all developed country Parties."[99]

In March 2008, James L. Conaughton, chairman of the White House Council on Environmental Quality under President Bush, announced that the United States was amenable to accepting a binding treaty to reduce emissions if China and other major developing countries were willing to do so also.[100] At the G8 Summit in Japan in July 2008, the G8 leaders agreed to "the goal of achieving at least 50% reduction of global emissions by 2050."[101] However, the leaders notably failed to agree to medium-term targets, and the emphasis on the need for both major developed and developing economies to make substantive commitments[102] may ultimately scupper the initiative.

The Obama administration has pledged to "engage vigorously" in climate change negotiations at the international level and has called for implementing a cap-and-trade program to reduce greenhouse gas emissions in the United States 14% below 2005 levels by 2020, and approximately 83% below 2005 levels by 2050.[103] These commitments hopefully will translate into greater willingness for the United States to make binding international commitments in the post-2012 treaty regime. Moreover, the U.S. Congress, with support from the executive branch, is considering major cap-and-trade legislation, and some U.S. cities, counties, and states are at the front end of innovative emissions reductions. But in the meantime, despite progress by particular smaller-scale governments, the United States as a whole continues to emit at troublingly high levels.[104]

---

[97] Peter Montague, *The Basket Our Eggs Are In*, 939 Rachel's Democracy & Health N., Dec. 27, 2007, *available at* http://www.rachel.org/en/newsletters/rachels_news/939 (last visited May 28, 2009).

[98] UNFCCC, Bali Action Plan, Decision – CP.13/ (2008), at para. 1.

[99] *Id.* at para. 1(b)(i).

[100] James Kanter & Andrew C. Revkin, *Binding Emissions Treaty Still a Possibility, U.S. Says*, N.Y. TIMES, Feb. 27, 2008, at A8.

[101] G8 Summit, *Statement on Environment and Climate Change* (2008), at para. 23, *available at* http://www.g8summit.go.jp/eng/doc/doco80709_02_en.html (last visited July 14, 2008).

[102] *Id.* at para. 24.

[103] Kim Chipman & Catherine Dodge, *Obama Plan Has $79 Billion From Cap-and-Trade in 2012*, Bloomberg News, Feb. 26, 2009, *available at* http://www.bloomberg.com/apps/news?pid=20601087& sid=aDT1Ybl.PccE&refer=home; Change.gov, *The Obama-Biden Plan, available at* http://change .gov/agenda/energy_and_environment_agenda (last visited Mar. 1, 2009); Jeff Mason, *Obama Vows Climate Action Despite Financial Crisis*, REUTERS, Nov. 18, 2008, *available at* http://www.reuters .com/article/vcCandidateFeed2/idUSN1827628 5 (last visited Mar. 1, 2009).

[104] *See* Kitty Bennett & Farhana Hossain, *The Presidential Candidates on Climate Change*, *available at* http://politics.nytimes.com/election-guide/2008/issues/climate.html (last visited May 25, 2008). Another potential positive development in the United States was the flurry of legislative activity in the 110th Congress to address climate change, with more than 180 bills, resolutions, and amendments introduced in the session through February 2008 to address climate change. Pew

Second, in developing the rules for implementing the Protocol, many concessions were made to wavering nations that substantially dilute the Parties' commitments. Thus, some analysts believe that implementation of the Protocol ultimately will

Center on Global Climate Change, *Legislation in the 110th Congress Related to Global Climate Change*, available at http://www.pewclimate.org/what_s_being_done/in_the_congress (last visited May 25, 2008). Several of the bills would have established cap-and-trade systems that would have dramatically reduced emissions over the course of the next few decades. In December 2007, the U.S. Senate Environment and Public Works Committee reported out the Lieberman-Warner Climate Security Act, S. 2191. This is the first greenhouse gas emissions cap-and-trade bill to be voted out of committee in the United States. Pew Center on Global Climate Change, *Legislation in the 110th Congress Related to Global Climate Change*, available at http://www.pewclimate.org/what_s_being_done/in_the_congress/110thcongress.cfm (last visited May 25, 2008). However, the bill ultimately died on the Senate floor in June 2008. Eric Pooley, *Why the Climate Bill Failed*, TIME, June 9, 2008, available at http://www.time.com/time/nation/article/0,8599,1812836,00.html (last visited Mar. 11, 2009). There is likely to be substantial activity related to climate change in the 111th Congress. President Obama has several powerful allies who advocate climate change legislation in the House, including Henry Waxman, the chairman of the House Energy and Commerce Committee, which has jurisdiction over climate change legislation in the House, and Representative Edward Markey, who heads up the committee's Subcommittee on Energy and Environment. Pew Center on Global Climate Change, *Climate Action in Congress*, available at http://www.pewclimate.org/what_s_being_done/in_the_congress (last visited Mar. 11, 2009). The Senate may also begin debate on climate change legislation again this summer. Ian Talley, *Sen. Reid: Aiming to Debate Climate Bill by Summer*, Wall St. J., Feb. 20, 2009, available at http://online.wsj.com/article/SB123516532284336065.html (last visited Mar. 12, 2009).

At the subnational level, there are also a number of regional and state initiatives to address climate change that may ultimately have a positive impact. For example, in 2006, California, which is the twelfth-largest emitter of carbon dioxide globally, Office of the Governor, Press Release, Gov. *Schwarzenegger Signs Landmark Legislation to Reduce Greenhouse Gas Emissions*, available at http://gov.ca.gov/index.php?/press-release/4111/ (last visited May 25, 2008), passed the California Global Warming Solutions Act, or AB32. California Legislature, Assembly Bill 32, Dec. 4, 2006, available at http://www.leginfo.ca.gov/pub/07-08/bill/asm/ab_0001-0050/ab_32_bill_20070501_amended_asm_v96.pdf (last visited May 25, 2008) [hereinafter AB32]. AB32 calls for the state to reduce its greenhouse emissions to 1990 levels by 2020. *Id.* at sec. 38550. The law provides for the establishment of additional targets thereafter, with the ultimate goal of reducing the state's emissions by 80% below 1990 levels by 2050. Office of the Governor, *supra*. It remains to be seen, however, whether the state can achieve this goal in the face of a projected doubling of its population in the next forty years and likely political pressure to downgrade the commitment if there is not ultimately a commensurate federal mandate. Bruce Murray, *Global Cooling in the Sunshine State*, available at http://www.analysisonline.org/energy/ab32.html (last visited Sept. 6, 2007). In the East, ten states have now joined the Regional Greenhouse Gas Initiative (RGGI), which sets a cap on power plant emissions at approximately current levels of 120 million tons of carbon dioxide between 2009 and 2015, and then 10% below this level by 2019. Regional Greenhouse Gas Initiative, *Frequently Asked Questions*, available at http://www.rggi.org/docs/mou_faqs_12_20_05.pdf (last visited May 25, 2008). Even assuming the states achieve this goal, this is an extremely modest commitment compared to what ultimately must be done, but at least RGGI establishes an institutional framework in the region that hopefully will both commit to further reductions in the future and help to pressure the federal government to establish national mandates. Moreover, a large number of states are taking actions to reduce greenhouse gas initiatives, including through renewable portfolio standards, greenhouse gas emissions targets, and tax incentives to reduce emissions. Pew Center on Global Climate Change, *Climate Change 101: State Action*, available at http://www.pewclimate.org/docUploads/101_States.pdf (last visited May 25, 2008). Similar initiatives have been established in the Midwest and South. *See* Western Climate Initiative, available at http://www.westernclimateinitiative.org/Useful_Links.cfm (last visited May 25, 2008); Midwestern Greenhouse Gas Accord, available at http://www.wisgov.state.wi.us/docview.asp?docid=12497 (last visited May 25, 2008).

result in substantially fewer reductions in emissions than originally contemplated, or perhaps even a net *increase* over 1990 levels.[105]

Third, it is far from clear that most of the industrialized State Parties to Kyoto will fulfill their obligations in the first commitment period. For example, Japan's emissions are currently more than 14% above its Kyoto targets.[106] Canada's emissions are now more than 30% above 1990 levels,[107] and the government recently acknowledged that it won't meet its commitments, but rather will seek to achieve less ambitious targets.[108] Even the European Union, the staunchest supporter of the Protocol, is struggling to meet its commitments. Greenhouse gas emissions in the EU rose in 2004 and 2005,[109] and seven of the EU-15 States are projected to exceed their individual emission limits set by the EU.[110] The European Commission projects that the bloc's Kyoto commitment will be met through the implementation of additional initiatives, but has emphasized that there is little room for error at this point.[111]

---

[105] Tom Athanasiou & Paul Baer, *Bonn and Genoa: A Tale of Two Cities and Two Movements*, Foreign Policy in Focus, Discussion Paper 3 (Aug. 2001) (Concessions made in negotiations to flesh out Kyoto Protocol could "render the protocol's nominal mandate of a 5.2% overall reduction in rich-world emissions (from their 1990 baseline) into a 0.3% increase); Miranda A. Schreurs, *Competing Agendas and the Climate Change Negotiations: The United States, the European Union, and Japan*, 31 ENVTL. L. REP. 11,218, 11,218 (2001).

[106] Ikuko Kao & Neil Chatterjee, *Japan's Kyoto Gap Widens as Emissions Rise*, Planet Ark, Oct. 18, 2006, *available at* http://www.planetark.com/dailynewsstory.cfm/newsid/38538/story.htm (last visited May 25, 2008). *See also Japan Emissions to Rise, Kyoto Target at Risk – Paper*, Aug. 9, 2007, http://www.planetark.com/dailynewsstory.cfm/newsid/43564/story.htm (last visited May 25, 2008) (Japanese government projects that Japan's greenhouse gas emissions will rise by 0.9% in the fiscal year ending in March 2011).

[107] Rob Gillies, *Canada Won't Meet Kyoto Emissions Targets*, Boston.com, Apr. 26, 2007, *available at* http://www.boston.com/news/world/canada/articles/2007/04/26/canada_wont_meet_kyoto_emission_target (last visited Sept. 6, 2007). The government's own new "Turning the Corner" climate change strategy would put Canada 39% above its Kyoto target in 2012. Environment News Service, *Canada Sued for Abandoning Kyoto Climate Commitment*, May 29, 2007, *available at* http://www.ecojustice.ca/media-centre/press-clips/canada-sued-for-abandoning-kyoto-climate-commitment/?searchterm=%22abandoning%20kyoto (last visited May 25, 2008).

[108] Gillies, *supra* note 107. Under the latest plan promulgated by the conservative Canadian government, Canada will not meet its commitments under the Kyoto Protocol until 2025 rather than 2012. *Environmentalists Pan Harper's Pitch on Climate*, CTV.ca, June 4, 2007, *available at* http://www.ctv.ca/servlet/ArticleNews/story/CTVNews/20070603/harper_g8_070604/20070604?hub=Canada (last visited May 25, 2008).

[109] Helena Spongenberg, *EU Falls Behind on Green Targets*, euobserver.com, June 23, 2006, *available at* http://euobserver.com/9/21944/?rk=1 (last visited May 25, 2008).

[110] Europa, *Climate Change: Member States Need to Intensify Efforts to Reach Kyoto Emission Targets*, Oct. 27, 2006, *available at* http://europa.eu/rapid/pressReleasesAction.do?reference=IP/06/1488&format=HTML&aged=0&language=EN&guiLanguage=en (last visited May 25, 2008).

[111] *Id.* Foreboding recent developments may make achievement of European's longer-range objective of reducing emissions by 20% below 1990 levels by 2020 increasingly unlikely. Italian Prime Minister Silvio Berlusconi recently announced his intention to veto the EU's proposal, and several Central and Eastern European countries, including Poland, Bulgaria, the Czech Republic, Hungary, Romania, and Slovakia, have expressed serious reservations about the proposal. Christian Spillmann, *Italy, Poland, Threaten to Veto EU Climate Change Plans*, Yahoo! News, Oct. 15, 2008, *available at* http://news.yahoo.com/s/afp/20081015/wl_afp/eusummitclimatewarmingenvironment (last visited on Oct. 22, 2008).

Finally, even if the Kyoto Protocol, as originally drafted, were faithfully implemented by all industrialized nations it would constitute only an extremely modest down payment on what ultimately must be done to stabilize atmospheric concentrations of greenhouse emissions. This is true for two primary reasons. First, as indicated above, the Kyoto Protocol in its first commitment period calls for Annex I Parties to reduce their overall greenhouse gas emissions by 5%. By contrast, stabilization of atmospheric greenhouse gases at levels that produce no more than a 2–3°C increase in temperatures from preindustrial levels, which many climate experts cite as a critical "climate tipping point that could lead to intolerable impacts on human well-being,"[112] will require the world community to reduce greenhouse gas emissions by 60–70%.[113] Moreover, industrialized countries might have to reduce their emissions by as much as 80% by the middle of the century if developing nations are to be permitted some growth in their emissions levels.[114]

Second, the Protocol currently does not impose emissions reductions commitments on developing countries, even though it is projected that by 2025 the developing world's share of global emissions will be approximately 55%.[115] Indeed, the Netherlands Environmental Assessment Agency recently concluded that China,

---

[112] Bierbaum et al., *supra* note 16, at xi. *See also* Paul Baer & Tom Athanasiou, *Honesty About Dangerous Climate Change*, EcoEquity, available at http://www.ecoequity.org/ceo/ceo_8_2.htm#dangerous (last visited May 25, 2008); B.C. O'Neill & M. Oppenheimer, *Climate Change - Dangerous Climate Impacts and the Kyoto Protocol*, 296 SCI. 1971–72 (2002). However, it needs to be emphasized that even lower temperature increases will have serious implications. For example, a 1°C increase in atmospheric temperatures will seriously imperil the world's coral reef ecosystems, as well as many other ecosystems in developing countries. *Id.* at 1971; Hadley Centre, *Avoiding Dangerous Climate Change* 14 (2005), available at http://www.stabilisation2005.com/Steering_Commitee_Report.pdf (last visited May 25, 2008).

[113] Jonathan Pershing & Fernando Tudela, *A Long-Term Target: Framing the Climate Effort*, in BEYOND KYOTO: ADVANCING THE INTERNATIONAL EFFORT AGAINST CLIMATE CHANGE (Joseph E. Aldy et al. eds., 2004), Pew Center on Global Climate Change, *Q&A: Kyoto Protocol* 23, available at http://www.pewclimate.org/docUploads/Long%2DTerm%20Target%2Epdf (last visited May 25, 2008) (Stabilization of atmospheric carbon dioxide levels at 550 parts per million, yielding an estimated 1.6–2.9°C increase in temperatures from preindustrial levels, necessitates 60% reduction in emissions). A recent study by Hare and Meinshausen suggests that the cutbacks may have to be even more dramatic. The study concludes that there is a 66% risk of overshooting a 2°C increase of temperatures from preindustrial levels even if atmospheric concentrations of carbon dioxide are held to 450 parts per million. Bill Hare & Malte Meinshausen, *How Much Warming Are We Committed to and How Much Can be Avoided?*, 75 CLIMATIC CHANGE 111, 129 (2006). The authors conclude that "[o]nly scenarios that aim at stabilization levels at or below 400 ppm CO_2 equivalence ($\sim$ 350ppm CO_2) can limit the probability of exceeding 2°C to reasonable levels . . . " *Id.* at 137. Even stabilization at 650ppm CO_2 equivalence would require reductions of approximately 50% by 2100. Detlef P. van Vuuren et al., *Stabilizing Greenhouse Gas Concentrations at Low Levels: An Assessment of Reduction Strategies and Costs*, 81 CLIMATIC CHANGE 119, 120 (2007).

[114] David D. Doniger, Antonia V. Herzog & Daniel A. Lashof, *An Ambitious, Centrist Approach to Global Warming Legislation*, 314 SCI. 764, 764 (2006); Ecofys GmbH et al., *WWF Climate Scorecards: Comparison of the Climate Performance of the G8 Countries* 4, available at http://www.panda.org/downloads/climate_change/g8scorecardsjun29light.pdf (last visited May 25, 2008).

[115] Kevin Baumert & Jonathan Pershing, *Climate Data: Insights and Observations*, Pew Center on Global Climate Change 16 (2004). Overall, the Parties to the Kyoto Protocol only generate approximately one third of the world's greenhouse emissions. Pew Center on Global Climate Change, *supra* note 16, at 36.

with fossil fuel consumption in recent years rising at a blistering pace of more than 9% annually,[116] surpassed the United States in 2006 in aggregate carbon dioxide emissions.[117]

Given the modest commitments undertaken under Kyoto, and the difficulties of controlling the rapid growth of emissions in the United States and large developing States, it is not surprising that the U.S. Energy Information Agency recently projected that worldwide emissions under Kyoto would be 43,000 MMT in 2030, only slightly below the business as usual projection of 43,676 MMT.[118] Overall, climate researchers have estimated that full implementation of Kyoto would reduce projected warming in 2050 by only about *one-twentieth of one degree* and projected sea level rise by a mere 5 *millimeters*.[119] By contrast, recent research indicates that if the world community wishes to have a reasonable prospect for avoiding temperatures increases beyond 2°C, global energy emissions must be stabilized by 2015 and rapidly decline by 6–8 per cent annually between 2020 and 2040, and full decarbonization by 2050.[120]

The glacial pace of progress under the UNFCCC and the Kyoto Protocol has led to growing despair by many actors, including nongovernmental organizations (NGOs), state and local governments in the United States, and in many nations, especially Southern States that are particularly vulnerable to the threat of climate change. Indeed, the trepidation of such stakeholders has been exacerbated over the course of the Bush administration by the failure of the United States to signal its willingness to reengage in the Kyoto process,[121] as well as tepid support for future commitments by other major greenhouse gas–emitting States, including China, Russia, and India.[122] Particularly disconcerting was the successful opposition by the

---

[116] Robert Collier, *China About to Pass U.S. as World's Top Generator of Greenhouse Gases*, SFGate.com, Mar. 5, 2007, *available at* http://sfgate.com/cgi-bin/article.cgi?file=/c/a/2007/03/05/MNG18OFHF21.DTL&type=printable (last visited May 25, 2008). China's carbon dioxide emissions over the period of 2001–2006 were almost 350% higher than the emissions of the United States, Canada, the European Union, South Korea, Australia, and New Zealand *combined. Id.*

[117] Press Release, Netherlands Environmental Assessment Agency, *Chinese CO2 Emissions in Perspective* (June 22, 2007), *available at* http://www.mnp.nl/en/service/pressreleases/2007/20070622ChineseCO2emissionsinperspective.html (last visited May 25, 2008).

[118] Michael Gerrard, *Introduction and Overview*, *in* GLOBAL CLIMATE CHANGE AND U.S. LAW 13 (Michael B. Gerrard ed., 2007).

[119] Martin Parry et al., *Buenos Aires and Kyoto Targets Do Little to Reduce Climate Change Impacts*, 8(4) GLOBAL ENVTL. CHANGE 285, 285 (1998). *See also* Mustafa H. Babiker, *The Evolution of a Climate Regime: Kyoto to Marrakech and Beyond*, 5 ENVTL. SCI. & POL'Y 195, 202 (2002).

[120] Kevin Anderson & Alice Bows, *Reframing the Climate Change Challenge in Light of Post-2000*, PHIL. TRANSACTIONS ROYAL SOC'Y A, Aug. 29, 2008, at 15.

[121] For example, at the most recent meeting of the Group of 8 industrialized nations, the United States refused to endorse carbon trading, one of the centerpieces of the Kyoto Protocol, as a means to reduce emissions. *U.S. Blocks Consensus of G8-plus-Five on Global Warming Issues*, GREENWIRE, Mar. 19, 2007, LEXIS-NEXIS, News.

[122] Alister Doyle, *U.N. Climate Talks Stagnate Despite Public Worries*, Environmental News Network, Mar. 2, 2007, *available at* http://uk.reuters.com/article/environmentNews/idUKL0223966020070303 (last visited May 25, 2008). Russia and India are, respectively, the third- and fourth-largest producers of greenhouse gas emissions globally, after China and the United States. Nita Bhalla, *India Says Its*

G77 countries and China to the European Union's efforts to insert language in negotiating documents that would have committed the Parties to seeking to keep temperature increases below 2°C.[123] Furthermore, the G77/China bloc expressed the view that developing countries should not be required to assume binding obligations to reduce emissions given their need for rapid economic growth and development.[124] Rather, the focus at COP12 was on adapting to climate change impacts that increasingly seem inevitable.[125]

More hopefully, the Bali Action Plan, adopted at the thirteenth Conference of the Parties, does call for the Ad Hoc Working Group on Long-term Cooperative Action to consider potential mitigation measures that could be taken by developing countries,[126] though this provision is freighted with ambiguous language, requiring that the measures be "nationally appropriate," and supported by "technology, financing and capacity building."[127] Burgeoning world emissions are setting us on a path that, absent a "radical reframing of both the climate change agenda, and the economic characterization of contemporary society," may ensure temperature increases of at least 4°C above preindustrial levels.[128] Even with the United States more committed to take action under President Obama, it is doubtful that the international climate treaty regime will be able to do enough to address this problem.

3. ENTER, CLIMATE ADJUDICATION

The consensus has become increasingly clear that meaningful reductions by major greenhouse gas–emitting nations must begin soon or we will inevitably cross the

---

Carbon Emissions not Harming the World, Environmental News Network, Dec. 14, 2006, *available at* http://www.enn.com/today.html?id=11845 (last visited May 25, 2008).

    The European Union in February 2007 did agree to reduce emissions to 20% below 1990 levels by 2020 and will push for a 30% commitment by industrialized states by that date. Europa, *Climate Change and the EU's Response*, MEMO/07/58, Feb. 15, 2007, *available at* http://europa.eu/rapid/pressReleasesAction.do?reference=MEMO/07/58&format=HTML&aged=0)#uage=EN&guiLanguage=enIan (last visited May 25, 2008); Ian Traynor & David Gow, *EU Promises 20% Reduction in Carbon Emissions by 2020*, GUARDIAN UNLIMITED, Feb. 21, 2007, *available at* http://environment.guardian.co.uk/climatechange/story/0,2017600,00.html (last visited May 25, 2008).

[123] *Id.*
[124] Chukwumerije Okereke et al., *Assessment of Key Negotiating Issues at Nairobi Climate COP/MOP and What it Means for the Future of the Climate Regime*, Tyndall Centre for Climate Change Research, Working Paper 106 (June 2006), *available at* http://www.oxfordclimatepolicy.org/publications/TyndallWorkingPaper2007.pdf (last visited May 25, 2008). More hopefully, the most recent Chinese Five Year Plan includes a commitment to reduce energy intensity by 20% by 2010. *Id.* at 19.
[125] UNFCCC, *Further Commitments for Annex I Parties and Programme of Work*, Ad Hoc Working Group (2006), *available at* http://unfccc.int/files/meetings/cop_12/application/pdf/awg__conclusions.pdf (last visited May 25, 2008); UNFCCC, *First In-Session Workshop of the Ad Hoc Working Group on Further Commitments for Annex I Parties under the Kyoto Protocol* (2006), *available at* http://unfccc.int/files/meetings/cop_12/application/pdf/awg2_in_sess__report_an.pdf (last visited May 25, 2008).
[126] *Bali Action Plan*, *supra* note 97, at ¶ 1(b)(ii).
[127] *Id.*
[128] Anderson & Bows, *supra* note 120, at 18.

critical thresholds that commit this world to centuries of potentially catastrophic impacts. As one recent study indicated, avoiding temperature increases of 2–2.5°C over preindustrial levels at this point would require carbon dioxide emissions to level off by 2015–2020 at not much above current levels, and to decline to no more than a third of those levels by 2100. In addition to these reductions, it would require substantial cuts in other potential greenhouse gases.[129]

As the foregoing discussion makes clear, the likelihood of the international legal regime achieving these goals seems low. Moreover, even without the existing political difficulties, climate change is not a problem that can be addressed at only one level of governance. Behavior that causes greenhouse gas emissions takes place and is regulated at the local, state, national, regional, and international levels.

This combination of the urgency of the problem and complexity of politico-legal solutions has caused many State and non-State actors to look beyond traditional international treaty mechanisms for solutions to anthropogenic climate change.[130] In this context, litigation and other legal actions at subnational, national, and international levels have evolved from innovative ideas to an emerging practice area over the last several years.[131] Although the U.S. Supreme Court's decision in *Massachusetts v. EPA* helped to bring these cases into the public consciousness,[132] actions have been pending in state, national, and regional and international tribunals for a number of years.

A representative sampling of this ever-growing list of cases includes the following:

Subnational and National Actions

- **United States**. More than a dozen actions related to climate change have been filed in state and federal courts in the United States, and more are anticipated.[133] A cross section of the actions filed to date include the following:
  - In *Massachusetts v. EPA*, twelve States and several cities and NGOs filed an action against the U.S. Environmental Protection Agency (EPA), challenging its denial of a petition to regulate greenhouse gas emissions from new motor vehicles under section 202(a)(1) of the Clean Air Act. The U.S. Supreme Court held for the plaintiffs, concluding that the EPA had authority to regulate such emissions and that the agency must ground its reason for action

---

[129] Bierbaum et al., *supra* note 16, at xi. See also Hansen, *supra* note 1, at 14,293.

[130] See Hari M. Osofsky, *The Geography of Climate Change Litigation: Implications for Transnational Regulatory Governance*, 83 WASH. U. L.Q. 1789, 1795–1800 (2005); Eric A. Posner, *Climate Change and International Human Rights Litigation: A Critical Appraisal*, 155 U. PA. L. REV. 1925 (2007).

[131] See Eric Torbenson, *Lawyers Preparing for Explosion of Climate-Related Work*, DALLAS MORNING NEWS, Business Section, June 24, 2007.

[132] Massachusetts v. EPA, 127 S. Ct. 1438; 167 L. Ed. 2d 248 (2007).

[133] Greenpeace, *History of Climate Change Litigation*, June 2007, *available at* http://www.greenpeace.org/raw/content/new-zealand/press/reports/history-climate-change-litigation.pdf (last visited May 25, 2008). For a good summary of current actions in U.S. courts, see Justin R. Pidot, *Global Warming in the Courts* (2006), Georgetown Environmental Law & Policy Institute, *available at* http://www.law.georgetown.edu/gelpi/current_research/documents/GlobalWarmingLit_CourtsReport.pdf (last visited May 25, 2008).

or inaction in the terms of the Act.[134] In April 2009, the EPA published a proposed finding that greenhouse gases in the atmosphere endanger the public welfare of current and future generations,[135] potentially paving the way for EPA regulation.

- In *Friends of the Earth, Inc. v. Watson*, a suit was brought by two NGOs and the City of Boulder, Colorado, against the Overseas Private Investment Corporation and Export-Import Bank, alleging that these entities' failure to conduct an environmental review of the impacts of their funding of fossil fuel projects violates the National Environmental Policy Act.[136] Cross-motions for summary judgment were denied by a U.S. district court in 2007.[137]
- In *State of Connecticut v. American Electric Power Co.*, several States, the City of New York, and several NGOs filed an action against five major power companies for "the public nuisance" of "global warming" under federal common law or state law. The case is currently on appeal to the Second Circuit Court of Appeals after a district court judge dismissed the case on the grounds that the action presented nonjusticiable political questions.[138]
- In *State of California v. General Motors Corp.*, the State of California filed an action against six auto manufacturers for public nuisance. The suit alleged that the greenhouse gas emissions associated with the defendants' production of automobiles "is harming California, its environment, its economy and the health and well-being of its citizens."[139] The suit seeks monetary damages and a declaratory judgment that each defendant was jointly and severally liable for future damages incurred by the state for the ongoing nuisance of climate change.[140] The case was dismissed in September 2007 by the U.S. District Court for the Northern District of California on the grounds that plaintiff's claims raised nonjusticiable political questions and is currently on appeal to the Ninth Circuit.[141]
- In *Comer v. Nationwide Mutual Insurance*,[142] fourteen individuals filed suit against a group of energy and refining companies for damages sustained to their property as a result of Hurricane Katrina. Plaintiffs contended that the greenhouse gas emissions of the defendants increased the damages suffered by plaintiffs by intensifying the hurricane.[143] The U.S. District Court

---

[134] *Massachusetts v. EPA*, 127 S. Ct. at 1438.
[135] 40 CFR Ch. 1, Part III, Proposed Endangerment and Cause or Contribute Findings for Greenhouse Gases Under Section 202(a) of the Clean Air Act, Proposed Rule, Apr. 24, 2009, *available at* http://www.epa.gov/climatechange/endangerment/downloads/EPA-HQ-OAR-2009-0171-0001.pdf (last visited on May 28, 2009).
[136] Friends of the Earth, Inc. v. Watson, 2005 U.S. Dist. LEXIS 42335 (2005).
[137] Friends of the Earth, Inc. v. Mosbacher, 488 F. Supp. 2d 889 (N.D. Cal. 2007).
[138] Connecticut v. Am. Elec. Power Co., 406 F. Supp. 2d 265, 273 (S.D.N.Y. 2005).
[139] State of California v. Gen. Motors Corp., Case No. C05–05755 (N.D. Cal. 2006).
[140] *Id.* at 3.
[141] State of California v. Gen. Motors Corp., Order Granting Defendants' Motion to Dismiss, No. C06–05755 MJJ (N.D. Cal. 2007).
[142] Comer v. Nationwide Mut. Ins., 2005 WL 1066645 (S.D. Miss. 2006) (unpublished opinion).
[143] Third amended complaint, Comer v. Murphy Oil, 2006 WL 147089 I (S.D. Miss. 2006).

for the Southern District of Mississippi dismissed the action without prejudice on the grounds that class action claims unrelated to climate change against insurance and mortgage companies were inappropriate.[144] The Court subsequently dismissed the amended complaint on the grounds that plaintiffs lacked standing and plaintiffs' claims were "non-justiciable pursuant to the political question doctrine."[145]
- The Center for Biological Diversity, a U.S. NGO, petitioned the U.S. National Marine Fisheries Service in 2004, seeking listing of elkhorn and staghorn corals under the Endangered Species Act.[146] The species were added to the official list of Threatened Species in 2006.[147] A petition by the Center for Biological Diversity to list the polar bear under the ESA[148] also resulted in its listing by the U.S. Department of Interior as a threatened species in May of 2008.[149]
- Canada
  - In 2007, the NGO Friends of the Earth Canada filed a lawsuit against the government of Canada alleging that the federal government is violating section 166 of the Canadian Environmental Protection Act, which requires compliance with international agreements to prevent pollution. The suit contends that Canada is failing to fulfill its commitments under the Kyoto Protocol and the United Nations Framework Convention on Climate Change.[150]
- New Zealand
  - *Greenpeace New Zealand Inc. v. Northland Regional Council and Mighty River Power Limited* involved the application by a power company to a regional council to develop a coal-fired facility. The High Court held that under the Resource Management Act of 1991 a consent authority could take into account whether the proposed project would enable a reduction in

---

[144] Id.
[145] Comer v. Murphy Oil USA, *Order Granting Defendants' Motion to Dismiss*, Civil Action No. 1:05-CV-436-LG-RHW I (S.D. Miss. 2007).
[146] Ctr. for Biological Diversity, Petition to List Acropora Palmata (Elkhorn Coral), Acropora Cervicornis (Staghorn Coral), and Acropora Prolifera (Fused-Staghorn Coral) as Endangered Species Under The Endangered Species Act (2004), *available at* http://www.biologicaldiversity.org/swcbd/SPECIES/coral/petition.pdf (last visited May 25, 2008).
[147] Endangered and Threatened Species: Final Listing Determinations for Elkhorn Coral and Staghorn Coral, 71 FED. REG. 26,852 (May 9, 2006); 50 C.F.R. § 223.102.
[148] Kassie Siegel & Brendan Cummings, *Petition to List the Polar Bear (Ursus Maritimus) as a Threatened Species under the Endangered Species Act*, Feb. 16, 2006, *available at* http://www.biologicaldiversity.org/swcbd/SPECIES/polarbear/petition.pdf (last visited May 25, 2008); Order Granting Plaintiffs' Motion for Summary Judgment and Injunction, Ctr. for Biological Diversity v. Kempthorne, No. C 08-1339 CW, Apr. 28, 2008, *available at* http://www.biologicaldiversity.org/species/mammals/polar_bear/pdfs/Order-Granting-Summary-Judgment-4-28-2008.pdf.
[149] Department of the Interior, Fish and Wildlife Service, Endangered and Threatened Wildlife and Plants; Determination of Threatened Status for the Polar Bear (Ursus maritimus) Throughout Its Range, 50 C.F.R. § 17 (2008), *available at* http://www.doi.gov/issues/polar_bears/Polar%20Bear%20Final%20Rule_to%20FEDERAL%20REGISTE%20-Final_05-14-08.pdf (last visited on June 2, 2008).
[150] *Canada Sued for Abandoning Kyoto Climate Commitment*, supra note 107.

greenhouse gas emissions by the use and development of renewable energy in determining whether to grant the application.[151]
- In *Genesis Power Ltd. v. Franklin District Council*, New Zealand's Environment Court allowed an appeal brought by the Energy Efficiency and Conservation Authority against the refusal for permission to build a wind farm, under the Resource Management Act of 1991. The Court cited reduction of emissions of greenhouse gases and climate change as factors supporting the case, and the project was subsequently approved.[152]

- **Australia**
  - In *Australian Conservation Foundation v. Latrobe City Council*, the owner of the Hazelwood coal-fired station in Victoria, one of Australia's largest producers of greenhouse gas emissions, applied to develop an alternative coalfield, which would prolong the plant's operation until 2031. Four environmental groups brought an action in the Victorian Civil and Administrative Tribunal alleging that a reviewing panel's failure to consider potential greenhouse gas emissions from the project violated the Victorian Planning and Environment Act of 1987. The Tribunal held that the panel should consider the environmental impacts of the greenhouse gas emissions associated with the project.[153]
  - In *Gray v. Minister for Planning*, an activist brought an action in the New South Wales Land and Environment Court, contending that the project's greenhouse gas assessment should have included greenhouse gas emissions from the combustion of coal bought from the project by third parties. The Court found for the plaintiff, holding that the failure to take into account the cumulative impacts of greenhouse gas emissions produced by the project violated the "environmentally sustainable development" principles of intergenerational equity and the precautionary principle.[154]

- **Germany**
  - In 2007, the NGO GermanWatch filed a complaint against Volkswagen with the Federal Ministry of Economics, contending that the auto companies "climate damaging product range" violates the OECD Guidelines for Multinational Enterprises by contravening principles of global sustainable development.[155] The Guidelines provide for a mediation process between the complainant and companies, and if this fails to resolve a complaint, requires

---

[151] Greenpeace New Zealand, Inc. v. Northland Reg'l Council and Mighty River Power Ltd., High Court of New Zealand, Auckland Registry, CIV 2006-404-004617 (2006).
[152] Genesis Power Ltd. v. Franklin Dist. Council, Decision No. A 148/2005, *available at* http://www.climatelaw.org/cases/elaw/wind.farms.decision.2005.pdf (last visited Aug. 8, 2007).
[153] Australian Conservation Found. v. Latrobe City Council, 140 LGERA 100 (2004).
[154] *Gray v. Minister for Planning* [2006] NSWLEC 720.
[155] Germanwatch, *Complaint against Volkswagen AG under the OECD Guidelines for Multinational Enterprises* (2000), submitted May 7, 2007, *available at* http://www.germanwatch.org/corp/vw-besche.pdf (last visited May 25, 2008); *see also* Cornelia Heydenreich, Gunda Züllich & Christoph Bals, *Germanwatch Raises Complaint Against Volkswagen 2*, GermanWatch Briefing Paper (2007), *available at* http://www.germanwatch.org/corp/vw-hg07e.pdf (last visited May 25, 2008).

a National Contact Point to make recommendations on the implementation of the Guidelines.[156] The complaint has not yet been resolved.
- In 2004, two NGOs, Germanwatch and BUND (the German branch of Friends of the Earth), brought an action in the Administrative Court in Berlin against the German Federal Ministry of Economics and Labor. The Applicants sought to compel the German government to disclose the contribution to climate change caused by projects supported by the German export credit agency Euler Hermes AG ("Hermes").[157] While the parties ultimately settled, the Applicants pressed for the settlement to set forth in the framework of a court order (Beschluss). The court order outlines the terms of the settlement, which included agreement by the defendant to disclose (1) all energy production projects of a certain value and duration – arranged by the kinds of energy – for which defendant had provided export credit since January 2003; (2) the total sum of credit provided; and (3) where available, specific information about the project, including kinds and origins of fuel, fuel output per ton, and projected period of operation of the plant. The Court found a legal basis for the Applicant's request in the German Access to Environmental Information Act (*Umwelt informations gesetz*).[158]
- **Nigeria**
  - Nigerian citizens living near oil production facilities that flare off natural gas filed a lawsuit against Royal Dutch Shell and other companies engaged in the practice. The plaintiffs alleged that the practice releases substantial amounts of greenhouse gases[159] and other pollutants into the atmosphere and violates their constitutional rights to life and dignity. The Federal High Court found for the plaintiffs, concluding that defendants' conduct constituted "a gross violation of [plaintiffs'] fundamental right to life (including healthy environment) and dignity of human person as enshrined in the Constitution," though it did not specifically address the impacts of greenhouse gas emissions. Defendants were also restrained from further flaring of gas in the plaintiffs' community.[160] The defendants were subsequently granted a "conditional stay of executive," permitting them to phase in the cessation of flaring; however, they have failed to comply with the conditions imposed by the Court to date.[161]

[156] *Complaint against Volkswagen AG under the OECD Guidelines for Multinational Enterprises, supra* note 155, at 6.
[157] Germanwatch & Bund, *German Government Sued over Climate Change: Briefing* 1 (2004), *available at* http://www.climatelaw.org/cases/case-documents/germany/export-credit-briefing.pdf (last visited June 9, 2009).
[158] Bund für Umwelt und Naturschutz Deutschland e.V. & Germanwatch e.V. v. the Federal Republic of Germany, Order, VG 10 A 215.04 (2004) (unofficial translation).
[159] The practice of gas flaring has contributed more greenhouse emissions than all other sources in sub-Saharan Africa. Friends of the Earth, *Shell Fails to Obey Court Order to Stop Nigeria Flaring, Again*, Media Advisor, May 2, 2007, *available at* http://www.foei.org/en/media/archive/2007/0502 (last visited May 25, 2008).
[160] Gbemre v. Shell Petroleum Dev. Co., Suit No. FHC/CS/B/153/2005, Order, Nov. 14, 2005.
[161] *Shell Fails to Obey Court Order to Stop Nigeria Flaring, Again, supra* note 159.

International Actions

- **Inuit Petition**
  - In 2005, a petition was filed with the Inter-American Commission on Human Rights on behalf of Inuit in Canada and the United States requesting relief for human rights violations associated with climate change "caused by actions and omissions of the United States."[162] The petition alleged that climate change threatened the lives, culture, and economy of the Inuit and constituted human rights violations under the American Declaration of the Rights and Duties of Man, as well as other human rights instruments. The Commission rejected the petition a year later, stating, "the information provided does not enable us to determine whether the alleged facts would tend to characterize a violation of the rights protected by the American Declaration."[163] However, it subsequently agreed to a hearing to more closely examine the nexus of human rights and climate change, which took place in March 2007; the Commission is currently deliberating.
- **World Heritage Committee Petitions**
  - Between 2004 and 2006, several petitions and a report were filed by NGOs with the World Heritage Committee to list World Heritage sites in Australia, Belize, Peru, Nepal, Canada, and the United States on the "List of World Heritage in Danger" under the World Heritage Convention on the grounds that they were threatened by climate change.[164] At its Thirtieth Session in 2006, the Committee decided not to list the sites and also rejected a request to encourage the Parties to draw on projections from the Intergovernmental Panel on Climate Change when assessing risks to World Heritage Sites.[165] The Committee did, however, adopt a "Strategy to Assist State Parties to Implement Appropriate Management Responses" to climate change and urged the Parties to the World Heritage Convention to implement the Strategy. Moreover, the Committee decided that World Heritage sites could be inscribed on the List of World Heritage in Danger on a case-by-case basis, but also called for a study on alternatives to such listings.[166]

The tremendous legal breadth of these cases is striking. Unlike efforts to regulate climate change through the international treaty regime, which clearly fall under

---

[162] Petition to the Inter American Commission on Human Rights Seeking Relief from Violations Resulting from Global Warming Caused by Acts and Omissions of the United States 1, Dec. 7, 2005, *available at* http://www.inuitcircumpolar.com/files/uploads/icc-files/FINALPetitionICC.pdf (last visited May 25, 2008).

[163] Letter from the Organization of American States to Sheila Watt-Cloutier et al. regarding Petition No. P-1413–05, Nov. 16, 2006, *available at* http://graphics8.nytimes.com/packages/pdf/science/16commissionletter.pdf (last visited May 25, 2008).

[164] For a summary of the petitions, *see* http://www.climatelaw.org/cases (last visited Aug. 9, 2007).

[165] *Heritage Body 'No' to Carbon Cuts*, BBC News, July 10, 2006, *available at* http://news.bbc.co.uk/2/hi/science/nature/5164476.stm (last visited May 25, 2008).

[166] World Heritage Convention, *World Heritage Committee Adopts Strategy on Heritage and Climate Change*, *available at* http://whc.unesco.org/en/news/262 (last visited May 25, 2008).

international environmental law, these suits and petitions employ a wide range of legal theories that intersect through their connection to the problem of climate change. However, despite their diversity, these cases generally involve two overarching themes: (1) disputes over the appropriate role of government in regulating greenhouse gas emissions and (2) efforts to force major corporate emitters to reduce their emissions. These dynamics reinforce the mixed public-private nature of anthropogenic climate change and the state-corporate regulatory dynamics that underlie both the problem and its solution.[167]

4. THE NEED FOR ASSESSMENT

As climate change litigation proliferates around the world, an assessment of what its role is and should be in transnational regulatory governance becomes important. This volume provides such an assessment by exploring representative examples at subnational, national, and supranational levels. Through employing the perspectives of academics and practitioners on a wide range of adjudication, the book explores the present and future of this litigation as part of multiscalar regulation of climate change.

The first part of this book focuses on subnational litigation. Stephanie Stern's chapter analyzes litigation in the mid-1990s over Minnesota's early efforts to include carbon dioxide in environmental cost valuation. In so doing, the chapter explores the role that even weak state regulation can play in addressing greenhouse gas emissions. Lesley McAllister's chapter describes several disputes in Australian courts over the greenhouse gas impacts of coal mining and discusses the role that such cases can play in encouraging the inclusion of emissions in environmental assessment. The chapter by Katherine Trisolini and Jonathan Zasloff considers a dispute over the siting of a wind farm in New Zealand and its implications for the involvement of localities in the climate regulation. Finally, Mary Wood's chapter on the public trust doctrine explores the potential use of these governmental responsibilities to the people to address emissions and impacts.

The second part of the book looks at national-level cases. Hari Osofsky's chapter on *Massachusetts v. EPA* examines the way in which the case involves disputes over the scale of climate regulation and the implications of viewing the case through a scalar lens. David Grossman's chapter on the use of tort law against greenhouse gas emitters discusses pending cases against the auto and power industries, as well as broader questions about the applicability of tort law to climate change. Jeff Stempel's analysis of climate change and insurance law analyzes the extent to which corporate liability insurance might apply to these suits. The chapter by Kassie Siegel and Brendan Cummings considers the ways in which the Endangered Species Act has and could be used to address climate impacts. Finally, Amy Sinden's chapter on a

---

[167] *See* Osofsky, *supra* note 130.

Nigerian human rights case over gas flaring engages the possibilities for applying a rights framework to the problem of climate change.

The third part of the book analyzes supranational cases. Erica Thorson's chapter on the World Heritage Convention petitions and report considers the role that this treaty and its danger listing process has and should play in addressing climate change. Hari Osofsky's chapter on the Inuit's petition to the Inter-American Commission on Human Rights analyzes the role of actions whose ability to affect direct formal change is limited. The chapter by Jennifer Gleason and David Hunter explores the possibility of actions using international financial mechanisms. William Burns's chapter discusses the way in which the U.N. Fish Stocks Agreement might be used to address climate change. Andrew Strauss's chapter considers the possibility of an action at the International Court of Justice. Finally, a chapter by David Hunter explores these petitions in the broader context of international environmental law.

The book concludes by synthesizing these individual accounts and returning to broader questions of governance. It argues that climate adjudication helps to provide policy dialogue across scales needed to address the regulatory challenges of climate change. Although litigation alone cannot solve this overwhelming problem, it serves as an important tool in encouraging much-needed innovation and action.

PART I

SUBNATIONAL CASE STUDIES

2

State Action as Political Voice in Climate Change Policy: A Case Study of the Minnesota Environmental Cost Valuation Regulation

Stephanie Stern*

INTRODUCTION

As the debate over global warming intensified during the Bush administration, state legislatures in the United States adopted regulations that conveyed their discontent with the failure of the national government to regulate carbon dioxide emissions adequately or to adopt the Kyoto Protocol. Even with the Obama administration's efforts at greater federal regulation, state activity continues. These subnational efforts by states range in stringency but often stop short of substantive regulation that burdens in-state business interests.[1] Such weak or "symbolic" regulation nonetheless plays an important role in the global climate change debate by fostering political voice, creating a threat of future regulatory action, and legitimating climate change as a legally redressable harm.[2] An individual state cannot make a significant impact on atmospheric carbon dioxide levels or arrest global warming. However, carbon dioxide regulation by states can make a strong statement about the political will to address global warming – a statement that has grown louder as individual state legislation encourages other states to act and in turn brings pressure to bear upon the federal government.[3]

---

* Associate Professor, Chicago-Kent College of Law. I would like to thank Greg Shaffer, Kirsten Engel, Barbara Freese, Fred Lebaron, and Annecoos Wiersema for their helpful comments and Jennifer Mongillo for her able research assistance.

[1] The trend of relatively weak state regulation may be slowly shifting. California recently passed the AB32 legislation, which requires a 25% reduction in the carbon dioxide produced within the state by 2020. Similarly, in 2001 the governors of the New England states and the eastern Canadian provinces signed a pact to reduce greenhouse gas emissions to 1990 levels by 2010 and to 10% below that by 2020.

[2] Symbolic regulation takes different forms. Legislators may impose only minor burdens on industry through legislation that is substantively weak or limited to information disclosure. Alternatively, lawmakers may enact unrealistically strict and sweeping regulatory measures that agencies cannot implement without great delay and compromise. See John P. Dwyer, *The Pathology of Symbolic Legislation*, 17 ECOLOGY L.Q. 233, 233–34 (1990) (discussing the harms from this latter type of symbolic legislation to the regulatory process and public debate).

[3] See Kirsten H. Engel & Scott R. Saleska, *Subglobal Regulation of the Global Commons: The Case of Climate Change*, 32 ECOLOGY L.Q. 183, 224–26 (2005) (describing how subnational levels of government, such as individual states, can motivate industry to support federal regulation by creating an inconsistent patchwork of state laws).

There has been increasing interest among state policymakers in regulatory models that force polluting entities to internalize the societal costs of their carbon dioxide emissions. Electricity generation, which is responsible for 38% of U.S. carbon dioxide emissions, has been one target of state regulatory efforts.[4] In 1993, well before the current flurry of climate change activity, Minnesota enacted an environmental cost valuation statute that requires utility companies to provide estimates of environmental costs associated with power generation. The statute empowers the Minnesota Public Utilities Commission (Commission) to consider these costs when approving resource plans or issuing permits. The statute delegated to the Commission the task of determining which environmental externalities to value and how to quantify those costs. The Commission in turn charged an administrative law judge with overseeing a contested case proceeding and drafting detailed recommendations for covered pollutants and cost value ranges (i.e., the lowest reasonable value and a midlevel value for the environmental costs of electricity generation).

The contested case proceeding, *In the Matter of Quantification of Environmental Costs*, created a dual role for the administrative law judge as both an interpreter and a creator of law.[5] The administrative law judge, Allan Klein, weighed expert testimony and proposals in light of the statutory mandate and crafted the implementing regulations that the Commission subsequently adopted. Judge Klein recommended requiring cost valuation for carbon dioxide emissions, but proposed cost value ranges that were too low to influence typical Commission decisions. Following the Commission's final order adopting these cost value ranges, industry and environmental interests petitioned the Minnesota Court of Appeals for a writ of certiorari.[6] The Court of Appeals upheld the cost value regulation, finding that the Commission was acting within the sphere of its administrative expertise and thus was entitled to significant judicial deference.[7]

The Minnesota environmental cost value regulation provides a case study of the linkages between judicial and regulatory dialogues and the multifaceted role of state judges in adjudicating global public goods problems. The regulation also offers a more nuanced view of the effects of weak or symbolic regulation on industry, government, and the public. At first glance, the environmental cost regulation appears to lack meaningful impact. Closer examination reveals a more subtle and complex dynamic. The relatively weak regulation and low cost values that the Commission ultimately adopted were a compromise between the state's desire to introduce carbon dioxide regulation and to safeguard in-state businesses. Although the carbon dioxide cost values had limited regulatory impact, the regulation nonetheless had important

---

[4] See U.S. Environmental Protection Agency, Human-Related Sources and Sinks of Carbon Dioxide, *available at* http://www.epa.gov/climatechange/emissions/co2_human.html.
[5] See In the Matter of the Quantification of Environmental Costs, Office of Administrative Hearings for the Minnesota Public Utilities Commission, Findings of Fact, Conclusions, Recommendation and Memorandum (Mar. 22, 1996), *available at* http://www.oah.state.mn.us/aljBase/25008632.rt3.htm.
[6] See In re Matter of Quantification of Environmental Costs, 578 N.W.2d 794, 799 (Minn. Ct. App. 1998).
[7] See id.

indirect effects on political awareness and industry action. First, the Minnesota regulation served as a form of political protest against national policy. The cost value regulation fostered political voice and encouraged regulatory dialogue and information sharing. Second, the statute created risk for utility companies that the Commission would delay or deny permits for high-polluting baseload facilities or that the Commission would substantially increase cost values in the future. Thus, while the regulation was weak overall, it held the threat of being significantly stronger in certain circumstances. Perhaps most importantly, the use and acceptance of cost valuation in Minnesota raised the specter of future regulation that would require utilities to pay for, rather than merely report, the social costs of electricity generation. The holdings of the administrative law judge and the Minnesota Court of Appeals amplified this regulatory threat. The contested case proceeding and appellate litigation paved the way for future regulation by substantiating the role of carbon dioxide in global warming and creating a precedent that scientific uncertainty would not bar climate change regulation.[8]

1. MINNESOTA ENVIRONMENTAL COST VALUATION: A TALE OF LEGISLATIVE, ADMINISTRATIVE, AND JUDICIAL INTERACTION

1.1. Legislative Efforts

In 1993, Minnesota enacted § 216B.2422, an environmental externality reporting statute, which requires state utility companies to submit information on the environmental costs of their electricity generation. This statute replaced an "adder approach"[9] where utilities compensated renewable energy providers for avoided costs by purchasing mandated percentages of renewable energy.[10] The adder approach was passed in 1991 without debate on the floor as an amendment to another bill and was unusual for the high burdens it imposed on industry (and thus consumers).[11] The statute required utility companies to pay renewable energy providers for the social costs avoided by the renewable energy (i.e., positive externalities) by buying specified amounts of renewable energy from those providers. Notably, the burdensome nature of the original adder approach departs from the public choice model, where politicians attempt to curry public favor while imposing few if any costs on

---

[8] Holly Doremus describes a similar pattern of judicial deference to agency science in the federal system but notes that in some contexts, such as the Endangered Species Act, courts have frequently taken a more interventionist approach. *See* Holly Doremus, *The Purposes, Effects, and Future of the Endangered Species Act's Best Available Science Mandate*, 34 ENVTL. L. 397, 430–31 (2004) ("[W]here there is substantial scientific uncertainty, such that experts disagree on the interpretation of the available data, the agency's interpretation will generally enjoy substantial deference.").
[9] Adder approaches add a unit externality cost to the standard resource cost to reflect costs to society. Adders may be useful in resource planning, raising awareness, or, as in the 1991 Minnesota statute, estimating environmental taxes and payments.
[10] *See* MINN. STAT. § 216B.164(4)(b) (repealed 1993).
[11] *See* Interview with Michael Noble, Executive Director, Fresh Energy, in St. Paul, MN (Oct. 27, 2006).

industry. The utilities apparently did not notice the adder provision until after it was signed into law, at which point they vigorously opposed it. The passage of the adder approach, and the threat that it would remain on the books, gave the legislature and environmental interests strong political leverage. In the face of political outcry and significant implementation concerns, the legislature repealed the adder in 1993 but was able to replace it with a second piece of legislation, the environmental cost valuation statute. The information disclosure approach of the cost valuation statute did not equal the regulatory bite of the adder approach. Nonetheless, the cost valuation statute was progressive for a time when climate change was just beginning to appear on the radar of political consciousness.

The Minnesota cost valuation statute requires the Commission to "quantify and establish a range of environmental costs associated with each method of electricity generation" so that utilities can report their environmental externality costs.[12] The statute instructs utilities to "use the values established by the commission in conjunction with other external factors, including socioeconomic costs, when evaluating and selecting resource options in all proceedings before the commission."[13] Minnesota law requires utilities to submit resource plans every two years that assess energy resources and forecast energy needs. The Commission has the power to approve, reject, or modify resource plans consistent with the public interest.[14] In addition, if a utility wishes to build a large electric power generating plant, it must apply to the Commission for a certificate of need. To gain approval, the utility must show that the demand for electricity cannot be met more cost-effectively through other measures and that the planned nonrenewable facility imposes lesser socioeconomic and environmental costs than a renewable energy facility.[15] The cost valuation statute requires utilities to disclose environmental costs, which the Commission then considers when deciding whether to approve resource plans or issue certificate of need permits.

The cost valuation statute provided no substantive guidance on how to implement its broad dictate that the Commission consider environmental costs in resource planning decisions. It did not specify how the Commission should weigh environmental costs against other concerns, such as consumer rates. The statute also left open such

---

[12] MINN. STAT. § 216B.2422(3)(a).

[13] Id. A prior agency rule passed in 1990 required the Commission to consider adverse effects on the environment, but section 216B.2422 added a new dimension by requiring utilities to actually *quantify* these costs. See Findings of Fact, Conclusions of Law and Order Adopting Rules, Minnesota Public Utilities Commission, Docket No. E999/R-89-201 (July 10, 1990) (agency rule).

[14] See MINN. STAT. § 216B.2422(2) ("As part of its resource plan filing, a utility shall include the least cost plan for meeting 50 and 75 percent of all new and refurbished capacity needs through a combination of conservation and renewable energy resources.").

[15] See MINN. STAT. §§ 216B.243(3) & (3)(a). Minnesota law also requires that, "The commission shall not approve a new or refurbished nonrenewable energy facility in an integrated resource plan or certificate of need... nor shall the commission allow rate recovery for such a nonrenewable energy facility, unless the utility has demonstrated that a renewable energy facility is not in the public interest." See MINN. STAT. § 216B.2422(4).

important questions as what types of environmental impacts are subject to valuation (air emissions, water contaminants, land use, etc.), the methodology utility companies should use to quantify costs, and whether environmental costs values should vary by geographic area.[16]

1.2. *The Administrative Process and Contested Case Proceeding*

In August 1993, the Minnesota Public Utilities Commission, the administrative agency with rulemaking authority under the statute, turned to the task of filling the legislative gaps.[17] The Commission noted that the overarching goal of the implementing regulations was to "enable utility planners to compare the costs of resource alternatives more accurately and facilitate the selection of the lowest-cost resources from a total societal cost perspective."[18] On March 1, 1994, the Commission adopted interim rules providing values for five types of emissions and determined that externality values are required only for proceedings to select new resources that replace or supplement existing facilities.[19]

The Commission then initiated a contested case proceeding for a final determination of the pollutants subject to cost valuation and the cost value range for each pollutant. Three groups participated in the contested case proceeding: (1) parties representing industrial interests, including Western Fuels and the major utility companies; (2) parties representing environmental interests, including the Izaak Walton League and Minnesotans for an Energy Efficient Environment; and (3) the Minnesota Pollution Control agency and other representatives of Minnesota government. Contested case proceedings in Minnesota are quasi-judicial hearings before an administrative law judge where testimony is under oath and subject to cross-examination. The administrative law judge's recommendations carry significant weight but do not bind the Commission. For controversial issues, such as carbon dioxide cost valuation, there is a strong incentive for the Commission to adopt the administrative law judge's recommendation and thus remove itself from the direct line of political fire.

On March 22, 1996, the administrative law judge, Allan Klein, determined that cost values should apply to the direct effects or byproducts of electricity generation rather than to various methods of electricity generation.[20] Judge Klein advised focusing on the byproducts of electricity generation that create the most significant environmental costs, are the easiest to quantify, and are most likely to be associated

---

[16] *See* Order Establishing Procedure for Establishing Interim Environmental Cost Values, 1993 WL 733124, at *3 (Minn. P.U.C. Aug. 17, 1993).
[17] *See id.*
[18] *See* Order Establishing Interim Environmental Cost Values for Air Emissions Associated with Electric Generation, 1994 WL 232372 (Minn. P.U.C. Mar. 1, 1994).
[19] *See id.*
[20] *See* In the Matter of the Quantification of Environmental Costs, *supra* note 5.

with Minnesota's future electricity resource planning decisions.[21] He then turned to the task of determining the specific pollutants that would be subject to environmental cost valuation. Carbon dioxide was the most controversial of the proposed pollutants. The environmental interests lobbied strenuously for the inclusion of carbon dioxide, while the utility companies presented expert testimony that denied climate change or disputed its significance.

In the Matter of Quantification of Environmental Costs was one of the first cases to test the science of climate change in a court setting. At the time, global warming trends were evident. However, there was substantial uncertainty about attribution and severity of consequences.[22] During the contested case proceeding, the utilities presented an A-team of five expert witnesses on climate change science. These witnesses included such well-known global warming skeptics as Richard Lindzen, Pat Michaels, and Robert Balling. Lindzen testified that global warming would only raise temperatures 0.3 degrees Celsius over the next fifty years.[23] Michaels testified that the research does not indicate that sea levels will increase, and Balling opined that temperatures would rise no more than a degree.[24] These experts criticized the global warming data from the Intergovernmental Panel on Climate Change (IPCC) as politically motivated and scientifically unsound. The IPCC, a joint venture of the United Nations Environment Programme and the World Meteorological Organization, analyzes peer-reviewed and published scientific, technical, and socioeconomic research to provide regular assessments of the state of knowledge on climate change.[25] The environmental coalition could not afford equally high-profile experts and offered a policy analyst from the state pollution control agency and two biologists from the University of Minnesota who volunteered their time.[26]

After hearing the evidence, Judge Klein recommended setting cost values for carbon dioxide, as well as particulates, sulfur dioxide, nitrogen oxides, lead, and carbon monoxide. He relied heavily on the IPCC reports in his decision to require environmental cost valuation for carbon dioxide.[27] However, in the eleventh hour of the contested case proceeding, the environmental interests suffered a huge defeat when Judge Klein determined the cost value range for carbon dioxide. He found

---

[21] See id.
[22] For example, in setting interim cost values the Commission considered the conflicting interpretations of the impact of CO_2, observing, "[A]lthough the scientific community does not unanimously endorse the purported connection between CO_2 and global warming, the international community and the federal government have established policies aimed at reducing CO_2 emissions on the chance that these emissions will, in fact, produce the climatic changes forecast by many." See Order Establishing Interim Environmental Cost Values for Air Emissions Associated with Electric Generation, *supra* note 18.
[23] See Ross Gelbspan, The Heat Is On 39 (1995).
[24] See id.
[25] See Intergovernmental Panel on Climate Change, *available at* http://www.ipcc.ch/about/about.htm.
[26] See Interview with Barbara Freese, Consultant to the Union of Concerned Scientists, in St. Paul, MN (Dec. 4, 2006).
[27] See Interview with Allan Klein, Minnesota Administrative Law Judge, in MN (Dec. 6, 2006).

that sufficient uncertainty existed to warrant a conservative approach of adopting low carbon dioxide cost values. He chose a damage rate of 1% of GDP coupled with a discount rate of 3 to 5% and set the cost range of carbon dioxide at $0.28 to $2.92 per ton.[28] The discount rate is the amount by which costs or damages in future years are reduced for comparison with present-day values. To put into context the conservative nature of these estimates, a 2006 report authored by Sir Nicholas Stern, a senior economist of the British government, estimates the damage rate at 5 to 20% of global GDP.[29] The environmental coalition had argued for a discount rate of zero or 1%, which would have valued carbon dioxide at $25 per ton.[30] The expert from the Minnesota Pollution Control Agency (MPCA), Peter Ciborowski, had advocated a discount rate of 1.5% and a damage rate of 1 to 2%, yielding a cost value range of $4 to $28 per ton.[31] In his final recommendation, Judge Klein decided that a discount rate of 3 to 5% and lower-end damage estimate of 1% was "consistent with the policy goal of using conservative values in the face of uncertainty."[32] Ultimately, the $0.28 to $2.92 cost value range for carbon dioxide was a hollow victory for the environmental interests; the cost valuation figure was set too low to have a significant impact on resource planning decisions.

On December 16, 1996, the Commission adopted Judge Klein's recommendations for covered pollutants and cost value ranges. The Commission updated the values to 1995 dollars and set carbon dioxide cost values at $0.30 to $3.10 per ton.[33] The Commission did not calculate global costs of CO_2 but rather focused on harms to four geographic ranges: urban, metropolitan fringe, rural areas, and areas within 200 miles of the Minnesota border.[34] The Commission also rejected the utility companies' argument that a broad spectrum of socioeconomic costs not related to environmental impact should be factored into the initial environmental cost valuation. Instead, the Commission said the statute called for a two-step procedure. Utilities must present environmental cost figures that allow the Commission to compare values at the low end and high end of the environmental cost range.[35] After disclosing the environmental cost values, a utility may then present additional evidence addressing socioeconomic costs, such as impacts on employment or consumer rates.[36]

---

[28] See id.
[29] See Sir Nick Stern, Stern Review: The Economics of Climate Change Summary of Conclusions, at http://www.hm-treasury.gov.uk/media/8A8/C1/Summary_of_Conclusions.pdf.
[30] See In the Matter of the Quantification of Environmental Costs, *supra* note 5, at 30.
[31] See id. at 32.
[32] See id.
[33] See Order Establishing Ranges of Environmental Cost Values for Certain Pollutants Associated with Electricity Generation, 1996 WL 773354 (Minn. P.U.C. Dec. 16, 1996).
[34] On July 7, 1997 the Commission removed CO_2 values for the 200-mile range by setting the environmental cost valuation for that area at zero.
[35] See Order Establishing Ranges of Environmental Cost Values for Certain Pollutants Associated with Electricity Generation, *supra* note 33.
[36] See Order Modifying Administrative Law Judge's Fifth Prehearing Order on the Consideration of Socioeconomic Factors, 1994 WL 777118, at *3 (Minn. P.U.C. Oct. 28, 1994).

1.3. Litigation before the Minnesota Court of Appeals

Following the Commission's order setting final environmental cost values, a non-profit trade association of fuel producers, users, and suppliers and an environmental coalition filed a certiorari appeal with the Minnesota Court of Appeals. The appeal alleged that the Commission's decision to set values for carbon dioxide was improper. The Minnesota Court of Appeals upheld the order of the Commission, noting throughout its opinion that administrative bodies are entitled to judicial deference.[37]

With respect to carbon dioxide cost valuation, the court rejected the claim that the Commission should not be entitled to deference because it was acting outside the realm of its expertise when it evaluated the environmental impacts of carbon dioxide. The court held that the Commission was the appropriate locus of regulatory decision making and legislative delegation was appropriate. The court also rejected the claim that the determination of carbon dioxide values was improper because of the speculative nature of the evidence underlying the environmental cost values. The Commission and administrative law judge had met the substantial evidence standard through a careful review that included consideration of expert testimony, the experiences of New York in setting environmental costs, and the IPCC's First Assessment Report and 1992 supplement.[38]

The Western Fuels Association and other industry representatives also challenged the Commission's threshold finding that carbon dioxide harms the environment. The court acknowledged uncertainty in the science of global warming, but found that the administrative law judge had explained the basis for his determination and that the Commission properly relied on expert testimony and the IPCC report. The court held that scientific uncertainty did not bar agency action: "While we acknowledge the concerns about the uncertain and speculative nature of the available data, we are disinclined to prohibit the state from directing its instrumentalities to engage in environmentally-conscious planning strategies."[39]

The appellate litigation and the contested case proceeding illustrate the power of adjudication to substantiate the importance of climate change as well as affirm state power to address global-scale harms. The litigation established that scientific uncertainty did not bar climate change regulation, validated reliance on the expertise of the IPCC, and upheld the Commission's discretion to regulate carbon dioxide pursuant to the cost valuation statute. These holdings reduced the legal stumbling blocks to future climate change regulation in Minnesota.

---

[37] See In re *Matter of Quantification of Environmental Costs*, supra note 6. The court also upheld the administrative law judge's decision to limit environmental valuation to six air pollutants rather than creating cost values for each method of generation and to set the carbon dioxide values at zero for pollution within 200 miles of Minnesota's borders.

[38] See IPCC 1990 First Assessment Report and IPCC 1992 supplement, *available at* http://www.ipcc.ch/pub/reports.htm.

[39] See In re *Matter of Quantification of Environmental Costs*, supra note 6.

1.4. *The Minnesota Environmental Cost Value Regulation Today*

Ten years later, environmental cost valuation remains a requirement for resource planning in Minnesota. The Commission has updated carbon dioxide values slightly to $0.36 to $3.76 to reflect inflation.[40] To date, the Commission has never denied a certificate of need or other resource approval based on environmental costs. During this time period, however, the Commission has not reviewed an application to site a high-emissions coal baseload facility in Minnesota.

The Minnesota environmental cost valuation regulation and ensuing litigation provide an interesting case study of the effects of seemingly weak regulation on public perceptions and industry behavior. Section 2 turns to an analysis of how state climate change regulation, filtered through quasi-judicial and judicial proceedings, can serve as political protest. Political action or voice in turn fosters regulatory dialogue.

2. STATE ACTION AS VOICE AND DIALOGUE

In the decade since the promulgation of the Minnesota cost valuation regulation, there have been a large number of initiatives nationwide at the state and local level to address climate change.[41] Although the amount of activity at the subnational level is substantial, the impact of these initiatives is less impressive. Environmentalists and scholars have criticized states for eschewing strict emissions reductions (such as the approach adopted by California in its AB32 legislation) and instead enacting symbolic regulation that is either toothless or infeasible.

The Minnesota environmental cost value regulation reveals a more complex picture of the subtle, yet significant, effects that may accrue from weak or symbolic regulation. Symbolic regulation may convey political messages and alter environmental norms by signaling opposition to federal climate change policy. Such regulation can raise awareness and facilitate interstate dialogue and innovation.

2.1. *Symbolic Regulation and Political Voice*

Scholars have long recognized the allure of symbolic regulation to legislators, agency officials, and the public. Murray Edelman argued that most regulatory programs are "symbolic campaigns" where legislators and regulators frame a public problem in the abstract and dramatic language of public interest.[42] The regulatory process assuages

---

[40] *See* Minnesota Public Utilities Commission, Environmental Externalities Values Updated Through 2005, *available at* http://www.puc.state.mn.us/docs/eeupdate.06.pdf.

[41] *See* Barry G. Rabe, *North American Federalism and Climate Change Policy: American State and Canadian Provincial Policy*, 14 WIDENER L.J. 121, 152–53 (2004) (discussing state initiatives and the potential for decentralized action to reduce emissions and serve as a testing ground for subsequent national or international policy approaches).

[42] *See* MURRAY EDELMAN, SYMBOLIC USES OF POLITICS 56 (1964); *cf.* Mark Fenster, *Polemicist of Public Ignorance*, 17 CRITICAL REV. 367, 379–83 (2005) (criticizing the lack of methodological rigor in Edelman's work).

public concern because it implies that the government shares the concerns of the citizenry and is acting to address important social problems. This interchange masks, or at least distracts the public from, the subsequent political capture of administrative processes by regulated parties.[43]

More recent work on symbolic regulation has refined these ideas in the context of environmental law. John Dwyer has observed that symbolic legislation occurs when politicians capitalize on the strong public support for the environment by enacting sweeping mandates that transfer the burden of reformulation (and blame) to agencies.[44] Symbolic regulation allows industry to escape substantive burdens or at least enjoy a period of reprieve while agencies struggle to find an acceptable middle-ground approach.[45] In their analysis of state climate change legislation, Kirsten Engel and Scott Saleska have described how states typically focus their efforts on weak regulation or information-gathering requirements that "appear largely motivated by legislators' symbolic desire to be seen as 'doing something' about the pressing global problem of climate change" without actually imposing regulatory costs on industry.[46] States generally justify these measures on grounds other than, or in addition to, climate change, such as price stability or protection of in-state alternative energy producers.[47]

The Minnesota case study offers insight into the circumstances that are likely to produce symbolic regulation. The familiar account is that symbolic regulation occurs when vote-hungry legislators attempt to appease public demand for environmental protection by enacting weak legislation or passing sweeping mandates that shift responsibility to agencies. The Minnesota experience suggests that this view is too narrow. Several factors prompted the weak cost value regulation: a prior failure with the ambitious adder approach, the public goods nature of climate change, uncertainty about the specific consequences of global warming and its societal costs, and a less powerful state market where strict regulation would disadvantage in-state business interests. These factors were critical not only to the legislative enactment but also to the administrative law judge's conservative recommendations for cost value ranges. This analysis suggests that symbolic regulation may be driven not only by self-serving legislators but also by historical context, economic concerns, public goods issues, and scientific uncertainty.[48]

The Minnesota experience also illustrates how symbolic regulation may affect political debate, industry practices, and future regulation. The Minnesota regulation was undeniably part of a national trend of state climate change regulation

---

[43] See EDELMAN, supra note 42.
[44] See Dwyer, supra note 2, at 231.
[45] See id.
[46] See Engel & Saleska, supra note 3, at 215.
[47] See id. at 218–19.
[48] This echoes the legal realists' focus on the social context of lawmaking. As Joseph William Singer writes, "Legal principles are not inherent in some universal, timeless logical system; they are social constructs, designed by people in specific historical and social contexts for specific purposes to achieve specific ends." Joseph William Singer, *Legal Realism Now*, 76 CAL. L. REV. 465, 474 (1988).

focused on information-gathering and disclosure. Yet the relatively weak nature of the Minnesota regulation did not render it useless. Even if the legislative or administrative intent was to create regulation in name only (which is unlikely given the strong representation of environmental interests in the Minnesota legislature and the initial passage of the stricter adder approach), the cost value regulation affected climate change debate on a political level.

The Minnesota regulation was a statement of political opposition to ineffective national and global climate change policies. Both the administrative law judge and the Commission acknowledged on the record that Minnesota's total carbon dioxide emissions represented 0.1% of global carbon dioxide emissions.[49] During the contested case proceeding, Judge Klein observed:

> [E]ven if Minnesota's utilities stopped emitting any carbon dioxide, the global problem would be virtually unaffected by our act, *except* as our action, and similar actions of others in this country and abroad, cause national governments to take the kind of actions that *will* make a difference.[50]

This statement by the administrative law judge framed carbon dioxide valuation as a matter of political voice. The cost value regulation was a political call meant to resonate in the national and even global arenas. Although the cost valuation statute could not benefit Minnesota or affect the global climate change problem, it could serve as a state protest of national government inaction. Judith Resnik has described local climate change legislation and initiatives as "expressive efforts [and] political speech aimed at changing ideas and policies."[51] The Minnesota cost valuation statute and regulations were a step toward changing perceptions and laying the groundwork for regional or federal responses. Higher cost values or substantive emissions limitations would have intensified the political protest, but such strong measures were not realistic in light of Minnesota's limited market power and the public goods nature of global warming.

In his seminal book, *Exit, Voice, and Loyalty*, Albert Hirschman describes two means of influencing organizations and political structures: exit and voice. "Exit" refers to expressing discontent solely through one's actions, such as leaving a firm, purchasing an alternative product, abandoning a political party, or even emigrating from one's country or state. "Voice" is the use of political protest, criticism, or outcry to encourage change in firms or governments.[52] In cases where exit is not an option, voice carries the sole burden of providing information on preferences.[53] Global warming is a public goods problem that nations or states cannot resolve through exit

---

[49] *See* Order Establishing Ranges of Environmental Cost Values for Certain Pollutants Associated with Electricity Generation, *supra* note 33; In the Matter of the Quantification of Environmental Costs, *supra* note 6.

[50] *See* In the *Matter of the Quantification of Environmental Costs*, *supra* note 5.

[51] *See* Judith Resnik, *Law's Migration: American Exceptionalism, Silent Dialogues, and Federalism's Multiple Ports of Entry*, 115 YALE L.J. 1564, 1654 (2006).

[52] *See* ALBERT O. HIRSCHMAN, EXIT, VOICE, AND LOYALTY 30 (1970).

[53] *See id.* at 34.

strategies. No community or individual can exit from the *effects* of climate change because of the global dispersion of carbon dioxide and other greenhouse gases. Exit from climate change *policy* is possible but often prohibitively expensive. For example, individuals can immigrate to a new country or state with stricter regulation of greenhouse gases but are typically dissuaded from doing so by the steep costs of relocation. In the same vein, individual states can adopt carbon taxes or mandatory emissions limitations in the absence of federal regulation. However, most states, including Minnesota, lack the market power necessary to adopt strict emissions requirements without placing their state at a competitive disadvantage.

Minnesota's imperceptible impact on global warming and the harm to in-state industry from strict substantive regulation explain the substantive weakness and political tone of the statute.[54] Environmental regulation by states with much larger markets, such as California, may prompt a "trading up" or "race to the top" phenomenon. Firms who must comply with strict regulations in a large market may voluntarily adopt, or even lobby for, stringent standards in other markets. Firms behave this way either because production standardization is more cost-effective or because their prior investment in regulatory compliance gives them an advantage over competitors.[55] A smaller state such as Minnesota lacks the market power to effect a ratcheting up of standards.[56] The Commission's decision to require cost valuation for carbon dioxide, but to set costs quite low, is consistent with a model of state regulation as political voice that stops short of onerous requirements.

2.2. Regulation as Dialogue

The environmental cost valuation legislation and subsequent legal proceedings reveal how even weak regulation can create discourse and information sharing.[57] Barry Rabe has described state climate change regulation as an interactive learning process where policy diffusion occurs through formal interstate organizations, informal relationships, or access to state policy documents.[58] Regulators may exchange information through interaction at national meetings or via more informal networks. Diffusion may also occur absent personal contact when regulators access another

---

[54] Public goods, such as emissions, "*can* be consumed by everyone, but... there is *no escape* from consuming them unless one were to leave the community by which they are provided." *See id.* at 101.

[55] *See* DAVID VOGEL, TRADING UP: CONSUMER AND ENVIRONMENTAL REGULATION IN A GLOBAL ECONOMY 254–62 (1995).

[56] *See, e.g.*, Gregory Shaffer, *Globalization and Social Protection: The Impact of EU and International Rules in the Ratcheting Up of U.S. Privacy Standards*, 25 YALE J. INT'L L. 1, 80–88 (2000) (discussing how the stricter data privacy requirements of the European Union increased U.S. regulation of data privacy).

[57] In the context of international information sharing, Anne-Marie Slaughter has discussed how "networks of bureaucrats responding to international crises and planning to prevent future problems are more flexible than international institutions and expand the regulatory reach of all participating nations." *See* Anne-Marie Slaughter, *The Real New World Order*, FOREIGN AFF. (Sept./Oct. 1997).

[58] *See* Rabe, *supra* note 41, at 156–60.

state's policy materials online or in a publication.[59] The Minnesota cost value regulation provides an early example of climate change policy diffusion.[60]

In the Minnesota case study, multiple channels of communication facilitated information sharing and regulatory dialogue. State utility regulators have a long-standing formal organization, the National Association of Regulatory Utility Commissioners (NARUC). NARUC disseminates state regulatory information and sponsors annual conferences. In addition, the Minnesota administrative law judge had access to interstate information during the contested case proceeding. There was no direct communication between the administrative law judge and other judges or organizations;[61] instead the parties to the contested case proceeding promoted information exchange through memoranda and expert witness testimony describing various state approaches.

Both the Commission and the administrative law judge considered externality valuation regulation from other states in crafting the Minnesota rules. When setting interim values prior to the contested case proceeding, the Commission noted the importance of looking to regulation in other states, such as California, Nevada, and Massachusetts.[62] The Commission noted that assigning cost value ranges to the externalities most commonly valued elsewhere "ensures that Minnesota's interim values represent the broadest possible consensus concerning which externalities pose the most significant risk to the environment and which lend themselves most to quantification."[63] Similarly, in the contested case proceeding, the administrative law judge carefully considered the damage and discount rates applied to carbon dioxide cost values in states that had already adopted externality valuation approaches.[64]

The Minnesota cost valuation proceedings also drew international bodies, such as the Intergovernmental Panel on Climate Change, into the regulatory discourse.[65] In the contested case proceeding, industry interests argued that the IPCC's data were biased and lacked credibility.[66] Judge Klein rejected these claims and relied on the IPCC data in his recommendations, noting that "the IPCC reports are the most authoritative sources available for information on climate change issues."[67] The Minnesota Supreme Court similarly validated the IPCC's credibility when it

---

[59] *See id.* at 157.
[60] *See generally* Hari M. Osofsky, *Climate Change Litigation as Pluralist Legal Dialogue?*, 26 STAN. ENVTL. L.J. & 43 STAN. J. INT'L L. 181 (2007).
[61] *See* Klein Interview, *supra* note 27.
[62] *See* Order Establishing Interim Environmental Cost Values for Air Emissions Associated with Electric Generation, *supra* note 18.
[63] *See id.* For example, in creating its interim values, the Commission relied heavily on values developed by Pace University and the Bonneville Power Association.
[64] The Minnesota cost valuation regulation has not had similar influence on other states' utility regulation because of the deregulation movement that began in the mid-1990s. *See* N. Edward Coulson et al., *The Effect of Electricity Deregulation on State Economies* 4, *available at* http://econ.la.psu.edu/~ecoulson/electric.pdf (Mar. 2005).
[65] *See supra* Section 1.
[66] *See* Klein Interview, *supra* note 27.
[67] *See In the Matter of the Quantification of Environmental Costs*, *supra* note 5, at 29.

held that the administrative law judge and the Commission based their decision on sufficient evidence, including careful review of the research reports of the IPCC.

In summary, the Minnesota environmental cost value regulation demonstrates how interlocking judicial and administrative processes can encourage political voice and foster regulatory dialogues. Through symbolic regulation, states can express political opposition to national or international inaction. Such regulation may also increase information sharing as judges and regulators look to existing state models and even international organizations.

3. THREATS TO INDUSTRY: BARGAINING IN THE SHADOW OF FUTURE REGULATION

Why did the cost valuation statute, with its low carbon dioxide values and weak substantive impact, draw such vigorous opposition from utilities? The reason is that the statute created multiple layers of risk for the utility industry. The regulations heightened the risk of adverse regulatory decisions for one type of high-emissions plant, coal baseload facilities,[68] and left open the possibility that the Commission could substantially increase cost values later in time. More broadly, the regulation raised the threat of more stringent state greenhouse gas legislation in the future. The statute also imposed costs by adding another law to the existing multiplicity of state initiatives. The Minnesota case study illustrates how regulatory threats may drive litigation, affect voluntary behavior, and increase support for federal legislation.

Because the statute lodged considerable discretion in the Commission, it generated uncertainty for utilities. During the contested case proceeding, the utilities were very concerned that the administrative law judge would recommend a high cost value range for carbon dioxide.[69] Steep cost values increase the likelihood that the Commission will deny permits under Section 216B.243(3), which requires a showing that a nonrenewable facility imposes fewer socioeconomic and environmental costs than a renewable facility.[70] In addition, because the cost value information is available to the public, high cost values generate negative publicity for utilities.

Following the contested case proceeding, the Commission allayed some of these fears when it adopted low CO_2 cost values. Substantial cost values for carbon dioxide would have undoubtedly posed a greater threat to utilities, and a stronger incentive to reduce carbon dioxide emissions. Even with modest CO_2 cost values, however, a risk remained that the Commission could reject applications for high-emissions coal baseload plants. Coal baseload plants are the most environmentally harmful type

---

[68] Coal baseload plants provide a steady flow of power, regardless of the energy demanded by the grid, by operating continuously rather than cycling on and off. Since the enactment of the cost valuation statute, the Commission has not considered an application to site a coal baseload facility in Minnesota. Currently, the Commission is considering an application to site such a facility in South Dakota, but because carbon dioxide costs are set at zero within the 200-mile range of Minnesota's borders, cost values won't influence this proceeding.

[69] See Klein Interview, *supra* note 27.

[70] See MINN. STAT. §§ 216B.243(3) & 3(a).

of facility and thus are the most likely to impose significant aggregate costs despite low cost values. In addition, because the Commission has the authority to update the cost value ranges, utilities assumed an ongoing risk of regulatory revision. If the Commission were to significantly increase carbon dioxide cost values in the future, the revised regulation could jeopardize resource planning and permit approval.

Perhaps most importantly, the environmental cost valuation statute and subsequent litigation raised the specter of more stringent state regulation in the future. The recognition by the administrative law judge and the state appellate judge of climate change as an environmental harm added a gloss of judicial endorsement to the scientific consensus on global warming. The appellate litigation also established that scientific uncertainty does not bar administrative action to address global warming, a key holding for future climate change regulation in Minnesota. The battle at the contested case proceeding, with the utility companies offering extensive expert testimony disputing climate change effects, suggests that the industry saw the cost value statute as a foundation for future regulation. Cost valuation poses risks to industry because of its variable methodologies, the potential for expansive definitions of social costs, and the scientific uncertainty regarding the magnitude of global warming harms.[71] The utilities recognized that legislative, judicial, and public acceptance of cost valuation disclosure requirements could pave the way for future regulation that requires industry to *pay* for environmental externalities.

Minnesota law gives present effect to these regulatory threats by requiring the Commission to consider future compliance costs. For new nonrenewable facilities, the Commission must consider "the applicant's assessment of the risk of environmental costs and regulation on that proposed facility over the expected useful life of the plant."[72] The Commission must also assess whether utility resource plans "limit the risk of adverse effects upon the utility and its customers from financial, social, and technological factors that the utility cannot control."[73] In a current proceeding before the Commission, environmental nonprofits are arguing that a utility application for a coal-fired generating facility must account for the costs of complying with *future* climate change legislation.[74] If future regulatory costs are taken into account under these standards, utilities may not be able to meet Minnesota's statutory requirement that a new nonrenewable facility must be less expensive, on the basis of socioeconomic and environmental costs, than a renewable energy facility.

The environmental cost value statute also increased costs for utilities by adding another law to the growing patchwork of state regulation across the United States.[75]

[71] *See* Interview with James Alders, Regulatory Projects Manager, Xcel Energy, in Minneapolis, MN (Jan. 30, 2007) (describing cost valuation as the "wrong tool" for addressing the important issue of global warming).
[72] *See* MINN. STAT. § 216B.243(3)(12).
[73] *See* Minn. R. 7843.0500(3)(D) & (E).
[74] *See* Supplemental Comments of the Izaak Walton League of America – Midwest Office, Fresh Energy, The Union of Concerned Scientists, and the Minnesota Center for Environmental Advocacy to the Minnesota Public Utilities Commission, No. E-017/RP-05–968 (Jan. 3, 2006).
[75] *See* Alders Interview, *supra* note 71.

When states adopt varying climate change measures, businesses are forced to meet multiple, conflicting requirements.[76] The Minnesota administrative law judge was strongly influenced by arguments from environmental groups, both in this litigation and in a previous case on state acid rain regulation, that state laws play an important role in forcing federal action.[77] Subnational regulation may trigger regional, national, or even international regulation as industrial interest groups lobby for coordinated regulation on more favorable terms or as alternative product producers search for a larger market.[78] This view of the state's role in climate change regulation now appears prescient. In a notable reversal, several large energy firms began lobbying Congress in 2006 for a national greenhouse gas cap-and-trade program that would eliminate the current multiplicity of state mandates.[79] The proliferation of state legislation, including the Minnesota cost value statute, is one factor that has decreased industry resistance to federal greenhouse gas regulation.[80] Other motivations have included the passage of stricter EU regulations on emissions and strategic preferences to press for national greenhouse gas legislation during a Republican presidency.

In summary, the cost value statute offers a different perspective on seemingly weak or informational regulation. If part of a state's goal is to encourage federal action or voluntary industry efforts, then it may not be necessary to have a strong statute – the threat of harsher regulatory action in the future or a looming patchwork of inconsistent state laws may suffice. For example, since the enactment of the cost value statute, smaller utility companies in Minnesota have voluntarily increased their energy from renewable sources. The reason for this appears to be the likelihood of stricter state or national regulation in the future as well as the cyclical demands for power, state policies supporting renewables such as windpower, and the advantage to utilities of incrementally and cost-effectively increasing their renewable energy in advance.[81] The Minnesota case study suggests that the risk of future regulation affects utilities' energy generation decisions as they invest in infrastructure in the shadow of regulatory threats.

CONCLUSION

The federal government's reluctance to create national legislation or ratify the Kyoto Protocol under the Bush administration gave states latitude to create their own climate change initiatives. States frequently responded by enacting weak or symbolic

---

[76] In recent years, business executives have made public statements criticizing the "patchwork quilt" of state climate change regulation. See Rabe, *supra* note 41, at 139–41.
[77] See Klein Interview, *supra* note 27.
[78] See Engel & Saleska, *supra* note 3, at 223–28.
[79] See Steven Mufson & Juliet Eilperin, *Energy Firms Come to Terms with Climate Change*, WASH. POST, Nov. 25, 2006, at A01.
[80] See Freese Interview, *supra* note 26.
[81] See Interview with Carol Casebolt, Counsel for Minnesota Public Utilities Commission, in St. Paul, MN (Oct. 6, 2006).

regulation that lacks regulatory bite. The dispute over Minnesota's environmental cost valuation statute provides the basis for a richer account of the indirect benefits that accrue from symbolic regulation. The cost value regulation and subsequent litigation fostered political voice and expressed dissatisfaction with federal and international climate change policy. The statute altered the regulatory dynamic by creating a risk of adverse decisions and more stringent legislation in the future. These regulatory threats encouraged voluntary efforts by utility companies and increased their support for federal legislation. The cost value regulation also affected regulatory norms and established global warming as a harm that could be redressed by state legislative and administrative action.

3

Litigating Climate Change at the Coal Mine

Lesley K. McAllister*

INTRODUCTION

In Australia, the most prominent climate change cases have involved attempts to stop greenhouse gas emissions from the burning of coal before that coal is even mined. Suing under state or national environmental impact assessment laws, Australian environmentalists have sought to compel government agencies responsible for approving coal mining projects to consider the very significant amounts of greenhouse gases that will be emitted at the time that the mined coal is burned to generate energy. With this approach, environmentalists have had some notable success in ensuring that such "indirect" or "downstream" greenhouse gas emissions are assessed as part of the decision-making process.

The Australian coal mining cases analyzed in this chapter are harbingers of a potentially large wave of legal actions that could arise under environmental impact laws in jurisdictions throughout the world. Environmentalists dissatisfied with their government's climate change policies are particularly likely to seek remedies through the judiciary where possible. Yet as they do, they will inevitably confront complex and difficult issues regarding how and the extent to which a particular project under consideration can be said to have an impact on the global climate. Whether and how greenhouse gas emissions should be assessed in the context of environmental impact studies for particular local projects is a key question for environmental policy in our new carbon-constrained world.

Australia is a fitting place for climate change activists to lodge novel claims about the climate change implications of coal mining. As discussed in Section 1 of this chapter, Australia is deeply reliant on coal both as an export commodity and for domestic energy production, and Australia has the highest greenhouse gas emissions per capita of any country in the world. Section 2 describes the three most significant legal cases involving climate change in Australia, all of which have concerned the

---

* Associate Professor of Law, University of San Diego; Assistant Adjunct Professor, School of International Relations and Pacific Studies, University of California, San Diego. J.D., Stanford Law School, 2000; Ph.D., University of California, Berkeley, 2004. The author's corresponding address is mcallister@sandiego.edu.

assessment of the environmental impacts of coal mines. Section 3 analyzes several legal barriers that may prevent favorable outcomes for environmental plaintiffs in such cases, with a focus on Australia and the United States.

1. KING COAL IN AUSTRALIA

Like the United States, Australia is rich in coal. Australia is the world's largest exporter of coal and the world's fourth-largest producer of coal after China, the United States, and India.[1] More than 70 percent of Australia's coal exports go to Asian countries, particularly Japan, India, Korea, and Taiwan.[2] Coal is also Australia's largest export commodity, accounting for about 16 percent of the value of total merchandise exports.[3]

Australia also relies heavily on coal for domestic energy production. Whereas coal accounts for 40 percent of electricity production worldwide, about 76 percent of electricity production in Australia comes from coal.[4] With its reliance on coal, Australia has the highest greenhouse gas emissions per capita in the world.[5] The stationary energy sector, which is fueled primarily by coal, accounts for 50 percent of Australia's greenhouse gas emissions, and its emissions rose by 47 percent between 1990 and 2006.[6]

Brown coal, of which Australia has extensive reserves, is a particularly egregious contributor to greenhouse gas emissions. Brown coal, also called lignite coal, is a low-rank form of coal that burns less efficiently than black coal because of higher moisture content.[7] Brown coal power plants typically emit about 37 percent more carbon dioxide per unit of power output than a black coal power plant.[8] Australia

---

[1] World Coal Institute, *The Coal Resource: A Comprehensive Overview of Coal*, at 14–15 (2005), available at http://www.worldcoal.org/assets_cm/files/PDF/thecoalresource.pdf (last visited February 27, 2009) [hereinafter WCI].

[2] Department of Resources, Energy and Tourism, *Australia's Coal Industry*, available at http://www.ret.gov.au/resources/mining/australian_mineral_commodities/Pages/australia_coal_industry.aspx (last visited February 27, 2009); Department of Foreign Affairs and Trade, *Composition of Trade Australia, 2005–06*, at 5 (November 2006), available at http://www.dfat.gov.au/publications/stats-pubs/cot_fy2006_analysis.pdf (last visited February 27, 2009) [hereinafter Composition of Trade Australia].

[3] Composition of Trade Australia, *supra* note 2, at 5.

[4] WCI, *supra* note 1, at preface.

[5] Pew Center on Global Climate Change, *Climate Data: Insights and Observations*, at 11 (December 2004), available at http://www.pewclimate.org/docUploads/Climate%20Data%20new.pdf (last visited February 27, 2009).

[6] Department of Climate Change, *National Greenhouse Gas Inventory 2006: Accounting for the Kyoto Target*, at 1 (2008), available at http://www.climatechange.gov.au/inventory/2006/index.html (last visited February 27, 2009).

[7] WCI, *supra* note 1, at 14–15.

[8] Institute for Sustainable Futures, *Why Brown Coal Should Stay in the Ground: Greenhouse Implications of the Proposed Expansion of Brown Coal Exploration and Mining in Victoria*, at 5 (2002), available at http://www.isf.uts.edu.au/publications/tarlo2002whybrowncoal.pdf (last visited February 27, 2009) [hereinafter ISF].

contains 20 percent of all demonstrated brown coal reserves in the world, and about 20 percent of all coal mined in Australia is brown coal.[9]

Australia's reliance on coal for export revenue and domestic energy production has influenced Australia's position in international climate change policy debates. Other than the United States, Australia was the only major industrialized country to reject the Kyoto Protocol.[10] As is also true in the United States, climate change activists have pursued remedies through litigation. The next section describes the most prominent climate change cases that have been filed in Australian courts.

2. THE COAL MINING CASES

The most significant legal cases relating to climate change in Australia have alleged inadequate assessments of the environmental impacts of proposed coal mining projects.[11] In the Hazelwood case, environmental groups alleged a violation of the land use planning statute of the state of Victoria. In the Isaac Plains and Sonoma Mines case, environmental groups alleged a violation of the national environmental impact law. Finally, in the Anvil Hill case, an environmentalist alleged a violation of the environmental planning statute of the state of New South Wales.

2.1. *The Hazelwood Case*

In 2004, the Victorian Civil and Administrative Tribunal (the Tribunal) held that the environmental planning studies required for the approval of an expansion of the Hazelwood coal mine must include an assessment of the greenhouse gas emissions that would result from the burning of the coal in the associated Hazelwood Power Station.[12] Although the greenhouse gas emissions were an indirect rather than a direct impact of the mining project, the Tribunal held that they had to be considered by the government because the plaintiff environmental organizations had submitted evidence about these impacts and this was a "relevant submission" under Victoria's Planning and Environment Act of 1987 (PE Act).[13]

---

[9] Australian Bureau of Agricultural Resources and Economics (ABARE)/Department of Industry, Tourism and Resources, *Energy in Australia 2005*, at 6 and 11 (2005), available at http://abareonlineshop.com/PdfFiles/energy2005_parta.pdf (last visited February 27, 2009).

[10] Kyoto Protocol to the U.N. Framework Convention on Climate Change, Dec. 10, 1997, art. 3, U.N. Doc. FCCC/CP/1997/L.7/Add.1, 37 I.L.M. 22 (1997), available at http://unfccc.int/resource/docs/convkp/kpeng.pdf (last visited February 27, 2009). Australia ultimately ratified the Kyoto Protocol on December 3, 2007.

[11] For further information about legal approaches to address climate change in Australia, see the website announcing the launching of the Australian Climate Justice Project in July 2003, available at http://www.cana.net.au/index.php?site_var=333 (last visited February 27, 2009).

[12] Australian Conservation Foundation v. Minister for Planning [2004] VCAT 2029 (October 29, 2004), available at http://www.austlii.edu.au/au/cases/vic/VCAT/2004/2029.html (last visited February 27, 2009) [hereinafter Hazelwood Decision].

[13] Full text of the PE Act is available at http://www.dms.dpc.vic.gov.au/Domino/Web_Notes/LDMS/PubLawToday.nsf/a12f6f60fbd56800ca256de500201e54/72df64b9bbadb89eca256ec3000084ef/$FILE/87-45a074.pdf (last visited February 27, 2009) [hereinafter PE Act].

International Power's Hazelwood Power Station is Australia's sixth-largest power station, and it supplies the state of Victoria with more than 20 percent of its baseload electricity.[14] The power station burns brown coal mined in the nearby Latrobe Valley, home to almost all of Australia's brown coal reserves.[15] Before it was purchased by International Power in 1996, the Hazelwood Power Station and the associated Hazelwood mine were owned and operated for more than thirty years by the State Electricity Commission of Victoria.[16]

In 2005, the environmental organization WWF labeled Hazelwood the dirtiest power station in the industrialized world.[17] Indeed, even as compared with other brown coal power stations in the state of Victoria, Hazelwood consistently emits larger quantities of greenhouse gases per unit of power produced.[18] The Hazelwood Power Station is Australia's largest single source of greenhouse gas emissions, accounting for 9 percent of Australia's total carbon dioxide pollution from power generators, an amount roughly equal to the greenhouse gas emissions of the entire fleet of 3.6 million cars in Victoria.[19]

In 2003, Hazelwood requested permission to open up and dredge a new part of the associated coal field referred to as the West Field. Hazelwood's operating mines had sufficient brown coal to fuel the power station only until 2009, and the mine expansion would provide enough coal for the station to remain operational until 2031.[20] Because the proposed project required the relocation of a highway and a river, International Power was required to seek an amendment to the Latrobe Planning Scheme under the PE Act.[21] International Power was also required to prepare an Environmental Effects Statement (EES) under Victoria's Environmental Effects

---

[14] Environmental Effects Act 1978, Hazelwood Mine West Field project (Phase 2) Assessment, at 1 (September 2005); The Australia Institute, *Victoria's Greenhouse Policy: The Moment of Truth*, at 1 (May 2005), available at http://www.tai.org.au/documents/downloads/WP75.pdf (last visited February 27, 2009) [hereinafter Victoria's Greenhouse Policy].

[15] Coal Mineral Fact Sheets, available at http://www.australianminesatlas.gov.au/education/fact_sheets/coal.jsp (last visited February 27, 2009). *See also* International Power Hazelwood, *International Power Hazelwood Business Report*, at 31 (2004), available at http://www.ipplc.com.au/_modules/Uploader/_uploaderBin/jbrinkworth/de491eb69a3126a3a352825375abfaec2004AnnualBusinessEnviroSocial Report.pdf (last visited February 27, 2009) [hereinafter IPH Business Report].

[16] Environmental Effects Act 1978, Hazelwood Mine West Field Project (Phase 2) Assessment, at 1 (September 2005), available at http://www.dse.vic.gov.au/CA256F310024B628/0/710B0D60BA19961 FCA2572F900136F87/$File/Hazelwood+Mine+West+Field+-+Ministers+Assessment.pdf (last visited February 27, 2009) [hereinafter Hazelwood Mine Assessment].

[17] WWF – Australia, *Hazelwood Tops International List of Dirty Power Stations* (July 13, 2005), available at http://wwf.org.au/news/n223 (last visited February 27, 2009) [hereinafter WWF].

[18] Charles Berger (Australian Conservation Foundation) and Tricia Phelan (Environment Victoria), *Greenhouse Pollution Intensity in the Victorian Brown Coal Power Industry* (May 2005), available at http://www.envict.org.au/file/Greenhouse_Brown_Coal_05.pdf (last visited February 27, 2009).

[19] Victoria's Greenhouse Policy, *supra* note 14, at 1–2; Greenpeace Australia Pacific, *Power Station Comes to Bracks* (September 5, 2005), available at http://www.greenpeace.org/australia/news-and-events/news/Climate-change/power-station-comes-to-bracks (last visited February 27, 2009).

[20] Hazelwood Mine Assessment, *supra* note 16, at 1.

[21] Hazelwood Decision, *supra* note 12, at paragraph 4.

Act (EE Act) of 1978.[22] The Minister for Planning appointed a panel to jointly consider the sufficiency of the EES under the EE Act and submissions regarding the amendment to the planning scheme under the PE Act.[23]

As part of the EES, International Power was required to discuss the direct effects of the coal mining operation on the atmosphere, including the emission of greenhouse gases from dredging the coal.[24] However, the Minister issued "terms of reference" to the panel for judging the sufficiency of the EES that instructed them that, when analyzing the environmental impacts of the project, they were not to consider greenhouse gas emissions from the burning of coal at the Hazelwood power plant.[25] The panel thereafter held a hearing and informed interested parties that it would abide by the terms of reference handed down by the Minister and would not consider the greenhouse gas emissions from the power plant in either the EES or in written submissions under the PE Act.[26] In August 2004, the panel held a hearing and considered a submission from the Australian Conservation Foundation (ACF) that presented testimony from an expert witness regarding the environmental impacts that would be caused by the burning of the mined coal.[27] Although the panel heard the submission, it said it would not consider it because it was outside the scope of the terms of reference.[28]

Soon after, four conservation groups, including the ACF, filed suit claiming that the panel failed to comply with the PE Act.[29] The Act states, in relevant part, "a panel appointed to consider submissions about an amendment to a planning scheme must consider all submissions referred to it and give a reasonable opportunity to be heard to any person who has made a submission referred to it."[30] In October 2004, the Tribunal issued its decision, agreeing with plaintiffs that the panel failed to comply with the PE Act. According to the Tribunal, the panel was obligated to consider "all relevant submissions," which included all submissions that raise "planning issues" and are "about an amendment." Applying this test, the Tribunal found that the plaintiffs' submissions regarding the environmental impacts of greenhouse gases from the coal to be mined at Hazelwood were relevant.

The Tribunal held that greenhouse gas emissions are planning issues because the PE Act includes as objectives of planning "the maintenance of ecological processes" and the balancing of the "present and future interests of all Victorians."[31] According

---

[22] Hazelwood Decision, *supra* note 12, at paragraph 5.
[23] *Id.* at paragraph 8.
[24] Hazelwood Decision, *supra* note 12, at paragraph 5.
[25] *Id.* at paragraph 10.
[26] *Id.* at paragraph 12.
[27] *Id.* at paragraph 20.
[28] *Id.* See also *Climate Change Litigation: Analysing the Law, Scientific Evidence & Impacts on the Environment, Health & Property* 60 (Joseph Smith & David Shearman eds., 2006).
[29] Other plaintiffs in the case included WWF Australia, Environment Victoria, and the Climate Action Network Australia.
[30] PE Act, *supra* note 13, at Section 24. *See also* Hazelwood Decision, *supra* note 12, at paragraph 23.
[31] Hazelwood Decision, *supra* note 12, at paragraph 38.

to the Tribunal, "ecological processes" include processes within the atmosphere of the earth, including its chemistry and temperature.[32] The Tribunal also acknowledged that the use of energy resources in the present may have a cost to future generations. Thus, submissions that deal with the emission of greenhouses gases are not only environmental issues but also planning issues that panels are required to consider under the PE Act.

The Tribunal determined, moreover, that the ACF's submission was "about an amendment" because the greenhouse gases emitted by the power plant are a sufficiently related effect of the amendment to the zoning ordinance. According to the Tribunal, a submission is about an amendment even if it "relates to an *indirect* effect of the amendment, if there is a sufficient nexus between the amendment and the effect."[33] The sufficiency of the nexus can be assessed by considering "whether the effect may flow from the approval of the amendment; and if so, whether, having regard to the probability of the effect and the consequence of the effect (if it occurs), the effect is significant in the context of the amendment."[34]

In finding that it was necessary to consider indirect effects, the Tribunal followed a 2004 decision of the Federal Court of Australia that held that indirect as well as direct effects should be considered under Australia's Environment Protection and Biodiversity Conservation Act (EPBC Act).[35] In this case, referred to as the Nathan Dam case, the Court considered whether an environmental impact study for the construction of a dam could exclude the effects on the Great Barrier Reef World Heritage Area that would likely result from the increase in irrigated agriculture that the dam would enable.[36] The Court held that the meaning of "all adverse impacts" was not confined to direct physical impacts but also included indirect impacts and effects "which are sufficiently close to the action to allow it to be said, without straining the language, that they are, or would be, the consequences of the action on the protected matter."[37] The Court thus held that the Minister of the Environment must consider the downstream pollution by irrigators as an impact of the dam.

Similarly, the Victorian Tribunal found that, although the greenhouse gases from Hazelwood Power Station are not a direct effect of the amendment enabling more coal mining, they were an indirect effect. As reasoned by the Tribunal, if the amendment was approved, there would be a greater likelihood that the power plant would

---

[32] *Id.* at paragraph 43.
[33] *Id.* at paragraph 41 (emphasis in the original).
[34] *Id.*
[35] Full text of the EPBC Act is available at http://www.austlii.edu.au/au/legis/cth/consol_act/epabca1999588/ (last visited February 27, 2009); for more information about the EPBC Act, see http://www.deh.gov.au/epbc/ (last visited February 27, 2009).
[36] Minister for the Environment and Heritage v. Queensland Conservation Council Inc. [2004] 139 FCR 24 (30 July 2004), available at http://www.austlii.edu.au/au/cases/cth/FCAFC/2004/190.html (last visited February 27, 2009) [hereinafter Nathan Dam Decision]. This decision affirmed the judgment of a single judge of the Federal Court of Australia, Queensland Conservation Council Inc. v. Minister for Environment and Heritage [2003] FCA 1463 (19 December 2003).
[37] Nathan Dam Decision, *supra* note 36, at paragraph 53.

remain operational past 2009.³⁸ If this happened, there would be a greater likelihood that more greenhouse gases would be emitted, which might constitute a significant environmental effect.³⁹ Thus, the greenhouse gases released by Hazelwood are not only planning issues, but are also a result of the amendment and must be considered before the panel can recommend that the planning amendment be enacted.⁴⁰

Once the Tribunal handed down its decision in 2004, the panel reconsidered whether to recommend that the government approve the West Field expansion.⁴¹ After hearing and considering submissions about greenhouse gas emissions from the burning of the coal, the panel released a report in April 2005 recommending that the Victorian government permit the extension of the coal mine.⁴² The panel placed some importance on the fact that there were ongoing negotiations for an agreement between the government and International Power Hazelwood that would provide for the reduction of greenhouse emissions from the Hazelwood Power Station. The panel voiced its support for the successful conclusion of this negotiation.⁴³

In September 2005, the Victorian government issued its approval of the minefield expansion, thus granting Hazelwood access to 43 million tons of coal that would allow the power station to remain operational through 2031. The government concurrently announced that International Power Hazelwood had signed the Greenhouse Gas Reduction Deed, the centerpiece of which was a cap on the total amount of carbon dioxide emissions that the Hazelwood Power Station could produce during the remainder of its operating life.⁴⁴ The cap was set at 445 million tons, an expected 7 percent reduction in the plant's emissions during the life of the deed, after which the plant must close down.⁴⁵

---

³⁸ Hazelwood Decision, *supra* note 12, at paragraph 47.
³⁹ *Id.*
⁴⁰ *Id.* at paragraph 49.
⁴¹ The Department of Premier and Cabinet for the Victorian Government, *Hazelwood Agreement to Secure Victoria's Energy Supply While Reducing Greenhouse Emissions* (September 6, 2005), available at http://www.legislation.vic.gov.au/domino/Web_Notes/newmedia.nsf/798c8b072d117a01ca256 c8c0019bb01/ce988ef03b5c71d4ca25707500082f68!OpenDocument (last visited February 27, 2009).
⁴² Victorian Department of Sustainability and Environment, *Final Panel Report – Hazelwood West Field EES La Trobe Planning Scheme Amendment C32* (March 2005) at 216–17, available at http://www.dse.vic.gov.au/CA256F310024B628/0/62EC957E3BFDFD77CA2572F900138DD6/$File/Hazelwood+Mine+West+Feild+-+Panel+Report+Ch1-12.pdf (last visited February 27, 2009).
⁴³ *Id.* at 177 and 185. These negotiations were also mentioned in the Tribunal's decision, which cited a letter written by the Minister for Planning on August 11, 2004 (shortly after the first panel convened and well before the litigation commenced), stating, "The full development of the West Field beyond the existing licence boundary will be subject to an agreement being reached between Government and IPHR. As is publicly known, the Minister's intention is that greenhouse gas emissions from the Power Station associated with any coal outside the existing licence boundary should be substantially reduced." Hazelwood Decision, *supra* note 12, at paragraph 19.
⁴⁴ The Greenhouse Gas Reduction Deed is available at http://www.dpi.vic.gov.au/dpi/dpinenergy.nsf/childdocs/-3f827e74c37e0836ca25729d00101eb0-866b51f390263ba1ca2572b2001634f9-3cd640176546d95fca2572b2008396f1?open (last visited February 27, 2009) [hereinafter Deed].
⁴⁵ Deed, *supra* note 44, at Section 2.1. However, under Section 8, this cap only applies to the boilers that are in existence at the time the Deed is entered into. As such, if Hazelwood obtains the government's approval to build new boilers, those boilers will not fall under the scope of the Deed. *See also Australian Power Plant Agrees to Reduce Emissions in Deal to Expand Lifespan*, 28 INT'L ENVTL. REP. 683 (September 21, 2005).

The Deed also established six-year milestones for cumulative greenhouse gas emissions and a system whereby Hazelwood could earn "emissions offset credits" if the company invested in wind power and other renewable sources of energy.[46] Earning these credits would allow Hazelwood to exceed the six-year emissions milestones but would not alter the total cap placed on its emissions. The power station was required to submit reports to the government every six years when the intermediate targets had been set regarding the total carbon dioxide emitted from the plant and the number of credits received for emission offsets.[47]

The Deed sought to spur emissions reductions at Hazelwood in a couple of other ways. In addition to capping the total emissions from the station, the Deed required that the station use the best "commercially viable" means to reduce its carbon dioxide intensity.[48] Also, Hazelwood was required to submit annual reports to the Environment Minister discussing any technological advances discovered through internal research and development that might reduce the station's carbon dioxide emissions.

Although it was the first legal agreement of its kind, the Deed was criticized by many for being too lenient. The ACF observed that because the six-year pollution milestones were not binding on the power plant, it could pollute at its current rate until 2030.[49] The ACF also expressed concern over language in the Deed to the effect that the plant would be treated "equitably" by the government in any future greenhouse gas legislation.[50] Finally, the group doubted Hazelwood's ability to comply with the deed, given its history as the worst performer of the five brown coal power plants in the Latrobe Valley.[51] Another critic of the Deed, Environment Victoria, referred to the deed as containing "trivial – and partly unenforceable – environmental commitments."[52] Yet, despite the environmentalists' ultimate dissatisfaction with the outcome, the Hazelwood decision was an important victory. It was the first time in Australia that a government agency was required to consider the implications of greenhouse gas emissions from burning coal as part of approving a mining project.

2.2. The Isaac Plains and Sonoma Mines Case

In the Isaac Plains and Sonoma Mines case, the environmental group Wildlife Preservation Society of Queensland (WPS) sought judicial review in Federal Court

---

[46] Deed, *supra* note 44, at Sections 2.1 and 2.4.
[47] *Id.*, at Section 2.1.
[48] *Id.*, at Section 2.2.
[49] Australian Conservation Foundation, *Help Slash Climate Pollution: Object to the Expansion of Australia's Dirtiest Power Station*, available at http://www.acfonline.org.au/news.asp?news_id=589 (last visited February 27, 2009).
[50] *Id.*
[51] Australian Conservation Foundation, *Victoria's Polluting Power Stations Revealed* (May 17, 2005), available at http://www.acfonline.org.au/news.asp?news_id=81 (last visited February 27, 2009).
[52] *See* website of Environment Victoria at http://www.envict.org.au/inform.php?menu=5&submenu=475&item=1019 (last visited February 27, 2009); for criticism by Greenpeace, see Greenpeace Australia Pacific, *Wrong Way, Go Back! Steve Bracks Condemns Victoria to Climate Change* (September 6, 2005), available at http://www.greenpeace.org/australia/news-and-events/media/releases/climate-change/wrong-way-go-back-steve-brac (last visited February 27, 2009).

of two decisions of the Federal Ministry for Environment and Heritage.[53] The Ministry had determined that the proposed Isaac Plains and Sonoma Coal mining projects did not constitute "controlled actions," and as a result they did not require approval under the federal Environment Protection and Biodiversity Conservation Act 1999 (EPBC Act). The Court found in favor of the Minister, declining to find a sufficient nexus between the proposed projects and the alleged climate change impacts.

The EPBC Act sets forth the environmental responsibilities of the federal or "commonwealth" government and specifies seven "matters of national environmental significance" with respect to which the federal government can legislate.[54] These matters include World Heritage properties, national heritage places, wetlands of international importance, threatened species and ecological communities, migratory species, commonwealth marine areas, and nuclear actions including uranium mining.[55]

The Act requires government assessment and approval of any actions that are likely to have a significant impact on a matter of national environmental significance. Any person proposing to take an action that will, or is likely to, have a significant impact on a matter protected by the EPBC Act must submit to the Minister a referral that contains information about the proposed action.[56] If the Minister determines that approval is required under the EPBC, the proposed action is called a "controlled action," and the proposal must go through a formal assessment and approval process before it can proceed.[57] Depending on the nature of the action and its likely significance, the Minister determines whether the preparation of an environmental impact statement or another assessment approach is most appropriate.[58] Once the required assessments are performed, the Minister has the power to deny approval to a project or grant approval subject to conditions that mitigate the environmental impact.

In October 2005, WPS brought suit against the Minister for the Environment and Heritage, Bowen Coal, and QCoal for their alleged failure to comply with the EPBC Act.[59] In April 2005, Bowen Coal had submitted to the Ministry a referral regarding

---

[53] Wildlife Preservation Society of Queensland Proserpine/Whitsunday Branch Inc. v. Minister for the Environment and Heritage & Ors [2006] FCA 736 (June 15, 2006); available at http://www.austlii.edu.au/au/cases/cth/federal_ct/2006/736.html (last visited February 27, 2009) [hereinafter Isaac Plains Decision].

[54] Department of the Environment and Water Resources, *EPBC Act – Environment Assessment Process*, available at http://www.environment.gov.au/epbc/publications/pubs/assessment-process.pdf (last visited February 27, 2009) [hereinafter EPBC Act summary]. Enacted in 1999, the Act replaced several major environmental statutes, including the Environment Protection (Impact of Proposals) Act 1974; the Endangered Species Protection Act 1992; the National Parks and Wildlife Conservation Act 1975; and the World Heritage (Properties Conservation) Act 1983. Richard B. Stewart, *A New Generation of Environmental Regulation?*, 29 CAP. U. L. REV. 21, 182 (2001).

[55] EPBC Act summary, *supra* note 54.

[56] *Id.*

[57] *Id.*

[58] Stewart, *supra* note 54, at 161. *See also* EPBC Act summary, *supra* note 54.

[59] Isaac Plains Decision, *supra* note 53.

the construction and operation of a coal mine, known as the Isaac Plains Project, which would produce an estimated 18 million tons of coal for export over a lifetime of nine years.[60] Also in April 2005, QCoal had proposed to construct and operate a coal mine, known as the Sonoma Coal Project, which would produce an estimated 30 million tons over a lifespan of fifteen years. This coal was primarily destined for export, but some would be used domestically.[61]

Soon after the companies' proposals were submitted, WPS made public submissions to the Ministry suggesting that the climate change–related impacts of each project required that they be considered controlled actions. WPS argued that the burning of the coal that would be mined would have climate change impacts that would adversely affect the Great Barrier Reef World Heritage Area and the Wet Tropics World Heritage Area. Studies were cited that showed that global warming could cause the collapse of coral populations and terrestrial biodiversity in tropical areas.[62]

In May 2005, the Minister issued decisions determining that neither project was a controlled action under the EPBC Act[63] because neither was likely to have a significant impact on matters of national environmental significance. In July, WPS applied for judicial review of both decisions, alleging that the Minister had not considered the effects of greenhouse gases generated in the mining, transportation, and burning of the coal extracted from the mines.[64] WPS based its argument in part on the absence of express references to climate change impacts in the Ministry's "statements of reasons" for its decisions.[65]

Several weeks before the October 2005 trial, the Ministry bolstered its defense by submitting to the Court an affidavit from the delegate of the Minister who had made the decision. The delegate stated that he had given detailed consideration to greenhouse gases from the mines.[66] He explained that he had viewed the greenhouse gas emissions that would be produced from the mining, shipping, and use of the coal from each project in relation to the "greenhouse gases currently in the atmosphere" and the "total annual global contributions from greenhouse gases from all global sources" and concluded that they represented only a "relatively small contribution" that was unlikely to have a "significant impact" on matters of national significance

---

[60] Applicant's Outline of Submissions and Summary of Relevant Facts (filed October 10, 2005) at paragraphs 11 and 12 [hereinafter Isaac Plains Applicant's Outline].
[61] *Id.* at paragraphs 26 and 27.
[62] *Id.* at paragraph 16.
[63] *Id.* at paragraphs 19 and 31.
[64] Application for an Order of Review, filed 20 July 2005, available at http://www.envlaw.com.au/whitsunday11.pdf (last visited February 27, 2009).
[65] Isaac Plains Applicant's Outline, *supra* note 60, at paragraph 6. See also Chris McGrath, *Federal Court Case Challenges Greenhouse Gas Emissions from Coal Mines*, at 2; available at http://www.envlaw.com.au/whitsunday19.pdf (last visited February 27, 2009) [hereinafter McGrath].
[66] *Id.*; Affidavit of Mark Flanigan, in the case of *Wildlife Preservation Society of Queensland Proserpine/Whitsunday Branch Inc. v. Minister for the Environment and Heritage* (October 2005), available at http://www.envlaw.com.au/whitsunday14.pdf (last visited February 27, 2009) [hereinafter Flanigan Affidavit].

protected under the Act.[67] The delegate stated that the link between the greenhouse gases produced by this coal and a "measurable or identifiable" change in climate "was speculative only and unlikely to be demonstrable." He also cited the "speculative and uncertain" nature of how the coal would be used at its export destinations.[68]

WPS argued that the affidavit should be given little or no weight because it "reads like a document prepared in response to the litigation and not as an accurate recounting of the true reasoning process that in fact occurred."[69] WPS further argued that the question of whether the mining projects were likely to have a "significant impact" under the Act should be addressed by asking whether the contribution to global warming emissions from the mines was significant at a national level in comparison with other actions in Australia contributing to global warming.[70] WPS calculated the amount of greenhouse gases emitted by the coal mined in these projects was roughly equivalent to 25 percent of Australia's greenhouse gas emissions and 0.6 percent of global emissions from fossil fuels in 2003.[71]

In a decision issued on July 15, 2006, the Court dismissed WPS's suit.[72] The Court gave credence to the Ministry's defense that climate change had been considered in the decision-making process. The Court found that the Ministry acted lawfully in considering greenhouse gas emissions from the projects and found no link between these emissions and any specific damage to the matters of national significance protected under the Act.

The Court thus arguably limited the principle announced in the Nathan Dam decision requiring the Minister to consider both the direct and indirect impacts of a project. The Court references the Nathan Dam decision only in the final paragraph of the opinion, stating that this case is "far removed from the factual situation" in that case.[73] The Court pointed out that the applicants in this case had not shown that the emissions from this coal would "directly affect" protected matter, nor had they identified the "extent (if any)" to which emissions from this coal would aggravate the climate change problem.[74] As the Court stated, "I am far from satisfied that the burning of coal at some unidentified place in the world, the production of greenhouse gases from such combustion, its contribution toward global warming and the impact of global warming upon a protected matter" could be said to be

---

[67] Flanigan Affidavit, *supra* note 66, at paragraphs 17–20.
[68] *Id.* at paragraphs 22–27.
[69] Isaac Plains Applicant's Outline, *supra* note 60, at paragraph 5–6.
[70] Environmental Defender's Office of North Queensland, *EDO Alert! Climate Change Case* (November 2, 2005), available at http://www.edo.org.au/edonq/images/stories/documents/climate_nov_2005_alert.pdf (last visited February 27, 2009) [hereinafter EDO Alert]. *See also* Application for an Order of Review (Version 3), in the case of *Wildlife Preservation Society of Queensland Proserpine/Whitsunday Branch Inc. v. Minister for the Environment and Heritage* (October 2005), available at http://www.envlaw.com.au/whitsunday15.pdf (last visited February 27, 2009).
[71] EDO alert, *supra* note 70.
[72] Isaac Plains Decision, *supra* note 53.
[73] *Id.* at paragraph 72.
[74] *Id.*

an impact of the proposed action.[75] Whereas the Nathan Dam case stands for the principle that indirect effects of a proposed action must be considered, this case stands for the principle that indirect effects that are not specifically identifiable and measurable need not be considered as impacts.

Some Australian commentators argue that the Court's decision in this case makes clear that the EPBC Act needs express greenhouse gas language to protect the Australian environment from the impacts of climate change. As explained by one advocate, "The decision in this case shows that the emissions from the use of the coal from the mines are effectively not regulated under the EPBC Act, which indicates an important gap in the ability of that Act to genuinely protect the matters of national significance it recognizes as warranting protection."[76] Environmental advocates suggest that the EPBC Act be amended to include a "greenhouse trigger" under which a proposed project would be considered to be a controlled action if its projected greenhouse gas emissions exceed a certain threshold amount.[77] Soon after the passage of the Act in 1999, the Australian government investigated a greenhouse trigger and drafted a regulation that would have created it.[78] But the regulation was never adopted. If it had been, these coal mining projects would likely have been classified as controlled actions, and thus been subject to requirements for more detailed assessments of their potential emissions and their implications.

2.3. The Anvil Hill Case

Another important climate change case arose in the Australian state of New South Wales.[79] In November 2006, the New South Wales Land and Environment Court decided that the state government's acceptance of an environmental assessment for a new coal mine, the Anvil Hill project, was invalid because the government had not required consideration of the greenhouse gas emissions from the burning of the mined coal.

Centennial Coal (Centennial) applied to the New South Wales Department of Planning to construct a new coal mine at Anvil Hill under the state's Environmental Planning & Assessment (EP&A) Act in January 2006.[80] Located in the Hunter Valley, the largest coal-producing region in New South Wales, Anvil Hill is the "largest

---

75 Id.
76 McGrath, *supra* note 65, at 4.
77 Id. at 2.
78 Id. at 5. In the draft regulation, the government included a trigger of 500,000 tons of CO_2-equivalent emissions in any twelve-month period. The Australian Network of Environmental Defenders Office recommended a trigger of 100,000 tons of CO_2-equivalent emissions. Id.
79 Gray v. Minister for Planning and Ors [2006] NSWLEC 720, decision available at http://www.lawlink. nsw.gov.au/lecjudgments/2006nswlec.nsf/61f584670edbfba2ca2570d40081f438/dc4df619de3b3f02ca 257228001de798?OpenDocument (last visited February 27, 2009) [hereinafter Anvil Hill Decision].
80 Environmental Planning and Assessment Act 1979, Part 3A, available at http://www.austlii.edu.au/ au/legis/nsw/consol_act/epaaa1979389/ (last visited February 27, 2009) (concerning the approval process for major infrastructure or other development of state or regional significance) [hereinafter EP&A Act].

intact stand of remnant vegetation" in the region, with significant biodiversity.[81] The proposed Anvil Hill project would produce up to 10.5 million tons of coal per year over its projected lifetime of twenty-one years, increasing the state's coal output by 20 percent.[82] The majority of it was destined for export.[83]

In April 2006, the Department issued to Centennial the environmental assessment requirements applicable to the Anvil Hill project.[84] Included was a requirement that the proponent address "Air Quality – including a detailed greenhouse gas assessment."[85] Centennial's completed environmental assessment was made public in August 2006.[86] The assessment included an analysis of the greenhouse gas emissions from the mining of the coal itself but not from the subsequent burning of the mined coal. On September 19, 2006, the Department directed that a "panel of experts" responsible for holding public hearings and providing expert advice on the proposal be constituted.[87]

Also on September 19, 2006, Peter Gray, a local environmentalist, filed a lawsuit against the Department in the New South Wales Land and Environment Court alleging a violation of the state EP&A Act. The applicant argued that the Department violated the Act by accepting as adequate an environmental assessment that failed to consider all potential greenhouse gas emissions and take into account principles of ecologically sustainable development such as the precautionary principle and intergenerational equity.[88] The following month, the Department requested and Centennial prepared an analysis of greenhouse gas emissions from the burning of the coal in response to submissions by various individuals and organizations both

---

[81] Anvil Hill Alliance website, available at http://www.anvilhill.org.au/ (last visited February 27, 2009).

[82] Executive Summary of the Environmental Assessment, Anvil Hill project, at 1 (August 2006), available at http://www.umwelt.com.au/anvil-hill/ (last visited February 27, 2009) [hereinafter Anvil Hill Environmental Assessment]. *See also* Anne Davies, *Appeal on Green Ruling Likely*, SYDNEY MORNING HERALD (November 29, 2006), available at http://www.smh.com.au/news/national/appeal-on-green-ruling-likely/2006/11/28/1164476204759.html (last visited February 27, 2009).

[83] Anne Davies, *Landmark Climate Change Ruling Puts Heat on Industry*, SYDNEY MORNING HERALD (November 28, 2006), available at http://www.smh.com.au/news/environment/landmark-climate-change-ruling-puts-heat-on-industry/2006/11/27/1164476140463.html (last visited February 27, 2009) (stating that 80 percent would be exported). *See also* Anvil Hill Decision, *supra* note 79, at paragraph 4 (stating that "about half the coal" is intended for export).

[84] *Director-General's Requirements, Section 75F of the Environmental and Planning Assessment Act 1979*, available at http://www.planning.nsw.gov.au/asp/pdf/anvil_hill_environmental_assessment_requirements.pdf (last visited February 27, 2009).

[85] Anvil Hill Decision, *supra* note 79, at paragraphs 16–17.

[86] *Id.* at paragraph 21.

[87] *Id.* at paragraph 25. *See also* Frank Sartor, Minister for Planning, *Direction: Section 75F of the Environmental and Planning Assessment Act 1979*, available at http://www.planning.nsw.gov.au/asp/pdf/06_0014_panel_of_experts_terms_of_reference.pdf (last visited February 27, 2009), and NSW Department of Planning, *Independent Panel to Review Anvil Hill Coal Mine Proposal* (October 6, 2006), available at http://www.planning.nsw.gov.au/mediarelplan/mr20061006_426.html (last visited February 27, 2009).

[88] Anvil Hill Decision, *supra* note 79, at paragraphs 35–45. Section 5 of the EP&A Act includes the encouragement of "ecologically sustainable development" as one of its objects, defined as in Section 6(2) of the New South Wales Protection of the Environment Administration Act 1991 to include implementation of, inter alia, the precautionary principle and intergenerational equity. *See* Anvil Hill Decision *supra* note 79, at paragraph 101.

before and after the assessment was made public. Centennial's analysis found that the coal mined at Anvil Hill would result in the emission of an average of 12.5 million tons of greenhouse gases per year, an annual amount equivalent to about 2 percent of Australia's total greenhouse gas emissions in 2004.[89]

The Court issued its decision on November 27, 2006.[90] Following the Nathan Dam case and distinguishing the Isaac Plains and Sonoma Mine case, the Court held that the EP&A Act required that the environmental assessment consider indirect greenhouse gas emissions that would result from the burning of the coal on the basis that there is "a sufficiently proximate link between the mining of a very substantial reserve of thermal coal in [New South Wales], the only purpose of which is for use as a fuel in power stations, and the emission of GHG which contribute to climate change."[91] The Court also held that the Department's failure to require Centennial to consider indirect greenhouse gas emissions in the environmental assessment violated the EP&A Act on the basis that it was inconsistent with the principle of intergenerational equity and the precautionary principle.[92]

The Court, however, did not rule that the Anvil Hill project's environmental assessment was "void and without effect" as requested by the applicant.[93] The Court took into account the fact that an analysis of the indirect greenhouse gas emissions had been submitted by Centennial in October 2006 and made publicly available by the Department.[94] The Court further observed that the panel of experts appointed to review the proposal had been authorized by the Department to consider indirect greenhouse gas emissions in its review of the project.[95] Despite the limited remedy granted by the Court, commentators consider the decision to be significant. Written broadly, the decision appears to require all projects seeking approval in New South Wales that directly or indirectly cause greenhouse gas emissions to include an assessment of their contribution to global warming.[96]

3. CLIMATE CHANGE AS AN ENVIRONMENTAL IMPACT

The coal mining cases raise the question of whether and how climate change impacts should be considered under Australian federal and state environmental and planning laws. The issue, however, is not specific to Australia; rather, it is an issue that

---

[89] Anvil Hill Environmental Assessment, Response to Submissions, Part A, at 32 (October 2006), available at http://www.planning.nsw.gov.au/asp/pdf/06_0014_response_to_submissions_parta.pdf (last visited February 27, 2009).

[90] According to the decision, the applicant conceded that if the analysis of greenhouse gas emissions that was prepared in October 2006 in response to submissions had been part of the original assessment released to the public as required under the EP&A Act, then he would not have had a legal claim. Anvil Hill Decision, *supra* note 79, at 28.

[91] *Id.* at paragraphs 90–93 and 97–100.

[92] *Id.* at paragraphs 126 and 135.

[93] *Id.* at paragraph 2.

[94] *Id.* at paragraph 150.

[95] *Id.*

[96] Davies, *supra* note 83. *See also* Matthew Warren, *Planning May Face Climate Test*, THE AUSTRALIAN (November 28, 2006), available at http://www.theaustralian.news.com.au/story/0,20867,20832685-30417,00.html (last visited February 27, 2009).

is likely to increasingly arise under subnational and national environmental impact laws in many countries. More than 100 countries have legal provisions relating to environmental impact assessment, and citizens in many countries may have legal recourse to compel or review their implementation.[97]

In the United States, the other major industrialized country not party to the Kyoto Protocol, several cases have been filed regarding the assessment of climate change–related impacts under the National Environmental Policy Act (NEPA),[98] which requires analysis of the environmental impacts of projects and policies that constitute "major federal actions."[99] Civil society organizations, states, and cities have sued federal agencies for not considering the greenhouse gas emissions resulting from new transmission lines that would connect new power plants in Mexico to the U.S. power grid;[100] a rulemaking on new federal fuel efficiency standards;[101] providing assistance in financing overseas fossil fuel projects;[102] and a new rail line to transport coal from mines in Idaho to power plants in the Midwest.[103]

Projects for which climate change impacts can be assessed are usefully categorized into two types: those projects that directly emit greenhouse gases and those that cause greenhouse gas emissions indirectly. Many projects that may require environmental assessments for governmental approval, such as new coal-fired power plants, directly emit greenhouse gases. One study suggested that the United States would add 72 new coal-fired power plants between 2005 and 2012, while China and India together would add 775 new plants.[104] Together they would emit more than five times the amount of greenhouse gases by which Kyoto countries are supposed to cut their emissions over the same period.[105] Other projects that directly emit significant amounts of greenhouse gases include iron and steel plants, cement plants, landfills, and cattle feedlots.[106]

---

[97] Annie Donelly, Barry Dalal-Clayton & Ross Hughes, *A Directory of Impact Assessment Guidelines* 3–4 (2nd ed., International Institute for Environment, 1998).

[98] 42 U.S.C. §§ 4321–4347 (2000).

[99] 42 U.S.C. § 4332(2)(c). Under applicable regulations, "[m]ajor Federal action" is defined to "includ[e] actions with effects that may be major and which are potentially subject to Federal control and responsibility." 40 C.F.R. § 1508.18 (2003). *Cf.* Justin R. Pidot, *Global Warming in the Courts: An Overview of Current Litigation and Common Legal Issues*, Georgetown Environmental Law & Policy Institute (2006), available at http://www.law.georgetown.edu/gelpi/current_research/documents/GlobalWarmingLit_CourtsReport.pdf (last visited February 27, 2009).

[100] Border Power Plant Working Group v. Department of Energy, 260 F. Supp. 2d 997 (S.D. Cal. 2003).

[101] Center for Biological Diversity v. National Highway Traffic Safety Administration, No. 06071891 (9th Cir. 2007).

[102] Friends of the Earth, Inc. v. Watson, 2005 U.S. Dist. LEXIS 42335 (N.D. Cal., 2005).

[103] Mayo Foundation v. Surface Transportation Board, No 06–2031 (8th Cir. 2006).

[104] Mark Clayton, *New Coal Plants Bury 'Kyoto': New Greenhouse-Gas Emissions from China, India, and the US Will Swamp Cuts from the Kyoto Treaty*, Christian Sci. Monitor (December 23, 2004), available at http://www.csmonitor.com/2004/1223/p01s04-sten.html (last visited February 27, 2009).

[105] *Id.*

[106] U.S. EPA, *Inventory of U.S. Greenhouse Gas Emissions and Sinks, 1990–2004*, at ES-6 to ES-10 (2006), available at http://www.epa.gov/climatechange/emissions/downloads06/06_Complete_Report.pdf (last visited February 27, 2009).

Many other projects do not directly emit greenhouse gases but can be viewed as leading to or causing greenhouse gas emissions. Indirect emissions are often temporally and spatially dislocated from the project under consideration and may be classified as either "downstream" or "upstream" indirect emissions.[107] Downstream emissions are those resulting from the products or processes that are outputs of the project under consideration. As discussed previously, the most significant greenhouse gas emissions related to coal mining occur when the mined coal is later burned for energy production rather than when it is mined. New highways and motor vehicle assembly plants can also be viewed as leading to downstream greenhouse gas emissions because they facilitate the use of motor vehicles, which constitute more than one-quarter of all greenhouse gas emissions in the United States.[108] Upstream emissions are emissions from products and processes that constitute necessary inputs to the project under consideration. For example, a new factory requiring large amounts of electricity to operate can be viewed as the cause of some of the greenhouse gas emissions of the power plant that provides it with electricity. The upstream emissions of a motor vehicle include the emissions associated with the collection, transport, storage, and refinement of the fuel it burns.

Litigants in cases based on governmental failures to assess climate-related impacts of project proposals encounter a host of barriers to favorable judicial resolution. "Standing to sue" is potentially a prominent barrier in some jurisdictions, including the United States. In Australia, standing has not emerged as a significant barrier, but other important barriers loom. Where plaintiffs seek the assessment and consideration of the impacts of greenhouse gas emissions directly caused by the proposed project, measuring and predicting local and cumulative impacts is likely to present a significant challenge. In cases such as the Australian coal mining cases where plaintiffs seek consideration of the impacts of greenhouse gas emissions indirectly caused by the proposed project, the difficulty of identifying such indirect emissions must also be confronted. The issues of standing, the difficulty of measuring and predicting cumulative and local impacts, and the difficulty of identifying indirect emissions are discussed next in turn.

3.1. *Standing to Sue*

In some jurisdictions, environmental litigants in climate change cases may have difficulty obtaining a favorable judgment because they are unable to establish that they have "standing to sue." In the Australia coal mining cases, standing did not emerge as a barrier to litigants because the applicable statutes established very broad standing requirements and the courts did not question the plaintiffs' standing.

---

[107] These terms have been used most extensively in the industrial ecology and energy efficiency literatures.
[108] David L. Greene & Andreas Schafer, Pew Center on Global Climate Change, *Reducing Greenhouse Gas Emissions from U.S. Transportation*, at 2 (May 2003), available at http://www.pewclimate.org/global-warming-in-depth/all_reports/reduce_ghg_from_transportation (last visited February 27, 2009).

In climate change litigation in the United States, in contrast, standing is a more controversial issue.[109]

Standing, in the words of the Supreme Court of the United States, is the question of "[w]hether the litigant is entitled to have the court decide the merits of the dispute."[110] It generally implicates the ability of a litigant to demonstrate to a court that he is sufficiently connected to or harmed by the law or action that he is challenging to justify his prosecution of the claim. Requirements for standing vary significantly among national jurisdictions. Many jurisdictions have virtually no standing requirements, while the U.S. judicial system has very extensive and detailed requirements.[111]

As noted earlier, in the Australian context, the statutes applicable to each case established very broad standing requirements. Under Victoria's PE Act, a person "who is substantially or materially affected by a failure of the Minister, a planning authority, or a panel" to comply with the Act's provisions regarding amendments to the planning scheme is entitled to judicial review.[112] New South Wales's EP&A Act is far more expansive in this context, stating that "[a]ny person may bring proceedings in the Court for an order to remedy or restrain a breach of this Act."[113] The EPBC Act sets forth "extended standing" for judicial review of administrative decisions pursuant to the Act.[114] Standing is extended to all Australian organizations that have engaged in a series of activities in Australia "for protection or conservation of, or research into, the environment" at any time in the preceding two years and whose "objects or purposes" included such protection, conservation, or research.[115] None of the judicial opinions in the coal mining cases raised the lack of standing as an issue.

U.S. courts, in contrast, have imposed standing requirements that are more likely to present barriers to plaintiffs in climate change litigation. To have standing, a plaintiff must show that he has suffered an actual or imminent injury that is "concrete and particularized," "fairly traceable to the challenged action of the defendant," and

---

[109] The U.S. Supreme Court's decision in the case of *Massachusetts v. Environmental Protection Agency*, 549 U.S. 497 (2007), is demonstrative. The majority ruled that the state of Massachusetts had standing to sue, but four justices vigorously dissented on this issue. *See also* David Hodas, *Standing and Climate Change: Can Anyone Complain about the Weather?* 15 J. LAND USE & ENVTL. L. 451 (2000). On standing in environmental cases generally, see Ann E. Carlson, *Standing for the Environment*, 45 U.C.L.A. L. REV. 931 (1998).

[110] Warth v. Seldin, 422 U.S. 490, 498 (1975).

[111] *See, e.g.*, Aharon Barak, *Foreword: A Judge on Judging: The Role of a Supreme Court in a Democracy*, 116, HARV. L. REV. 16, 106-10 (2002); Matt Handley, *Comment: Why Crocodiles, Elephants, and American Citizens Should Prefer Foreign Courts: A Comparative Analysis of Standing to Sue*, 21 REV. LITIG. 97 (2002).

[112] PE Act, *supra* note 13, at Section 39.

[113] EP&A Act, *supra* note 80, at Section 123.

[114] EPBC Act, Section 487, "Extended Standing for Judicial Review," available at http://www.frli.gov.au/ComLaw/Legislation/ActCompilation1.nsf/framelodgmentattachments/7274EE4D4BF1FB60CA257000000B1F19 (last visited February 27, 2009).

[115] *Id.*

likely to be "redressed by a favorable decision."[116] Defendants in climate change cases may argue that plaintiff's alleged injury is not sufficiently imminent because it may not occur for many years; that it is general to society rather than particular to the plaintiff because it may be experienced by so many people; and that it is not sufficiently traceable to the defendant's actions or would not be redressed by a favorable judicial decision because the defendant's actions constitute a small percentage of all greenhouse gas emissions and the injury could occur regardless of a change in the defendant's actions.[117]

In the landmark case of Massachusetts v. EPA, the U.S. Supreme Court ruled in favor of Massachusetts in deciding that the Environmental Protection Agency (EPA) had the authority to regulate the emissions of greenhouse gases under the Clean Air Act and that the EPA had not acted properly in declining to do so.[118] Significantly, the Court determined that the state of Massachusetts had standing to bring these claims. As the Court explained, "the rise in sea levels associated with global warming has already harmed and will continue to harm Massachusetts. The risk of catastrophic harm, though remote, is nevertheless real. The risk would be reduced by some extent if petitioners received the relief they seek."[119] However, while this case determined that the state of Massachusetts had standing, it left open the question of whether private parties such as individuals and environmental groups would have standing to bring such claims.[120]

Where plaintiffs challenge the government's adherence to procedural laws such as NEPA, standing requirements may be easier for plaintiffs to satisfy.[121] In *Friends of the Earth v. Watson*, plaintiffs alleged a violation of NEPA based on the failure of the Export-Import Bank and the Overseas Private Investment Corporation to prepare an environmental impact statement (EIS) before engaging in the financing of overseas oil and gas extraction and energy generation projects that emitted large quantities of greenhouse gases.[122] The court held that to demonstrate standing in cases raising procedural issues, environmental plaintiffs must show only that "it is reasonably

---

[116] Friends of the Earth, Inc. v. Laidlaw Environmental Services, Inc., 528 U.S. 167, 180–81 (2000).

[117] See the briefs for the respondents in the case of *Massachusetts v. EPA*, available at http://supreme.lp.findlaw.com/supreme_court/docket/2006/november/05-1120-massachusetts-v-environmental-protection-agency.html (last visited February 27, 2009).

[118] Massachusetts v. EPA, 549 U.S. 497 (2007).

[119] *Id.* at 526.

[120] The majority opinion emphasizes that because Massachusetts is a sovereign state, it deserved "special solicitude" in resolving the standing issue. *Id.* at 536. In dissent, Chief Justice Roberts interprets the majority's decision as "an implicit concession that petitioners cannot establish standing on traditional terms." *Id.* at 540.

[121] See Bradford C. Mank, Standing and Global Warming: Is Injury to All Injury to None? 35 ENVTL. L. 1, 45–63 (2005) (discussing the current split among circuits on this question, wherein the Ninth and Tenth Circuits have explicitly rejected a more stringent standard for NEPA standing set forth by the D.C. Circuit).

[122] Friends of the Earth, Inc. v. Watson, 2005 U.S. Dist. LEXIS 42335 (N.D. Cal., 2005). Environmental groups and several cities sued the Overseas Private Investment Corp. (OPIC) and Export-Import Bank of the United States for their alleged failure to comply with the National Environmental Protection Act (NEPA).

e that the challenged action will threaten their concrete interests," not that substantive environmental harm is imminent.[123] Moreover, the court held that causation and redressability standards are relaxed for plaintiffs that assert procedural challenges and make the required showing as to injury.[124] The court thus rejected the defendants' arguments that their role with respect to the overseas projects producing greenhouse gas emissions was too "limited or attenuated" to demonstrate causation.[125] Plaintiffs satisfied the redressability standard by showing that a decision of an agency "could be" influenced by further environmental studies, rather than showing that it necessarily "would" be influenced.[126]

The question of whether standing presents a barrier to climate change litigants thus depends upon where claims are filed and the nature of those claims. In Australia and many other countries, environmental plaintiffs in climate change cases do not encounter standing as a barrier to the same degree as in the United States. Further, in some U.S. jurisdictions, plaintiffs suing under environmental impact laws for alleged procedural violations encounter relaxed standing requirements under which they are likely to prevail.

3.2. Cumulative and Local Impacts

Litigants seeking favorable judicial resolution in cases alleging an illegal governmental failure to consider greenhouse gas emissions as environmental impacts are likely to be plagued by the difficulty of assessing the climate change impacts of a given project. Project-related climate change impacts are difficult to measure because of the cumulative nature of climate change: it is the combined emissions of many sources together over a long period of time that lead to elevated atmospheric levels of greenhouse gases.[127] Moreover, how climate change will be manifested in terms of local impacts remains very difficult to predict. Although much may depend on the wording of the particular statute and regulations that a national court is applying, courts are likely to be reluctant to order that a governmental agency assess climate change impacts that are difficult or impossible to predict and measure.

Although the quantity of greenhouse gases that a particular project will produce may be ascertainable, any ultimate climate change impact is inherently a "cumulative" impact. A single source of emissions would be unlikely to sufficiently increase the concentration of greenhouse gases in the atmosphere to lead to a detectable degree of global warming. Because climate change impacts are caused by the cumulative activities of many sources, and because the concrete manifestations of climate change remain uncertain, courts may conclude that the greenhouse gas emissions

---

[123] *Id.* at 8–9.
[124] *Id.* at 12.
[125] *Id.* at 14–15.
[126] *Id.* at 16–17.
[127] Unlike many pollutants, greenhouse gases may not be harmful to human health or the environment at all in small quantities. Rather, it is *only* when viewed cumulatively that there are potentially significant adverse impacts.

of a single project cannot be said to cause global warming alone or to cause any particular local environmental impact.[128]

The environmental and planning laws of national and subnational jurisdictions are likely to vary with respect to the need to assess and consider cumulative impacts. In the United States, NEPA regulations define cumulative impact as "the impact on the environment which results from the incremental impact of the action [being analyzed] when added to other past, present, and reasonably foreseeable future actions regardless of what agency (Federal or non-Federal) or person undertakes such other actions. Cumulative impacts can result from individually minor but collectively significant actions taking place over a period of time."[129] Despite a clear requirement that they be analyzed, cumulative impacts have presented many difficulties for project-based environmental assessments in the United States.[130] Lacking information about the impacts of past projects, as well as reasonably foreseeable future actions, agencies struggle to assess both the cumulative impact and the significance of the project's contribution.[131] And although the language is broad enough to encompass climate change–related impacts, evidence suggests that many environmental assessments and environmental impact assessments do not contain cumulative impact analyses related to a project's greenhouse gas emissions.[132]

Other jurisdictions may have statutory or regulatory language requiring that global warming impacts be examined if the greenhouse gas emissions of a project are above

---

[128] A similar issue arose in *Massachusetts v. EPA* within arguments over petitioners' standing to sue. Respondents argued that petitioners failed to demonstrate that the regulation they sought was likely to affect climatic or environmental conditions in Massachusetts. *Brief for the Federal Respondent*, available at http://supreme.lp.findlaw.com/supreme_court/briefs/05–1120/05–1120.mer.resp.fed.pdf (last visited February 27, 2009). The Supreme Court rejected this argument, explaining that "[a]gencies, like legislatures do not generally resolve massive problems in one fell regulatory scoop... That a first step might be tentative does not by itself support the notion that federal courts lack jurisdiction to determine whether that step conforms to law." 549 U.S. 497.

[129] 40 C.F.R. § 1508.7 (1989).

[130] *Cf.* Robert L. Fischman, *The EPA's NEPA Duties and Ecosystem Services*, 20 STAN. ENVTL. L.J. 497, 512 (2001); *See also* Terence L. Thatcher, *Understanding Interdependence in the Natural Environment: Some Thoughts on Cumulative Impact Assessment under the National Environmental Policy Act*, 20 ENVTL. L. 611 (1990) (discussing the importance of cumulative impact analysis and the confusion about it that stemmed from the Supreme Court case *Kleppe v. Sierra Club*, 427 U.S. 390 (1976)).

[131] Fischman, *supra* note 130, at 513.

[132] In 1997, the Council on Environmental Quality (CEQ) issued draft guidelines on how global climate change should be treated under NEPA, but final guidelines were never issued. For the text of the draft guidelines, see http://www.mms.gov/eppd/compliance/nepa/procedures/climate/considerations.htm (last visited February 27, 2009). *See also* letter from Sierra Club to Western Area Power Administration, Re: Big Stone II Expansion Proposal, DEIS Comments, dated July 24, 2006 (commenting that the Draft EIS fails to address indirect and cumulative impacts of the project's carbon dioxide emissions), available at http://www.northstar.sierraclub.org/campaigns/air/coal/bigStoneProposal.html (last visited February 27, 2009); letter from Center on Biological Diversity to BLM, Re: Notice of Intent (NOI) to Prepare a Programmatic EIS and Plan Amendments for Oil Shale and Tar Sands Resources Leasing on Lands Administered by the BLM in Colorado, Utah, and Wyoming, dated January 31, 2006 (commenting that the NOI omitted the direct, indirect and cumulative impacts of greenhouse gas emissions as an issue to be addressed), available at http://www.biologicaldiversity.org/swcbd/Programs/policy/energy/BLM-Tar-Sands-Scoping-FINAL.pdf (last visited February 27, 2009).

threshold amount. Such provisions would be useful in preventing a court from having to make a judgment call regarding the significance of a single project's emissions. Indeed, as discussed previously, Australian environmentalists have called for the inclusion of such a trigger in the EPBC Act.[133] Taking this approach, however, raises the possibility that large projects would be burdened by the regulation while even a larger number of separate small projects would be able to escape the regulatory requirements.

Aside from the problem of cumulative impacts, there is a great deal of uncertainty regarding how elevated levels of greenhouse gases will manifest as local environmental impacts. Many subnational and national environmental and planning laws are likely to be focused upon such impacts rather than general global impacts. In the Isaac Plains and Sonoma case, for example, the judge interpreted the EPBC Act to require plaintiffs to show impacts on specific national protected areas in order to prevail in their claim. The Court stated the plaintiffs "paid little or no attention to the actual effect on any identified protected matter" and concluded that "[t]here has been no suggestion that the mining, transportation or burning of coal from either proposed mine would directly affect any such protected matter." General assertions pertaining to issues such as sea level rise, increasing or decreasing rainfall, and changing average temperatures were not sufficient in this case.

In sum, the difficulty of ascertaining the impacts of global warming has been, and is likely to continue to prove to be, an imposing barrier to plaintiffs in climate change cases. The extent to which climate change is inherently a cumulative impact distinguishes it from many of the traditional environmental impacts that environmental impact laws were designed to deal with. Moreover, laws that require environmental impact assessments may explicitly focus on adverse local impacts, which often remain impossible to predict in the case of climate change. Even where the legislative language clearly requires agencies to assess cumulative impacts and to consider environmental impacts at all scales, courts may view the task of identifying and quantifying such impacts as very burdensome and be reluctant to require agencies to undertake it.

3.3. Indirect Impacts

A third set of barriers to favorable judicial resolution arises in the subset of potential cases where greenhouse gas emissions are an indirect or "downstream" impact of the project under assessment. In these cases, courts are likely to hesitate in ordering an assessment of the impacts of such indirect emissions because of the difficulty or impossibility of identifying such emissions with certainty.

Where the emissions that were allegedly required to be assessed and considered are indirect emissions, as in the coal mining cases discussed earlier, a definitive quantification of the amount of greenhouse gases that will be emitted may prove

---

[133] *See supra* notes 76–78 and accompanying text.

elusive. In both the Isaac Plains and Sonoma Mines case and the Anvil Hill case, the mined coal was destined primarily for export. Given that fact, there is uncertainty with respect to where, when, and how the coal will be used, and thus as to the ultimate emissions.[134] In the example of the downstream impacts of a new highway, it is similarly difficult to predict how much it will be used and the resulting emissions that will be generated. Moreover, when dealing with indirect impacts, there is the distinct possibility that if the project under consideration does not go forward, the indirect emissions would occur anyway. In other words, if Australia does not mine its coal for export to Japan, Japan will acquire and burn coal from elsewhere. Similarly, if the proposed new highway is not constructed, drivers will emit the greenhouse gases using other highways and roads.

Perhaps most problematically, the chain of causation with respect to indirect emissions is potentially infinite. Australian advocates pointed to the Nathan Dam case to advocate that all indirect effects be considered, but the assessment of all possible indirect effects is likely to be infeasible. Judges thus confront the difficult questions of how predictable or certain indirect emissions must be. In the Hazelwood case, the indirect emissions were quite certain because all the mined coal was destined to be burned in the nearby Hazelwood power plant. The judge found a "sufficient nexus" between the mining and the indirect effect.[135] In the cases where the coal was being mined primarily for export, the judges espoused disparate views. The judge in the Isaac Plains and Sonoma case stated that he had "proceeded on the basis that greenhouse gas emissions consequent upon the burning of coal mined in one of these projects might arguably cause an impact upon a protected matter, which impact could be said to be an impact of the proposed action," but then expressed serious doubt with respect to this proposition.[136] As discussed earlier, he remained unsatisfied that the production of greenhouse gases from the burning of coal in "some unidentified place in the world" should be considered an impact of the proposed project at all.[137] The judge in the Anvil Hill case, by contrast, easily found "a sufficiently proximate link."[138] The difficulty of establishing a bright-line rule with respect to when indirect emissions are sufficiently related to the project under consideration to govern such cases could lead to extensive litigation on the issue.

In the United States, NEPA regulations require the analysis of both direct and indirect impacts. Direct impacts are caused by the action and occur at the same time and place, whereas indirect impacts are "caused by the action and are later in time or

---

[134] It can be argued that although it is not clear where, when, and how the coal would be burned, it can very reasonably be assumed that it will be burned somewhere and that the ultimate emissions can be predicted with reasonable certainty. The greenhouse gas intensities of various coal burning production facilities fall within a known range, and an assessment could assume the lowest pollution intensity as a conservative estimate.

[135] See supra notes 33–34 and associated text.

[136] Isaac Plains Decision, supra note 53, at paragraph 72.

[137] Id. See also supra note 68 and associated text. On this point, the judge appears to be criticizing the Department's acceptance that such greenhouse gases should be considered an impact of the project.

[138] See supra note 91 and associated text.

farther removed in distance, but are still reasonably foreseeable."[139] According to the regulations, examples of indirect impacts or effects include "growth inducing effects and other effects related to induced changes in the pattern of land use, population density, or growth rate, and related effects on air and water and other natural systems, including ecosystems."[140] The regulations thus appear sufficiently broad to include the emissions that would result from burning mined coal as well as many other indirect emissions. Evidence suggests, however, that there are many environmental assessments and environmental impact statements prepared under NEPA that do not adequately address such indirect impacts.[141]

For coal mining projects, highway projects, and many other types of projects that are subject to environmental impact laws, indirect greenhouse gas emissions may be much more significant than direct greenhouse gas emissions. Governmental agencies, and ultimately courts, are thus faced with the question of how far "downstream" to look when analyzing emissions and their impacts. Australian courts have pioneered the way by focusing on the sufficiency of the "link" or "nexus" between the project and the downstream emissions and thereby determining in some cases that such indirect impacts must indeed be assessed.

4. CONCLUSION

Legal approaches to addressing climate change have experienced a significant degree of success in Australia. In well-publicized cases, Australian courts have required that governmental agencies examine not just the greenhouse gases directly produced by proposed coal mines but also those that will be produced when the coal is burned. As a result, governmental agencies are increasingly required to analyze and consider a project's climate change impacts, both direct and indirect.

Yet the requirement to examine climate change impacts of a particular project, especially where those impacts are indirect, may be difficult to attain under environmental impact laws in other cases and in other countries. In the United States, some courts might dismiss such a claim on the procedural basis that environmental plaintiffs lack standing. Even where standing is not a barrier, courts are challenged by the problems that the climate change impacts of particular projects are only significant when viewed cumulatively and their local manifestations are difficult or impossible to ascertain. In claims regarding indirect emissions of particular projects, legal claims are likely to be even more difficult to substantiate and prevail upon, as the emissions that the plaintiff seeks to have assessed are dislocated in time and space from the project under consideration.

In Australia, the obstacles to new coal mining projects that arise from environmentalists' success in the courts may slow the development of Australia's immense

---

[139] 40 C.F.R. § 1508.8 (1989).
[140] Id.
[141] See references contained in *supra* note 132.

coal resources. Environmentally and legally, these cases represent a success for climate change activists. Yet many questions remain as to whether project-based environmental assessment is a form of regulation that makes sense in confronting the problem of climate change. Environmental assessment laws tend to work best when a project can be clearly viewed as directly causing one or more environmental impacts that are well defined and understood. Other policy instruments are likely to be much better suited to addressing the problem of climate change in a more coordinated and coherent manner.

In the absence of such alternative policies, however, litigation is a critical mechanism by which citizens can force their governments to take climate change seriously. Soon after the decision in the Anvil Hill case was handed down, Ian Campbell, Australia's Environment Minister, referred to the decision as "fatally flawed."[142] In his view, "What we need to do as a world is keep mining coal. In fact, mine more coal for energy security but invest in the technologies to make sure that when we burn that coal, we have the technology to capture the carbon and stop it going into the atmosphere."[143] Perhaps the greatest significance of Campbell's statement is that he invoked the need for an alternative policy approach. By its nature, project-based assessment of climate change impacts may not be able to address the problems of climate change holistically, but legal decisions such as that in the Anvil Hill case may well play an important role in leading reluctant governments and industries toward developing policies that can.

---

[142] *Australian Judge Blocks Coal Mine on Climate Grounds*, ENVTL. NEWS SERVICE (November 29, 2006), available at http://www.ens-newswire.com/ens/nov2006/2006-11-29-03.asp (last visited February 27, 2009).

[143] *Law Changes Won't Solve Climate Change: Campbell*, ABC NEWS ONLINE (Nov. 28, 2006), available at http://www.abc.net.au/news/australia/nsw/newcastle/200611/s1799315.htm (last visited February 27, 2009).

4

Cities, Land Use, and the Global Commons: *Genesis* and the Urban Politics of Climate Change

Katherine Trisolini* and Jonathan Zasloff**

INTRODUCTION

Ruin is the destination toward which all men rush, each pursuing his own best interest in a society that believes in the freedom of the commons. Freedom in a commons brings ruin to all.[1]

By quoting this pessimistic philosophy, the New Zealand Environment Court, in *Genesis Power Ltd. v. Franklin District Council* (*"Genesis"*),[2] demonstrated its lack of faith in local governments as protectors of the Earth from climate change. Little wonder, then, that it mandated that a local district government set aside its concern for local environmental impacts and indigenous peoples' cultural resources, and permit the construction of a wind farm that could reduce carbon emissions. In describing the legal basis for its decision, the Environment Court echoed prevailing assumptions that climate change can only be dealt with from the top down. That is, nation-states implement international treaties by imposing concern on local governments that would otherwise, "in pursuit of their well-being, destroy existing stock of natural and physical resources so as to improperly deprive future generations of their ability to meet their needs."[3]

This chapter explores such a view of localities and its alternatives. We observe the seemingly contrary behavior of cities in the United States, which appear to have taken up the charge for climate protection despite the complete absence of the national influence identified as necessary by the New Zealand court and others. We then

---

* Associate Professor, Loyola Law School, Los Angeles; J.D., Stanford Law School, M.A., Political Science, University of California at Berkeley. Professor Trisolini teaches environmental law, property, and climate change courses and has litigated local state, and federal environmental and land use cases. The author thanks UCLA School of Law for providing a teaching and research fellowship during which this chapter was completed.
** Professor of Law, UCLA School of Law, and Associate Director, Richard S. Ziman Center for Real Estate at UCLA; J.D. Yale Law School, Ph.D. Harvard University, zasloff@law.ucla.edu.
[1] Garrett Hardin, *The Tragedy of the Commons*, 162 SCIENCE 1243, 1244 (1968).
[2] [2005] N.Z.R.M.A. 541 (Env. C.) (We cite the case as *"Genesis"* following the Environment Court's own format.).
[3] *Id.* 225 (citing Canterbury Reg. Council v. Selwyn Dist. Council, [1997] N.Z.R.M.A. 25 (Env. C.)).

initiate an inquiry into the possible basis of these apparently against interest actions, drawing upon loose analogies to international relations theory.

Unlike the focus of most other climate change lawsuits, which have predominantly challenged actions of either large polluters or national agencies, the Environment Court in *Genesis* reviewed a local land use decision regarding the siting of a small wind farm. The Franklin District Council, responsible for the only required discretionary approval, refused consent to the wind farm's application because of the adverse impacts to the coastal landscape, the cultural resources of indigenous people, and local equestrian facilities.[4] Under New Zealand's Resource Management Act,[5] Genesis Power appealed to the Environment Court, a body invested with authority to conduct a de novo review of the Council's decision.[6] The Environment Court reversed the decision, finding that the proposal's benefits, "when seen in the national context, outweigh the site-specific effects, and the effects on the local surrounding area."[7]

The case turns our attention to the impact of local land use decisions on climate change and the constraints under which local decision-makers act. Taken cumulatively, local governments' land use decisions – their determinations of which categories of activities go where – have a substantial impact on greenhouse gas production. Local governments may control the availability of sites for alternative energy production, and their proximity to consumers, which influences transmission efficiency, as in *Genesis*. Among other things, local governments also determine the design of cities and towns that either require carbon-heavy auto use or encourage alternatives.

The New Zealand Environment Court expressed extreme skepticism about local governments' ability to incorporate global concerns about climate change into their decision making. Rather, it envisioned them as trapped by parochial concerns analogous to those famously described by Garrett Hardin as producing the "tragedy of the commons,"[8] a phrase that has become a trope in environmental law. Hardin described herdsmen following rational incentives as tragically overgrazing the commons necessary for their livelihoods.[9] Many scholars following Hardin have further illuminated the traps precluding rational actors from preserving shared or "common pool" resources.[10] The scholarship on this problem largely presumes that there

---

[4] *Id.* 3–5, 39–41.
[5] 1991 S.N.Z. No. 69.
[6] *See id.* § 120; *see also* New Zealand Ministry of Justice, Environment Court, http://www.justice.govt.nz/environment/index.html#jurisdiction (last visited Jan. 3, 2008) (describing the Environment Court's jurisdiction).
[7] *Genesis*, [2005] N.Z.R.M.A. 541, ¶ 228.
[8] *Id.* 223 (quoting REPORT OF THE BOARD OF INQUIRY: PROPOSED TARANAKI POWER STATION – AIR DISCHARGE EFFECTS 7.103 (Feb. 1995)); Hardin, *supra* note 1.
[9] Hardin, *supra* note 1, at 1244.
[10] ELINOR OSTROM, GOVERNING THE COMMONS: THE EVOLUTION OF INSTITUTIONS FOR COLLECTIVE ACTION 2–8 (1990) (summarizing the literature). For a summary of scholarship criticizing Hardin's theory, see DANIEL H. COLE, POLLUTION AND PROPERTY: COMPARING OWNERSHIP INSTITUTIONS FOR ENVIRONMENTAL PROTECTION 15–16 (2002). As Cole explains, none of the scholarship criticizing

are only two ways out of this trap – privatization of the resource or imposition of resource protection by an external sovereign.[11] Few are sanguine about the prospects of resolving the commons problem without one of these two solutions. On this view, we would not expect local governments – as small, individual users of the global commons – to tackle climate change constructively.

Yet, contrary to this tragic vision of resource users, we observe an apparent movement among U.S. cities[12] to tackle climate change even when it appears to be against their immediate interest – at least as those interests have traditionally been understood.[13] Thus, *Genesis* symbolizes a cluster of issues far broader than one dispute. It presents, rather, an opportunity to consider how local governments' actions on climate change can both inform theories of urban governance and enrich our understanding of relationships among varying international actors.

Although ostensibly categorized as climate change litigation, *Genesis* inspires comparison with political decision making by city governments in the United States rather than with judicial determinations for several reasons. *Genesis* fits uneasily into a general discussion of climate change litigation because it was not really a piece of litigation at all, at least not in the way that American observers might think of it. Rather than a judicial proceeding, it more closely resembles an appeal from a lower to a higher level within an administrative agency empowered to make substantive policy. It also bears a striking similarity to an administrative appeal from a city's planning commission to its city council in the United States. The

---

Hardin undermines his "chief insight" that "open access resources tend to be unsustainably exploited" absent the imposition of a regime for their protection. *Id.* One recent work, however, challenges the common assumption that climate change presents a tragedy of the commons scenario at all. Rather, the authors propose that subglobal governments may unilaterally regulate climate change without behaving irrationally because regulation provides some benefit relative to no regulation at all. *See* Kirsten H. Engel & Scott R. Saleska, *Subglobal Regulation of the Global Commons: The Case of Climate Change*, 32 ECOLOGY L.Q. 183, 188 (2005).

[11] *See* OSTROM, *supra* note 10, at 8–13; Engel & Saleska, *supra* note 10, at 191.

[12] Cities, of course, do not comprise all local governments. The Franklin District Council itself represented a different type of local government. For simplicity of expression, however, we will use "cities" and "local governments" interchangeably unless there is a particular reason to be more specific.

[13] *See infra* Section 3. Note that many cities around the world have become involved in this movement to combat climate change. We discuss the role of U.S. cities because they provide a model of local governments with apparently clear disincentives to take this action and because, given the federal government's refusal to implement mandatory carbon reductions, their decision to do so particularly conflicts with the accepted model of local action.

We note, however, that although an increasing number of cities have autonomously initiated policies to reduce their carbon footprints, some local governments have resisted implementation of carbon reduction plans. San Bernardino County, California, for example, initially refused to incorporate robust climate change analysis and mitigation into revisions to its General Plan, a required document under California law that governs the physical development of land within the county's jurisdiction. After being sued by the California attorney general for failing to adequately assess and mitigate the climate change impacts of updates to its plan, the county and the attorney general reached a settlement which requires the county to develop a Greenhouse Gas Emissions Reduction Plan. *See* California v. County of San Bernardino, No. CIVSS 0700329 (Cal. Sup. Ct. Aug. 28, 2007). Order Regarding Settlement, *available at* http://ag.ca.gov/cms_pdfs/press/2007-08-21_San_Bernardino_settlement_agreement.pdf.

Environment Court's specialized jurisdiction empowers it to vet applications for water permits, subdivision approvals, zoning and planning designations, and to conduct enforcement actions – matters usually dealt with by administrative bodies such as planning commissions, water boards, or even city councils in the United States.[14] Unlike the district courts in New Zealand that try common civil cases, the Environment Court is not bound by the same rules of evidence and its hearings occur in a much less formal environment, somewhat like a local agency hearing in the United States.[15]

We consider the theoretical assumptions underlying the Environment Court's dim view of local governments through the lens of land use planning by municipalities in the United States, which presented nearly a mirror image of the context of the *Genesis* case in two important respects. First, while New Zealand's central government has expressly set out to address climate change through national policy, the U.S. government expressly rejected the Kyoto Protocol and refused to adopt mandatory emissions reductions during George Bush's presidential administration (2000–2008), the precise period during which U.S. mayors rapidly signed up to address the issue. Second, while the New Zealand Resource Management Act provides the Environment Court with de novo review of district council decisions, local governments in the United States enjoy substantial discretion in land use matters, which are generally conceived of as matters of eminently local concern on which they are the final arbiter.[16]

The relationship between local government and climate change policy is thus particularly critical in the United States. As of this writing, the federal government has yet to adopt emission reduction legislation despite the inauguration of President

---

[14] The Environment Court's work includes:

- "Designations authorising public works such as energy projects, hospitals, schools, prisons, sewage works, refuse landfills, fire stations, major roads and bypasses; and also major private projects, for example, dairy factories, tourist resorts, timber mills and shopping centres.
- Classifications of waters, water permits for dams and diversions, taking of geothermal fluids, discharges from sewage works, underground mines; maximum and minimum levels of lakes and flows of rivers, and minimum quality standards; and water conservation orders.
- Land subdivision approvals and conditions, development levies, car parking contributions, reserve contributions, development levy fund distributions, road upgrading contributions, regional roads, limited access roads, and stopping roads.
- Environmental effects of prospecting, exploration, and mining, including underground, open pit and alluvial mining.
- Enforcement proceedings (including interim enforcement orders), declarations about the legal status of environmental activities and instruments, existing and proposed, and appeals against abatement notices."

http://www.justice.govt.nz/environment/index.html#jurisdiction (last visited Jan. 2, 2008).

[15] *See id.*

[16] The U.S. judiciary reviews land use actions, of course, but defers to local bodies. Even in "quasi-judicial" actions, where local decision-makers are theoretically applying preexisting standards, review is confined to whether "substantive evidence" in the record supports the decision-maker's judgment. For legislative acts, courts accord to local bodies' judgments a strong presumption of validity. Both notions are hornbook law. *See* JULIAN CONRAD JUERGENSMEYER & THOMAS E. ROBERTS, LAND USE PLANNING AND DEVELOPMENT REGULATION LAW § 5.33 (2003).

Obama who favors such action. And even state legislation regarding global warming leaves a large policy space open for municipal action.[17] But most importantly, local control over land use policy means that cities will play a major role in determining if the United States can reduce its emissions sufficiently to mitigate climate change. Scholars thus are in need of a theory to explain local governments' policy affecting this global common pool resource. We cannot provide such a theory at this point, but we do set forth a framework for developing one. We suggest that analogies to international relations theory may help expand upon current theories of urban politics. We also find that the Environment Court's conclusion concerning the competence of local governments to grapple with climate change is less certain than the Environment Court assumed.

1. GENESIS – CASE BACKGROUND

The *Genesis* case originated as a matter of local concern in the Franklin District on the North Island of New Zealand. Genesis Power's proposal to build nineteen wind turbines on the Awhitu Peninsula begat an unusual opposition alliance of horse trainers and representatives of New Zealand's indigenous population. The Awhitu Peninsula land was an important element of the cultural heritage of New Zealand's Tangata Whenua (literally, people of the land), the first people to settle in New Zealand.[18] The Te Iwi O Ngati Te Ata people objected because the project would adversely affect an area of cultural importance and because, prior to Genesis Power's application for approval, no survey had been performed to discern the areas of cultural significance.[19] Owners of local, decades-old equestrian facilities feared that vibrations, visual stimulation, and noise from the construction and operation of the wind farm would spook the horses and undermine their training, racing, and other horse-related businesses.[20] The project objectors argued that they had no objection to the project per se, but *not here*. The location of this particular project, they

---

[17] *See, e.g.*, California Global Warming Solutions Act of 2006, CAL. HEALTH & SAFETY CODE §§ 38500–38599 (West Supp. 2007). This Act directs the State Air Resources Board to achieve emission reductions through energy conservation, increased use of renewables, cap-and-trade programs, and directions to state regulators to reduce emissions from motor vehicles. *Id.* But it says nothing about local land use authority. For a summary of state efforts, see PEW CENTER ON GLOBAL CLIMATE CHANGE & PEW CENTER ON THE STATES, CLIMATE CHANGE 101: STATE ACTION (Oct. 2006), *available at* http://www.pewclimate.org/docUploads/101_States.pdf (last visited Dec. 2, 2008); PEW CTR. ON GLOBAL CLIMATE CHANGE, LEARNING FROM STATE ACTION ON CLIMATE CHANGE (Mar. 2007), *available at* http://www.pewclimate.org/docUploads/States%20Brief%20Template%20_March%202007_jgph.pdf (last visited Jan. 2, 2008).

[18] *Genesis*, [2005] N.Z.R.M.A. 541, 7 (Env. C.) ("The Awhitu Peninsula has been described as possibly one of the most densely populated areas of the Auckland Province, prior to European contact. Te Iwi O Ngati Te Ata are the tangata whenua. They have a long and close association with the Peninsula. It accordingly is very special to them as part of their cultural heritage.") (citation omitted). For background on Tangata Whenua, see Tangata Whenua, People of the Land, http://www.enzed.com/tw.html (last visited Jan. 2, 2008).

[19] *Genesis*, [2005] N.Z.R.M.A. 541, 38.

[20] *Id.* 37, 129–63.

argued, would adversely impact the "visual, landscape, natural character, amenity and cultural values in the environs of the site... and the surrounding rural area; and the current, lawfully established, use of the properties adjacent to the proposed wind farm site."[21]

As noted previously, the Franklin District Council refused consent because of the adverse impacts to the landscape, the Tangata Whenua, and the equestrian facilities.[22] The Council's decision reminds us that local governments must directly answer to constituents whose way of life often may rely on existing land use patterns and whose cultural heritage may be disrupted by changes to the landscape. *Genesis* suggests potential limitations on local governments' ability to promote land use changes that help reduce greenhouse gas emissions when those changes disturb settled expectations that existing land uses will continue and will not be compromised by unfamiliar uses.

Project proponents sought review before the Environment Court, which reviews the Council's decision pursuant to the Resource Management Act. On appeal, the parties' stipulated statement of facts sets forth the numerous environmental benefits of the proposed wind farm. They agreed that it would create security of supply by diversifying New Zealand's generating base and providing 18 MW of power, enough to supply 7500 households per year, or 37% of the homes in Franklin District and 0.18% of New Zealand's annual electricity consumption; reduce greenhouse gas emissions by generating electricity without emitting greenhouse gases during operation and emit 40,000 fewer tons of CO_2 per year than a comparable coal-fired power plant; reduce dependence on the national grid because of the proximity to the source of the demand; reduce transmission losses; provide a reliable energy resource; provide development benefits of wind energy generally; and contribute to New Zealand's renewable energy target.[23]

The Environment Court identified four potential negative impacts: "(i) effects on the visual amenity of the area – including effects on the landscape and natural character; (ii) noise effects on areas of recreation and workplaces; (iii) various horse-related effects; and (iv) effects on tangata whenua."[24] The Environment Court acknowledged that the Te Iwi O Ngati Te Ata "have a long and close association with the [Awhitu] Peninsula," making it a very special part of their cultural heritage.[25] Nonetheless, it dismissed the Tangata Whenua concerns because it found that most of the sites had been degraded or were of questionable cultural value, and that the project conditions would be sufficient to protect cultural artifacts.[26] The Environment Court found most of the other negative impacts to be minimal as well.

---

[21] *Id.* 43(i)–(ii).
[22] Subsequent to the lodging of the appeal, Genesis amended the project to address several of the Franklin District Council's concerns by removing one turbine of the original nineteen proposed and relocating two. At that point, the Council amended its position to "not opposing" the project. *Id.* ¶ 42.
[23] *Id.* 64(vi)(a)–(g).
[24] *Id.* 67.
[25] *Id.* 7.
[26] *Id.* 212(v).

The Environment Court acknowledged one major adverse impact of the project, that the "scale of the turbines is such that they would dominate the surrounding area and undermine the visual integrity of the natural character and landscape of the coastal environment."[27] However, the Environment Court emphasized that the Resource Management Act's mandate to preserve the coast's "natural character" and protect it from "inappropriate development" must be subordinated to the Act's general purpose to provide for "sustainable management."[28] Among the factors to be considered in assessing appropriate development in the context of sustainable management, the Resource Management Act required the Environment Court to consider "the effects of climate change and the benefits to be derived from the use and development of renewable energy."[29] The Court concluded that the latter outweighed impacts to the coastal environment.

Given the Environment Court's factual findings dismissing the majority of impacts and its conclusion that sustainable management outweighed the mandate to preserve the coast's natural features, it could have based its decision entirely on these factual elements. Instead, the Environment Court includes a theoretical discussion suggesting that it acts as an agent of the national government in ensuring protection of common pool resources that would otherwise be managed ineffectively by individuals and localities.

The Environment Court stated that Parliament's amendment of the Resource Management Act in 2004 to include explicit consideration of climate change provided "a clear recognition by Parliament of both the importance of the use and development of renewable energy and the need to address climate change, both of which are key elements in the proposed the wind farm."[30] In response to the project opponents' contention that the project did not warrant the environmental cost because reduction in greenhouse gas emissions attributable to the project would be de minimis, the Environmental Court quoted at length from a passage authored by the Board of Inquiry in a report rejecting claims that a power station's contribution to worldwide CO_2 emissions, and hence climate change, would be negligible:

> An implication could be taken from this statement that, as the contribution of the proposed power station to the total world emissions of CO_2 would be miniscule, then it would make no difference to any global warming effects whether the power station were to be built or not.
>
> We do not accept the argument. To do so would imply that as the world's CO_2 emission is composed of a great number of small emissions, the effect of any one of them could be discounted. But if one, why not more, or many, or, indeed, all? Without the Convention, and united efforts toward compliance, the situation becomes another example of what the economist Garret Hardin called the "tragedy of the Commons" in his famous article bearing that title. Each man is locked into

---

[27] Id. 215.
[28] Id. 215–16.
[29] Id. 224 (citing Resource Management (Energy and Climate Change) Amendment Act, 2004 S.N.Z. No. 2 (amending Resource Management Act, 1991 S.N.Z. No. 69, § 7)).
[30] Id. 220.

> a system that compels him to increase his herd without limit in a world that is limited. Ruin is the destination toward which all men rush, each pursuing his own best interest in a society that believes in the freedom of the commons. Freedom in a common brings ruin to all.
>
> Here because there is no one owner of an exploitable common resource, in this case the air as a receiver of carbon dioxide, the resource becomes overused and ill-used or even destroyed.
>
> Furthermore, even though the emission from the proposed power station is small by world standards, nevertheless the harm or potential for harm, throughout the world is very large. A small proportion of a very large amount may itself be large.[31]

The Environment Court emphasized that Parliament, through its 2004 amendments to the Resource Management Act, affirmed the Board of Inquiry's view that New Zealand must address climate change.[32] Moreover, Parliament had "reinforced the intention" of requiring the Environment Court to pay particular attention to climate change.[33] The Environment Court took this mandate as its authorization to answer the Hardin problem by assuming the role of outside sovereign "to ensure present people and communities do not, in pursuit of their well-being, destroy existing stock of natural and physical resources so as to improperly deprive future generations of the ability to meet their needs."[34]

It is noteworthy that, even prior to the 2004 amendments, the Environment Court presumed that climate change could only be addressed effectively at the national level. The Environment Court's 2002 decision in *Environmental Defence Society (Inc.) v. Auckland Regional Council*[35] recognized that climate change is a "serious concern" that "is likely to result in significant changes to the global environment, including New Zealand and the Auckland region."[36] Despite recognizing the scientific reality of climate change, however, the Environment Court in 2002 refused to grant the Environmental Defence Society's request to impose a mitigating condition on the Auckland Regional Council's air discharge approval of a 400 MW gas-fired combined cycle power station. In refusing to require the power plant owner to offset its carbon dioxide emissions by planting trees to act as carbon sinks, the Environment Court stated that New Zealand had a "clear preferred policy . . . to address greenhouse gas emissions . . . at a national level to ensure consistency of approach to guarantee an efficiency compatible with achieving the best social environmental and economic outcome."[37] Although not ruling on the District Council's claim that such a condition was outside of its jurisdiction, the Environment Court emphasized the difficulty the Auckland Regional Council would have enforcing and monitoring such a

---

[31] *Id.* 223 (quoting REPORT OF THE BOARD OF INQUIRY, *supra* note 8, 7.102–7.104).
[32] *Id.* 220.
[33] *Id.* 224.
[34] *Id.* 225.
[35] [2002] N.Z.R.M.A. 492 (Env. C.).
[36] *Id.* 65, *quoted in Genesis*, [2005] N.Z.R.M.A. 541, 221.
[37] *Id.* 88.

condition if trees were planted outside of the Auckland region.[38] Thus, the Environment Court has twice presumed that national action constitutes the only effective and appropriate way to address the climate commons on the grounds that local governments will be either unwilling or unable to successfully address this problem.

2. CITIES AND CLIMATE CHANGE ACTIVISM: UNLIKELY BEDFELLOWS

The widely accepted understanding of the difficulties in resolving commons problems, as exemplified by the Environment Court's decision in *Genesis*, suggests that local governments will not address climate change effectively in their decision making on land use matters. This would be particularly true of cities in the United States given the lack a federal mandate. Yet a number of municipalities in the United States are interjecting themselves into the national and international policy arena by tackling climate change despite apparently strong institutional incentives to avoid this issue. This is particularly surprising because of the diffuse nature of the benefits and the localized nature of the costs. Although Elinor Ostrom has observed users of common pool resources autonomously generating successful allocation systems from the bottom up without private ownership, her analysis focuses on smaller-scale resources such as fisheries, grazing meadows, and irrigation institutions.[39] Thus, while her work inspires us to recognize that local users may be capable of addressing commons problems in the absence of private ownership or imposition of regulation from a superseding sovereign, it does not help us understand the actions of municipalities tackling the *global* common resource implicated in climate change.

The following section discusses our reasons for finding it unlikely that municipalities in the United States would address climate change and then discusses the surprising indications that a number of them seem to be doing so. In order to understand these actions, we then turn to analogies with international relations theory to expand on current theories of urban politics and begin working on a model of municipal action on climate change.

Local governments' control over many land use decisions in the United States can have a monumental impact on climate change. For example, their planning approach has a dramatic impact on transportation choices. One-third of all the carbon dioxide that enters the atmosphere in the United States comes from the transportation sector.[40] This sector causes more CO_2 emissions than any other, and since 1980 its emissions have also been growing the fastest,[41] consuming seven out of every ten barrels of oil that the United States uses.[42]

[38] *Id.* 92.
[39] *See* OSTROM, *supra* note 10.
[40] *See* Eileen Claussen, *Foreword* to DAVID L. GREENE & ANDREAS SCHAFER, PEW CENTER ON GLOBAL CLIMATE CHANGE, REDUCING GREENHOUSE GAS EMISSIONS FROM U.S. TRANSPORTATION ii (May 2003), *available at* http://www.pewclimate.org/docUploads/ustransp.pdf.
[41] GREENE & SCHAFER, *supra* note 40, at 2–3.
[42] *Id.* 3.

Efforts to lessen transportation's role in global climate change have focused largely on making vehicles more fuel efficient and the fuel they run on cleaner. Yet Americans spend more and more time behind the wheel every year. As the Center for Clean Air Policy warns, growth in vehicle miles traveled in the United States "has outpaced population growth and is projected to continue to outstrip improvements in vehicle efficiency."[43] Sprawling residential and commercial development is the chief problem. For many Americans, cars are the most practical, and often the only way, to get to work, stores, entertainment, social gatherings, or grandmother's house for the holidays.

Moreover, vehicle miles traveled (or "VMT") can increase drastically at the smallest level of urban planning. Traditional Euclidean zoning – the type preferred by most suburbs, possibly because of its salutary impacts on property values – requires the radical separation of uses. Homeowners cannot even walk to the supermarket; they must drive there because commercial and residential neighborhoods are usually separated by major arterials.[44] This separation is highly significant, as nearly 40% of vehicle miles traveled are for local trips, not commuting.[45]

The automobile's significant contribution to emissions has put a distinctly green cast on the smart growth and New Urbanist views of planning. To completely define these views would require an article in itself, but for our purposes, they hold that sprawling development increases automobile use, leading to greater congestion and pollution. Their solution is more compact, higher-density development, which allows for greater use of mass transit. Moreover, the possibility of residents walking to amenities and essential services is seen as the critical test of appropriate neighborhood design. New Urbanists also favor mixed-use developments with narrower streets to allow for a pedestrian-friendly character.[46]

---

[43] Progressive Policy Institute, Driving Down Carbon Dioxide (Nov. 24, 2003), http://www.ppionline.org/ppi_ci.cfm?knlgAreaID=116&subsecID=900039&contentID=252224 (quoting Center for Clean Air Policy); see also GREENE & SCHAFER, supra note 40, at 6 ("Transportation energy use and greenhouse gas emissions are increasing because the growth of transportation activity exceeds the rate of improvement in energy efficiency and because little low-carbon fuel is used.") (italics omitted).

[44] See Peter Calthorpe, Land Use and Building the American Community, Presentation at the Fourth Annual Land Use Conference, The Rocky Mountain Land Use Institute, University of Denver College of Law (1996) (videotape, on file with authors).

[45] Id.

[46] A good characterization of New Urbanist development is found at the website of the Congress for the New Urbanism:

- Rule out any project that is gated, that lacks sidewalks, or that has a tree-like street system, rather than a grid network. The project as a whole should connect well with surrounding neighborhoods, developments, or towns, while also protecting regional open space.
- Rule out "single-use" projects that are just housing, retail, or office. The various types of building should all be seamlessly integrated – from different types of housing, to workplaces, to stores.
- The project should have a neighborhood center that is an easy and safe walk from all dwellings in the neighborhood. Buildings should be designed to make the street feel safe and inviting, by having front doors, porches, and windows facing the street – rather than having a streetscape of garage doors.
- The project, and particularly the neighborhood center, should include formal civic spaces and squares.

Whatever the other pros and cons of the New Urbanist/smart growth paradigm,[47] any actual decrease in VMT requires it. That need creates a significant problem. Unlike in New Zealand, where the Environment Court could force consideration (and prioritization) of national concerns, land use in the United States has a nearly sacrosanct position as a local concern.[48] But the most powerful incentives to cities all appear to point against a low-VMT planning policy. Indeed, as explored subsequently, cities might encourage auto dependency as a way of attracting commerce and capital.[49]

The benefits of mitigating climate change are about as widely diffused as possible; the city incurring the costs of adaptation bears them by itself – a highly unfavorable calculus. And those costs of adaptation are precisely ones that cities should be loath to endure. Consider the most basic element of local public finance: the "capitalization" of public services into home values. Although no one likes paying local property taxes, those taxes eventually lead to higher property values because of the public services they pay for. This is hardly an earth-shattering insight; everyone knows that – all things being equal – homes in a town with excellent public schools will cost more than those in its neighbor with poor ones.

Not everyone lives in high-tax, high-service jurisdictions, of course. Instead, people vote with their feet. In the 1950s, economist Charles Tiebout famously hypothesized that local governments could efficiently provide public goods because individuals could "shop" among local jurisdictions, choosing the jurisdiction that provides their optimal mix of taxes and services. Local political entrepreneurs, the theory states, will compete to attract mobile consumer-taxpayers, offering distinct tax-service packages to suit consumer demand.[50]

Yet simply relying on property taxes to maintain the desired mix of taxes and services is highly unstable. Under the Tiebout scheme, lower-income individuals have a great incentive to migrate to wealthy communities to free ride on the larger

- Finally, there is the "popsicle test." An eight-year-old in the neighborhood should be able to bike to a store to buy a popsicle without having to battle highway-size streets and freeway-speed traffic.

  http://www.cnu.org/charter (last visited Jan. 3, 2008).

[47] Some scholars see sprawling development as beneficial, and thus object to New Urbanism. *See, e.g.,* Peter Gordon & Harry W. Richardson, *Are Compact Cities a Desirable Planning Goal?*, 63 J. AM. PLAN. ASS'N 1 (1997). This position is effectively critiqued in Reid Ewing, *Is Los Angeles-Style Sprawl Desirable?*, 63 J. AM. PLAN. ASS'N 107 (1997).

[48] *See, e.g.,* ROBERT C. ELLICKSON & VICKI L. BEEN, LAND USE CONTROLS: CASES AND MATERIALS 29 (3d ed. 2005) ("Public land use regulation in the United States traditionally has been mainly the province of local governments."); Richard L. Briffault, *Our Localism: Part I – The Structure of Local Government Law*, 90 COLUM. L. REV. 3 (1990) ("Land use control is the most important local regulatory power.... [In land use], state-delegated power, supported by judicial attitudes sympathetic to local control, has resulted in real local legal authority, notwithstanding the nominal rules of state supremacy.").

[49] *See, e.g.,* Greg Hise, MAGNETIC LOS ANGELES: PLANNING THE TWENTIETH-CENTURY METROPOLIS 130–31 (1997); THOMAS J. SUGRUE, THE ORIGINS OF THE URBAN CRISIS: RACE AND INEQUALITY IN POSTWAR DETROIT 129–30 (1996).

[50] *See* Charles M. Tiebout, *A Pure Theory of Local Expenditures*, 64 J. POL. ECON. 416, 419–20 (1956). Our discussion of the Tiebout-Hamilton framework relies primarily on Kirk Stark & Jonathan Zasloff, *Tiebout and Tax Revolts: Did Serrano Really Cause Proposition 13?*, 50 UCLA L. REV. 801, 811–13 (2003).

tax base. Bruce Hamilton, writing twenty years after Tiebout, termed this possibility "musical suburbs" – the poor chasing the rich in a "never-ending quest for a tax base." Hamilton's insight was that cities would use land use controls to block the free riders and increase their own fiscal base.[51] As William Fischel has bluntly noted, "The family of eight that wants to rent part of a lot in Scarsdale and park two house trailers on it and send their kids to Scarsdale's fine schools is apt to find a few regulations in their way."[52]

Localities interested in enhancing property values, however, often will attempt to increase VMT. They will adopt large lot-size requirements and generally low densities, making it extremely difficult to support a public transportation system financially. They also will resist providing sites for affordable housing, because these residents will be the classic "free riders" of the Tiebout-Hamilton system. But no matter how property wealthy a city might be, it will still need its working class: the police officers, firefighters, teachers, nurses, secretaries, janitors, and clerks who provide critical services but are rarely highly compensated. And zoning them out (and killing the density necessary for transit) means that they will have to drive to their jobs, further increasing vehicle miles traveled and greenhouse gas emissions.

These structural problems are compounded by literally decades of local land use practices and bureaucratic culture, combined with developer practices and business models, which presume and foster high-VMT development. In order to retool, cities must completely overhaul their zoning codes, general plans, road and street designs, parking requirements, block-length specifications, and virtually every aspect of what they have done since the end of the Second World War.[53] These changes also will require developers to change alongside the cities, uprooting established business models and thereby incurring large new design costs, a prospect likely to lead to some resistance.

All of these patterns and incentives would lead the observer to anticipate that cities in the United States would continue to promote land use policies that exacerbate climate change. One would not expect cities to be on the front lines of tackling this issue.

3. UNDERSTANDING NEW INTERNATIONAL ACTORS: WHY ARE U.S. CITIES TACKLING CLIMATE CHANGE?

Despite the pressures just identified, an increasing number of municipalities throughout the country appear to be confronting the specter of climate change. The

---

[51] See Bruce W. Hamilton, *Property Taxes and the Tiebout Hypothesis: Some Empirical Evidence*, in FISCAL ZONING AND LAND USE CONTROLS 13, 15 (Edwin S. Mills & Wallace E. Oates eds., 1975); see also Bruce W. Hamilton, *Zoning and Property Taxation in a System of Local Governments*, 12 URB. STUD. 205 (1975).

[52] William A. Fischel, *Property Taxation and the Tiebout Model: Evidence for the Benefit View from Zoning and Voting*, 30 J. ECON. LITERATURE 171, 171 (1992).

[53] A good demonstration of how postwar planning has undermined New Urbanism is found in Michael Lewyn, *New Urbanist Zoning for Dummies* (George Washington University Legal Studies Research Paper No. 183, 2006), *available at* http://ssrn.com/abstract=873903.

2005 U.S. Mayors Climate Change Protection Agreement provides that signatories agree on a common goal: to meet or beat Kyoto Protocol targets within their own communities, that is, to reduce greenhouse gas emissions to 7% below 1990 levels by 2012.[54] As of March 12, 2009, 916 mayors have signed the Agreement.[55] The Sierra Club estimated that if the first 230 signatory cities succeed, their reductions would equal those expected from the combined Kyoto commitments of the United Kingdom, the Netherlands, and all Scandinavian countries.[56] The city of Portland, Oregon, already claims to have substantially reached its Kyoto-"mandated" levels.[57] Salt Lake City, Utah, states that it has reduced emissions from its municipal operations by 31% since 2001, exceeding its target.[58]

Why are these cities embracing this initiative despite the pressures against them? A clear answer is not available in the existing literature. Determining the best explanation for the apparent urban leadership on climate change obviously awaits more detailed empirical research on implementation of the Agreement – and of course the track record that such research would investigate.[59] But any research outcome promises to illuminate two scholarly disciplines generally not associated with each other: urban theory and international relations theory. Considering the role of cities in the politics of climate change suggests that bridging these two fields could enrich them both. It also may provide critical information to advocates seeking to mitigate climate change.

The traditional Westphalian model of international law and international relations focuses exclusively on nation-states as international actors.[60] A burgeoning

---

[54] They also agree to urge their state governments and the federal government to do the same. *See* U.S. CONFERENCE OF MAYORS, THE U.S. MAYORS CLIMATE PROTECTION AGREEMENT (2005), *available at* http://usmayors.org/climateprotection/documents/mcpAgreement.pdf.

[55] Press Release, The U.S. Conference of Mayors, *600 Mayors in All 50 States and Puerto Rico Take Action to Reduce Global Warming* (July 13, 2007), *available at* http://usmayors.org/climateprotection/climateagreement_071307.pdf.

[56] Jennifer Hattam, *Green Streets: Where Great Ideas Are Transforming Urban Life*, SIERRA CLUB MAG., July/Aug. 2006, at 36.

[57] *See* A PROGRESS REPORT ON THE CITY OF PORTLAND AND MULTNOMAH COUNTY LOCAL ACTION PLAN ON GLOBAL WARMING 1 (June 2005), *available at* http://www.portlandonline.com/shared/cfm/image.cfm?id=112118 ("Despite rapid population and economic growth, local greenhouse gas emissions in 2004 were only *slightly above* 1990 levels, the benchmark year established in the Kyoto Protocol.") (emphasis added). We note that this figure comprises all emissions from the area, not simply those produced by governmental activities.

[58] *See Salt Lake City Green, Current and Completed Climate Change Initiatives*, *available at* http://www.slcgreen.com/CAP/current.htm (last visited March 16, 2009) ("Salt Lake City has reduced carbon dioxide in its municipal operations energy use by 31% since 2001, surpassing our goal to meet the Kyoto Protocol standard by 148%, seven years early.")

[59] One excellent project by Harriet Bulkeley and Michelle M. Betsill reviews the influence of the Local Governments for Sustainability's (ICLEI) Cities for Climate Protection program on six cities, including two in the United States assessing the success of this transnational network in influencing climate change policy. This gives us a good start on the empirical work necessary to understand local governments' actions. *See* HARRIET BULKELEY & MICHELLE M. BETSILL, CITIES AND CLIMATE CHANGE: URBAN SUSTAINABILITY AND GLOBAL ENVIRONMENTAL GOVERNANCE (2003). For further discussion on this topic, see *infra* Sections 3.1, 3.4.

[60] *See* MARK W. JANIS, AN INTRODUCTION TO INTERNATIONAL LAW 162–63 (4th ed. 2003) ("Sovereignty was the crucial element in the peace treaties of Westphalia, the international agreements intended to

literature recognizes, however, that other actors, such as nongovernmental organizations, subnational governments, transgovernmental networks, and multinational corporations profoundly affect the course of international law and politics;[61] and thus, the traditional Westphalian model is becoming obsolete. These new players in world affairs include local governments. Yishai Blank, for example, argues that "globalization is not only imposed on passive localities by their states or international institutions – it is also advanced from the ground up by localities themselves. Localities are thus doing their part to further disaggregate the waning Westphalian concepts of the unitary state."[62]

Knowledge of how precisely local governments behave in such a highly complex global politics, however, remains rudimentary. We suggest that one potentially fruitful way to frame the issue lies in applying the insights of international relations theory to municipal behavior on the global stage. Indeed, the very origin of international relations theory derives from Thucydides, who attempted to explain the actions of city-states.[63] More importantly, international relations theory represents a useful series of frames of the behavior of governmental units competing for survival without centralized, sovereign-created rules.[64] As cities increasingly interject themselves into the international dialogue on climate change both as policymakers and as litigants, they step out of the domestic realm in which they are governed by a set of well-established legal rules. Moreover, in a competitive global economic environment, cities, like nations, must be concerned for their survival. To be sure, the "destruction" of a city may not be political but economic, but few can examine the status of, say, Detroit, and claim that it has "survived" in anything but the most nominal terms. While analogizing competition between cities to the anarchical global order is far from perfect, the management of common pool resources such as the Earth's atmosphere lends itself well to international analogies.

We most emphatically do *not* argue that cities will behave in world politics in the same way as nation-states, if for no other reason than cities are embedded within a set of domestic laws far denser and more powerful than the anarchy of global politics. Rather, we believe that enough similarities exist between the incentives for

---

end a great war and to promote a coming peace ... [T]he key actor on the world's stage was the sovereign state to which all loyalty was due internally and which was unrestrained externally."). If cities have an independent role in global politics, this development clearly cuts against the Westphalian grain.

[61] See, e.g., Earl H. Fry, The Expanding Role of State and Local Governments in U.S. Foreign Affairs 23 (1998) ("While predictions concerning the rapid demise of the nation-state are premature, an evolutionary process is certainly under way, in terms of both the distribution of governing authority within the nation-state and the constant interaction among governments, international organizations, and citizens groups."); Hari M. Osofsky, Climate Change Litigation as Pluralist Legal Dialogue?, 26 Stan. Envtl. L.J. & 43A Stan. J. Int'l L. 181 (2007) (joint issue); Kal Raustiala, The Architecture of International Cooperation: Transgovernmental Networks and the Future of International Law, 43 Va. J. Int'l L. 1 (2002).

[62] Yishai Blank, Localism in the New Global Legal Order, 47 Harv. Int'l L.J. 263, 268 (2006).

[63] Laurie M. Johnson Bagby, The Use and Abuse of Thucydides in International Relations, 48 Int'l Org. 132 (Winter 1994).

[64] To the extent that cities globally begin to make strong efforts toward combating climate change, the international relations analogy becomes stronger, because no central enforcement mechanism exists for relations between cities in different nations.

cities and those for nation-states that international relations theory can provide a framework that helps to generate useful research hypotheses for investigating the motivations behind cities' actions on climate change. Moreover, broader theories can help clarify discrete data, detecting broader behavior patterns that might appear at first to be driven by idiosyncratic or highly localized factors.

As will be explained in greater detail herein, the turn to international relations theory also might help to explain an anomaly in urban theory, particularly the regnant models that analyze cities' motivation and function as attracting capital or serving as neutral "markets" or "bankable" locations. In sum, then, connecting two streams of scholarship both broadens and deepens our understanding of the increasingly complex global political environment.

3.1 *Urban Theory and Quasi-Realism: Cities as Markets*

We should begin, then, with the most influential model in modern international relations theory, "Structural Realism," and its less demanding counterpart that we term "Quasi-Realism."[65] Structural Realism posits that policy outcomes are principally shaped by the international system. That system, Structural Realists contend, is anarchical, and thus threatens the viability of all the states within that system. In such a competitive environment, states measure their own success by their power *relative* to other states in the system.[66]

Structural Realism parallels the most significant trends in urban theory, in particular the significant expansion of scholarship seeking to understand the role of cities in the globalizing economy. Michael Porter's work, which stressed that cities must compete for scarce capital in order to survive economically,[67] has been applied worldwide. Porter's work echoed an older model developed by Paul Peterson,[68] which argued that the weakness of cities within a federalist framework requires them to avoid redistributive politics and privilege business elites as a way to promote economic development and secure their tax base. Cities must compete "with one another so as to maximize their economic position."[69] To achieve this objective, "the city must use the resources its land area provides by attracting as much capital and as high a quality labor force as possible."[70]

---

[65] The standard work outlining such a theory is KENNETH N. WALTZ, THEORY OF INTERNATIONAL POLITICS (1979). Since Waltz's study appeared, the literature commenting on and reacting to it has been vast.

[66] An excellent summary of Structural Realism's implications for states' foreign policies is found in Fareed Zakaria, *Is Realism Finished? (Method of Analyzing International Relations)*, 30 NAT'L INT. 21 (Winter 1992).

[67] Michael Porter, *The Competitive Advantage of the Inner City*, 73 HARV. BUS. REV. 55 (1995). In order to advance this model, Porter established the Initiative for a Competitive Inner City, http://www.icic.org (last visited Jan. 3, 2008). Porter's framework derived from his earlier work, which focused on national competitiveness. *See* MICHAEL PORTER, THE COMPETITIVE ADVANTAGE OF NATIONS (1990).

[68] PAUL E. PETERSON, CITY LIMITS 12 (1981).

[69] *Id.*

[70] *Id.*

Urban sociologists and geographers whose politics sharply differ from Porter's (and to a lesser extent, Peterson's) have also gotten into the game. Saskia Sassen's influential writings focus on a narrower band of urban centers, which she terms "world cities." Sassen finds fierce competition among larger cities to attract the "command-and-control functions of the global economy," which we understand to comprise the network of critical business services – such as banking/finance, accounting, advertising, and law – that international capital relies upon to maintain the health of the global capitalist system. Recent scholarship by Gerald Frug and David Barron on "international local government law" discusses how international law implicitly endorses a vision of the "private city."[71] This urban form, promoted by the World Bank among others, emphasizes the city's role as a market location that can facilitate economic growth. Although not limited to a narrow range of command centers, the normative focus of international law on cities as private markets again emphasizes the economic function of cities and deemphasizes their governmental role.[72]

Despite highly divergent disciplinary backgrounds and political outlooks, a powerful theme underlies the cities-as-markets view: globalization and the mobility of capital have sharply curtailed urban autonomy, forcing cities to compete for capital and driving a convergence of urban politics around the attraction of business. If not zero sum, this model certainly has little room for mutually beneficial cooperation between cities.

Not surprisingly, other urban theorists take exception to – or at least highly qualify – this picture. H.V. Savitch and Paul Kantor contend that "cities need not be leaves in the wind," and argue forcefully that cities have varying bargaining positions with national and international capital, thus allowing them wiggle room to develop their own independent policies.[73] (Nonetheless, they still acknowledge that the relationship with capital is a critical factor in shaping modern cities.) In similar fashion, Peter Newman and Andy Thornley find regional differences among North America, Europe, and Pacific Asia, based largely on differences in national-local relationships.[74] Newman and Thornley also find that cities within a region often adopt very different economic development strategies.

Cities, of course, do not function in an entirely zero-sum world, which is why Structural Realism in international relations theory provides an imperfect analogy for examining city behavior[75] and why we reference its application to cities as

---

[71] Gerald Frug & David Barron, *International Local Government Law*, 38 URB. L. 57 (2006).

[72] To be clear, Frug and Barron do not endorse the model; rather, they argue that international law appears to do so. See id.

[73] See generally H.V. SAVITCH & PAUL KANTOR, CITIES IN THE INTERNATIONAL MARKETPLACE: THE POLITICAL ECONOMY OF URBAN DEVELOPMENT IN NORTH AMERICA AND WESTERN EUROPE (2002).

[74] See generally PETER NEWMAN & ANDY THORNLEY, PLANNING WORLD CITIES: GLOBALIZATION AND URBAN POLITICS (2005).

[75] Another obvious difference is that cities do not anticipate military conflict with other cities. Their primarily economic conflict with each other, however, contains large implications for their overall health and well-being.

"Quasi-Realism." However, the analogy is not far off. As Porter, Peterson, Sassen, and others have emphasized, however, cities do compete for vital goods like investment capital, which does yield a zero-sum result. Moreover, there can only be so many "centers" of critical command functions; once these functions are too spread out, no city can serve as a center. Thus, we find the analogy useful for identifying a set of presumptions in urban theory and for framing hypotheses about municipal action.

Despite the theoretical presumption that cities compete for capital, any serious municipal attack on climate change would, upon first impression, appear to be *adverse* to business interests. While global economic command-and-control functions might not necessarily require a particular urban land use and environmental strategy, one would not expect cities to focus on environmental, energy, and land use policies to prevent climate change as a means of attracting capital. As for cities' own competitive advantage, the business groups pursued as part of a competitive-advantage strategy likely would be wary of a city whose policies were focused on global warming.

Such local governments might advance regulations and policies that could substantially curtail capital's ability to conduct business in the way that it wants. They could restrict or penalize energy use, or require energy use from renewable sources, increasing its cost. They could zone for high-density development, thereby curtailing the construction of low-rise office parks popular in both the commercial and the industrial sectors.[76] They could insist on restrictions on employee automobile usage or require and enforce parking cash-outs.[77] They could attempt to set urban growth boundaries and thus reduce business' options for development and expansion. And they could restrict free parking, thus forcing retail businesses to internalize the costs of automobile dependence.[78] Indeed, the very unpredictability of what regulatory steps they might take could deter capital.

Scholars taking a Quasi-Realist view, then, would hardly be surprised if they found that municipal rhetoric on climate change remains just that. They would expect cities to be engaging in mere window dressing, signing memoranda that sound impressive – and enhancing the profile of ambitious local politicians[79] – without really taking the difficult steps required to actually reduce their emissions. Inaction or limited action, then, would suggest that the pessimistic theories set forth here actually do explain cities' actions.

A 2003 review of municipal policies on climate change found some evidence for this pessimistic Quasi-Realist position. Harriet Bulkeley and Michele M. Betsill

---

[76] Many industries prefer sprawling development, as low-rise office parks are considered particularly suitable for many commercial and industrial users. *See* HISE, *supra* note 49; SUGRUE, *supra* note 49; PETERSON, *supra* note 67.

[77] *See* Jennifer Dill, *Mandatory Employer-Based Trip Reduction: What Happened?*, 1618 TRANSP. RES. REC. 103–10 (1998).

[78] *See generally* DONALD SHOUP, THE HIGH COST OF FREE PARKING (2005).

[79] This certainly could explain the actions of some prominent mayors. *See, e.g., Governor, L.A. Mayor Doing Power-Inspired Duet,* DAILY BREEZE (Torrance, Cal.), July 10, 2006, at A13 ("It's no secret that [Los Angeles Mayor Antonio] Villaraigosa thirsts to be Governor.").

assessed the climate change policies in six cities on three continents, including two cities in the United States.[80] They found that although some of these cities did take proactive measures on in-house energy management, such as municipally owned vehicle fleets and public property, they took few measures in the critical areas of planning, transportation, and other land use controls such as building codes.[81] They found that the more serious attempts to promote climate change policies usually derived from preexisting agendas; in other words, urban policymakers pursued measures that reduced carbon emissions not for their own sake but for other policy reasons or to foster their own images as pro-environmental politicians.[82] (We have more to say about these reasons later.)[83] Bulkeley and Betsill's work was published two years prior to the Mayor's Agreement and was directed toward international efforts to shape cities' actions rather than domestic ones; nonetheless, it does suggest the difficulty of implementing climate change mitigation policies that impose costs and inconveniences on a city's residents or businesses.

But what if cities actually mean it? Quasi-Realism would not be without explanations. Cities competing for economic development resources might very publicly fight climate change, casting doubt on the Environment Court's thesis.

First, Sassen argues that in order to attract the command-and-control functions of the international economy, cities must also attract the highly educated professionals who provide critical services to global capital. These professionals might care a great deal about living in a sustainable city.[84] Moreover, at some point, cities might pass a tipping point, where pollution, congestion, and livability get so bad that the municipality is unable to attract capital investment.

Second, the mere fact that cities might be competing within the Tiebout equilibrium hardly implies that they would compete in the same way. Many local governments simply cannot hope to compete with the fast-growing outer suburbs and exurbs: they lack the huge tracts of land necessary for sprawling development.[85] No matter how much San Francisco tries, it cannot provide space for large office parks with free employee parking. Thus, such cities will move toward more compact development and New Urbanist form as a way of finding a market niche.

Simply because Quasi-Realism could explain both mere rhetoric on climate change and genuine municipal policy changes does not imply that it is incoherent; rather, it shows that the framework raises a set of new research questions. If cities prove to be only paying lip service to climate change politics, then this would seem to

---

[80] See generally BULKELEY & BETSILL, supra note 59. We should note, however, that Bulkeley and Betsill do not identify themselves as realists.
[81] Id. at 171–85.
[82] Id.
[83] See infra Section 3.4.
[84] See NEWMAN & THORNLEY, supra note 74, at 44 ("There is increasing awareness that the success of the city also lies in maintaining... environmental sustainability."). Newman and Thornley do not, however, provide any direct evidence of this awareness or that it has been translated into policy.
[85] See Matthew P. Drennan & Michael Manville, Lagging Behind: California's Interior Metropolitan Areas (May 31, 2006) (unpublished manuscript, on file with authors).

demonstrate the pessimistic version of Quasi-Realist theory. If, however, cities are actually taking the kinds of serious, difficult measures required to reduce overall emissions, then establishing the validity of Quasi-Realism becomes more complex. Researchers would have to determine whether the strategy of attracting key professionals actually represents a pattern of municipal policy across different governments: can we point to any actual and continuing[86] governmental action to promote economic development through sustainability policies that would otherwise cut against the interests of capital?[87] If not, then it would be hard to represent it as a function of a global system since presumably many cities face the same incentives. But if so, then such a pattern could provide important evidence for a more positive Quasi-Realism.

3.2 Liberalism: All Politics Is Local

Broadly speaking, Liberal theories of international politics argue that the actions of states are driven by domestic political conditions.[88] These conditions can be ideological (states behave according to cultural or religious traditions), material (states behave according to the economic interests of powerful political actors), or institutional (states behave according to the specific mode of aggregating preferences within them, such as democracy or dictatorship). Using the lens of Liberal theory, we would consider how individuals or political groups use the climate change issue for either idealistic or material purposes. Moreover, we would look to the internal politics of the city, not its relative relationship to other cities, as the impetus behind its regulatory stance on climate change. From this perspective, we must consider that, as a distinct political milieu, large American cities are generally left of center,[89] and in the United States, the Bush administration's refusal to take climate change seriously intensely politicized the issue. For the domestic opposition, the administration's attitude symbolized an arrogant, parochial government hostile

---

[86] We emphasize the continuing nature of such policies as a way to avoid confusing temporary fads – which might be best described as constructivist, see infra Section 3.3 – from genuine systemic incentives.

[87] There is some sketchy evidence that this may be happening. See, e.g., TERRY NICHOLS CLARK, THE CITY AS AN ENTERTAINMENT MACHINE (2003), L.T. Ker, Towards a Tropical City of Excellence, in CITY AND THE STATE: SINGAPORE'S BUILT ENVIRONMENT REVISITED (Ooi Giok Ling & Kenson Kwok eds., 1997).

[88] See Andrew Moravcsik, Taking Preferences Seriously: A Liberal Theory of International Politics, 51 INT'L ORG. 513–53 (1997); Anne-Marie Slaughter Burley, International Law and International Relations Theory: A Dual Agenda, 87 AM. J. INT'L L. 205, 228 (1993). Because a broad range of theoretical approaches use this label, forests have been felled attempting to define "liberalism." The way we use it here has become accepted in international relations theory, despite the seemingly endless variety of contexts and meanings that the word holds. We adopt it here to follow the international relations literature, not as an assessment of any other use of the term.

[89] See, e.g., Jill Lawrence, Democratic Gains in Suburbs Spell Trouble for GOP, USA TODAY, Nov. 26, 2006, at 6A (noting Democratic candidates won 60% of the vote in inner suburbs); Posting of Ezra Klein, Places, Not States, http://ezraklein.typepad.com/blog/2007/03/places_not_stat.html (Mar. 26, 2007) ("It's long been understood that urban centers go Democratic....").

to scientific data – indeed, data of all kinds.[90] It should hardly surprise observers that mayors representing their democratic constituencies would take pleasure in highlighting an opposing administration's abdication of leadership. These kinds of Democratic-leaning cities are also the most likely to have local environmental activists who make campaign contributions and show up at the polls in low-turnout municipal elections.

If, by analogy to Liberal theory, we presume that all politics is local, we would expect Democratic-leaning cities to take stronger anti–climate change actions. But this signals more than a reaction to the Bush administration. Environmentalism has become a principal hallmark of the modern Democratic Party, so we would naturally anticipate that cities dominated by Democrats would also have a distinctly green cast. This is particularly true because land use policies to mitigate climate change hardly signal a new direction in planning theory. They represent the "smart growth" policies that progressive planners and New Urbanists have been advocating for nearly two decades. Left-of-center Democrats were advocating "smart growth" and energy conservation long before climate change appeared on the national political horizon, and they will continue to do so even if the climate change problem miraculously disappears tomorrow. Thus, Democratic-leaning cities will adopt climate change policies not only because of the politicization of the climate debate nationally but also because those policies conform to a preexisting political agenda.[91]

At the same time, the Agreement's signatory list includes many traditionally conservative cities. This could indicate that alarm over climate change is crossing partisan lines. Polling suggests that this, indeed, is happening. One poll found that 63% of Americans believe environmental hazards such as climate change present a threat equivalent to that of terrorism,[92] a supermajority that reveals strong bipartisan preferences. Similarly, 70% of respondents to a January 2007 poll stated that global warming is having a serious impact now.[93] Sharper evidence is available from polling data broken down along party lines. Although a higher percentage of Democrats perceive a threat from climate change, a substantial majority of Republican primary voters do as well; thus, a December 2006 poll of New Hampshire Republican primary voters (whose viewpoints figure to be highly conservative) found that 70% believe

---

[90] For a discussion on this latter point, see the superb analysis in Joshua Micah Marshall, *The Post-Modern President*, WASH. MONTHLY, Sept. 2003, at 22.

[91] For example, much of Los Angeles mayor Antonio Villaraigosa's announcement concerning his "climate change" policy actually concerned several items detached from the climate change agenda, such as restoring the Los Angeles River, reducing pollution at the Port of Los Angeles, cleaning up the Santa Monica Bay, and increasing open space. *See* Antonio R. Villaraigosa, L.A. Mayor, Remarks before the Latino Congreso Conference (Sept. 8, 2006), *available at* http://www.earthday.net/news/Remarks-LatinoCongresoConference.pdf.

[92] *See* Memorandum from Global Strategy Group to Yale Center for Environmental Law & Policy (Mar. 7, 2007), *available at* http://research.yale.edu/envirocenter/uploads/epoll/YaleEnvironmentalPoll2007 Keyfindings.pdf (visited Jan. 3, 2008).

[93] *See* Press Release, CBS News Poll, The President, the State of the Union and the Troop Increase (Jan. 22, 2007), *available at* http://www.cbsnews.com/htdocs/pdf/012207_bush_poll.pdf.

climate change to be a "serious threat"[94] and 75% think that the United States should take action to reduce greenhouse gas emissions.[95] Congressional Republicans, however, seem out of step with their constituents and with local elected Republican officeholders: Only 13% of GOP members of Congress believe "beyond a reasonable doubt" that human activity is causing climate change, compared with 95% of their Democratic counterparts.[96]

Such data suggests that preferences form differently depending upon the level of government: Republican voters will back climate change skeptics at the national level while supporting action against climate change at home. They may be more willing to cross party lines for environmental causes when the stakes are local. This in turn suggests unique aspects of local politics that will translate into unexpected municipal behavior on climate change – precisely the sort of outcome that would be anticipated by Liberal theory. All politics may be local, as Tip O'Neill famously stated, but at least with climate change, local politics may be becoming global.

The linkages between smart growth policies and the prevention of climate change shows that predictions of Liberal theory also might derive from the material realm. New Urbanism fulfills the needs of some social groups more than others, particularly young professionals, empty nesters, and (to a lesser extent) families where all the adults work. None of these household types needs the standard suburban form of sprawling single-family homes with lawns.[97] Liberal theory might suggest that the degree to which a local government adheres to anti–climate change policies would vary with these kinds of critical demographic factors. And this trend could have a ratchet effect: Promoting compact development would attract more members of these demographic groups, which could in turn strengthen political support for New Urbanist form.[98] The theory might also suggest an alternative explanation if we find that older cities and inner-ring suburbs prefer compact development over sprawl; local elites with fixed investments in these localities, Liberals will suggest, will serve as the driving force behind local economic development policies in order to buttress their own assets.

Thus, in the same way that Quasi-Realism frames a research agenda, Liberalism does as well. It directs us to consider that the impetus for cities' actions may not lie in purely economic terms, but may rather stem from the ideological bent of its inhabitants. (Of course, these may often extend beyond environmental issues to other matters such as religion, etc.)

---

[94] See THE MELLMAN GROUP, LARGE MAJORITIES CLAIM AT LEAST SOME KNOWLEDGE ABOUT GLOBAL WARMING (Dec. 2006), *available at* http://www.cleanair-coolplanet.org/information/nhpollresults.pdf.
[95] See *id.*
[96] See *Congressional Insiders Poll*, NAT'L J., Feb. 3, 2007, at 6 (asking the question, "Do you think it's been proven beyond a reasonable doubt that the Earth is warming because of man-made problems?").
[97] See Calthorpe, *supra* note 44.
[98] Liberal skeptics would wind up on the same side as the Realists, although for different reasons: cities will not respond, Liberals will argue, because of the Homevoter Hypothesis. There is an important (and potentially unanswerable) question as to whether the operation of the Homevoter Hypothesis is better described as a Liberal or Realist development. Moravcsik suggests that internal politics driven by external system effects are Liberal. See Moravcsik, *supra* note 88, at 523.

Most obviously, does any relevant difference exist between municipal climate change policies based upon the strength of Democratic partisanship or other indicators of left-leaning politics? If not, do we find evidence that environmental issues generally or climate change policy particularly may influence voters at the local level to cross partisan lines in a manner that would not occur on national-level votes? Similarly, can we point to differences based upon the demographic characteristics mentioned here? If we find central cities and inner-ring suburbs pushing more compact development, can we point to particular elites driving such a decision, with variations in local elites affecting the degree of New Urbanism adopted?

3.3 Constructivism: International Discourse

Constructivist international relations theory suggests that, by producing a set of discursive practices that shape knowledge and ideas, international interaction creates the international system itself.[99] Thus, constructivist theory focuses on dialogue between actors as the primary source of their relationship. For constructivists, language creates the international system by constructing parties' self-definitions. The most extreme position sees it as a matter of discourse creating states, not states engaging in discourse. A more moderate position holds that notions of national interest are altered fundamentally through the interactions of states and other actors in world politics.[100]

Margaret Keck and Kathryn Sikkink's influential account of "transnational advocacy networks" (TANs), which they define as "networks of activists distinguishable largely by the centrality of principled ideas or values in motivating their formation,"[101] serves as a recent outstanding example of the constructivist turn. The overall point is that strong forces outside a polity influence its politics on principled grounds.[102]

---

[99] It is worthwhile to distinguish between "hard" and "soft" constructivism. See Richard H. Steinberg & Jonathan M. Zasloff, *Power and International Law*, 100 Am. J. Int'l L. 64, 82–85 (2006) (making this distinction). Hard constructivists argue that the most basic building blocks of the international order are social constructs. See, e.g., Friedrich V. Kratochwil, Rules, Norms, and Decisions: On the Conditions of Practical and Legal Reasoning in International Relations and Domestic Affairs (1989). Soft constructivists explain the international social processes that affect the international system.

[100] The more extreme position is sometimes associated with the work of Friedrich Kratochwil. See Kratochwil, *supra* note 99. The more moderate approach is taken by Alexander Wendt. See Alexander Wendt, Social Theory of International Politics (1999); Alexander Wendt, *The Agent-Structure Problem in International Relations Theory*, 41 Int'l Org. 335 (1987).

[101] Margaret E. Keck & Kathryn Sikkink, Activists Beyond Borders: Advocacy Networks in International Politics (1998).

[102] Anne-Marie Slaughter has also suggested that these international networks run between governmental officials as well as among activists. See Anne-Marie Slaughter, *The Real New World Order*, Foreign Aff., Sept./Oct. 1997, at 183. It is not clear whether this represents a Constructivist turn or an Institutionalist one. Slaughter herself describes her work as Liberal because she sees the individual rather than the state as the repository of preferences.

Although Bulkeley and Betsill also highlight the function of transnational networks in their discussion, they focus almost exclusively on transnational networks of government officials as opposed to nongovernmental advocacy networks. See Bulkeley & Betsill, *supra* note 58, at 186–93. This is largely because their study concerns the ICLEI CCP program, which by definition is restricted to

The intense national and international focus on climate change suggests that viewing cities' actions as primarily a reaction to local pressure (as Liberalism would posit) may be missing the forest for the trees. The Mayors' Agreement itself was launched at the U.S. Conference of Mayors,[103] and when Los Angeles Mayor Antonio Villaraigosa declared his adherence to the Mayors' Agreement, he did so not at City Hall but at an international conference at the University of California, Los Angeles (UCLA). On the dais, he was accompanied not by members of the City Council, but rather by Bill Clinton, Tony Blair, San Francisco Mayor Gavin Newsom, and London Mayor Ken Livingstone.[104]

Thus, the politics of local climate change may be driven by powerful advocacy networks whose origins lie far from the cities in which they operate. Within this framework, national and international environmental organizations raise the issue's salience, propose specific actions, and command press attention, making it useful – and at times necessary – for local politicians to embrace the initiative. Indeed, they may shape politicians' own values. TANs may provide financial, informational, and political support to local organizational chapters and new cadres of activists energized by the climate change issue. Thus, they not only create connections between cities on this policy issue but also help policymakers and their constituents believe that climate change requires a significant policy response from all levels of government.[105]

The growth of local climate change policy advocacy even could suggest that Keck and Sikkink's framework has more power than they gave it credit for. Keck and Sikkink hypothesize that TANs are most effective in dealing with "issues involving bodily harm to vulnerable individuals, and legal equality of opportunity."[106] Neither category includes climate change.[107] Moreover, Keck and Sikkink argue that TANs achieve prominence because they target repressive governments (such as human rights abusers) where domestic political opposition is practically close to impossible. Here, however, we might see the influence of powerful TANs even in open and

---

government officials. But this focus, in our view, means that their discussion of transnational networks fits more precisely into an Institutionalist framework, because it concerns the use of international institutions to facilitate cooperation between governments (in this case, local governments). We recognize that the borders of the international relations theory paradigms blur in this case, because it is not clear whether the institution is reducing the transactions costs of cooperative behavior (the Institutionalist account) or changing the discourse and thereby altering governments' perceptions of their self-interest (the Constructivist account). In international legal scholarship, such hybrid theories are becoming more common as empirical work becomes more detailed. See Steinberg & Zasloff, *supra* note 99, at 86–87.

[103] *See* Office of the Mayor, Seattle, Wash., *What Is the U.S. Mayors' Climate Protection Agreement?*, http://www.seattle.gov/mayor/climate (last visited Jan. 3, 2008).

[104] *See* Carla Marinucci, *In L.A. Speech, Blair Talks Tough in Defense of Israel*, S.F. CHRON., Aug. 2, 2006, at A10.

[105] For further discussion of this approach, see generally Osofsky, *supra* note 61.

[106] *Id.* at 204.

[107] Catastrophic climate change could, of course, cause severe bodily harm, but Keck and Sikkink emphasize that it involves specifically vulnerable groups (e.g., women subjected to circumcision) or the grotesque human rights abuses of the Pinochet government.

tolerant U.S. cities. Such developments could point to a wider and more powerful role for TANs in the emerging global order.

3.4 Institutionalism: Evolving Cooperation

Institutionalists accept the rationalist framework offered by Structural Realists, but suggest that transgovernmental institutions can transcend the Prisoner's Dilemma by providing a framework for cooperation. The Prisoner's Dilemma is a two-player game structured so that even though cooperation between the two parties would yield overall benefits for both, each player's "dominant strategy" (i.e., the course that it will take regardless of the other player's moves) is to "defect," or take uncooperative action. As Douglas G. Baird and his colleagues note, the Prisoner's Dilemma "is emblematic of some collective action problems in the law in which individual self-interest leads to actions that are not in the interest of the group as a whole."[108] Many international problems can be conceptualized as Prisoner's Dilemmas: Nations have powerful temptations to break arms control agreements, for example, for fear that the other party is doing the same.[109]

By considering a series of crosscutting yet related matters, international institutions effectively make nations "repeat players" in these areas. This, in turn, allows for the repetition, or "iteration" of the Prisoner's Dilemma, which as Robert Axelrod has shown, can facilitate cooperative outcomes.[110] In addition, international institutions reduce information and transactions costs by generating otherwise-costly monitoring information, thereby reducing uncertainty about compliance and assisting the production of international stability.[111] Finally, institutions reduce the usually high transaction costs of achieving international agreements and cooperation.[112]

By way of analogy with institutionalism, we would look for the formation of transmunicipal institutions, which could generate information about compliance with the Mayor's Climate Change Protection Agreement as well as reducing the transactions costs of multilateral action. The Agreement itself might represent cities' attempts to create such institutions. Organizations such as the International Council on Local Environmental Initiatives (ICLEI), the Clinton Foundation's Climate Change Initiative, and the Large Cities Climate Change Leadership Group claim

---

[108] Douglas G. Baird, Robert H. Gertner & Randal C. Picker, Game Theory and the Law 312–13 (1994).

[109] An excellent summary of the Prisoner's Dilemma game with implications can be found at *Prisoner's Dilemma*, in Stanford Encyclopedia of Philosophy (Aug. 11, 2003), available at http://plato.stanford.edu/entries/prisoner-dilemma.

[110] *See* Robert Axelrod, The Evolution of Cooperation (1984).

[111] Two leading accounts of Institutionalist theory are Robert O. Keohane, After Hegemony: Cooperation and Discord in the World Political Economy (1984); Robert O. Keohane, *Neoliberal Institutionalism: A Perspective on World Politics*, in International Institutions and State Power: Essays in International Relations Theory (1988).

[112] Theoretically, TANs could be providing the same sorts of services for local governments, providing an intriguing possible link between constructivist and Institutionalist theory.

to be operating in a similar fashion, creating international institutions to facilitate cooperation.[113]

Institutional existence, of course, hardly guarantees effectiveness. Bulkeley and Betsill's study questions ICLEI's ability to significantly promote robust municipal climate change policies. The United Nations' Environment Programme provides little assistance to local governments and is itself a weak link in an overmatched U.N. framework.[114] The Clinton Foundation states that it is developing a worldwide purchasing cooperative to lower the prices of energy-saving products and to catalyze the development and deployment of low-energy and low-greenhouse-gas-producing products and services.[115] Such plans, however, while valuable in garnering cooperation to surmount fiscal barriers to efficiency improvements, may not provide an iterative or monitoring process for overcoming the competitive pressures that drive unsustainable land use practices.

But this might not be the end of the story. As suggested earlier, one major obstacle preventing New Urbanist land use is not political but inertial.[116] Sprawling development occurs because of encrusted layers of obsolete zoning and planning codes, which in turn create business models designed for traditional suburban growth, which in turn leads developers to propose and advocate for these models. And while New Urbanist designers have crafted codes in response to suburbia, these documents have nowhere near the specificity and applicability for cities to deploy them, especially since cities are generally not working with empty land, but rather with suburbs needing retrofitting. Hiring the extra staff and costly consultants necessary to achieve low-VMT land use may be well beyond the range of most cities. To the extent that institutions such as the Mayors' Conference, ICLEI, the Clinton Foundation, and others can provide such services at lower cost to urban areas, they would represent the kind of transaction-cost-reducing bodies envisioned by institutionalist theory.

Institutionalist theory, then, suggests a potentially fruitful research agenda in investigating these groups and others like them.[117] Most importantly, to what extent has institutional involvement allowed individual cities to accelerate the pace and depth of their initiatives to address climate change? Can we detect the use of institutional information in augmenting monitoring capabilities? Can we point to any actual

---

[113] *See President Clinton Launches Large Cities Climate Initiative*, ENV'T NEWS SERVICE, Aug. 2, 2006, *available at* http://www.cns-newswire.com/ens/aug2006/2006-08-02-05.asp (last visited Jan. 3, 2008).

[114] *See* PHILIP SHABECOFF, A NEW NAME FOR PEACE: INTERNATIONAL ENVIRONMENTALISM, SUSTAINABLE DEVELOPMENT, AND DEMOCRACY 49–50 (1996).

[115] *See* William J. Clinton Foundation, *Clinton Climate Initiative*, *available at* http://www.clintonfoundation.org/cf-pgm-cci-home.htm (last visited Jan. 3, 2008).

[116] More precisely, one might say that the problem derives not from inertia but rather from fixed costs.

[117] Its applicability suffers because of the relative dearth of enforcement measures that cities have against other cities. Institutional monitoring is useful because it allows states to retaliate against each other if they do not comply. But whereas states can retaliate against each other in several ways, usually through trade and other economic barriers, cities have far fewer levers. Nevertheless, cities have some weapons available to them – most notably tax incentives, planning and zoning codes, and tit-for-tat strategies – which could provide at least some basis for reciprocity.

reduction in transactions costs by the provision of these institutional services? And can we determine whether institutional involvement has led to outcomes that would not have otherwise occurred?

Our own suspicion is that *all* of the theories will turn out to explain different aspects of municipal behavior. After all, in the context of international relations, "[n]one of the metatheories of the last century ha[s] been able to deliver the knockout blow that some may have once thought possible. No one trying to understand international relations can ignore power, *or* law, *or* the state, *or* civil society, *or* norms, *or* language."[118] Now, "midlevel analysis of international legal and political developments using hybrid theories"[119] is the best course. Creating and applying these hybrids represents the challenge for those seeking to understand the local politics of climate change. And the better the understanding, the more effective can be advocacy strategies designed to influence policy.

4. CONCLUSION: IS A THEORY NECESSARY?

As cities increasingly become international players, particularly on climate change, the development of a theory modeling the local politics of climate change could be of great value to both the scholarly and policymaking community. Understanding what cities do and why they do it is imperative. Whether or not the United States enacts comprehensive (or even piecemeal) climate change regulation, local governments will control the land use process for the foreseeable future, and thus they will play a central role in the mitigation of or adaptation to global warming – or the failure of such efforts. Practically, the better the understanding of local politics, the more effective can be advocacy strategies designed to influence land use policies. And even if state and federal policy eventually manages to somehow regulate land use for the purpose of mitigating climate change, these policies will have to be implemented at the local level. Will local governments be able to respond constructively? Or is "sustainable development" at the urban level only an oxymoron in the face of fierce international competitive pressures? We still do not know, but we should find out.

What does seem clear is that the pace of international climate change litigation will surely quicken as impacts become increasingly apparent. Cities will find it advantageous for a host of reasons to serve as climate change plaintiffs. Even if some cities are hypocritical about climate change, hypocrisy is the tribute that vice pays to virtue, and in the case of lending a few city attorneys to the effort, it is not a very big tribute at all. The seconding of attorneys to international climate change litigation, however, will mean a great deal to the cash-strapped NGOs that now must devote a major portion of their resources to the effort. To the extent that city leaders gain domestic support for climate change reduction efforts, serving as a plaintiff certainly will promise political benefits. Plaintiff cities such as Oakland,

---

[118] Steinberg and Zasloff, *supra* note 99, at 86.
[119] *Id.*

California, and Boulder, Colorado, have already been lionized by transnational advocacy networks,[120] which figure to pressure others to join in. And as attorneys from different jurisdictions develop connections representing co-plaintiffs in climate change litigation, they could provide ballast and organizational heft for the growth of cooperative institutions that further environmental goals.[121] Moreover, from a risk perspective, cities ignore climate change at their peril; cities and other local governments are beginning to find themselves defending against climate change actions as well.[122] Thus, recalcitrant jurisdictions may find it costly to miss jumping on this policy bandwagon.

Perhaps the Environment Court was right, and perhaps cities rush headlong into disaster. But enough evidence exists, both inside and outside the climate change arena, to suggest that the Environment Court may have seriously overstated the matter. If local and regional policy continues to play a central role in efforts to forestall and adapt to climate change, neglecting it would represent a failure not only of scholarship but of the world community's effort to contend with the greatest environmental threat that humanity has ever faced.

---

[120] *See* Press Release, Friends of the Earth, *City of Oakland, Calif., Joins Global Warming Lawsuit in Unanimous City Council Vote* (Dec. 18, 2002), *available at* http://www.foe.org/new/releases/12020akland.html ("We congratulate the cities of Oakland and Boulder for their leadership in holding the Bush administration accountable for failing to take action on global warming" (quoting Gary Cook, Coordinator of Greenpeace's Global Warming Campaign)). The statements refer to the lawsuits challenging the Overseas Private Investment Corporation's and the Export-Import Bank's investment decisions on National Environmental Policy Act grounds. On March 30, 2007, the federal district court in San Francisco partially denied defendants' motions for summary judgment, allowing the case to proceed. *See* Order Denying Plaintiffs' Motion for Summary Judgment and Granting in Part and Denying in Part Defendant's Motions for Summary Judgment, Friends of the Earth, Inc. v. Mosbacher, No. C 02–04106 JSW (N.D. Cal. Mar. 30, 2007). It is not known as of this writing precisely how this litigation will be affected by the Supreme Court's decision in *Massachusetts v. EPA*, 549 U.S. 497 (2007).

[121] We derive this idea from the work of Anne-Marie Slaughter, who has argued that the state is "disaggregating" into its component functions, thereby unbundling into its separate, functionally distinct parts. These courts, regulatory agencies, executives, and legislatures, says Slaughter, are then networking with their counterparts abroad, creating a new, transgovernmental order. *See generally* ANNE-MARIE SLAUGHTER, A NEW WORLD ORDER (2004).

[122] *See, e.g.*, Ctr. for Biological Diversity v. City of Banning, No. RIC 46097 (Cal. Super. Ct. Nov. 21, 2006) (challenging Banning's approval of a large housing development because it failed to disclose, analyze, or mitigate greenhouse gas emissions from the project); Ctr. for Biological Diversity v. City of Desert Hot Springs, No. RIC 464585 (Cal. Super. Ct. Jan. 24, 2007) (similar issues). The California attorney general's suit against the County of San Bernardino, discussed *supra* note 13, presents another example of this phenomenon.

5

Atmospheric Trust Litigation

Mary Christina Wood*

INTRODUCTION

This chapter outlines the contours of potential "atmospheric trust litigation," designed to provide a means by which courts can hold governments at the national and subnational level accountable for reducing carbon emissions. Such litigation rests on the premise that all governments hold natural resources in trust for their citizens and bear the fiduciary obligation to protect such resources for future generations. The trust is embedded in the law as an attribute of sovereignty itself. While most frequently applied to state governments, public trust theory applies with equal force to the federal government, and seemingly indeed, to any sovereign.[1] Atmospheric trust litigation would characterize the atmosphere as one of the assets in the trust, shared as property among all nations of the world as co-tenants.

Protection of the trust through judicial oversight lies at the heart of the public trust jurisprudence in this country. As this chapter explains, the courts have the ability to enforce a fiduciary obligation to reduce carbon at all levels of government. Whether they will do so or not depends largely on individual judges' perception of the urgency of climate crisis, their belief as to whether the political system will address the issue, and their view of the role of the judiciary in confronting climate change. While atmospheric trust litigation bears the risk of any untested strategy, it is perhaps the only macro approach that can empower courts to effectuate the reductions in emissions within the limited time frame afforded to us before critical climate thresholds are exceeded.

---

* Philip H. Knight Professor of Law, Wayne Morse Center for Law and Politics Resident Scholar, 2006–2007, Luvaas Faculty Fellow, 2007–2008, University of Oregon School of Law, mwood@law.uoregon.edu. Generous support for this project came from the Luvaas Faculty Fellowship Endowment Fund. The author is a member of the Consultants Working Group, Climate Legacy Initiative, http://www.vermontlaw.edu/cli/, and a contributor to the Presidential Climate Action Project, http://www.climateactionproject.com. The author wishes to thank Dawn Winalski, Kelly Fahl, Erin Roach, Rachel Black-Maier, and Matt Rykels for valuable research assistance; Heather Brinton, Marianne Dellinger, and Tim Ream for contributing analysis; and Professors Patrick Parenteau and Deepa Badrinarayana for reviewing an earlier draft of this chapter. For additional scholarship and speeches by the author on climate crisis, see http://www.law.uoregon.edu/faculty/mwood/.

[1] See *infra* note 19 and accompanying text.

Section 1 of this chapter describes the body of public trust law and presents the atmosphere as an asset in the "res" that all governments have the duty to protect. It asserts that the United States' fiduciary obligation is measured according to the targets recently set forth by scientists. This fiduciary obligation applies to every level of government as an organic, uniform responsibility – an approach formulated to leave no orphan shares of carbon reduction. Section 2 then offers a framework for judicial enforcement of government's trust obligation to protect the atmosphere.

1. PUBLIC TRUST AS A MACRO APPROACH

Atmospheric trust litigation is premised on the generic and inherent fiduciary obligation of all governments to protect a shared atmosphere that is vital to human welfare and survival. The judicial role is to compel the political branches to meet their fiduciary standard of care through whatever measures and policies they choose, as long as such measures sufficiently reduce carbon emissions within the required time frame. The courts' role is not to supplant a judge's wisdom for a legislature's approach, but rather to police the other branches to ensure fulfillment of their trust responsibility in accordance with the climate imperatives of Nature.[2]

As a strategy, atmospheric trust litigation is geared toward enforcing planetary carbon reduction requirements, formulated to hold each government accountable for its share of the necessary reduction. The carbon reduction regime prescribed by scientists serves as the yardstick for determining whether government is carrying out its fiduciary obligation to protect the atmosphere.[3] Put another way, the scientific prescription is the expression of whether public trustees are meeting the "reasonable care" standard in protecting the trust.[4] As explained later, the remedy of a carbon accounting provides courts and the public with the necessary information as to whether governmental fiduciaries are adequately recovering the atmospheric trust.

By taking a macro approach, public trust litigation seeks not only to impose concrete, quantitative carbon requirements on all levels of government but also to

---

[2] See Lake Mich. Fed'n v. U.S. Army Corps of Eng'rs, 742 F. Supp. 441 (D. Ill. 1990) ("The very purpose of the public trust doctrine is to police the legislature's disposition of public lands. If courts were to rubber stamp legislative decisions the doctrine would have no teeth. The legislature would have unfettered discretion to breach the public trust as long as it was able to articulate some gain to the public."). While beyond the scope of this chapter, defenses based on the "political question" doctrine should carry far less weight in public trust litigation than in climate nuisance litigation, where they have presented barriers to actions brought by states against carbon polluters. See California v. Gen. Motors, Order Granting Defendants' Motion to Dismiss, 2007 U.S. Dist. LEXIS 68547, at *29 (N.D. Cal. Sept. 17, 2007) (dismissing nuisance lawsuit brought by California against auto manufacturers, stating, "the Court finds that injecting itself into the global warming thicket at this juncture would require an initial policy determination of the type reserved for the political branches of government."); see also Connecticut v. Am. Elec. Co., 406 F. Supp. 2d 265, 272 (S.D.N.Y. 2005) (dismissing climate nuisance lawsuit brought by state against electric company on basis that it raised nonjusticiable political questions).

[3] See infra note 56 (discussing TARGET FOR U.S. EMISSIONS REDUCTIONS).

[4] RESTATEMENT (SECOND) OF TRUSTS § 176 (1957) ("The trustee is under a duty to the beneficiary to use reasonable care and skill to preserve the trust property.").

invoke the full bureaucratic capacity of the United States and all of its subdivision governments to accomplish the rapid transformation of infrastructure necessary to achieve requisite carbon reduction.[5] In these ways, trust litigation is much different from claims that may be brought under environmental statutes including the National Environmental Policy Act (NEPA), the Endangered Species Act (ESA), and the Clean Air Act (CAA). Such claims are directed toward discrete actions that have carbon consequences. Grounded in more traditional litigation, these types of claims are not geared toward assuring the sum total of carbon reduction needed.[6] That is not to say they are unimportant; they do provide a vital check on government policies that contribute to major individual sources of carbon.

The following sections explore the contours of a potential public trust claim. They provide an overview of the doctrine and discuss its applicability to climate change.

1.1. *Public Trust Law*

Deriving from the common law of property, the public trust doctrine is the most fundamental legal mechanism to ensure that government safeguards natural resources necessary for public welfare and survival.[7] In the context of the climate crisis, which threatens the lives of innumerable human beings into the future, the public trust doctrine functions as a judicial tool to ensure that the political branches of government protect the basic right to life held by citizens.[8] An ancient yet enduring

---

[5] It is clear that carbon reduction will have to occur across all sectors of society. Different levels of government bring different resources and regulatory authority to the task. It would be ill considered to expect the federal government alone to solve the carbon problem. Cities and counties, for example, have primary jurisdiction over local transportation infrastructure and land use planning, both of which account for significant carbon emissions.

[6] By the same token, trust claims are broader than nuisance claims, which are directed against single sources of emissions. Two district courts have dismissed global warming nuisance claims partly due to the vast nature of the problem and the courts' sense that the political branches should make initial policy determinations on how to regulate the various sources. *See supra* note 2 and cases cited therein. Trust litigation is geared toward forcing action in the political branches without invading the province of those branches to decide how to accomplish atmospheric recovery.

[7] For sources and materials on the public trust doctrine, see JAN G. LAITOS, SANDRA B. ZELLMER, MARY C. WOOD & DAN H. COLE, NATURAL RESOURCES LAW, chap. 8.II (2006). For discussion of the public trust concept, see Joseph L. Sax, *The Public Trust Doctrine in Natural Resource Law: Effective Judicial Intervention*, 68 MICH. L. REV. 471, 558–66 (1970); Harrison Dunning, *The Public Trust: A Fundamental Doctrine of American Property Law*, 19 ENVTL. L. 515 (Spring 1989); Allen Kanner, *The Public Trust Doctrine, Parens Patriae, and the Attorney General as the Guardian of the State's Natural Resources*, 16 DUKE ENVTL. L. & POL'Y F. 57 (2005).

[8] Perhaps the best expression of this organic concept comes from the Philippines Supreme Court's opinion in Juan Antonio Oposa v. Fulgencio S. Factoran, Jr., G.R. No. 101083 (S.C., 1993) (Phil), *excerpted in* LAITOS, ZELLMER, WOOD & COLE, *supra* note 7, at 441–44:

> [T]he right to a balanced and healthful ecology belongs to a different category of rights altogether for it concerns nothing less than self-preservation and self-perpetuation the advancement of which may even be said to predate all governments and constitutions.
>
> As a matter of fact, these basic rights need not even be written in the Constitution for they are assumed to exist from the inception of humankind. If they are now explicitly mentioned it is because of the well-founded fear of its framers that unless the right to a balanced and healthful

legal principle, it underlies modern statutory law.[9] At the core of the doctrine is the principle that every sovereign government holds vital natural resources in "trust" for the public.[10] The doctrine invokes the sovereign's property powers and obligations, distinct from the police powers of a state.[11] Its fulfillment depends largely on judicial enforcement through injunctive relief. In the United States, the doctrine is redolent in hundreds of judicial decisions, including landmark Supreme Court opinions.[12]

As trustee, government must protect the natural trust for present and future generations.[13] It may not allow irrevocable harm to critical resources by private interests. As the Supreme Court said in *Geer v. Connecticut*:

> [T]he power or control lodged in the State, resulting from this common ownership, is to be exercised, like all other powers of government, as a trust for the benefit of the people, and not as a prerogative for the advantage of the government, as distinct from the people, or for the benefit of private individuals as distinguished from the public good.... [T]he ownership is that of the people in their united sovereignty.[14]

---

ecology and to health are mandated as state policies by the Constitution itself the day would not be too far when all else would be lost not only for the present generation, but also for those to come – generations which stand to inherit nothing but parched earth incapable of sustaining life.

[9] *See* National Environmental Policy Act of 1969 (NEPA), 42 U.S.C. § 4331(b)(1) (2006) (declaring a national duty to "fulfill the responsibilities of each generation as trustee of the environment for succeeding generations"). Federal pollution laws also designate sovereigns (federal, tribal, and state governments) as trustees of natural resources for purposes of collecting natural resource damages. For discussion, see Mary Christina Wood, *The Tribal Property Right to Wildlife Capital (Part II): Asserting a Sovereign Servitude to Protect Habitat of Imperiled Species*, 25 VT. L. REV. 355, 443 (2001). The public trust is also expressed in many state constitutions. *See, e.g.*, HAW. CONST. art. XI, § 1; PA. CONST. art. I, § 27; R.I. CONST. art. I, § 17. For discussion, see Robin Kundis Craig, *A Comparative Guide to the Eastern Public Trust Doctrine: Classifications of States, Property Rights, and State Summaries*, 16 PENN ST. ENVT'L L. REV. 1 (2007), *available at* http://ssrn.com/abstract=1008161 (last visited June 8, 2008).

[10] Ill. Cent. R.R. Co. v. Illinois, 146 U.S. 387, 455 (1892); Geer v. Connecticut, 161 U.S. 519, 525–29 (1896) (detailing ancient and English common law principles of sovereign trust ownership of air, water, sea, shores, and wildlife); *see also* Charles L. Wilkinson, *The Public Trust Doctrine in Public Land Law*, 14 U.C. DAVIS L. REV. 269, 315 (1980) ("The public trust doctrine is rooted in the precept that some resources are so central to the well-being of the community that they must be protected by distinctive, judge-made principles.").

[11] *See* LAITOS, ZELLMER, WOOD & COLE, *supra* note 7, at 623 ("Because the public trust doctrine emanates from property ownership on behalf of the public, the duties and powers to preserve the trust are distinct from the states' legislative police powers."); *see also* Gerald Torres, *Who Owns the Sky?* 19 PACE ENVTL. L. REV. 515, 525 (2002) (distinguishing sovereign's police power and property interests in context of air pollution).

[12] *See* discussion at Kanner, *supra* note 7 at 71–72; Torres, *supra* note 11, at 521.

[13] *Geer*, 161 U.S. at 534 ("The ownership of the sovereign authority is in trust for all the people of the state; and hence, by implication, it is the duty of the legislature to enact such laws as will best preserve the subject of the trust, and secure its beneficial use in the future to the people of the state.").

[14] *Id.* at 529. *See also* Lake Mich. Fed'n v. U.S. Army Corps of Eng'rs, 742 F. Supp. 441, 445 (D. Ill. 1990) ("[T]he public trust is violated when the primary purpose of a legislative grant is to benefit a private interest."); Gail Osherenko, *New Discourses on Ocean Governance: Understanding Property Rights and the Public Trust*, 21 J. ENVTL. L. & LITIG. 317, 327 (2006).

The lodestar public trust opinion is *Illinois Central Railroad Co. v. Illinois*, where the Supreme Court announced that the shoreline of Lake Michigan was held in public trust by the State of Michigan and could not be transferred out of public ownership to a private railroad corporation. In broad language encompassing the public's fundamental right to natural resources, the Court stated:

> [T]he decisions are numerous which declare that such property is held by the state, by virtue of its sovereignty, in trust for the public. The ownership of the navigable waters of the harbor, and of the lands under them, is a subject of public concern to the whole people of the state. The trust with which they are held, therefore, is governmental, and cannot be alienated.[15]

Public trust jurisprudence makes clear that government is not at liberty to disclaim its fiduciary obligation to protect crucial natural resources. As the Court said in *Illinois Central*:

> The state can no more abdicate its trust over property in which the whole people are interested than it can abdicate its police powers in the administration of government and the preservation of the peace.
>
> Every legislature must, at the time of its existence, exercise the power of the state in the execution of the trust devolved upon it.[16]

As a federal district court said more recently in applying the doctrine to both the federal and state governments, "The trust is of such a nature that it can be held only by the sovereign, and can only be destroyed by the destruction of the sovereign."[17] The public trust is appropriately viewed as a fundamental attribute of sovereignty itself, applicable to all governmental bodies.[18]

---

[15] *Ill. Cent. R.R. Co.*, 146 U.S. at 455 (but noting that parcels could be alienated "when parcels can be disposed of without detriment to the public interest in the lands and waters remaining."). *Id.* at 453.

[16] *Id.* at 460.

[17] *United States v. 1.58 Acres of Land*, 523 F. Supp. 120, 124 (D. Mass. 1981).

[18] *See Geer*, 161 U.S. at 528 (referring to the trust over wildlife as an "attribute of government" and tracing its historical manifestation "though all vicissitudes of government."); *State v. Bartee*, 894 S.W.2d 34, 41 (Tex. App. 1994) ("attribute of government"); *see also* Gary D. Meyers, *Variation on a Theme: Expanding the Public Trust Doctrine to Include Protection of Wildlife*, 19 ENVTL. L. 723, 728 (1989) (noting "[t]he ownership of wildlife, like water, historically has been treated as an aspect of sovereignty").

While most public trust cases involve states, the doctrine, as an attribute of sovereignty, logically applies to the federal government as well. *See* Complaint of Steuart Transp. Co., 495 F. Supp. 38, 40 (E.D. Va. 1980) (applying doctrine to federal government); *1.58 Acres of Land*, 523 F. Supp. at 124 (same); *see also* ZYGMUNT J.B. PLATER ET AL., ENVIRONMENTAL LAW AND POLICY: NATURE, LAW, AND SOCIETY 1103 (Erwin Chemerinsky et al. eds., Aspen Publishers, 3d ed. 2004) ("In several cases, courts have asserted that the federal government is equally accountable and restricted under the terms of the public trust doctrine.... [Since] the federal government is a creature of the states by delegation through the Act of Union and the federal Constitution[,] the federal government is therefore exercising delegated powers [and] cannot have greater rights and fewer limitations than the entities that created it."). For further discussion, see Mary Christina Wood, *Protecting the Wildlife Trust: A Reinterpretation of Section 7 of the Endangered Species Act*, 34 ENVTL. L. 605, n.38 and accompanying text (2004).

As a property law concept, the trust has an orientation very different, and in fundamental ways more exacting, than the body of statutory law. Though the prescriptions of statutory law are vastly more detailed, the discretion afforded to bureaucratic managers and the widespread cultural tolerance for behind-the-scenes political decisions often defeat the purposes of such statutes. Agencies typically use their statutory discretion to allow depletion and degradation of resources.[19] Trust law, by contrast, holds trustees to the "most exacting fiduciary standards."[20]

1.2. The Res of the Trust

At its core, the public trust doctrine defines certain natural resources as quantifiable assets that the government holds for the benefit of present and future citizen beneficiaries. Those assets are the "res" or "corpus" of the trust. The beneficiaries of the trust are present and future generations.[21]

The assets constituting the res of the public trust have been expanded by courts to meet society's changing needs.[22] The original cases focused on submersible lands as trust assets.[23] As society industrialized, a much broader array of resources became critical. Over time, the doctrine has reached new geographic areas including water, wetlands, dry sand beaches, and nonnavigable waterways.[24] The doctrine has also pushed beyond the original societal interests of fishing, navigation, and commerce to protect modern concerns such as biodiversity, wildlife habitat, aesthetics, and recreation.[25] Such expansion is well within the function of common law to adapt to emerging societal needs.[26]

---

[19] *See generally* Mary Christina Wood, *Nature's Trust: Reclaiming an Environmental Discourse*, 25 VA. ENVTL. L.J. 243 (2007).

[20] Seminole Nation v. United States, 316 U.S. 286, 297 (1942). *See also* Jicarilla Apache Tribe v. Supron Energy Corp., 728 F.2d 1555, 1563 (10th Cir. 1984) (where "the Secretary is obligated to act as a fiduciary ... his actions must not merely meet the minimal requirements of administrative law, but must also pass scrutiny under the more stringent standards demanded of a fiduciary.").

[21] *See supra* note 13; Ariz. Ctr. for Law in the Pub. Interest v. Hassell, 837 P.2d 158, 169 (Ariz. Ct. App. 1991) ("The beneficiaries of the public trust are not just present generations but those to come"); *see also* Deborah G. Musiker et al., *The Public Trust and Parens Patriae Doctrines: Protecting Wildlife in Uncertain Political Times*, 16 PUB. LAND L. REV. 87 (1995); Peter H. Sand, *Sovereignty Bounded: Public Trusteeship for Common Pool Resources*, 4 GLOBAL ENVTL. POL. 47, 55 (2004), available at http://www.mitpressjournals.org/doi/pdfplus/10.1162/152638004773730211?cookieSet=1 (defining beneficiaries, on the global level, as "future humanity").

[22] As the New Jersey Supreme Court said, "[W]e perceive the public trust doctrine not to be 'fixed or static,' but one to be 'molded and extended to meet changing conditions and needs of the public it was created to benefit.'" Matthews v. Bay Head Improvement Ass'n, 471 A.2d 355, 365 (N.J. 1984) (citation omitted); *see also* Marks v. Whitney, 491 P.2d 374, 380 (Cal. 1971) ("In administering the trust the state is not burdened with an outmoded classification favoring one mode of utilization over another."); Kanner, *supra* note 7, at 72 ("United States judges have broadened the geographic protections and widened the range of activities under the public trust.").

[23] *See Ill. Cent. R.R. Co.*, 146 U.S. at 453.

[24] *See* LAITOS, ZELLMER, WOOD & COLE, *supra* note 7, at 651.

[25] *Matthews*, 471 A.2d at 363; Nat'l Audubon Soc'y v. Sup. Ct., 658 P.2d 709, 719–22 (Cal. 1983).

[26] *In re* Hood River, 227 P. 1065, 1086–87 (Or. 1924):

The governmental trustee bears a fiduciary obligation to protect the assets of the trust from damage.[27] Scores of cases emphasize this duty of protection,[28] and many hold that the duty imposes an affirmative obligation on government.[29] Under well-established principles of private trust law, trustees may not sit idle and allow damage to occur to the trust.[30] Moreover, where trust assets have been damaged, the trustee

> The very essence of the common law is flexibility and adaptability. It does not consist of fixed rules, but is the best product of human reason applied to the premises of the ordinary and extraordinary conditions of life, as from time to time they are brought before the courts.... If the common law should become so crystallized that its expression must take on the same form whenever the common-law system prevails, irrespective of physical, social, or other conditions peculiar to the locality, it would cease to be the common law of history, and would be an inelastic and arbitrary code. It is one of the established principles of the common law, which has been carried along with its growth, that precedents must yield to the reason of different or modified conditions.

[27] GEORGE T. BOGERT, TRUSTS § 99, at 358 (6th ed. 1987) ("The trustee has a duty to take whatever steps are necessary... to protect and preserve the trust property from loss or damage."); 76 Am. Jur. 2D TRUSTS § 331, at 404 ("[T]he trustee must make the trust property productive, and must not suffer the estate to waste or diminish, or fall out of repair."); RESTATEMENT (SECOND) OF TRUSTS, *supra* note 4, at § 176 ("The trustee is under a duty to the beneficiary to use reasonable care and skill to preserve the trust property."); United States v. White Mountain Apache Tribe, 537 U.S. 465, 475 (2003) (fundamental common law duty of a trustee is to maintain trust assets); State v. City of Bowling Green, 313 N.E.2d 409, 411 (Ohio 1974) ("[W]here the state is deemed to be the trustee of property for the benefit of the public it has the obligation to bring suit... to protect the corpus of the trust property."); State Dep't of Envtl. Prot. v. Jersey Cent. Power & Light Co., 336 A.2d 750, 758–59 (N.J. Super. Ct. App. Div. 1975) (finding both right and duty to recover damages for harm to natural resources held in public trust), *rev'd on other grounds*, 351 A.2d 337 (N.J. 1976); Fort Mojave Indian Tribe v. United States, 23 Cl. Ct. 417, 426 (1991) (finding federal trust duty to protect Indian water rights because "the title to plaintiffs' water rights constitutes the trust property, or the res, which the government, as trustee, has a duty to preserve.").

[28] *See, e.g.*, Geer, 161 U.S. at 534 ("[I]t is the duty of the legislature to enact such laws as will best preserve the subject of the trust, and secure its beneficial use in the future to the people of the state."); Nat'l Audubon Soc'y v. Sup. Ct. of Alpine County, 658 P.2d 709, 724 (Cal. 1983) (expressing the "duty of the state to protect the people's common heritage of streams, lakes, marshlands and tidelands"); *White Mountain Apache Tribe*, 537 U.S. at 475; *Fort Mojave Indian Tribe*, 23 Cl. Ct. at 426.

[29] *See* City of Milwaukee v. State, 214 N.W. 820, 830 (Wis. 1927) ("The trust reposed in the state is not a passive trust; it is governmental, active, and administrative [and] requires the lawmaking body to act in all cases where action is necessary, not only to preserve the trust, but to promote it...."). For discussion, see Kanner, *supra* note 7, at 75–77; Torres, *supra* note 11, at 549 (government has an obligation to act to preserve the atmospheric trust); *City of Bowling Green*, 313 N.E.2d at 411 ("[W]here the state is deemed to be the trustee of property for the benefit of the public it has the obligation to bring suit... to protect the corpus of the trust property."); *see also* Musiker et al., *supra* note 21, at 96 ("The [government], as trustee, must prevent substantial impairment of the wildlife resource so as to preserve it for the beneficiaries – current and future generations.").

[30] *See* BOGERT, *supra* note 27, § 99, at 358 (duty to protect and preserve property); *id.* at § 107, at 391 ("The trustee is liable for damages if he should have known of danger to the trust, could have protected the trust, but did not do so."); Am. Jur. 2D TRUSTS, *supra* note 27, at § 656 (noting the "power, and a duty of the trustee, to initiate actions... for the protection of the trust estate"). Courts have imported principles of protection from the private realm of trust law to govern public trustee duties in state lands management. *See* Idaho Forest Indus. v. Hayden Lake Watershed Improvement Dist., 733 P.2d 733, 738 (Idaho 1987) (noting the administration of public trust is governed by the same principles applicable to the administration of trusts in general).

has the affirmative duty to recoup damages and restore the corpus, or res.[31] Common law has vested sovereigns with the right and obligation to sue third parties to recoup natural resource damages for destruction of public trust assets.[32]

1.3. Co-tenancy Sovereign Interests

A singular failing of statutory law is its confinement to jurisdictional boundaries. A notable strength of the trust doctrine's property framework is that it creates logical rights to shared assets that are not confined within any one jurisdictional border. It is well established that, with respect to transboundary trust assets, all sovereigns with jurisdiction over the natural territory of the asset have legitimate property claims to the resource.[33] States that share a waterway, for example, have correlative rights to the water.[34] Similarly, states and tribes have coexisting property rights to share in the harvest of fish passing through their borders.[35] Shared interests are best described as a sovereign co-tenancy. A co-tenancy is "the ownership of property by two or more persons in such manner that they have an undivided... right to possession."[36] The Ninth Circuit has invoked the co-tenancy model to describe shared sovereign rights to migrating salmon.[37]

---

[31] See Jersey Cent. Power & Light, 336 A.2d at 758–59 (finding duty to seek damages for harm to natural resources held in public trust); City of Bowling Green, 313 N.E.2d at 411 (noting public trustee's "obligation... to recoup the public's loss occasioned by... damage [to] such property"); Wash. Dep't of Fisheries v. Gillette, 621 P.2d 764, 767 (Wash. Ct. App. 1980) (noting right and "fiduciary obligation of any trustee to seek damages for injury to the object of its trust"); Mary Christina Wood, The Tribal Property Right to Wildlife Capital (Part I): Applying Principals of Sovereignty to Protect Imperiled Wildlife Populations, 37 IDAHO L. REV. 1, 58–59, 92–93 (2000) (discussing duty); Musiker et al., supra note 21, at 107–08 (discussing trust obligations as parens patriae); Susan Morath Horner, Embryo, Not Fossil: Breathing Life into the Public Trust in Wildlife, 35 LAND & WATER L. REV. 23, 27–28 (2000) (discussing rights and duties).

[32] See Charles B. Anderson, Damage to Natural Resources and the Costs of Restoration, 72 TUL. L. REV. 417, pt. III (1997); Thomas A. Campbell, The Public Trust, What's It Worth?, 34 NAT. RESOURCES J. 73, 82–86 (1994); Carter H. Strickland, Jr., The Scope of Authority of Natural Resource Trustees, 20 COLUM. J. ENVTL. L. 301, pt. III.A (1995) (some common law claims may be preempted if they fall within a comprehensive program established by federal statutory law). See generally Md. Dep't of Nat. Res. v. Amerada Hess Corp., 350 F. Supp. 1060 (D. Md. 1972); City of Bowling Green, 313 N.E.2d at 411; State of North Dakota v. Dickinson Cheese Co., 200 N.W.2d 59, 61 (N.D. 1972). Modern statutes also provide the right to recover damages to public trust assets. See Anderson, supra. Natural resource damages must be applied to restoration of the trust. See Comprehensive Environmental Response, Compensation, and Liability Act of 1980 (CERCLA), 42 U.S.C. § 9607(f).

[33] See Idaho ex rel. Evans v. Oregon, 462 U.S. 1017, 1031 n.1 (1983) (O'Connor, J., dissenting) (noting "recognition by the international community that each sovereign whose territory temporarily shelters [migratory] wildlife has a legitimate and protectible interest in that wildlife").

[34] Arizona v. California, 373 U.S. 546, 601 (1963).

[35] See Washington v. Wash. State Commercial Passenger Fishing Vessel Ass'n, 443 U.S. 658, 676–79 (1979); see also Minnesota v. Mille Lacs Band of Chippewa Indians, 526 U.S. 172 (1999).

[36] 20 Am. Jur. 2d COTENANCY AND JOINT OWNERSHIP § 1 (1995). A co-tenancy typically implies each party's right to full possession of the asset. JOSEPH WILLIAM SINGER, PROPERTY LAW: RULES, POLICIES, AND PRACTICES 711 (2d ed. 1997).

[37] Puget Sound Gillnetters Ass'n v. U.S. Dist. Ct., 573 F.2d 1123, 1126 (9th Cir. 1978) ("We held that [the treaty] reserved an exclusive right to fish on the reservation and that [the treaty] established something analogous to a cotenancy, with the tribes as one cotenant and all citizens of the Territory (and later of

Co-tenancy relationships give rise to correlative duties not to waste the common asset.[38]

Within the United States, layered sovereign interests in natural resources arise from the constitutional configuration of states and the federal government. Where the federal government has a national interest in the resource, it is a co-trustee along with the states.[39] The concurrence of federal and state trust interests is reflected in statutory provisions that provide natural resource damages to both sovereign trustees.[40] As one court has made clear in the context of streambed ownership, the federal government and states are held to identical trust obligations but must carry them out in accordance with their unique constitutional roles:

> This formulation recognizes the division of sovereignty between the state and federal governments.... [T]hose aspects of the public interest... that relate to the commerce and other powers delegated to the federal government are administered by Congress in its capacity as trustee of the jus publicum, while those aspects of the public interest in this property that relate to nonpreempted subjects reserved to local regulation by the states are administered by state legislatures in their capacity as co-trustee of the jus publicum.[41]

the state) as the other."); United States v. Washington, 520 F.2d 685, 686, 690 (9th Cir. 1975) (applying co-tenancy construct, by analogy, to Indian fishing rights). Of course, a co-tenancy framework for sovereign management of natural resources differs in some ways from a private co-tenancy in land among individuals. For example, a sovereign co-tenancy in natural resources may not be capable of partitioning. *See Puget Sound Gillnetters, supra,* at 1134–35 (Kennedy, J., concurring). Nevertheless, the basic co-tenancy construct is helpful and instructive in the sovereign context. *See id.* at 1128, n.3 (stating, in the treaty fisheries context, "We refer to the cotenancy analogy only because it is helpful in explaining the rights of the parties, not because all the rights and incidents of a common law cotenancy necessarily follow.... Obviously, not all the rules of cotenancy in land can apply to an interest of the nature of a profit.").

[38] Acts that amount to permanent damage to the common property are held to constitute waste. E. HOPKINS, HANDBOOK ON THE LAW OF REAL PROPERTY § 214, at 342 (1896); 2 W. WALSH, COMMENTARIES ON THE LAW OF REAL PROPERTY § 131, at 72 (1947). *See also Washington,* 520 F.2d at 685 (stating, in context of fisheries shared between states and tribes: "Cotenants stand in a fiduciary relationship one to the other. Each has the right to full enjoyment of the property, but must use it as a reasonable property owner. A cotenant is liable for waste if he destroys the property or abuses it so as to permanently impair its value. A court will enjoin the commission of waste.... By analogy, neither the treaty Indians nor the state on behalf of its citizens may permit the subject matter of these treaties to be destroyed.").

[39] For an extensive discussion of these co-trustee interests, see United States v. 1.58 Acres of Land, 523 F. Supp. 120, 121 (D. Mass. 1981) (discussing tidelands: "Since the trust impressed upon this property is governmental and administered jointly by the state and federal governments by virtue of their sovereignty, neither sovereign may alienate this land free and clear of the public trust."). *See also* Wood, *The Tribal Property Right to Wildlife Capital (Part I), supra* note 31, at 79 (describing concurrent federal, state, and tribal trust interests in wildlife).

[40] *See, e.g.,* CERCLA, 42 U.S.C. § 9607(f).

[41] *1.58 Acres of Land,* 523 F. Supp. at 121. Recently, in *Massachusetts v. U.S. Environmental Protection Agency,* the Supreme Court echoed this division of authority with respect to the air assets in the trust:

> When a State enters the Union, it surrenders certain sovereign prerogatives. Massachusetts cannot invade Rhode Island to force reductions in greenhouse gas emissions, it cannot negotiate an emissions treaty with China or India, and in some circumstances, the exercise of its police powers to reduce in-state motor-vehicle emissions might well be pre-empted.
>
> These sovereign prerogatives are now lodged in the Federal Government....

549 U.S. 497, 519 (2007) (citation omitted).

1.4. The Atmosphere as a Public Trust Asset

Guided by the essential doctrinal purposes expressed by the Supreme Court in foundational public trust cases, it is no great leap to recognize the atmosphere as one of the crucial assets of the public trust. At the time of the *Illinois Central* case, the Court made clear that the essence of the doctrine is to protect resources of "special character" that serve purposes "in which the whole people are interested."[42] The Court was presented with a novel situation – the conveyance of a major shoreline to a private party. The Court noted:

> We cannot, it is true, cite any authority where a grant of this kind has been held invalid, for we believe that no instance exists where the harbor of a great city and its commerce have been allowed to pass into the control of any private corporation. But the decisions are numerous which declare that such property is held by the state, by virtue of its sovereignty, in trust for the public.[43]

Climate crisis presents the courts with an equally novel, yet necessary, application of the public trust. In the crisis at hand, the public interests at stake are leagues beyond the traditional interests at the forefront of *Illinois Central*: fishing, navigation, and commerce. Atmospheric health is essential to all facets of civilization and human survival. As such, it falls within the core of the purpose of the public trust doctrine: to protect natural assets crucial to human survival and welfare. While air has not previously been the subject of trust litigation, the Roman origins of the public trust doctrine classified air – along with water, wildlife, and the sea – as "res communes."[44] In a landmark public trust decision, *Geer v. Connecticut*, the Supreme Court relied on this ancient Roman classification of "res communes" to find the public trust applicable to wildlife.[45] Since then, the Court has also recognized the states' sovereign interests in air as a basis upon which to bring an interstate nuisance suit. In *Georgia v. Tennessee Copper Co.*, the Court upheld an action brought by the state of Georgia against Tennessee copper companies for discharging noxious gas that drifted across state lines, stating: "This is a suit by a state for an injury to it in its capacity of quasi-sovereign. In that capacity the state has an interest independent of and behind the titles of its citizens, in all the earth and air

---

[42] *Ill. Cent. R.R. Co.*, 146 U.S. at 453.
[43] *Id.* at 455.
[44] See *Geer*, 161 U.S. at 525 ("These things are those which the jurisconsults called 'res communes'... the air, the water which runs in the rivers, the sea, and its shores.... [and] wild animals."); *id.* at 524. See also Torres, *supra* note 11, at 529–30 ("The evolution of the public trust doctrine is complex, but it is essentially rooted in Roman law and from those laws through the various commentators on Roman law.... If a resource were excluded from private ownership because by its nature it could only be used in common, it was called res communes.... The principle of res communes was expressed in the English common law and in 19th century American law as jus publicum. The beneficial interest in any res communes is held by the people in common.").
[45] See *Geer*, 161 U.S at 523 (quoting Ponthier treatise on property) (citations omitted): "Among other subdivisions [in property], things were classified by the Roman law into public and common. The latter embraced animals ferae naturae, which, having no owner, were considered as belonging in common to all the citizens of the state."

within its domain."[46] Recently, in *Massachusetts v. EPA*, the Supreme Court drew upon *Georgia v. Tennessee Copper Co.* to underscore the state's unique interest in air, alluding to the state's position as trustee in its discussion of standing.[47]

Given the essential nature of air, it is unsurprising that numerous state court decisions, constitutions, and codes have recognized air as part of the res of the public's trust,[48] and commentators have urged a focus on the atmosphere as a trust asset.[49] Moreover, federal statutory law includes air as a trust asset for which the federal government, states, and tribes can gain recovery for natural resource damages.[50] On the international level, the United Nations Framework Convention on Climate Change declares an atmospheric trust obligation by calling upon nations to "protect the climate system for the benefit of present and future generations of humankind."[51]

1.5. *The Carbon Fiduciary Obligation*

The trust construct positions all nations of the Earth as sovereign co-tenant trustees of a shared atmosphere.[52] In addition to a fiduciary obligation owed to their own citizens to protect the atmosphere, all nations have duties to prevent waste arising from their co-tenancy relationship with each other.[53] Courts are positioned to define these duties by tying them directly to scientists' concrete prescription for carbon

---

[46] Georgia v. Tenn. Copper Co., 206 U.S. 230, 237 (1907).
[47] *See Massachusetts v. EPA* 549 U.S. 497, 519 (2007) (finding that Massachusetts had standing to sue the federal government over its inaction to prevent carbon emissions from new automobiles).
[48] *See, e.g.*, Her Majesty v. City of Detroit, 874 F.2d 332, 337 (6th Cir. 1989) (citing Michigan act that codifies public trust to include "air, water, and other natural resources"); HAW. CONST. art. XI, § 1 (stating, "All public natural resources are held in trust by the State for the benefit of the people," and "the State and its political subdivisions shall conserve and protect Hawaii's natural resources, including land, water, air, minerals and energy resources."); LA. CONST. art. IX, § 1 ("natural resources of the state, including air and water shall be protected"); R.I. CONST. art. I, § 16 (duty of legislature to protect air), interpreted as codification of Rhode Island's public trust doctrine in State ex. Rel. Town of Westerly v. Bradley, 877 A.2d 601, 606 (R.I. 2005); Nat'l Audubon Soc'y v. Super. Ct. of Alpine County, 658 P.2d 709, 720 (1983) ("purity of the air" protected by the public trust); *c.f.* PA. CONST. art. I, § 27 (declaring public trust duty to conserve natural resources, and expressing citizens' right to clean air); *see also* WILLIAM RODGERS, JR., ENVIRONMENTAL LAW: AIR AND WATER § 2.20, at 162 (1986) ("It is eminently clear now that trust properties not only can, but must, be administered to protect birdlife and to prevent air and water pollution.").
[49] *See* Torres, *supra* note 11, at 533, 526: ("Properly understood the traditional rationale for the public trust doctrine provides a necessary legal cornerstone to protect the public interest in the sky."); *id.* at 532 ("The public trust doctrine supplies a broad framework that supports the establishment of a mechanism to supervise the government dealings in relationship to the carrying capacity of the atmosphere."); PETER BARNES, WHO OWNS THE SKY: OUR COMMON ASSETS AND THE FUTURE OF CAPITALISM (2006).
[50] CERCLA, 42 U.S.C. § 9601 (16) (defining air as among the natural resources subject to trust claims for damages).
[51] United Nations Framework Convention on Climate Change, Article 3, Principle 1 (1992).
[52] *See* Sand, *supra* note 21, at 51–54 (discussing concept of global trusteeship for common resources vital to humanity). For an analysis applying the trust to the analogous global oceans resource, see Osherenko, *supra* note 14.
[53] *See* discussion at *supra* note 38.

reduction.[54] The Union of Concerned Scientists has developed such a prescription – called the *Target for U.S. Emissions Reductions* – based on the extensive body of climate science.[55] Courts often rely on independent scientific recommendations in assessing liability and formulating injunctive relief.[56] The *Target* maps a climate stabilization pathway whereby the industrialized nations on Earth must collectively: (1) arrest the rising trajectory of carbon emissions by 2010; (2) reduce emissions an average of 4% per year starting in 2010; and (3) reduce carbon by at least 80% below 2000 levels by 2050.[57]

The scientifically established structure reflected in the *Target for U.S. Emissions Reductions*, as adapted to comport with changed scientific understanding,[58]

---

[54] While beyond the scope of this chapter, courts may invoke several procedural tools to gain the scientific expertise necessary to define the fiduciary standard of care. Increasingly, judges use court-appointed experts, technical advisers, and special masters to resolve difficult scientific questions in environmental, toxic torts, and product liability cases. See FEDERAL JUDICIAL CENTER, REFERENCE MANUAL ON SCIENTIFIC EVIDENCE (1994); THE CARNEGIE COMMISSION ON SCIENCE, TECHNOLOGY, AND GOVERNMENT, SCIENCE AND TECHNOLOGY IN JUDICIAL DECISION MAKING: CREATING OPPORTUNITIES AND MEETING CHALLENGES (1993). For discussion of these various judicial tools, see Samuel H. Jackson, *Technical Advisors Deserve Equal Billing with Court Appointed Experts in Novel and Complex Scientific Cases: Does the Federal Judicial Center Agree?*, 28 ENVTL. L. 431 (1998); Karen Butler Reisinger, *Court-Appointed Expert Panels: A Comparison of Two Models*, 32 IND. L. REV. 225 (1998).

[55] A. LUERS, M. D. MASTRANDREA, K. HAYHOE & P. C. FRUMHOFF, HOW TO AVOID DANGEROUS CLIMATE CHANGE: A TARGET FOR U.S. EMISSIONS REDUCTIONS 5 (UNION OF CONCERNED SCIENTISTS 2007) (hereinafter TARGET FOR U.S. EMISSIONS REDUCTIONS), *available at* http://www.ucsusa.org/assets/documents/global_warming/emissions-target-report.pdf (last visited June 16, 2008).

[56] *See* Pac. Coast Fed'n of Fishermen's Ass'ns v. U.S. Bureau of Reclamation, 138 F. Supp. 2d 1228, 1250 (N.D. Cal. 2001). The scientifically established target for carbon reduction is similar in concept to scientifically established river flows for endangered fish, which courts have relied upon in structuring injunctions for federal water project operations. *Id.* at 1250 (relying on independent scientist's report in setting river flows for fish); *see* discussion at Wood, *Restoring the Abundant Trust*, *infra* note 105, at 10,178. For examples of injunctions in the river context, see Am. Rivers v. U.S. Army Corps of Eng'rs, 271 F. Supp. 2d 230, 262 (D. D.C. 2003) (ordering more natural flow regime in Missouri River); Rio Grande Silvery Minnow v. Keys, 333 F.3d 1109, 1119 (10th Cir. 2003) (affirming injunction requiring flows in Rio Grande and reservoir release pending compliance with the ESA), *vacated as moot by* 355 F.3d 1215 (10th Cir. 2004); *see also* Wash. Toxics Coal. v. U.S. EPA, 413 F.3d 1024 (9th Cir. 2005) (banning application of pesticides near salmon streams).

[57] TARGET FOR U.S. EMISSION REDUCTIONS, *supra* note 55. The report groups the United States with other industrialized nations and then sets forth specific U.S. targets. The first part of the prescription, arresting emissions growth by 2010, is by far the most urgent and important because the world is dangerously close to climate thresholds, or a "tipping point" that will cause runaway heating. For discussion, see DAVID SPRATT & PHILIP SUTTON, CLIMATE CODE RED: THE CASE FOR A SUSTAINABILITY EMERGENCY (Friends of the Earth, 2008), *available at* http://www.climatecodered.net/ (last visited June 16, 2008) (hereinafter CLIMATE CODE RED). The call for arresting U.S. emissions growth by 2010 is in line with a call by the United Nations to arrest the growth of worldwide emissions by 2015. *See* Cahal Milmo, *"Too Late to Avoid Global Warming," Say Scientists*, INDEPENDENT UK, Sept. 19, 2007, *available at* http://www.independent.co.uk/environment/climate-change/too-late-to-avoid-global-warming-say-scientists-402800.html (last visited June 16, 2006). The worldwide date is set out five years beyond the U.S. date because the developing nations like China and India are going to take more time to arrest emissions.

[58] The assumptions underlying these target levels may already be outdated by more recent data showing accelerated polar ice melting, suggesting that a lower atmospheric level of carbon may be necessary

can be invoked as a generic standard of fiduciary obligation applicable to each industrialized nation. Such targets can also be "scaled down" to each subnational jurisdictional level as well.[59] Established liability principles create a rational structure for apportioning responsibility among sovereigns and subsovereigns. The law has often imposed proportionate liability on multiple contributors to a problem.[60] Moreover, in co-tenancies, each tenant is responsible for paying his share of the expenses proportionate to his interest in the property.[61]

The carbon loading of the atmosphere may be thought of as a sort of "natural lien." Having created the debt, all industrialized nations have the duty to pay it off. Invoking these basic principles, each nation is responsible for its proportionate share, or "fair share," of carbon reduction.[62] If each industrialized sovereign reduces

   to achieve climate stability. For discussion, see CLIMATE CODE RED, supra note 57, at 26–28. The TARGET delineates a "reasonable emissions pathway" for the United States calibrated to the goal of not exceeding 450 ppm carbon equivalent in the atmosphere. TARGET FOR U.S. EMISSIONS REDUCTIONS, supra note 55, at 3, 8, 14. This scientific prescription should be subject to change if data shows that 450 ppm is too high to achieve climate equilibrium. Courts must necessarily adjust the fiduciary standard of care to emerging science. Recently, NASA scientist Jim Hansen suggested that a lower goal may be necessary to avoid dangerous climate feedbacks that would trigger runaway heating. See Bill McKibben, Remember This: 350 Parts per Million, WASH. POST (Dec. 28, 2007), available at http://www.washingtonpost.com/wp-dyn/content/article/2007/12/27/AR2007122701942.html (last visited June 16, 2008); see also CLIMATE CODE RED, supra note 57 (climate stability may require reducing atmospheric carbon dioxide to 320 ppm). Courts may incorporate new scientific understanding into litigation management through use of the judicial tools described in supra note 4. For an example of a court's use of a technical advisor to resolve complex and rapidly changing science involving species survival, see Nat'l Wildlife Fed'n v. Nat'l Marine Fisheries Serv., 2005 U.S. Dist. LEXIS 16658, slip op. at 15–18 (Mar. 2, 2005) (upholding use of technical adviser in case brought under Endangered Species Act).

[59] See Hari M. Osofsky, The Geography of Climate Change Litigation Part II: Narratives of Massachusetts v. EPA, 8 CHI. J. INT'L L. 573, 583 (2008) (concept of "scaling up and down" in climate strategies); Sand, supra note 21, at 57 (discussing "transfer of the public trust concept from national to the global level").

[60] The RESTATEMENT OF TORTS provides: "Damages for harm are to be apportioned among two or more causes where there is a reasonable basis for determining the contribution of each cause to a single harm." RESTATEMENT OF THE LAW (SECOND) TORTS (ALI 1965); § 433 A (1). As an illustration, it states:

   Such apportionment is commonly made in cases of private nuisance, where the pollution of a stream, or flooding, or smoke or dust or noise, from different sources, has interfered with the plaintiff's use or enjoyment of his land. Thus where two or more factories independently pollute a stream, the interference with the plaintiff's use of the water may be treated as divisible in terms of degree, and *may be apportioned among the owners of the factories, on the basis of evidence of the respective quantities of pollution discharged.* . . . Id. (emphasis added).

For broad discussions of proportionate liability, comparative fault, and contribution in tort law, see generally Gerald W. Boston, Apportionment of Harm in Tort Law: A Proposed Restatement, 21 DAYTON L. REV. 267 (1996); Gary T. Schwartz, The Beginning and the Possible End of the Rise of Modern American Tort Law, 26 GA. L. REV. 601 (2001).

[61] See, e.g., Willmon v. Koyer, 143 P. 694, 695 (Cal. 1914) ("In proportion to their interests all tenants in common are in duty bound to pay taxes. . . . ").

[62] This was the general premise of the Kyoto Protocol, which called for each industrialized nation that was a party to Annex I to reduce its greenhouse gas emissions by an average of 5% below 1990 levels by 2008–2012. Kyoto Protocol, Art. 3.1. Courts have quantified and enforced a "fair share" concept among governmental jurisdictions in other contexts. See infra note 108 (land use planning context).

proportionately its carbon emissions – ultimately 80% or more below 2000 levels by 2050 – the total industrialized carbon share on a planetary level will be reduced by that amount. The developing nations, in turn, have the corresponding duty not to waste the common asset.[63]

Atmospheric trust litigation calls upon United States courts to translate this "fair share" concept into a streamlined obligation that applies within this nation evenly to each of the fifty states and all of their subdivisions (cities and counties), as well as to the federal government. Each jurisdictional level, in other words, bears a uniform responsibility of meeting the *Targets for U.S Emissions Reductions*. For example, the states of New Jersey and Idaho, though they have vastly different carbon footprints, will each bear the same *proportionate reduction* of carbon emissions over the same time frame.

This approach will have its critics, as it does not take into account certain state differences. Some states are more carbon heavy than others in particular sectors.[64] Some, for example, rely primarily on coal for energy, and others rely more heavily on hydropower.[65] But because carbon emissions are spread across all sectors, the differences between states in any one sector are likely to be compensated for in other sectors. All of the fifty states are similarly situated in terms of the carbon-dependent lifestyles of their citizens. In other words, there is no "developed versus undeveloped world" within the United States as there is on the international level that would justify differentiating between states in terms of the burden they carry. It would be a nearly impossible judicial task to arrive at any approach that accounts for all state differences. Courts should reject liability distinctions that encumber a swift governmental response appropriate to the scale of the crisis facing society.[66]

In other environmental contexts, courts and Congress have rejected a fine-lined fairness approach in order to carry out paramount public objectives. In imposing liability for hazardous waste pollution, Congress and the courts adopted strict liability principles out of a concern that fairness-based principles such as negligence would vastly complicate the role of the courts and impede swift judicial recovery.[67] These concerns are manifold in the context of global heating. Furthermore, in the area

---

[63] See discussion at *supra* note 38 and accompanying text.
[64] See Seth Borenstein, *Texas, Wyoming Take Lead in Emissions*, USA TODAY (June 2, 2007), *available at* http://www.usatoday.com/weather/climate/globalwarming/2007-06-02-emissions_N.html (last visited June 16, 2008).
[65] *Id.*
[66] Moreover, courts should reject arguments that liability percentages should be reduced to account for carbon "sinks" within the jurisdiction that provide natural offsets. Not only is the inquiry far too complex, but the sinks are not reliable. Forests, for example, have traditionally operated as sinks. Global warming, however, can make vegetation less able to absorb carbon dioxide from human activities. *See* David Adam, *Ten-Year Warming Window Closing*, SYDNEY MORNING HERALD (May 12, 2007) (reporting scientific findings that plants take up less carbon dioxide under unusually hot and dry conditions); *see also* FRED PEARCE, WITH SPEED AND VIOLENCE Ch. 13 (Beacon Press 2007) (explaining process of natural carbon sinks turning to carbon sources).
[67] *See* CERCLA § 107; Veolia Es Special Servs., Inc. v. Techsol Chem. Co., 2007 U.S. Dist. LEXIS 88127 at *29–30 (S.D. W. Va. 2007).

of regulatory takings, the Supreme Court has said that individuals may bear the burden of exactions that are "roughly proportional" to the harm that they cause, emphasizing that "no precise mathematical calculation is required."[68]

1.6. The Inexcusability of Carbon Orphan Shares

The needed emissions reductions will be achieved only if the apportioned responsibility definitively adds up to the required "carbon math." Each industrialized nation must carry out its proportion of the overall planetary carbon reduction, or it leaves an "orphan share" on the doorstep of the world. An orphan share is a share of liability for which the liable party does not take responsibility. In the context of carbon reduction, any significant orphan share is likely to defeat efforts to reduce emissions adequately in the short time frame needed. Because the required carbon reduction is as steep as it is, no sovereign is positioned to take on more than its share, at least at the outset. In other words, no industrialized nation is likely equipped to adopt an orphan share left by another sovereign. Doing so would mean that citizens of the adoptive sovereign must decrease their emissions multifold, beyond what their national liability would require – simply so that citizens of the "deadbeat" sovereign can continue living a lifestyle in carbon excess with no regard to meeting their planetary obligations. The concept applies to fractional orphan shares as well. If a sovereign has liability to decrease its emissions 80% but actually decreases its carbon emissions by only 50%, it will leave a 30% orphan remainder. A bedrock principle of atmospheric trust liability must be the inexcusability of orphan shares and partial orphan shares.[69]

The concept of orphan shares applies at every jurisdictional level, from national to local. For example, on the state level, if all California cities except for San Diego were to meet their fiduciary obligation to reduce carbon emissions, San Diego's orphan share would nevertheless sink the state's ability to meet its fiduciary obligation unless some other city or county also took on San Diego's share, which, as previously discussed, is an unlikely scenario and certainly not one that courts can rely on. The difficulty of adopting any significant orphan shares in the carbon context means, as a practical matter, that virtually all levels of government, including cities, counties, states, and national governments, must be held to their fiduciary duty to meet their fair, proportionate share of the planetary carbon liability. As a legal reality, this means that each sovereign and subsovereign must have a clear, generic fiduciary obligation that courts are capable of strictly enforcing. This does not mean that atmospheric trust litigation must be brought against every jurisdiction. More than likely, a few

---

[68] Dolan v. City of Tigard, 512 U.S. 374, 391 (1994). Moreover, in the context of approving consent decrees for treaty fishing rights, courts have emphasized that mathematical precision is not required and the courts' approval is an "amalgam of delicate balancing, gross approximations and rough justice." U.S. v. Oregon, 913 F.2d 576, 580–81 (9th Cir. 1990).

[69] This does not necessarily prevent sovereigns from using cap-and-trade mechanisms to fulfill their proportionate share of reduction.

precedent-setting lawsuits that create a clear liability framework can spur necessary action on the political level nationwide, without the need for massive litigation. A press strategy can magnify the public impacts of atmospheric trust litigation long before any judicial decision is rendered.

2. ENFORCING THE ATMOSPHERIC FIDUCIARY OBLIGATION

The trust framework presents two causes of action, available to different classes of parties, to enforce the atmospheric fiduciary obligation. The first is an action by citizen beneficiaries against their governmental trustees for failing to protect their natural trust. It is well settled that beneficiaries may sue the trustee to protect their property.[70] Citizens are positioned to bring trust actions against their cities, counties, states, or the federal government.[71] The second is an action brought by one sovereign trustee against another for failure to maintain common property. Co-tenants have a right against other co-tenants for waste and for failing to pay necessary expenses.[72] States may bring an action for waste against other states or the federal government. Tribal sovereigns may also bring actions. Waste and breach of trust claims find grounding within the same basic property framework.

As with any claim, a myriad of issues may bar recovery. Litigants must navigate potential barriers such as standing, sovereign immunity, preemption, the political question doctrine, ripeness, jurisdiction, and intervention. This chapter does not discuss such hurdles, charting instead the broad terrain of atmospheric trust litigation. It should be noted, however, that courts recognizing the enormity of climate crisis and the crucial role of the judiciary may approach these barriers with a leniency that is not characteristic of past decisions. At its core, the unparalleled force of the public trust doctrine is its mandate to preserve resources for future generations – and the role of the court in policing the legislature and agencies in their management of trust assets. The substantive underpinning of the doctrine thus creates powerful arguments in defense of many potential barriers.[73]

---

[70] See BOGERT, supra note 27, § 154 at 551 ("If the trustee is preparing to commit a breach of trust, the beneficiary need not sit idly by and wait until damage has been done. He may sue in a court of equity for an injunction against the wrongful act.").

[71] Marks v. Whitney, 491 P.2d 374, 381 (Cal. 1971) (private citizens have standing to sue under public trust though a court may raise the issue on its own). Of course issues of sovereign immunity may arise in such suits, and general Constitutional requirements of standing apply.

[72] Willmon v. Koyer, 143 P. 694, 695 (Cal. 1914) (each co-tenant has a right to maintain an action against co-tenants "to have refunded to him by the other his proportion of any expenditures made for the benefit of common property."); 63C Am. Jur. 2D PROPERTY § 31; Chosar Corp. v. Owens, 370 S.E.2d 305 (Va. 1988) (co-tenants who allowed mining without consent of all other co-tenants were liable for waste); Anders v. Meredith, 1839 WL 525 (N.C. 1839); see also supra note 38 (discussing waste in context of sovereign co-tenancy in migrating fishery).

[73] While procedural issues are beyond the scope of this chapter, they are considered in Mary Christina Wood, Courts as Guardians of the Global Trust (work in progress).

2.1. Declaratory Relief

It is important to design a remedy with a view toward providing the macro relief imperative to addressing the climate crisis. A declaratory judgment setting forth the trust framework for atmospheric obligations will greatly advance society's task of clarifying the responsibilities of governments worldwide. Amid the present political chaos surrounding climate change, such clarification may bring results reaching far beyond the courtroom because it infuses citizens with the conceptual tools they need to hold their respective governments accountable in quantifiable terms at all jurisdictional levels. In that sense, a declaratory judgment could become a yardstick for political action.

A declaratory judgment should clearly iterate the following principles: (1) all governments have a fiduciary obligation, as trustees, to protect the atmosphere as a commonly shared asset; (2) all governments bear liability for reducing carbon; (3) the fiduciary obligation among industrialized nations and subjurisdictions is to: (a) arrest the growth of emissions no later than 2010; (b) reduce carbon levels by at least 4% each year; and (c) ultimately bring carbon levels down to 80% or more below 2000 levels by 2050; (4) this fiduciary obligation is organic to government and permits no orphan shares or partial orphan shares; (5) the fiduciary obligation is enforceable by the citizen beneficiaries of the trust representing present and future generations; and (6) the fiduciary obligation and the concomitant duty to prevent waste are enforceable by co-tenant trustees.[74]

Declaratory relief should be accompanied by suitable injunctive relief that allows courts to provide a remedy on a macro level without invading the province of the political branches.[75] Courts have emphasized that the core purpose of the public trust doctrine is to police the other branches of government in their disposition of public assets.[76] By drawing on traditional relief available against co-tenants and trustees for misuse of property, courts may require carbon accountings and enforceable carbon budgets as remedies for sovereign breach of the atmospheric fiduciary obligation without reaching into the lawmaking purview of the other branches.

2.2. A Carbon Accounting

An accounting is a traditional remedy springing from the equitable powers of the court in both the co-tenancy and trust contexts.[77] It is a judicial process whereby

---

[74] However, a declaratory judgment should not be a "general admonition" but must be narrowly crafted to define a duty according to "concrete facts presented by a particular dispute." United States v. Washington, 2007 U.S. Dist. LEXIS 61850, at *23 (W.D. Wash. 2007). Courts have rejected overly broad declaratory judgments. See id.

[75] Winberger v. Romero-Barcelo, 456 U.S. 305, 312 (1982) (the basis for injunctive relief is a finding of irreparable injury and the absence of an adequate legal remedy) (citations omitted).

[76] See supra note 2.

[77] See, e.g., Evans v. Little, 271 S.E. 2d 138, 141 (Ga. 1980) (co-tenancy); Koyer, 143 P. at 695 ("As an incident to a cotenancy relationship, either cotenant has a right to demand of the other an accounting

co-tenants or trustees must account for expenses and/or profits in connection with the property.⁷⁸ The basic premise of an accounting in the co-tenancy context is that each co-tenant is responsible for his share of the expenses and is due his share of the profit from the property.⁷⁹ An accounting is the procedural method by which this "fair share" principle is enforced by courts. In the trust context, an accounting is the method by which beneficiaries may ensure proper management of their property.⁸⁰ Accordingly, courts have held that "any beneficiary, including one who holds only a present interest in the remainder of a trust, is entitled to petition the court for an accounting."⁸¹ The scope of an accounting must include "all items of information in which the beneficiary has a legitimate concern."⁸² In the financial context, this means a statement "in clear and concise terms of the nature and value of the corpus of the trust and the amount and location of any balance or remainder."⁸³

In the context of atmospheric trust litigation, an accounting would take the form of quantifying carbon emissions and tracking their reduction over time.⁸⁴ This form of accounting is an extrapolation from the traditional remedy in two ways. First, it is applied against a sovereign trustee, not a private trustee. It is

---

as to rents and profits of the cotenancy, which of course, involves the right of one cotenant to have refunded to him by the other his proportion of any expenditures made for the benefit of the common property."); Zuch v. Conn. Bank & Trust Co., 500 A.2d 565, 568 (Conn. App. 1985) ("As a general matter of equity, the existence of a trust relationship is accompanied as a matter of course by the right of the beneficiary to demand of the fiduciary a full and complete accounting at any proper time.") (citations omitted); Cobell v. Norton, 240 F.3d 1081 (D.C. Cir. 2001) (*Cobell VI*) (accounting against federal government for mismanagement of Indian trust funds).

78 *Evans*, 271 S.E.2d at 141.
79 *See, e.g.*, Garber v. Whittaker, 174 A. 34, 37 (Super. Ct. Del. 1934) ("Tenants in common of the legal title to land are ordinarily entitled to the use, benefit and possession of such land, including their just and proper shares of the rents and profits therefrom."); *Koyer*, 143 P. at 695–96 ("The rule is that when one tenant in common has paid a debt or obligation for the benefit of the joint property, or has discharged a lien or assessment imposed upon it as a common burden, he is entitled as a matter or right to have his co-tenant, who has received the benefit of it, refund to him his proportionate share of the amount paid."); *see also* WILLIAM B. STOEBUCK & DALE A. WHITMAN, THE LAW OF PROPERTY 205 (3d ed. 2000) (where a cotenant derives income from a use of land that permanently reduces its value the cotenant must account to the other cotenants); White v. Smyth, 214 S.W.2d 967, 978 (Tex. 1948) ("When it is claimed that a cotenant in possession of property has become liable to his cotenants for profits accruing from his productive operations, the usual mode of settling the account is to charge him with all his receipts and credit him with all his expenses, thereby ascertaining the net profits available for distribution [among cotenants].").
80 *See Zuch*, 500 A.2d at 567 ("The fiduciary relationship is in and of itself sufficient to form the basis for the [accounting].") (citations omitted); Faulkner v. Bost, 137 S.W. 3d 254, 259 (Tex. App. 2004) (beneficiaries may file suit to compel a trustee to provide an accounting, citing Texas Property Code).
81 *In re* Estate of Ehlers, 911 P.2d 1017, 1021 (Wash. App. 1996) (*citing* Nelsen v. Griffiths, 585 P.2d 840, 843 (Wash. App.1978)).
82 *Zuch*, 500 A.2d at 568.
83 *Id.*
84 On a broader level, NASA scientist Jim Hansen has suggested keeping an annual public scorecard of measured changes of (1) fossil fuel CO_2 emissions; (2) atmospheric CO_2 amount; (3) human-made climate forcing; and (4) global temperature. James E. Hansen, *A Brighter Future*, 52 CLIMATIC CHANGE 435, 438 (2002). The carbon accounting suggested herein would track all greenhouse gas emissions.

well established, however, that a sovereign defendant may be subject to an accounting for mismanagement of a trust. In the Indian law context, for example, the federal government is currently subject to a multibillion-dollar accounting action for its mismanagement of tribal trust funds.[85] Second, a carbon accounting invokes a tool developed in the financial context and extends it to the natural context. Such a leap should be well within the imagination of judges. Modern natural resource management increasingly imports concepts from the financial world. Approaches recognizing "natural capital" and "environmental services" draw upon financial constructs to organize human demands on a natural resource.[86] Courts are also quite familiar with assigning monetary value to resources through the process of awarding natural resource damages to governmental trustees.[87] Moreover, courts have essentially engaged in natural "accountings" in the environmental context before, without using the label. In determining rights to fish runs shared between states and tribes, for example, courts have delved into the quantitative aspects of beneficial use of a resource held in common by sovereign co-tenants.[88]

In the climate context, the accounting consists of a judicially supervised periodic quantification of the amount of greenhouse gas pollution emitted by the sovereign defendant. Such pollution is an overriding factor affecting the productivity of the atmospheric trust.[89] The accounting establishes the current carbon pollution emitted on the particular jurisdictional level (local, state, or federal) so as to define a baseline and then tracks progressive reduction over time. Modern modeling is capable of quantifying a carbon footprint on virtually any scale, from individual to global.[90] Much of the necessary data has been developed and is already accessible. The U.S. Department of Energy, for example, has released overall carbon emissions of all fifty states and will continue the reporting on an annual basis.[91] Several cities,

---

[85] See Cobell v. Kempthorne, 455 F.3d 317, 319–21 (D.C. Cir. 2006) (describing background of litigation).
[86] See generally PAWL HAWKEN, AMORY LOVINS & L. HUNTER LOVINS, NATURAL CAPITALISM: CREATING THE NEXT INDUSTRIAL REVOLUTION (1999); PETER BARNES, CAPITALISM 3.0 (Barrett-Koehler 2006); Costanza et al., *The Value of the World's Ecosystem Services and Natural Capital*, NATURE 387, 253 (May 15, 1997).
[87] See generally Coeur D'Alene Tribe v. Asarco, Inc., 280 F. Supp. 2d 1094 (D. Idaho 2003).
[88] See generally Wood, *The Tribal Property Right to Wildlife Capital (Part 1)*, supra note 31, at 16.
[89] See Torres, supra note 11, at 547 (calling for accounting).
[90] U.S. Environmental Protection Agency, *Personal Emissions Calculator*, available at http://www.epa.gov/climatechange/emissions/ind_calculator.html (last visited Sept. 18, 2007) (allows individuals to calculate their emissions); The Climate Registry, available at http://www.theclimateregistry.org/index.html (last visited Sept. 18, 2007) (tracks emissions from private industry); Borenstein, supra note 64 (chart with all state emissions in million metric tons); Australian Government, *National Carbon Accounting System*, available at http://www.greenhouse.gov.au/ncas/background.html (last visited Sept. 18, 2007) (describing National Carbon Accounting System for Australia); UNFCCC, *Counting Emissions and Removals: Greenhouse Gas Inventories Under the UNFCCC*, available at http://unfccc.int/resource/docs/publications/counting.pdf (last visited June 16, 2008).
[91] See Borenstein, supra note 64. Raw data for state carbon dioxide emissions is available from the Energy Information Administration, *Energy Emissions Data & Environmental Analysis of Energy Data*, available at http://www.eia.doe.gov/environment.html (last visited June 16, 2008).

such as Seattle, Washington, have already quantified their carbon footprint.[92] A new carbon registry has formed to account for releases from private sources as well.[93] While inevitably there will be areas of dispute regarding some emissions sources, particularly mobile sources, the methodology for measuring jurisdictional carbon footprints will continue to be refined as professional standards emerge in the field of carbon accounting.[94]

Carbon accounting allows co-tenants and beneficiaries of the trust to evaluate government's measures to protect the atmospheric trust. The accounting would determine jurisdictional compliance with the *Target for U.S. Emissions Reductions*, which, as explained previously, is the quantitative standard of government's fiduciary obligation. A court must maintain ongoing jurisdiction over the case to receive periodic progress reports, a common procedure in accounting cases. Accountings for trust management are often performed on a regular basis, such as quarterly, biannually, or annually, and contain an inventory of the trust fund at the end of the accounting period.[95] The narrow window of time remaining before climate thresholds are crossed seemingly justifies carbon accounting reports every quarter.

2.3. Nested Jurisdictions

Unprecedented as it may seem, atmospheric trust litigation may call forth municipal judges, state court judges, and federal judges to enforce the fiduciary obligation against various levels of government. This is because, for greenhouse gas reductions to truly add up to the "carbon math" in time, each jurisdiction must be accountable for reducing carbon. The unavailability of judicial relief at any jurisdictional level risks orphan shares at that level. Accordingly, the atmospheric trust obligation must be viewed as a general mandate capable of multilevel jurisdictional enforcement. This in turn presents a need for coordination among various courts. Cross-judicial coordination is a familiar challenge for many judges. Complex class action cases, such as those involving asbestos and tobacco claims, have often involved more than one court. Water appropriation cases also frequently involve simultaneous proceedings at both the state and federal level. The dynamic nature of environmental resources typically creates a need for transjurisdictional judicial coordination.

---

[92] City of Seattle, Climate Action Plan, http://www.seattle.gov/climate/carbonfootprint.htm (last visited June 16, 2008); City of Seattle, *Our Carbon Footprint*, available at http://www.seattle.gov/climate/PDF/Our_Carbon_Footprint.pdf ("Any serious initiative to reduce global warming pollution must begin with a very challenging first step: A greenhouse gas emissions inventory that establishes the baseline against which progress will be measured, and identifies the major sources of pollution that will be the focus of the program.").

[93] The Climate Registry, *supra* note 90.

[94] Australia, for example, has developed a carbon accounting system with uniform standards. Australian Government, *National Carbon Accounting System*, *supra* note 90.

[95] *See, e.g.*, Fraser v. Se. First Bank of Jacksonville, 417 So.2d 707, 708 (Fla. App. 1982) (citing Florida statutes); Cobell v. Norton, 240 F.3d 1081, 1086 (D.C. Cir. 2001) (*Cobell VI*) (quarterly reports in Indian trust litigation against the federal government).

Coordination in atmospheric trust litigation is made possible using the "nested jurisdiction" concept. Greenhouse gas reductions achieved on a subjurisdictional level (i.e., cities and counties) are readily and easily attributable to the umbrella jurisdiction (the state). For the same reason, reductions at the subnational (state) level are easily accounted for at the federal level. Through open accounting processes, carbon reduction can simultaneously be attributed to the most immediate jurisdictional level as well as the broadest jurisdictional level.

Courts can facilitate the process by posting accounting results on the website that houses litigation documents.[96] In this way, carbon accountants nationwide can quickly obtain information and incorporate it into ongoing analysis for any jurisdiction of concern. For example, as carbon reduction takes place as a result of a suit against the city of Phoenix, such reduction may be accounted for in any suit against the state of Arizona. The judicial process would be greatly streamlined by uniform reporting schedules as well as accounting templates and processes. Like any emerging field, atmospheric trust litigation would benefit from procedural uniformity among various courts and advisers to the court.[97]

2.4. An Enforceable Carbon Budget and Recovery Plan

While an accounting remedy provides the means whereby a beneficiary or cotenant can measure the performance of a governmental trustee, additional injunctive relief is necessary to enforce the sovereign duty to restore the natural trust where it has been damaged. At a very simple level, the fiduciary obligation to reduce carbon pollution can be carried out through a "budget" for carbon reduction over time that sets forth quantifiable mileposts. The jurisdiction must also develop an asset recovery plan containing measures calibrated to bring about such reduction.[98]

Seattle, Washington, has undertaken an initiative that provides an example of a template for such action. Under the leadership of Mayor Greg Nickels, the city set a goal for reducing its greenhouse gas emissions to 7% below 1990 levels by 2012.[99]

---

[96] Many courts post documents related to current cases online through a case management and electronic filing system. See, e.g., U.S. District Court for the Eastern District of California, CM/ECF, http://www.caed.uscourts.gov/caed/StaticOther/page_1581.htm (last visited June 16, 2008); U.S. District Court for the District of Columbia, Electronic Case Filing, http://www.dcd.uscourts.gov/ecf.html (last visited June 16, 2008).

[97] To that end, it is imperative that carbon professionals develop a uniform protocol for measuring the carbon footprint at each jurisdictional level. Such protocol must address, for example, attributing mobile sources and origins of electricity consumption. The lack of such protocol should not deter courts in enforcing the carbon fiduciary obligation. Litigation will most certainly create a demand for professional assistance that will rapidly be met by a cadre of professional carbon accountants.

[98] Proposed British legislation provides an example of a "carbon budget." *Britain Proposes Bold Environmental Legislation that Could Pave Way for Post-Kyoto Pact*, INT'L HERALD TRIB. Mar. 13, 2007, available at http://www.iht.com/articles/ap/2007/03/13/europe/EU-GEN-Britain-Climate-Change.php.

[99] City of Seattle, *Our Carbon Footprint*, supra note 92.

This goal requires reducing current emissions by 680,000 metric tons a year.[100] The city then created a plan that divided the overall emissions into sectors such as city lighting, coal, heating, cars and trucks, airports, maritime, and other.[101] The plan sets forth specific action items designed to reduce carbon from the various sectors.[102]

Courts can require governmental trustees at any jurisdictional level to establish a budget and asset recovery plan calibrated to the uniform fiduciary standard set forth in the *Target for U.S. Emissions Reductions*. The contemplated injunctive relief does not invade the prerogatives of the other branches because it does not dictate to the trustee how to accomplish the carbon reduction. It simply spurs action where the political branches neglect to carry out their fiduciary responsibility. Cities, counties, and states have wide latitude in devising plans that are tailored to the unique circumstances of their jurisdiction. Periodic reports provided to the court through the accounting process inform the court and the beneficiaries whether the trustee is making adequate progress in accordance with the budget and plan.[103] In this respect, the trust remedy may strike the ideal balance between necessarily potent, macro judicial enforcement and traditional deference to the political branches.

While some judges may be overwhelmed by the novel and all-encompassing context of carbon reduction, it is important to bear in mind that the envisioned judicial role is much the same as in other natural resource contexts where courts enforce management and/or recovery of diminished natural assets. In the treaty fishing wars of the late 1960s and 1970s, the district courts of Oregon and Washington became, for a time, "fishmasters," tasking themselves with detailed supervision of tribal and state salmon harvests.[104] The courts created a consent decree structure whereby the states and tribes developed a judicially supervised and enforceable plan for future harvest of the salmon.[105] More recently, in the ESA lawsuits over the imperiled Columbia River salmon, the federal district court of Oregon has assumed a rigorous role overseeing the development of a fish recovery plan pursuant to a process of multisovereign consultation structured by the court.[106] Courts have supervised broad plans to address exclusionary zoning[107] and racial desegregation.

---

[100] City of Seattle, A Climate of Change: Meeting the Kyoto Challenge, *Climate Action Plan Highlights* (Sept. 2006), *available at* http://www.seattle.gov/climate/docs/SeaCAP_summary.pdf (last visited June 16, 2008).

[101] *Id.*

[102] Action items include converting city governmental fleets to more efficient cars, creating bicycle paths, imposing parking taxes, creating residential electrical efficiency programs, and other measures. *See id.*

[103] As carbon reduction measures come "on line," progress may be quantified in the accounting process.

[104] *See* Puget Sound Gillnetters Ass'n v. U.S. Dist. Court, 573 F. 2d 1123, 1133 (9th Cir. 1978); United States v. Washington, 520 F.2d 685, 686, 693 (9th Cir. 1975); *see also* Mary Christina Wood, *Restoring the Abundant Trust: Tribal Litigation in Pacific Northwest Salmon Recovery*, 36 ENVTL. L. REP. 10,163, 10,176–77 (2006).

[105] For discussion, see Mary Christina Wood, *Reclaiming the Natural Rivers: The Endangered Species Act Applied to Endangered River Ecosystems*, 40 ARIZ. L. REV. 197, 233 (1998).

[106] *See* Wood, *Restoring the Abundant Trust*, *supra* note 104, at 10, 175–76.

[107] In Southern Burlington County, NAACP v. Township of Mount Laurel, 336 A.2d 713 (N.J. 1975) (Mt. Laurel I), the New Jersey Supreme Court found that housing, along with food, is one of the

While courts must be cognizant of appropriate judicial boundaries in structuring relief for trust violations,[108] they seemingly have wide latitude in requiring sovereigns to develop enforceable plans for proper trust management.[109]

2.5. Backstops through Injunctive Relief

Enforcing the carbon reduction budget and plan presents a challenge for the courts. On one hand, it is clear that enforcement is necessary, because the political branches may lack the will to institute the measures necessary to carry out their fiduciary obligation. On the other hand, a carbon reduction plan is likely to contain a set of measures beyond the power of courts to enforce – measures such as carbon taxes, infrastructure projects, and transfer of public investment. In structuring enforcement mechanisms, courts must walk a fine line between affording meaningful relief and overstepping their role. It is unlikely that carbon reduction plans may be enforced in their entirety down to the last detail.

Nevertheless, courts have it well within their power to force carbon reduction through discrete injunctive measures tailored toward obvious carbon sources. An injunction may contain "backstops" that consist of measures that the court will mandate if the budget is not carried out. The broad realm of environmental and land use litigation provides precedent for measures that may serve as effective backstops. Such measures might include, for example, injunctions prohibiting new coal-fired plants[110] and injunctions against large-scale logging, recreational vehicle use on public lands, airport expansions, sewer hookups, issuance of air pollution permits, and

---

"most basic human needs" and interpreted affordable housing as a right implicitly guaranteed by the State's constitution. The Court held that towns must bear their "fair share" of providing housing needed on a regional level and ordered a town to amend its zoning law to fulfill its fair share, noting, "The municipality should first have full opportunity to itself act without judicial supervision." Id. at 734. However, a second challenge was brought after the town failed to provide adequate housing, and the Court devised a detailed remedy structure that included ordering affirmative remedies involving government subsidies, incentive zoning, mandatory set-asides and other measures. S. Burlington County, NAACP v. Township of Mount Laurel, 456 A.2d 390 (N.J. 1983) (*Mt. Laurel II*); *see* discussion in JOSEPH WILLIAM SINGER, PROPERTY LAW: RULES, POLICIES, AND PRACTICES 908 (4th ed. 2006). The Court authorized the appointment of special masters to rewrite the zoning ordinances to provide constitutionally sufficient housing. Id. (discussing remedy aspects of the case). The Court also provided for the appointment of regional trial judges to handle all zoning cases in order to generate consistent definitions of regions and to "determine in an orderly way each community's fair share of the regional housing need." Id. The *Mt. Laurel II* case seems particularly helpful to the global warming context, where courts must allocate a fair share of carbon reduction liability on a regional basis and devise innovative approaches to enforcing that share.

[108] *See* Cobell v. Kempthorne, 455 F.3d 317, 330–31 (D.C. Cir. 2006) (reviewing reversals of district court remedies in an Indian trust accounting case).

[109] Cobell v. Norton, 283 F. Supp. 2d 66 (D. D.C. 2003), *rev'd on other grounds*, 392 F.3d 469 (D.C. Cir. 2004).

[110] The world's preeminent climate scientists are clear that climate stability cannot be achieved if humanity uses the remaining coal reserves. James Hansen et al., *Climate Change and Trace Gases*, PHIL. TRANS. ROYAL SOC'Y A, 1925, 1939 (2007), *available at* http://www.planetwork.net/climate/Hansen2007.pdf (last visited June 16, 2008) ("Given the estimated size of fossil fuel reservoirs, the chief implication is that we, humanity, cannot release to the atmosphere all, or even most, fossil fuel

a myriad of other activities.[111] While most of the precedent for such injunctions is grounded in claims brought under statutory law, the relief awarded is typically not statutorily mandated but rather devised by a court to afford a meaningful remedy. It is within the traditional province of courts of equity to devise relief to remedy the harm.[112] Of course, the ultimate enforcement mechanism is to hold government officials personally in contempt of court for failure to carry out court-ordered fiduciary duties.[113]

2.6. Collateral Benefits of Atmospheric Trust Litigation

Statutory avenues of litigation offer the advantage of detailed frameworks and helpful precedent. Atmospheric trust litigation carries the risk of any novel strategy. Professor David Hunter, however, emphasizes the "awareness-building" and other positive impacts of climate litigation apart from court victories.[114] Atmospheric trust litigation provides two substantial collateral benefits missing from other strategies.

CO_2. To do so would guarantee dramatic climate change, yielding a different planet than the one on which civilization developed and for which extensive physical infrastructure has been built."); James Hansen, *Dangerous Human-Made Interference with Climate*, Testimony Before Select Committee on Energy Independence and Global Warming, U.S. House of Representatives 18 (Apr. 26, 2007), available at http://www.columbia.edu/~jeh1/testimony_26april2007.pdf (last visited June 16, 2008) ("Thus the most critical action for saving the planet at this time, I believe, is to prevent construction of additional coal-fired power plants without CO_2 capture capability.").

[111] *See, e.g.*, United States v. Metro. Dist. Comm'n, 757 F. Supp. 121, 128–29 (D. Mass 1991), aff'd, 930 F.2d 132 (1st Cir. 1991) (moratorium against sewer hookup); Jeffery J. Matthews, *Clean Water Act Citizen Suit Requests for Municipal Moratoria: Anatomy of a Sewer Hookup Moratorium Law Suit*, 14 J. ENVTL. L. & LITIG. 25 (1999) (discussing injunctions imposing moratoria against sewer hookups); Am. Motorcyclist Ass'n v. Watt, 543 F. Supp. 789, 798 (C.D. Cal. 1982) (enjoining off-road vehicle use because agency plan did not comply with the statute); Pac. Rivers Council v. Thomas, 30 F.3d 1050 (9th Cir. 1994) (enjoining the U.S. Forest Service from proceeding with projects under land resource management plans prior to ESA consultation); Lane County Audubon Soc'y v. Jamison, 958 F.2d 290, 294 (9th Cir. 1992) (enjoining the BLM from new timber sales until ESA consultation was completed); Thomas v. Peterson, 753 F.2d 754 (9th Cir. 1975) (enjoining construction of road until agency prepared biological assessment); Or. Nat. Desert Ass'n v. Singleton, 75 F. Supp. 2d 1139 (permanently enjoining grazing in all "areas of concern").

[112] *See* Alaska Ctr. for the Env't v. Browner, 20 F.3d 981, 986 (9th Cir. 1994) ("The district court has broad latitude in fashioning equitable relief when necessary to remedy an established wrong."); Weinberger v. Romero-Barcelo, 456 U.S. 305, 311 (1982) ("The essence of equity jurisdiction has been the power of the [Court] to do equity and to mould each decree to the necessities of the particular case. Flexibility rather than rigidity has distinguished it.").

[113] One district court recently threatened U.S. Agriculture Undersecretary Mark Rey with contempt of court and jail time for the agency's "systematic disregard of the rule of law." The agency failed to conduct environmental analysis required by statute in connection with the use of fire retardant that kills fish. *See* Matt Gouras, *Judge: Ag Undersecretary Avoids Jail Time*, ASSOCIATED PRESS, Feb. 28, 2008, available at http://hosted.ap.org/dynamic/stories/B/BUSH_OFFICIAL_CONTEMPT?SITE=AP&SECTION=HOME&TEMPLATE=DEFAULT&CTIME=2008-02-28-00-41-37.

[114] David Hunter, *The Implications of Climate Change Litigation for International Environmental Law-Making*, in ADJUDICATING CLIMATE CHANGE: STATE, NATIONAL, AND INTERNATIONAL APPROACHES 1 (William C. G. Burns & Hari M. Osofsky eds., 2009).

First, as a macro strategy, trust litigation speaks directly to government's obligation to address climate crisis – in quantitative terms applicable to any jurisdictional level. So far, government's approach to climate crisis has been perceived as a matter of political discretion, not obligation.[115] The trust approach applies a logical, obligatory framework to a situation dangerously devoid of any standards for government behavior. Through the mere act of "preparing, announcing, filing, advocating and forcing a response,"[116] trust litigants will be positioned to change the public's expectation of government on all levels. Statutory claims, by contrast, fail to address government's full obligation in the face of climate crisis. Such claims are geared toward isolated instances of government action, such as approvals of air permits or programs, or listing decisions under the ESA. While valuable in many other ways and worthy of pursuit, they nevertheless embrace an approach of incremental change rather than the rapid overhaul necessary to combat climate change.

Second, atmospheric trust litigation harnesses strength from the economic, moral, and political realms. Perhaps the most dangerous aspect of global warming is that there is no overarching paradigm to turn governmental, economic, or individual choices away from the business-as-usual approach that has led the world to the threshold of climate catastrophe. In order to accomplish the massive shift that society must make in the short time frame remaining, there needs to be an encompassing moral, political, spiritual, economic, and legal framework that draws from a common wellspring of human thought and experience. From this perspective, the major drawback of most statutory legal claims is that they are often divorced from any unifying framework that reaches across other realms. Claims brought under the ESA, CAA, and various other statutes are mired in complexity and beyond the understanding of most citizens. Such claims do not serve as good vehicles for expression of values and do not speak to the experience of citizens in the nonlegal realm. They therefore often lack much-needed fortification on economic or moral grounds.

Trust litigation, by contrast, draws on fundamental principles that are increasingly invoked by today's visionaries. In economic terms, the trust dovetails with principles of natural capitalism, which leading thinkers present as a paradigm of business and industrial reform.[117] These principles urge business to structure operations using the Earth's interest, not its capital. Emphasis on renewable energy is an example of this approach. On a moral level, trust principles reflect an ethic toward children and underscore the strong urge of human beings to pass estates along to future generations.[118] The atmosphere is an endowment to which future generations

---

[115] See Mary Christina Wood, *Government's Atmospheric Trust Responsibility*, 22 J. ENVTL. L. & LITIG. 369, 374 (2008).

[116] Hunter, *supra* note 114, at 1.

[117] See NATURAL CAPITALISM, *supra* note 86.

[118] Addressing climate crisis has been presented by civic and religious leaders as a moral obligation toward future generations. See AL GORE, Introduction, in AN INCONVENIENT TRUTH: THE PLANETARY EMERGENCY OF GLOBAL WARMING AND WHAT WE CAN DO ABOUT IT (Rodale Books 2006); Al Gore, Op-Ed., *Moving Beyond Kyoto*, N.Y. TIMES, July 1, 2007, *available at* http://www.nytimes.com/2007/07/01/opinion/01gore.html?ex=1341115200&en=beob465c91dbcaaf&ei=5124&partner=permalink&

have a legitimate moral claim, and failure to safeguard it amounts to generational theft. Moreover, on the political level, by defining the atmosphere as common property, the trust positions all nations of the world in a logical relationship toward each other and toward Nature.[119] The trust framework defines respective sovereign obligations in quantifiable, straightforward terms.[120] Once presented in U.S. courts, citizen litigants of other countries may invoke the trust concept. India, for example, already has a robust public trust doctrine that citizens there can draw upon to establish atmospheric trust responsibility.[121] In short, by defining the trust paradigm of sovereign obligation in a litigation venue, courts may play a tremendous role in harnessing the collective momentum from various other realms in which a paradigm shift is necessary and already taking place.

3. CONCLUSION

Atmospheric trust litigation challenges lawyers and judges to take fundamental principles of public trust law and apply them in coherent fashion to a new and urgent context so as to arrive at a uniform, quantifiable measure of governmental responsibility to reduce carbon. The task is made easier by the fact that these principles are logical, compelling, and seemingly applicable to all governments. However, judges have become so accustomed to issuing rulings within the detailed confines of statutory law that many may have lost the imagination to construct

---

exprod=permalink ("Our children have a right to hold us to a higher standard when their future – indeed, the future of all human civilization – is hanging in the balance."); Colin Woodard, *In Greenland, An Interfaith Rally for Climate Change*, CHRISTIAN SCI. MONITOR, Sept. 12, 2007, *available at* http://www.csmonitor.com/2007/0912/p06s01-woeu.html?page=1 (Shiite, Buddhist, Hindu, Jewish, Christian, and Shinto leaders join in commitment at Greenland interfaith climate rally to leave the planet "in all its wisdom and beauty to the generations to come."). The legal arguments setting forth an obligation to future generations have been compiled and analyzed in Burns Weston, *Climate Change and Intergenerational Justice: Foundational Reflections*, 9 VT. J. ENVTL. L. (2008).

[119] *See* Ved P. Nanda & William K. Ris, Jr., *The Public Trust Doctrine: A Viable Approach to International Environmental Protection*, 5 ECOLOGY L.Q. 291, 306 (1976) (inventorying trust concepts in other countries and concluding, "The principles of public trust are such that they can be understood and embraced by most countries of the world."); Sand, *supra* note 21, at 57–58 (suggesting trust principles as framework for international law, stating, "[A] transfer of the public trust concept from the national to the global level is conceivable, feasible, and tolerable.... The essence of transnational environmental trusteeship... is the democratic *accountability* of states for their management of trust resources in the interest of the beneficiaries – the world's 'peoples'....") (emphasis in original).

[120] Absent such quantification, nations are settling for inadequate measures. *See* Reuters, *G8 Agreement on Climate Change a 'Disgrace' – Al Gore*, June 14, 2007, *available at* http://www.alertnet.org/thenews/newsdesk/L14317004.htm (last visited June 16, 2008) (G8 leaders agreeing only to "substantial" reduction in greenhouse gases and failing to adopt concrete numerical commitments).

[121] M.C. Mehta v. Kamal Nath, 1 SCC 388 (India 1997); Karnataka Indus. Areas Dev't Bd. v. C. Kenchappa, AIRSCW 2546 (India 2006); T.N. Godavaman Thirumalpad v. Union of India, WP 202 1995, CDJ 2005 SC 713 (India 2005); Perumatty Grama Panchayat v. State of Kerala, 2004 (1) KLT 731 (India 2003). For discussion of India's public trust doctrine, see Deepa Badrinarayana, *The Emerging Constitutional Challenge of Climate Change: India in Perspective*, 19 FORDHAM ENVTL. L.J. 1, n. 137 (2009).

meaningful remedies under traditional common law. At a time in history when thinkers across the world are calling for new, innovative technologies and practices to address climate crisis, lawyers should pioneer promising, if untested, legal constructs to address carbon loading of the atmosphere. If one thing is clear, it is that the body of statutory environmental law is a product of an altogether different era, formulated to respond to circumstances far less urgent, less dangerous, and less encompassing than those now confronting society. The environmental statutes were not crafted to address a planetary emergency.[122] The trust claim defines a binding fiduciary obligation organic to all levels of government – one that is calibrated mathematically to scientific understanding. In that way, it is perhaps the only claim that speaks directly to the sovereign's duty at this pivotal time in the history of human civilization.

[122] CLIMATE CODE RED, *supra* note 57, at 63–66.

PART II

NATIONAL CASE STUDIES

6

The Intersection of Scale, Science, and Law in *Massachusetts v. EPA*

Hari M. Osofsky*

INTRODUCTION

Justice Scalia: But I always thought an air pollutant was something different from a stratospheric pollutant, and your claim here is not that the pollution of what we normally call "air" is endangering health.... [Y]our assertion is that after the pollution leaves the air and goes up into the stratosphere it is contributing to global warming.

Mr. Milkey: Respectfully, Your Honor, it is not the stratosphere. It's the troposphere.

Justice Scalia: Troposphere, whatever. I told you before I'm not a scientist.

(Laughter).

Justice Scalia: That's why I don't want to deal with global warming, to tell you the truth.[1]

The above exchange occurred between Justice Scalia and James Milkey, Assistant Attorney General of Massachusetts, during the oral argument in *Massachusetts v. EPA*,[2] the first case heard by the U.S. Supreme Court on governmental regulation of greenhouse gas emissions. It not only illustrates the complexities of judicial

---

* Associate Professor, Washington and Lee University School of Law; B.A., J.D., Yale University. The author can be contacted through osofskyh@wlu.edu. This chapter is an edited version of an identically titled symposium piece published in Volume 9 of the *Oregon Review of International Law*. A truncated version of this piece, based on my remarks at the ASIL conference, was published in the *ASIL Proceedings of the 101st Annual Meeting*. The initial version of this chapter was prepared for Bart Bartlein's course Climatological Aspects of Global Change, and I am very grateful to him for his many insights on climate science, which have fundamentally shaped my analysis. The piece also benefited from feedback during the *Oregon Review of International Law* Symposium, as well as from its presentation at the 2007 American Society of International Law (ASIL) Annual Meeting. In addition, Wil Burns, Holly Doremus, Alexander Murphy, and Eve Vogel provided very helpful comments that greatly improved the piece. I would like to thank my research assistants – Will Cooksey, Michelle Platt, and Jeff Richards – for their helpful work in exploring the initial impacts of *Massachusetts v. EPA*, as well as Stefanie Herrington for her meticulous review of drafts. As always, the piece would not have been possible without the loving support of Joshua, Oz, and the newly arrived Scarlet Gitelson.

[1] Transcript of Oral Argument at 22–23, Massachusetts v. EPA, 549 U.S. 497 (2007) (No. 05-1120), 2006 WL 3431932 at 22–23.
[2] Massachusetts v. EPA, 549 U.S. 497 (2007).

engagement with the science of global warming but also provides a window into one of the greatest obstacles to effective regulatory approaches to the problem of climate change, which the Obama administration must grapple with as it embarks upon its ambitious climate and energy initiatives. Namely, greenhouse gas emissions and their impacts are foundationally multiscalar; they range from the most individual to global levels.

Referencing climate change as a multiscalar problem, however, only serves as a starting point for further discussion. "Scale" is a complex and contested concept in both the geography and the ecology literatures. Geographers have defined it as (1) "a nested hierarchy of bounded spaces of differing size"; (2) "the level of geographical resolution at which a given phenomenon is thought of, acted on or studied"; (3) "the geographical organizer and expression of collective social action"; and (4) "the geographical resolution of contradictory processes of competition and cooperation."[3] Ecologists supplement this understanding with additional concepts. They define scale as being composed of grain ("the finest level of spatial or temporal resolution available within a given data set") and extent ("the size of the study area or the duration of the study").[4]

This chapter analyzes the interaction of scale (in its many guises), science, and law in the Supreme Court briefs, oral argument, opinion, and dissents in *Massachusetts v. EPA* as a window into the complex dynamics at play in climate change litigation. Its focus is not on the complexities of implementing the decision, another important topic beyond the scope of this chapter, but rather on what can be learned from the interactions that took place in this lawsuit. Formally, the case primarily occurs at a national level; the parties dispute the interpretation of federal law in an action that was heard by federal courts at every level. It is precisely this apparently "national" character of the case, however, that makes it a good example of the multiscalar dynamics of international decision making. Despite the formal federal level of this case, both its actors and its arguments have subnational and supranational dimensions that are deeply intertwined with the science of climate change.

Section 1 draws from Holly Doremus's work on the use of science as a tool in, and an obstacle to, regulatory approaches and from Nathan Sayre's analysis of the concept of scale to consider the particular challenges posed by the multiscalar context of climate change. The section intertwines their theories to argue that both sides in *Massachusetts v. EPA* use scientific uncertainty together with the scale of the problem to forward their version of appropriate regulation. Section 2 then supports this argument through a detailed analysis of the interaction of scale, science, and law in the briefs and opinions. Section 3 examines the implications of that interaction for how this case should be fit into a model of international legal decision making

---

[3] Neil Brenner, New State Spaces: Urban Governance and the Rescaling of Statehood 9 (2004) (internal quotations omitted).
[4] Nathan F. Sayre, *Ecological and Geographical Scale: Parallels and Potential for Integration*, 29 Progress Hum. Geography 276, 281 (2005).

with respect to climate change. The chapter concludes with broader reflections on strategies for improving the way in which courts engage the scale-science confluence.

1. THE SCALE-SCIENCE INTERSECTION AS AN ARGUMENTATIVE TOOL

This section interweaves the work of two California-based academics: Holly Doremus, a law professor whose scholarship explores the way in which science is used in natural resource regulation,[5] and Nathan F. Sayre, a geographer whose recent scholarship has compared the analysis of scale in geography and ecology literatures.[6] This section summarizes each of their approaches and then interweaves them in the context of *Massachusetts v. EPA*.

1.1. *Defensive Uses of Scientific Uncertainty*

Holly Doremus's article *Science Plays Defense: Natural Resource Management in the Bush Administration* explains that the biggest difficulty regarding science and politics in natural resources management is not the politicization of science, but rather the scientizing of politics. Both conservationists and those who seek to block regulation can use science as a tool. Doremus notes: "The combination of actual uncertainty and public expectations of certainty makes the rhetoric of science equally available to the regulatory offense and defense."[7] She traces offensive and defensive uses of science and then explores four main ways in which the Bush administration used science defensively: high burden of proof, value choices in the face of ambiguity,

---

[5] See Holly Doremus, *Science Plays Defense: Natural Resource Management in the Bush Administration*, 32 Ecology L.Q. 249 (2005) [hereinafter *Science Plays Defense*]; Holly Doremus & A. Dan Tarlock, *Science, Judgment, and Controversy in Natural Resource Regulation*, 26 Pub. Land & Resources L. Rev. 1 (2005); Holly Doremus, *The Purposes, Effects, and Future of the Endangered Species Act's Best Available Science Mandate*, 34 Envtl. L. 397 (2004); Holly Doremus, *Listing Decisions under the Endangered Species Act: Why Better Science Isn't Always Better Policy*, 75 Wash. U. L.Q. 1029 (1997). For additional analyses of the intersection between law and science in public policymaking, see Adaptive Governance: Integrating Science, Policy, and Decision Making (Ronald D. Brunner et al. eds., 2005); Rescuing Science from Politics: Regulation and the Distortion of Scientific Research (Wendy Wagner & Rena Steinzor eds., 2006); Donald T. Hornstein, *Accounting for Science: The Independence of Public Research in the New, Subterranean Administrative Law*, 66 Law & Contemp. Probs. 227 (2003).

[6] See Sayre, *supra* note 4. A substantial geography literature engages these questions of scale and science. See, e.g., Louis Lebel, Po Garden & Masao Imamura, *The Politics of Scale, Position, and Place in the Governance of Water Resources in the Mekong Region*, 10 Ecology & Soc'y 18 (2005); James McCarthy, *Scale, Sovereignty, and Strategy in Environmental Governance*, 37(4) Antipode 731 (2005); Erik Swyngedouw, *Scaled Geographies: Nature, Place, and the Politics of Scale*, in Scale and Geographic Inquiry: Nature, Society, and Method 129 (Eric Shepard & Robert B. McMaster eds., 2004). An extensive review of that literature is beyond the scope of this brief chapter, though I plan to address it in more depth in future work. I choose to focus on Nathan Sayre's approach here, however, because of the particular way in which he interweaves ecological and scalar issues.

[7] Doremus, *Science Plays Defense*, *supra* note 5, at 258.

resolution of scientific certainty issues at the agency level, and limits to information gathering.[8]

These offensive and defensive strategies around science are apparent in the debates over climate change regulation in the United States. In fact, Doremus even quotes a memorandum from communication professional Frank Luntz on the topic to illustrate the defensive approach:

> The most important principle in any discussion of global warming is your commitment to sound science. Americans unanimously believe all environmental rules and regulations should be based on sound science and common sense. Similarly, our confidence in the ability of science and technology to solve our nation's ills is second to none. Both perceptions will work in your favor if properly cultivated.[9]

If Luntz is correct, a reinforcement of current scientific uncertainty and of the importance of waiting for future technological and scientific developments can serve as a powerful tool in blocking more stringent regulation of greenhouse gas emissions.

Moreover, as Doremus has explained, in judicial decision making, the framing of science is often outcome determinative.[10] The climate change context is no exception. The regulatory debates at the core of the arguments in *Massachusetts v. EPA*, discussed in detail in Section 2, exemplify the offensive and defensive uses of science that she has highlighted.

1.2. Debates over Scale

The arguments over science in *Massachusetts v. EPA*, however, consistently have a particular geographic dimension to them: scale. Both the geography and the ecology literatures, which engage scientific issues very much interconnected with climate change, have their own distinct discourses about scale. Sayre's recent article, *Ecological and Geographical Scale: Parallels and Potential for Integration*, attempts to interweave the two debates. He explains:

> In both ecology and human geography, the adequacy of research at any single scale is clearly in question, but the concept of scale itself remains unclear. Most participants in the debates acknowledge the need for studies that span multiple scales, and most conceive of different scales as being organized in some sort of hierarchical fashion. Within human geography, recent contributions have established several further points of general agreement: that scale is socially constructed and thus historically contingent, that it is politically contested, and that it is centrally important to understanding a variety of political, sociocultural, economic and environmental

---

[8] *Id.* at 266–95.
[9] *Id.* at 255 (quoting Luntz Research Companies, *Straight Talk, The Environment: A Cleaner, Safer, Healthier America* 138, available at http://www.luntzspeak.com/graphics/LuntzResearch.Memo.pdf (last visited June 16, 2008)).
[10] Doremus, *Science Plays Defense*, *supra* note 5.

phenomenon. The debate has foundered on basic conceptual and methodological questions, however. What exactly is scale? How should researchers theorize and use it?[11]

He goes on to argue that human geographers should draw three primary lessons from the ecologists' work on scale: that it is critical to distinguish between scale and level,[12] that rescaling processes are about "shift[ing] the level at which some process occurs" within "an existing structure of social organization,"[13] and that hierarchical models of scale can be misleading at times.[14]

Sayre's analysis of scale, like Doremus's discussion of the scientizing of politics, is reflected in the arguments of *Massachusetts v. EPA*. As discussed in detail in Section 2, both sides consistently try to (1) rescale, that is, change the relevant level for the argument, and (2) create hierarchies among levels – that is, assert the primacy of a particular level – in order to accomplish their goal of proving the appropriateness or inappropriateness of the EPA exercising its discretion to regulate motor vehicle greenhouse gas emissions.

1.3. Scale as a Lens on Science and the Law

The key point of this chapter is not simply that both scientizing and rescaling occur in this case, but rather that they are being used together to accomplish litigative goals. The large scale – both spatially and temporally – of climate change, and the resulting scientific uncertainties about subnational contributions to it and impacts from it, are combined by the respondents in an attempt to block regulatory behavior. In contrast, petitioners assert the appropriateness of nation-level regulation of supranational phenomenon and certainty around subnational contributions and effects to try to push for EPA action.[15]

These dynamics suggest that offensive and defensive strategies around science have particular nuances in multiscalar contexts in which relevant levels range from the individual to the global. Namely, the existence of multiple levels to jump and many possible arrangements of hierarchy allows for intersecting efforts at rescaling that place judges in a particularly difficult decision-making position. Moreover, the nexus of uncertainty around both science and scale creates additional judicial discretion and opportunities for litigants to attempt to manipulate the outcome.[16]

---

[11] Sayre, *supra* note 4, at 277–78.
[12] See id. at 283–85.
[13] See id. at 285.
[14] See id. at 286.
[15] See *infra* Section 2. As Holly Doremus has noted, the dynamics of this case represent only one variation of the intersection of scale, science, and regulation. In other contexts, such as debates over critical habitat, scaling down also can be an antiregulatory strategy because scientific uncertainty is often magnified at smaller scales. E-mail from Holly Doremus, Professor, UC Davis School of Law, to Hari Osofsky, Assistant Professor, University of Oregon School of Law (Mar. 20, 2007) (on file with author).
[16] Frederic Kirgis has explored similar issues in the context of legal formulas that contain two elements. In particular, he notes that courts and other decision makers are often unaware, or at least do not

2. THE COLLISION OF SCALE AND SCIENCE IN *MASSACHUSETTS V. EPA*

Massachusetts v. EPA involves the appropriateness of the U.S. EPA's denial of a petition requesting that it regulate motor vehicles' greenhouse gas emissions under section 202(a)(1) of the Clean Air Act.[17] The case is just one of many petitions and lawsuits engaging global climate change that have been filed around the world in subnational, national, and supranational fora. These litigative efforts tend to take two main approaches: (1) claims against governmental entities to force or limit regulatory behavior and (2) claims against corporate emitters to limit emissions directly. *Massachusetts v. EPA* falls into the first category. In both forms, the cases serve as part of state-corporate regulatory interactions around climate change.[18]

This Section explores the dynamics among scale, science, and law in the case. It considers the scales represented by the petitioners and respondents in the case, the use of science and scale in the claims by petitioners and respondents, and the implications of these approaches for efforts to use science as a tool for and against regulation.

2.1. Actors

The parties to *Massachusetts v. EPA* constitute a diverse group that cuts across scales. Twelve states, three cities, a U.S. territory, and thirteen nongovernmental organizations brought the petition. Ten other states and nineteen industry and utility groups – organized into six conglomerate entities – and the U.S. EPA served as respondents.[19]

These petitioners and respondents span numerous geographic regions at multiple levels of governance. The state and local level governmental petitioners tend to be located toward the coasts and respondents mostly are based in the middle of the

---

articulate an awareness, that they are using a sliding scale – "[t]he greater the degree to which one element is satisfied, the lesser the degree to which the other need be" – in such situations. Frederic L. Kirgis, *Fuzzy Logic and the Sliding Scale Theorem*, 53 ALA. L. REV. 421, 422–23 (2002).

[17] Massachusetts v. EPA, 549 U.S. 497 (2007).

[18] For a discussion of the geography of many of these suits, see Hari M. Osofsky, *The Geography of Climate Change Litigation: Implications for Transnational Regulatory Governance*, 83 WASH. U. L.Q. 1789 (2005) [hereinafter *Geography of Climate Change Litigation*]. For other analyses of climate change litigation, see, for example, JOSEPH SMITH & DAVID SHEARMAN, CLIMATE CHANGE LITIGATION: ANALYSING THE LAW, SCIENTIFIC EVIDENCE & IMPACTS ON THE ENVIRONMENT, HEALTH & PROPERTY (2006); RODA VERHEYEN, CLIMATE CHANGE DAMAGE AND INTERNATIONAL LAW, PREVENTION DUTIES AND STATE RESPONSIBILITY (2005); William C.G. Burns, *The Exigencies That Drive Potential Causes of Action for Climate Change Damages at the International Level*, 98 AM. SOC'Y INT'L L. PROC. 223 (2004); Richard W. Thackeray, Jr., Note, *Struggling for Air: The Kyoto Protocol, Citizens' Suits under the Clean Air Act, and the United States' Options for Addressing Global Climate Change*, 14 IND. INT'L & COMP. L. REV. 855, 884–98 (2004).

[19] A complete list of parties is available at International Center for Technology Assessment (ICTA), Global Warming Petitioners, http://www.icta.org/doc/global%20warming%20petitioners%20final.pdf (last visited June 16, 2008) [hereinafter *ICTA Parties Listing*].

country. The national-level governmental respondent, the U.S. EPA, is based in Washington, D.C., but has ten regional offices located in major cities throughout the country; it thus engages national policy issues through interacting in multiple places with various levels of government.[20] The nongovernmental entities similarly have a mix of local, state, national, and international ties.[21] And the preceding lists do not even include the many who filed amicus briefs or other actors engaged in responding to the Supreme Court's ruling.

From a scalar perspective, then, this case interacts with far more than simply the federal level at which it occurs. The actors reveal *Massachusetts v. EPA* as a situs for contestation across levels of governance between a wide variety of interested actors. As I have analyzed elsewhere, these dynamics pose difficult questions about how to locate this case in an analysis of transnational regulatory governance of climate change.[22]

2.2. *Claims*

The facts in this case involve the U.S. EPA's denial of a national-level rulemaking petition under a national-level law, the Clean Air Act, to address emissions by vehicles in places around the United States. These localized emissions contribute to the supranational phenomenon of climate change, which produces varying specific effects in particular places at a subnational level. The substantive and procedural claims made by the petitioners rely upon national-level statutes to address a situation that occurs across spatial and temporal scales. Moreover, this intersection of scalar issues and scientific data was at the core of both the standing and substantive issues debated in the U.S. Supreme Court.

2.2.1. Standing

Although standing was not one of the issues initially before the court,[23] the respondents raised it in their briefing and the Supreme Court justices discussed it

---

[20] EPA Organizational Chart, http://www.epa.gov/epahome/organization.htm (last visited June 16, 2008). Massachusetts v. EPA, 549 U.S. 497 (2007).
[21] For an in-depth discussion of those ties, see *Geography of Climate Change Litigation*, supra note 18, at 1830–34.
[22] *See* Hari M. Osofsky, *The Geography of Climate Change Litigation Part II: Narratives of Massachusetts v. EPA*, 8 CHI. J. INT'L L. 573 (2008); Hari M. Osofsky, *Climate Change Litigation as Pluralist Legal Dialogue?*, 26 STAN. ENVTL. L.J. & 43 STAN. J. INT'L L. 181 (2007) (Joint Issue).
[23] The questions presented in the petition for writ of certiorari were: "1. Whether the EPA Administrator may decline to issue emission standards for motor vehicles based on policy considerations not enumerated in section 202(a)(1). 2. Whether the EPA Administrator has authority to regulate carbon dioxide and other air pollutants associated with climate change under section 202(a)(1)." Petition for Writ of Certiorari at i, Massachusetts v. EPA, 549 U.S. 497 (2007) (No. 05-1120), 2006 WL 558353 at i.

extensively in oral argument. The Brief of the Federal Respondent claimed that the supranational and extended time scales of climate change limited the impact of national-level decisions to limit reductions:

> Global climate change is, by definition, a global phenomenon. The greenhouse gases at issue here are "fairly consistent in concentration, everywhere along the surface of the earth." The vast majority – as much as 80 percent – of all greenhouse gas emissions emanate from countries other than the United States. For that reason, reducing greenhouse gas emissions within the United States is unlikely, as a general matter, to have a significant long-term impact on climatic conditions in this country without reductions of greenhouse gas emissions in other parts of the world.[24]

The respondents further argue with respect to standing that the impacts at state and local levels are too speculative because of the extent of both the space and time involved. As the Brief for Respondents Alliance of Automobile Manufacturers, Engine Manufacturers Association, National Automobile Dealers Association, and the Truck Manufacturers Association (Brief for Respondents AAA) put it:

> [B]ecause they do not face any imminent injury, petitioners are forced to rely on predictions of harm decades in the future, the occurrence of which is largely (if not entirely) dependent on actions other nations take in their own regulation of greenhouse gas emissions. Petitioners' hypotheses, each of which is the subject of an active scientific debate, are reduced to conjecture by the inherent uncertainty of global events that will unfold between now and the time of the predicted injury.[25]

These claims by respondents thus use scientific uncertainty together with the alleged global scale of the problem to argue against the appropriateness of the petitioners being allowed to be before the Supreme Court.

The petitioners' reply to the standing argument rescales the issue back to the state and local levels and the present time. They note:

> Rising temperatures have injured petitioners in the following specific and concrete ways: coastal States have lost and are losing land to rising sea levels; ground-level ozone (smog) is exacerbated by rising temperatures, leading to adverse health effects and costly efforts on the part of States to address the problem; glaciers are melting, causing distinct injuries to particular individuals. These injuries span a broad range, from the Commonwealth of Massachusetts losing coastal land to Frank Keim no longer being able to hike on the Alaskan glaciers he used to enjoy.
>
> Petitioners' injuries are not "some day" injuries, as respondents contend; they are injuries in the here and now. Nor do petitioners' declarations describe mere

---

[24] Brief for Federal Respondent at 13, Massachusetts v. EPA, 549 U.S. 497 (No. 05–1120), 2006 WL 3043970, at *13 (citation omitted).
[25] Brief for Respondents Alliance of Auto. Mfrs., Engine Mfrs. Ass'n, Nat'l Auto. Dealers Ass'n, Truck Mfrs. Ass'n at 13, Massachusetts v. EPA, 549 U.S. 497 (No. 05–1120), 2006 WL 3023028 at *13 (citations omitted).

"generalized grievances"; they attest to harms being visited – right now – upon particular individuals and particular States.[26]

This reply relies on the same scientific data set but, by scaling down the argument, engages the alleged injuries in ways that tie them more easily to legal standing requirements.

The Supreme Court's opinion sides with the petitioners and indicates that the "widely shared" character of climate-change risks does not prevent Massachusetts from having an interest in the case's outcome.[27] It concludes the standing analysis:

> In sum – at least according to petitioners' uncontested affidavits – the rise in sea levels associated with global warming has already harmed and will continue to harm Massachusetts. The risk of catastrophic harm, though remote, is nevertheless real. That risk would be reduced to some extent if petitioners received the relief they seek. We therefore hold that petitioners have standing to challenge the EPA's denial of their rulemaking petition.[28]

Although the Court's holding on standing narrowly focuses on the interests of state parties, its approach to them scales down the problem of climate change and its regulation; this "global" phenomenon can cause harm at a state level and choices at a federal level influence the risks faced by states.

The dissenters, unsurprisingly, side with the respondents. Chief Justice Roberts's dissent, for example, explains how, in his view, the multiscalar nature of the problem defeats standing.

> The Court's sleight-of-hand is in failing to link up the different elements of the three-part standing test. What must be *likely* to be redressed is the particular injury in fact. The injury the Court looks to is the asserted loss of land. The Court contends that regulating domestic motor vehicle emissions will reduce carbon dioxide in the atmosphere, *and therefore* redress Massachusetts's injury. But even if regulation *does* reduce emissions – to some indeterminate degree, given events elsewhere in the world – the Court never explains why that makes it *likely* that the injury in fact – the loss of land – will be redressed.[29]

In so doing, Chief Justice Roberts articulates his concerns about whether the occurrence of emissions around the world (essentially, local emissions taking place at a global scale) makes the impact of U.S. national-level regulatory behavior less clear at a subnational scale.

At the core of this battle over standing lies scientific data. Both sides acknowledge the problem of climate change, but they part ways in how to map the scientific information, and its uncertainties, onto existing legal structures. As emissions and their impacts connect to multiple levels of governance, the parties and Court are

---

[26] Reply, Massachusetts v. EPA, 549 U.S. 497 (No. 05–1120), 2006 WL 3367871, at *2–*3 (citations omitted).
[27] Massachusetts v. EPA, 549 U.S. 497, 522 (2007).
[28] *Id.* at 526.
[29] *Id.* at 546.

forced to grapple with how to apply the more simply structured standing doctrine to this problem.

2.2.2. Substantive Claims

The substantive arguments reveal a similar dynamic of scaling climate change and regulatory authority over it up and down. For example, the respondents claim that states cannot implement National Ambient Air Quality Standards (NAAQS) in this context because their regulatory level fails to match the global level at which the problem was occurring. The brief of respondent CO_2 litigation group argues:

> None of these regulatory authorities makes sense if the "air pollutant" to which they are applied is CO_2 or another greenhouse gas being regulated for the purpose of mitigating potential global climate change. Since the projected effect of greenhouse gas emissions is a function of changes in the global atmosphere, rather than local or regional air quality, and it is the aggregate contribution of all greenhouse gas emissions around the world to global atmospheric greenhouse gas contributions that is believed by many to cause global climate change, notions of attaining or not attaining an ambient air quality standard within a state or air quality control region are inapplicable.[30]

The theme of scientific uncertainty is intertwined with the claim of scalar mismatch, as represented in language like "believed by many" in that statement. As with the standing argument, respondents are portraying climate change as something occurring at a supranational level and over a long period of time with substantial deficits in current understanding about how anthropogenic greenhouse gas emissions fit into that model.

The petitioners' argument on this point, in contrast, relies upon the various levels at which the Clean Air Act provides regulatory authority. They note in their opening brief:

> Whatever question exists about the applicability of the NAAQS program to the air pollutants at issue here cannot excuse the failure to adopt emission standards under section 202. Section 202 *does* provide a perfectly feasible mechanism for regulating emission of these pollutants from motor vehicles: the establishment of the same sort of limits on these pollutants that EPA has already imposed on pollutants such as carbon monoxide and hydrocarbons.[31]

In other words, regulation can work according to the petitioners if one changes levels – to the national one – and type of regulatory approach.

A similar debate among the parties takes place over whether Congress's specific action with respect to ozone limits EPA's ability to regulate prior to a similar type of action regarding global climate change. The Brief for Respondents AAA argues:

---

[30] Brief for Respondent CO_2 Litigation Group at 20, Massachusetts v. EPA, 549 U.S. 497 (No. 05–1120), 2006 WL 3043971, at *20 (citation omitted).

[31] Brief for Petitioners, Massachusetts v. EPA, 549 U.S. 497 (No. 05–1120), 2006 WL 2563378, at *29 (emphasis original).

Congress has previously dealt with emissions issues relating to non-localized gases that implicate global environmental concerns. For example, when Congress addressed stratospheric ozone depletion it used an express delegation under a new regulatory framework: Title VI of the Clean Air Act. The addition of Title VI to address global issues reflects Congress's views about the regulatory limits of Titles I and II of the Act.

Much like carbon dioxide, anthropogenic substances that deplete ozone are emitted around the world and are very long-lived. Their upper-atmosphere ozone depleting effects – and the consequences of those effects – occur on a global scale.[32]

This approach indicates a presumption that similarities in the scale and time frame of two problems, as described in the existing scientific literature, means that a Congressional approach to one of them limits regulatory discretion with respect to another.

The petitioners, unsurprisingly, resist such an interpretation of the ozone legislation. Beyond arguing that the ozone provisions have been used to regulate "air pollutants associated with climate change," they note:

EPA cannot seriously maintain that "coordination with the international community" is a prerequisite for regulating pollutants that "are emitted around the world and are very long-lived," the consequences of which "occur on a global scale." Congress directed EPA to regulate ozone-depleting substances themselves without awaiting such coordination.[33]

The petitioners thus use the same analogy between ozone and global climate change to indicate that national-level regulation of multiscalar problems is appropriate.

As with the standing issue, the majority opinion substantively sides with the petitioners over a vigorous dissent. It holds that Clean Air Act section 202(a)(1), read together with the Act's broad definition of "air pollutant," gives the EPA statutory authority to regulate greenhouse gas emissions from motor vehicles.[34] Moreover, the Court rejects the EPA's alternative argument that even if it has statutory authority, it should not exercise it.[35] In so doing, the opinion notes that the agency cannot avoid its regulatory responsibilities simply by invoking scientific uncertainty. Rather, the EPA must address the statutory question of whether "sufficient information exists to make an endangerment finding."[36]

Although Chief Justice Roberts's dissent engages only the standing question, Justice Scalia's dissent – joined by the other three dissenting judges – addresses the merits. Justice Scalia's dissent begins by arguing that EPA's discretion is broader than the majority holds,[37] but then further indicates that the majority is wrong on

---

[32] Brief for Respondents Alliance of Auto. Mfrs., Engine Mfrs. Ass'n, Nat'l Auto. Dealers Ass'n, Truck Mfrs. Ass'n, *supra* note 25, at 38–39 (emphasis and citation omitted); *accord* Brief for Federal Respondent, *supra* note 24, at 27–30.
[33] Brief for the Petitioners, Massachusetts v. EPA, 549 U.S. 497 (No. 05–1120), 2006 WL 2563378, at *27 (citation omitted).
[34] Massachusetts v. EPA, 549 U.S. at 527–32.
[35] *Id.* at 532–35.
[36] *Id.* at 534.
[37] *Id.* at 549–53.

its own terms because of the EPA's statements on scientific uncertainty.[38] Its final argument addresses scale even more clearly through arguing against the majority's broad interpretation of "air pollutant."[39] In particular, the dissent focuses quite literally on the question of the part of the atmosphere in which "pollution" resides. Because greenhouse gases build up in the upper atmosphere, the dissent claims that the EPA's exclusion of them through focusing on "ambient air at ground level or near the surface of the earth" is statutorily consistent.[40]

Together, the actors and arguments in this case demonstrate the judicial challenge that the collision of scientific uncertainty and multiscalar regulatory problems poses. Although the parties used particular conceptions of that intersection in their argument, the briefs and arguments are not explicit about the fact that the U.S. Supreme Court's selection of scalar perspective would influence how the scientific data should be viewed. Similarly, choices about the scale of climate change and its regulation run through the discourse among the majority and dissenting opinions, but those decisions are often buried in the legal analysis.

3. IMPLICATIONS FOR INTERNATIONAL LEGAL DECISION MAKING

The strategic use of science with scale in *Massachusetts v. EPA*, especially when not explicit, suggests dangers for the way in which decision making that has supranational dimensions tends to be dichotomized. In particular, the Balkanization of both scalar and identity categories allows for distorting efforts at rescaling. This section focuses on three types of divisions that not only are inaccurate descriptors in a multiscalar, multiactor framework but also provide the basis for the political games being played in the case.

3.1. Domestic vs. International

Is *Massachusetts v. EPA* domestic or international?[41] The case clearly was brought under domestic law and many of the petitioners are domestic governmental actors, but simply characterizing it as a domestic case does not encompass all of the scales involved. As was repeatedly expressed by parties on both sides of the litigation, the case involves a problem and broader law and policy discourse that have international dimensions.[42]

---

[38] *Id.* at 553–55.
[39] *Id.* at 555–60.
[40] *Id.* at 560 (internal quotation marks omitted).
[41] For an explication of the traditional Westphalian perspective on international law, see IAN BROWNLIE, PRINCIPLES OF PUBLIC INTERNATIONAL LAW 287–88 (6th ed. 2003); *see also* Michael J. Kelly, *Pulling at the Threads of Westphalia: "Involuntary Sovereignty Waiver," Revolutionary International Legal Theory or Return to Rule by the Great Powers?*, 10 UCLA J. INT'L L. & FOREIGN AFF. 361 (2005).
[42] For an interesting analysis of the increasingly blurry boundaries between domestic and international, see Judith Resnik, *Law's Migration: American Exceptionalism, Silent Dialogues, and Federalism's Multiple Ports of Entry*, 115 YALE L.J. 1564 (2006).

Neither "domestic" nor "international" conveys fully the multiscalar character of the case, and a notion that there is an appropriate regulatory level, either domestic or international, fails to capture the many levels at which climate change must be regulated. Moreover, the domestic/international distinction privileges the national level at which the case is taking place by using it as the fulcrum point between relevant categories. Using this dichotomy as a frame thus plays a distorting role in a discourse over problems like climate change.[43]

3.2. Local vs. State vs. Federal

Similarly, if domestic, is *Massachusetts v. EPA* simply federal?[44] The case was brought in a federal court and involved the regulatory discretion of a federal actor, but in both its actors and its claims, it involved many other scales and places associated with them in the United States.[45] After all, a good portion of the previously described debate involved state and local actors, regulatory decisions, and impacts. Moreover, the distinction – local vs. state vs. federal – fails to capture the nuances of the levels involved or the fact that multiple levels are involved in every aspect of the discourse.

This point becomes even clearer if this case is viewed in the broader context of climate change litigation and policy. For example, as I have discussed in depth elsewhere,[46] California is not only a plaintiff in *Massachusetts v. EPA* but also a plaintiff or defendant in several other resolved and pending cases involving climate change, some of which specifically focus on motor vehicle emissions.[47] Moreover,

---

[43] I explore these issues in more depth and in additional contexts in Hari M. Osofsky, *Is Climate Change "International"?: Litigation's Diagonal Regulatory Role*, 49 VA. J. INT'L L. 585 (2009).

[44] For examples of broader federalism debates in the context of environmental regulations, see Kirsten H. Engel, *State Environmental Standard-Setting: Is There a "Race" and Is It "to the Bottom"?*, 48 HASTINGS L.J. 271 (1997); Daniel C. Esty, *Revitalizing Environmental Federalism*, 95 MICH. L. REV. 570 (1996); Joshua D. Sarnoff, *The Continuing Imperative (But Only from a National Perspective) for Federal Environmental Protection*, 7 DUKE ENVTL. L. & POL'Y F. 225 (1997); Peter P. Swire, *The Race to Laxity and the Race to Undesirability: Explaining Failures in Competition Among Jurisdictions in Environmental Law*, 14 YALE J. ON REG. 67 (1996); Henry N. Butler & Jonathan R. Macey, *Externalities and the Matching Principle: The Case for Reallocating Environmental Regulatory Authority*, 14 YALE L. & POL'Y REV. 23 & 14 YALE J. ON REG. 23 (1996); Richard L. Revesz, *Rehabilitating Interstate Competition: Rethinking the "Race-to-the-Bottom" Rationale for Federal Environmental Regulation*, 67 N.Y.U. L. REV. 1210 (1992); Richard L. Revesz, *The Race to the Bottom and Federal Environmental Regulation: A Response to Critics*, 82 MINN. L. REV. 535 (1997); Richard B. Stewart, *Environmental Regulation and International Competitiveness*, 102 YALE L.J. 2039 (1993).

[45] For an interesting analysis of the complexities of regulation at multiple scales, see William W. Buzbee, *Recognizing the Regulatory Commons: A Theory of Regulatory Gaps*, 89 IOWA L. REV. 1 (2003).

[46] See Osofsky, *Climate Change as Pluralist Legal Dialogue?*, supra note 22.

[47] See, e.g., First Amended Complaint for Declaratory and Injunctive Relief, Cen. Valley Chrysler-Jeep v. Witherspoon, 456 F. Supp. 2d 1160 (E.D. Cal. 2006), 2004 WL 5001055; Complaint, Connecticut v. Am. Elec. Power Co., 406 F. Supp. 2d 265 (S.D.N.Y. 2005) (Nos. 04 Civ. 5669(LAP), 04 Civ. 5670(LAP)), available at http://caag.state.ca.us/newsalerts/2004/04-076.pdf; Complaint for Damages and Declaratory Judgment, State of California v. Gen. Motors Corp., No. C06-05755 MJJ (N.D. Cal. Sept. 17, 2007), available at http://ag.ca.gov/newsalerts/cms06/06-082_oa.pdf (last visited June 17, 2008); Petition for Review, State of California v. NHTSA, No. 06-2654 SC (N.D. Cal. June 12,

California's representatives in Congress are playing leadership roles in efforts to regulate emissions more aggressively,[48] and its cities are engaging in both litigation[49] and their own regulatory efforts.[50] Divorcing *Massachusetts v. EPA* from that multiscalar context decontextualizes the case in ways that portray its significance inaccurately.

3.3. Public vs. Private

Finally, is this litigation about public or private decision making?[51] Because this case involves the behavior of a federal regulator, one could argue that it is a public law case. But such a view of the case would suffer from some of the same flaws as the other two efforts to categorize it.

2007), *available at* http://ag.ca.gov/newsalerts/cms06/06-046_oa.pdf (May 2, 2006) (last visited June 17, 2008); Non-Binding Statement of Issues of Petitioners, Coke Oven Envtl. Task Force v. EPA, Case No. 06-1131 (Sept. 3, 2003).

[48] *See, e.g.*, Press Release, Pelosi and Reid: We Should Work Together to Take America in a New Direction (Jan. 27, 2007), *available at* http://www.speaker.gov/newsroom/pressreleases?id=0047 (last visited June 17, 2008); NPR Talk of the Nation: Is U.S. Energy Independence a Pipe Dream? (Jan. 24, 2007), *available at* http://www.npr.org/templates/story/story.php?storyId=7002504 (last visited June 17, 2008) ("Today Speaker of the House Nancy Pelosi upped the ante and called for energy independence within 10 years."); Press Release, *Boxer, Bingaman and Lieberman Ask President to Commit to Working with Congress to Fight Global Warming* (Nov. 15, 2006), *available at* http://boxer.senate.gov/news/releases/record.cfm?id=265906&& (last visited June 17, 2008).

[49] *See* Complaint for Declaratory and Injunctive Relief (Second Amended), Friends of the Earth, Inc., v. Watson, No. 02-4106 (N.D. Cal. Sept. 3, 2002), *available at* http://www.climatelawsuit.org/documents/Complaint_2Amended_Declr_Inj_Relief.pdf (last visited June 7, 2008).

[50] *See* ICLEI website, Regional Membership Lists by Country, http://www.iclei.org/index.php?id=1387&region=NA (last visited June 17, 2008); ICLEI website, CCP: Participants, http://www.iclei.org/index.php?id=1121 (last visited June 17, 2008); Fact Sheet, California Climate Activities, http://www.climatechange.ca.gov/climate_action_team/factsheets/2005-06_CLIMATE-ACTIVITIES_FS.PDF (last visited June 17, 2008); City of Los Angeles website, Council Actions, http://www.lacity.org/ead/EADWeb-AQD/council_actions.htm (last visited June 17, 2008); City of Los Angeles website, Awards Received, http://www.lacity.org/ead/EADWeb-AQD/awards_received.htm (last visited June 17, 2008); Tomas Alex Tizon, *Mayor Is on a Mission to Warm U.S. Cities to the Kyoto Protocol*, L.A. TIMES, Feb. 22, 2005, at A15. For scholarly analysis of the state and local dimensions of climate change regulation, see BARRY G. RABE, STATEHOUSE AND GREENHOUSE: THE EMERGING POLITICS OF AMERICAN CLIMATE CHANGE POLICY (2004); Donald A. Brown, *Thinking Globally and Acting Locally: The Emergence of Global Environmental Problems and the Critical Need to Develop Sustainable Development Programs at State and Local Levels in the United States*, 5 DICK. J. ENVTL. L & POL'Y 175 (1996); Ann E. Carlson, *Federalism, Preemption, and Greenhouse Gas Emissions*, 37 U.C. DAVIS L. REV. 281 (2003); David R. Hodas, *State Law Responses to Global Warming: Is It Constitutional to Think Globally and Act Locally?*, 21 PACE ENVTL. L. REV. 53 (2003); Laura Kosloff & Mark Trexler, *State Climate Change Initiatives: Think Locally, Act Globally*, 18 NAT. RESOURCES & ENV'T 46 (Winter 2004); Robert B. McKinstry, Jr., *Laboratories for Local Solutions for Global Problems: State, Local and Private Leadership in Developing Strategies to Mitigate the Causes and Effects of Climate Change*, 12 PENN ST. ENVTL. L. REV. 15 (2004); Hari M. Osofsky, *Local Approaches to Transnational Corporate Responsibility: Mapping the Role of Subnational Climate Change Litigation*, 20 PAC. MCGEORGE GLOBAL BUS. & DEV. L.J. 143 (2007); Barry G. Rabe, *North American Federalism and Climate Change Policy: American State and Canadian Provincial Policy Development*, 14 WIDENER L.J. 121 (2004); Resnik, *supra* note 42, at 1643–47.

[51] For a historical perspective on the evolution of the public/private distinction in a local government context, see Gerald Frug, *A Legal History of Cities*, *in* THE LEGAL GEOGRAPHIES READER 154 (Nicholas Blomley, David Delany & Richard T. Ford eds., 2001).

A mix of public and private actors appears on both sides of the lawsuit in *Massachusetts v. EPA*, and in other instances of climate change litigation. Moreover, some of the cases over vehicle emissions focus on governmental regulatory decisions, and others focus on emissions decisions of private actors directly.[52] To fail to see these cases as involving a state-corporate regulatory dynamic would be just as flawed as ignoring California's critical role in the multiscalar dialogue about climate change.

As a wide range of actors operate across scales and play multidimensional roles in the policy and lawmaking debate, *Massachusetts v. EPA* becomes one step in a complex dance. This reality creates a risk that traditional ways of categorizing the case – which might focus on it as simply a public environmental regulatory dispute – will miss critical elements of what it is.

4. CONCLUDING REFLECTIONS: STRATEGIES FOR MANAGING THE CONFLUENCE

The confluence of scale, science, and law in *Massachusetts v. EPA* does not simply challenge our conventional approaches to categorization but also suggests important strategies for managing these ever-more-common convergences better. In particular, the lack of explicit acknowledgment of battles at this intersection has troubling implications for judicial decision making. When petitioners and respondents are scaling up and scaling down without acknowledging it, judicial discretion is increased tremendously. The scalar lens that the court chooses may well be outcome-determinative and may not reflect a great deal of consciousness about the ways in which the framing influenced the decision.

Unfortunately, efforts to engage scale more directly may not actually fix this problem. If parties spotlight the way in which scale and science are being used, the court may make its decision more reflectively. However, the outcome will not necessarily vary much. Judges may well choose the same scalar framing that they were initially inclined toward selecting. As Holly Doremus's work makes clear, even if scale and science are approached more consciously, the scientizing and scaling of politics may be unavoidable.

Under such a view, explicit acknowledgment of the ways in which scale, science, and law interact would simply change the words that advocates and judges use. Both sides likely will continue to use scale and science in tandem both offensively and defensively, and the arguments about why a particular framing is appropriate would simply become more explicit and nuanced. Moreover, the lack of university-level

---

[52] I have discussed this dynamic in depth in Osofsky, *supra* note 18, at 1796–97; *see also* Robert Dufresne, *The Opacity of Oil: Oil Corporations, Internal Violence, and International Law*, 36 N.Y.U. J. INT'L L. & POL. 331 (2004). For an interesting analysis of corporate responsibility in the context of indigenous peoples' land rights, see Lillian Aponte Miranda, *The Hybrid State-Corporate Enterprise and Violations of Indigenous Land Rights: Theorizing Corporate Responsibility and Accountability under International Law*, 11 LEWIS & CLARK L. REV. 135 (2007); *see also* Hari M. Osofsky, *Learning from Environmental Justice: A New Model for International Environmental Rights*, 24 STAN. ENVTL. L.J. 71, 72–76 (2005).

exposure to geography among many elites in the United States[53] may cause resistance to a deeper engagement of the nuances of scale and feed the politicization of its confluence with science and law.

Even so, I think that an active effort to engage this intersection more systematically would be an improvement over the status quo. When assumptions are allowed to control discourse without conscious acknowledgment, the possibilities for political manipulation of science are heightened. If courts and litigants engage the confluence of law, scale, and science more thoughtfully, the potential for an adequate regulatory discourse over complex issues like climate change improves. At the very least, an explicit dialogue about these issues might help lawyers and judges become more comfortable with the dynamics underlying cases like *Massachusetts v. EPA*. Given the growing climate crisis and the Obama administration's commitment to greater federal regulatory efforts, which include rapidly responding to the *Massachusetts v. EPA* decision, further exploration of these scalar questions is critical.[54]

---

[53] Beginning with Harvard in 1948, many U.S. universities eliminated their geography departments or failed to constitute them. Alexander Murphy, *Geography's Place in Higher Education in the United States*, 31 J. GEOGRAPHER HIGHER ED. 121, 122–23 (2007); William A. Koelsch, *Academic Geography, American Style: An Institutional Perspective, in* GEOGRAPHY: DISCIPLINE, PROFESSION AND SUBJECT SINCE 1870: AN INTERNATIONAL SURVEY 245, 270 (Gary S. Dunbar ed., 2001); *see also* Thomas J. Wilbanks & Michael Libbee, *Avoiding the Demise of Geography in the United States*, 31 PROF. GEOGRAPHER 1, 1 (1979). A recent study indicates that 93% of U.S. liberal arts institutions lack geography departments. Mark D. Bjelland, *A Place for Geography in the Liberal Arts College?*, 56 PROF. GEOGRAPHER 326, 326 (2004). I have discussed this issue in depth in Hari M. Osofsky, *A Law and Geography Perspective on the New Haven School*, 32 YALE J. INT'L L. 421 (2007).

[54] For the Obama administration's efforts in response to *Massachusetts v. EPA*, see Proposed Endangerment and Cause or Contribute Findings for Greenhouse Gases Under Section 202(a) of the Clean Air Act; Proposed Rule, 74 Fed. Reg. 18885 (proposed Apr. 24, 2009) (to be codified at 40 C.F.R. ch. 1); John M. Broder, *E.P.A. Clears Way for Greenhouse Gas Rules*, N.Y. TIMES, Apr. 17, 2009 at A15, *available at* http://www.nytimes.com/2009/04/18/science/earth/18endanger.html. I am exploring these issues of scale in more depth in Osofsky, *Is Climate Change "International"?*, *supra* note 43; Hari M. Osofsky, *Diagonal Climate Regulation: Implications for the Obama Administration* (draft on file with author); and Hari M. Osofsky, *Scales of Law: Rethinking Climate Change, Terrorism, and the Global Financial Crisis* (draft précis of monograph on file with author).

7

Biodiversity, Global Warming, and the United States Endangered Species Act: The Role of Domestic Wildlife Law in Addressing Greenhouse Gas Emissions

Brendan R. Cummings* and Kassie R. Siegel**

INTRODUCTION

On May 15, 2008, the U.S. Fish and Wildlife Service, an executive branch agency within the Department of Interior, issued a regulation[1] listing[2] the polar bear under the Endangered Species Act (ESA),[3] a federal statute designed to prevent the extinction of imperiled plants and animals, due to global warming[4] and the melting of the bear's sea-ice habitat. This highly publicized milestone firmly cemented the polar bear as the iconic example of the devastating impacts of global warming on the planet's biodiversity.[5] While the polar bear is the most well-known species imperiled by global warming, and the first to be listed under the ESA *solely* due to this factor, it was not the first species protected under the statute in which global warming played a significant role. Two years earlier, on May 9, 2006, the National Marine Fisheries Service, declared two species of Caribbean coral Threatened under the ESA.[6]

---

* Center for Biological Diversity, P.O. Box 549, Joshua Tree, CA 92252, 760–366-2232, bcummings@biologicaldiversity.org.
** Center for Biological Diversity, P.O. Box 549, Joshua Tree, CA 92252, 760–366-2232, ksiegel@biologicaldiversity.org.
[1] Endangered and Threatened Wildlife and Plants, Determination of Threatened Status for the Polar Bear (Ursus maritimus) throughout Its Range, 73 FED. REG. 28,212 (May 15, 2008) (to be codified at 50 C.F.R. § 17).
[2] A species does not receive the protections of the ESA until, following an administrative rulemaking process, a regulation is promulgated adding the species to the official List of Threatened and Endangered Species. 50 C.F.R. § 17.11 (2007) (list of "threatened" and "endangered" species; *see also* 16 U.S.C. § 1533 (ESA listing process).
[3] 16 U.S.C. §§ 1531–1544.
[4] Throughout this chapter we use the phrase "global warming" to refer to anthropogenic climate change resulting from greenhouse gas emissions, recognizing that the resultant climate impacts to a particular region or ecosystem are often more complex and may involve other factors (e.g., changes in rainfall, cloud cover, storm frequency, etc.) in addition to a rise in ambient air temperature.
[5] *See, e.g.*, Jeffrey Kluger, *Polar Ice Caps Are Melting Faster Than Ever . . . More and More Land Is Being Devastated by Drought . . . Rising Waters Are Drowning Low-Lying Communities . . . By Any Measure, Earth Is at . . . the Tipping Point*, TIME, Apr. 3, 2006, at cover page (Cover photograph of polar bear).
[6] Endangered and Threatened Species: Final Listing Determinations for Elkhorn Coral and Staghorn Coral, 71 FED. REG. 26,852 (May 9, 2006) (to be codified at 50 C.F.R. § 223).

The ESA has been declared by the U.S. Supreme Court to be "the most comprehensive legislation for the preservation of endangered species ever enacted by any nation."[7] The statute is widely considered to be one of the strongest environmental laws in the United States, and hence one of the most controversial.[8] The listing under the ESA of species threatened by global warming raises the possibility of applying this law, which "admits of no exception,"[9] and which affords endangered species "the highest of priorities,"[10] to the seemingly intractable issue of reducing U.S. greenhouse gas emissions.

In this chapter we place the listings of the corals and polar bear in context. We begin by providing an overview of the ESA, including its overarching objectives and key provisions. We then discuss how the ESA should operate to protect species imperiled by global warming and create an obligation on U.S. federal agencies and corporations to reduce greenhouse gas emissions. We use the examples of coral species and the polar bear to explore the possibilities and limitations of using domestic wildlife law such as the ESA to bring the subject of global warming into the courtroom and address otherwise unregulated greenhouse gas emissions.

1. THE ENDANGERED SPECIES ACT AS A MECHANISM TO ADDRESS GREENHOUSE GAS EMISSIONS

1.1. *The Endangered Species Act: Overview*

In the seminal ESA case, *Tennessee Valley Authority v. Hill* (TVA),[11] the Supreme Court held that the ESA's unequivocal mandate that federal agencies "insure" that their actions do not "jeopardize" any species protected by the statute, meant that a multimillion-dollar dam project already near completion could not proceed because its completion threatened the existence of the snail darter, a small endemic fish of no known economic value.[12] In so doing, the Supreme Court elevated a little-known statute that had passed Congress with near unanimity into one of the most powerful and controversial environmental statutes in the United States. In the three decades since TVA was decided, courts enforcing the ESA have halted such activities as logging, to protect threatened owls;[13] commercial fishing, to protect Hawaiian monk seals;[14] military activities, to protect endangered whales;[15] oil and gas development,

---

[7] Tenn. Valley Auth. v. Hill (TVA), 437 U.S. 153, 180 (1978).
[8] THE ENDANGERED SPECIES ACT AT THIRTY: RENEWING THE CONSERVATION COMMITMENT (Dale D. Goble, J. Michael Scott & Frank W. Davis eds.) (2006).
[9] TVA, 437 U.S. at 173.
[10] Id. at 174.
[11] Id.
[12] Id. at 171–93.
[13] Silver v. Babbitt, 924 F. Supp. 972, 976 (D. Ariz. 1995); Lane County Audubon Soc'y v. Jamison, 958 F.2d 290, 295 (9th Cir. 1992).
[14] Greenpeace Found. v. Mineta, 122 F. Supp. 2d 1123, 1138 (D. Haw. 2000).
[15] Natural Res. Def. Council v. Evans, 279 F. Supp. 2d 1129 (N.D. Cal. 2003).

to protect grizzly bears;[16] off-road vehicles, to protect imperiled plants;[17] pesticide authorizations, to protect imperiled salmon;[18] and numerous other habitat-damaging activities that threatened a particular protected species. In granting such injunctive relief, courts have repeatedly found that "[g]iven a substantial procedural violation of the ESA in connection with a federal project, the remedy must be an injunction of the project pending compliance with the ESA."[19] Regardless of the economic consequences of halting a given project, protection of the species must receive precedence.[20]

The two primary mechanisms by which the ESA protects listed species are contained in sections 7 and 9 of the statute. Section 7 directs all federal agencies to "insure through consultation with the Secretary"[21] that all actions authorized, funded, or carried out by such agencies are "not likely to jeopardize the continued existence" or "result in the destruction or adverse modification" of "critical habitat" of any listed species.[22] In contrast to the National Environmental Policy Act (NEPA)[23] – which requires only informed agency decision-making and not a particular result,[24] and is therefore strictly procedural – section 7 of the ESA contains both procedural ("through consultation") and substantive ("insure" the action does not "jeopardize") mandates for federal agencies. As such, the statute, and litigation under it, can force analysis through the consultation process of the environmental effects of a given project and, if the project is determined to jeopardize a listed species or adversely modify its critical habitat, trigger modification or cancellation of the project so as to avoid such impacts.

While section 7 applies only to the actions of federal agencies, the prohibitions of section 9[25] apply to "any person," including federal, state, and local agencies and

[16] Conner v. Burford, 848 F.2d 1441 (9th Cir. 1988).
[17] Ctr. for Biological Diversity v. Bureau of Land Mgmt., 422 F. Supp. 2d 1115 (N.D. Cal. 2006).
[18] Washington Toxics Coal. v. Envtl. Prot. Agency, 413 F.3d 1024 (9th Cir. 2005).
[19] Thomas v. Peterson, 753 F.2d 754, 764 (9th Cir. 1985).
[20] Sierra Club v. Marsh, 816 F.2d 1376, 1387 (9th Cir. 1987).
[21] The "Secretary" refers to either the Secretary of Interior or the Secretary of Commerce depending on the species at issue. 16 U.S.C. § 1532(15). The Secretaries have delegated authority to the U.S. Fish and Wildlife Service ("FWS") and National Marine Fisheries Service ("NMFS"), respectively. 50 C.F.R. § 402.01(b). FWS has authority over all terrestrial species, while NMFS manages most marine species. However, several marine mammals, including the polar bear, are managed by FWS.
[22] 16 U.S.C. § 1536(a)(2). To "jeopardize" a species is defined by regulation as "to engage in an action that reasonably would be expected, directly or indirectly, to reduce appreciably the likelihood of both the survival and recovery of a listed species in the wild by reducing the reproduction, numbers, or distribution of that species." 50 C.F.R. § 402.02. A similar regulatory definition of "destroy or adversely modify critical habitat" has been struck down by several courts as not properly encompassing recovery of the species. See, e.g., Gifford Pinchot Task Force v. U.S. Fish and Wildlife Serv., 378 F.3d 1059, 1069–71 (9th Cir. 2004). While no replacement regulatory definition has yet been promulgated, courts have made clear that an agency action in critical habitat cannot compromise the species' recovery and still rationally be deemed to have not "adversely modified" that habitat. See, e.g., Ctr. for Biological Diversity v. Bureau of Land Mgmt., 422 F. Supp. 2d 1115, 1136 (N.D. Cal. 2006).
[23] 42 U.S.C. §§ 4321–4347.
[24] Robertson v. Methow Valley Citizens Council, 490 U.S. 332, 350 (1989).
[25] 16 U.S.C. § 1538.

entities, individuals, and corporations.[26] Section 9 prohibits, inter alia, the "taking" of any endangered species in the United States or upon the high seas.[27] Regulations promulgated pursuant to section 4(d) apply most of the take prohibitions applicable to endangered species to threatened species as well.[28] "Take" means "to harass, harm, pursue, hunt, shoot, wound, kill, trap, capture, or collect, or to attempt to engage in any such conduct."[29] "Harass" is further defined as any "act or omission which creates the likelihood of injury to wildlife by annoying it to such an extent as to significantly disrupt normal behavior patterns which include, but are not limited to, breeding, feeding or sheltering."[30] "Harm" includes "significant habitat modification or degradation where it... injures wildlife by significantly impairing essential behavioral patterns, including breeding, feeding or sheltering."[31]

The ESA's legislative history supports "the broadest possible" reading of "take,"[32] and courts have consequently found violations of section 9 from activities ranging from direct intentional killing of a listed species,[33] to harm resulting from habitat degradation,[34] to government authorizations of activities that inevitably would result in prohibited take, such as pesticide use[35] or fishing.[36] In perhaps the most expansive reading of section 9's reach to date, one appellate court found that "inadequate regulation" of light pollution could make a local government liable for the take of listed sea turtles.[37]

In addition to the prohibitions against jeopardy and take provided by sections 7 and 9, the ESA mandates an array of affirmative conservation actions for listed species. These include the designation of "critical habitat,"[38] the development and implementation of recovery plans,[39] the acquisition of land,[40] and the release of

[26] 16 U.S.C. § 1532(13) (definition of "person"). The statute contains an exception for hunting by residents of Alaskan native villages. 16 U.S.C. § 1539(e).
[27] 16 U.S.C. § 1538. In contrast to Section 7, which should apply to any federal agency action, no matter where it occurs, Section 9 is explicitly limited in its geographical scope to the United States and high seas.
[28] 50 C.F.R. § 17.31(a) (FWS rule applying Section 9 prohibitions to all Threatened species); 50 C.F.R. §§ 17.40–17.48 (FWS rules modifying take prohibitions for certain species); 50 C.F.R. § 223 (NMFS rules applying take prohibitions to Threatened species on a species by species basis). As discussed in Section 1.4.3, infra, the lack of a blanket 4(d) rule applying Section 9 prohibitions to all Threatened species under NMFS's jurisdiction has significant consequences for the two listed coral species, which, as of this writing, are subject of no specific 4(d) rule, and hence receive none of the protections against "take" provided by Section 9.
[29] 16 U.S.C. § 1532(19).
[30] 50 C.F.R. § 17.3 (FWS definition of "harass"). NMFS has no corresponding definition of "harass."
[31] 50 C.F.R. § 17.3 (FWS definition of "harm"); see also 50 C.F.R. § 222.102 (NMFS's near-identical definition of "harm").
[32] Babbitt v. Sweet Home Chapter of Cmtys. for a Great Or., 515 U.S. 687, 704–05 (1995).
[33] United States v. Billie, 667 F. Supp. 1485 (S.D. Fla. 1987).
[34] Palila v. Haw. Dep't of Land & Natural Res., 852 F.2d 1106 (9th Cir. 1988).
[35] Defenders of Wildlife v. Envtl. Prot. Agency, 420 F.3d 946 (9th Cir. 2005).
[36] Strahan v. Coxe, 127 F.3d 155 (1st Cir. 1997).
[37] Loggerhead Turtle v. County of Volusia, 120 F. Supp. 2d 1005 (M.D. Fla. 2000).
[38] 16 U.S.C. § 1533(a)(3).
[39] 16 U.S.C. § 1533(f).
[40] 16 U.S.C. § 1534.

federal funding for domestic[41] and international[42] conservation programs. The ESA contains a "citizen suit" provision allowing interested parties, such as nongovernmental organizations (NGOs), to bring suit against both private and government entities to enjoin violations of the statute.[43]

However, no matter how imperiled a species might be, none of the protections of the ESA apply to it unless it is officially listed, via regulation, as Threatened or Endangered under the statute. A species is "Endangered" if it "is in danger of extinction throughout all or a significant portion of its range."[44] A species is "Threatened" if it is "likely to become an endangered species within the foreseeable future."[45]

The listing process for a given species may be initiated either by FWS or NMFS on the agency's own volition or by petition from an interested party.[46] Under either scenario, once the listing process is initiated, strict timelines apply.[47] As discussed in more detail here, these timelines have played a crucial role in ESA actions related to climate change. All listing decisions are to be made "solely on the basis of the best scientific data available."[48] The failure to list a petitioned species is subject to judicial review.[49]

1.2. Case Study I: Elkhorn and Staghorn Corals

1.2.1. Global Warming and Coral Reefs

Coral reefs are among the first ecosystems to show the significant adverse impacts of global warming.[50] An estimated 30 percent of coral reefs globally are already severely degraded and 60 percent may be lost by 2030.[51] The primary cause of coral reef degradation on a global scale is bleaching, the expulsion of symbiotic algal zooxanthellae from coral triggered, inter alia, by elevated sea temperatures.[52]

---

[41] 16 U.S.C. § 1535(d).
[42] 16 U.S.C. § 1537(a).
[43] 16 U.S.C. § 1540(g).
[44] 16 U.S.C. § 1532(6).
[45] 16 U.S.C. § 1532(20). The ESA does not define "foreseeable future." As discussed *infra*, interpretation of this phrase is likely to be key to any listing decision (and litigation over that decision) based upon the projected impacts of global warming on the species.
[46] In practice, virtually all listing actions since the mid-1980s have been initiated by petition rather than independent agency initiative. *See* D. Noah Greenwald, Kieran F. Suckling & Martin F. Taylor, *The Listing Record*, in THE ENDANGERED SPECIES ACT AT THIRTY: RENEWING THE CONSERVATION COMMITMENT (Dale D. Goble, J. Michael Scott & Frank W. Davis eds., 2006).
[47] 16 U.S.C. § 1533(b).
[48] 16 U.S.C. § 1533(b)(1)(A).
[49] 16 U.S.C. § 1533(b)(3)(C)(ii).
[50] Ove Hoegh-Guldberg, *Climate Change, Coral Bleaching and the Future of the World's Coral Reefs*, 50 MARINE & FRESHWATER RES. 839 (1999).
[51] Terence P. Hughes et al., *Climate Change, Human Impacts, and the Resilience of Coral Reefs*, 301 SCI. 929 (2003).
[52] Hoegh-Guldberg, *supra* note 50, at 861.

In 1998, which at the time was the warmest year on record, bleaching occurred in every ocean, ultimately resulting in the death of 10 to 16 percent of the world's living coral.[53] In 2005, which eclipsed 1998 as the warmest year on record,[54] a major bleaching event swept through the Caribbean, bleaching over 90 percent of live coral in some areas and resulting in the ultimate death of about 20 percent of living coral regionwide.[55] Before this unprecedented single-year die-off even began, the Caribbean contained the world's most degraded coral reefs, having already lost as much as 80 percent of live coral over the preceding thirty years.[56] It will not take many more episodes like the 2005 bleaching event before living coral reefs in the Caribbean disappear entirely.[57]

While bleaching is perhaps the most widespread and worrisome impact of global warming on coral reefs, it is far from being the only such impact. As the authors of an authoritative review in 2003 put it:

> The link between increased greenhouse gases, climate change, and regional-scale bleaching of corals, considered dubious by many reef researchers only 10 to 20 years ago, is now incontrovertible. Moreover, future changes in ocean chemistry due to higher atmospheric carbon dioxide may cause weakening of coral skeletons and reduce the accretion of reefs, especially at higher latitudes. The frequency and intensity of hurricanes (tropical cyclones, typhoons) may also increase in some regions, leading to a shorter time for recovery between recurrences. The most pressing impact of climate change, however, is episodes of coral bleaching and disease that have already increased greatly in frequency and magnitude over the past 30 years.[58]

The regional or global loss of coral reefs will have a devastating impact on global biodiversity. Coral reefs are the oldest and most diverse ecological communities on Earth[59]; they occupy less than 0.1 percent of the area of the world's oceans; yet harbor about a third of all described ocean species.[60] Moreover, most coral reef species remain undescribed, with estimates of an additional 1 million or more reef-dependent species yet to be cataloged.[61] The impending loss of coral reef ecosystems, if allowed to proceed, will perhaps be the greatest anthropogenic extinction event in history.

---

53 Ove Hoegh-Guldberg, *Marine Ecosystems and Climate Change*, in CLIMATE CHANGE & BIODIVERSITY 256, 264 (Thomas E. Lovejoy & Lee Hannah eds., 2005).
54 James Hansen et al., *Global Temperature Change*, 103 PROC. NAT'L ACAD. SCI. 14288, 14290 (2006).
55 Federal Response to the 2005 Caribbean Bleaching Event, *available at* http://coralreefwatch.noaa.gov/caribbean2005/docs/2005_bleaching_federal_response.pdf (last visited Dec. 1, 2006).
56 Toby A. Gardner, Isabelle M. Côté, Jennifer A. Gill, Alastair Grant & Andrew R. Watkinson, *Long-Term Region-Wide Declines in Caribbean Corals*, 301 SCI. 958 (2003).
57 Hoegh-Guldberg, *supra* note 53, at 264.
58 Hughes, *supra* note 51, at 929.
59 James W. Porter & Jennifer I. Tougas, *Reef Ecosystems: Threats to Their Biodiversity*, in 5 ENCYCLOPEDIA OF BIODIVERSITY 73, 74 (Simon Asher Levin ed., 2001).
60 *Id.* at 74–75.
61 *Id.* at 75–77.

1.2.2. The Decline of Elkhorn and Staghorn Corals

Elkhorn coral (*Acropora palmata*) and staghorn coral (*Acropora cervicornis*) were, for at least the past 3,000 years, the dominant reef-building corals in the Caribbean.[62] Virtually every reef from the Florida Keys across the Caribbean to the Mesoamerican Reef in Belize was largely comprised of one or the other (or both) of these formerly ubiquitous species.[63] However, over the past thirty years the two species have declined by upwards of 90 percent.[64] The primary drivers of the decline have been disease and temperature-induced bleaching.[65] Additionally, the period of decline coincided with an ongoing period of increased hurricane activity, with intense storms destroying entire reef tracts in certain areas.[66] The cumulative result was that by the beginning of the twenty-first century, elkhorn and staghorn corals had been reduced to a scattering of mostly small colonies amid a large sea of coral rubble.[67]

The factors causing the decline of elkhorn and staghorn coral either are directly correlated with global warming or represent examples of events likely to be more prevalent and more intense in a warming climate.[68] And while the link between coral bleaching and global warming is relatively intuitive, even the outbreaks of coral disease that ravaged the two species have been linked to elevated water temperatures.[69] Similarly, scientific evidence indicates that global warming increases the probability of severe weather events like the series of intense hurricanes that have so impacted Caribbean reefs in recent decades.[70] Finally, there is clear evidence that the record-setting ocean temperatures of 1998 and 2005 that triggered widespread bleaching and mortality are the product of global warming.[71]

In sum, over the course of less than three decades, elkhorn and staghorn corals went from being the most visible and ecologically most important corals of Caribbean reefs, a position they have held for at least 3,000 years, to species whose continued existence beyond the next few decades is now in serious doubt. Global warming helped bring them to this state and, absent significant reductions in greenhouse gas emissions, is poised to be the final nail in their coffin.

---

[62] Terence P. Hughes, *Catastrophes, Phase Shifts and Large-Scale Degradation of a Caribbean Coral Reef*, 265 SCI. 1547 (1994).
[63] *Id.*
[64] *Id.*
[65] *Id.*
[66] William F. Precht & Richard B. Aronson, *Climate Flickers and Range Shifts of Reef Corals*, 2 FRONTIERS ECOLOGY & ENV'T 307 (2004).
[67] On top of these regionwide factors, the two corals virtually disappeared in the face of a diverse array of threats, the primary ones being overfishing and nutrient runoff, both of which promote algal overgrowth, which smothers living coral. *See* Hughes, *supra* note 51, at 861.
[68] *Id.*
[69] C. Drew Harvell et al., *Climate Warming and Disease Risks for Terrestrial and Marine Biota*, 296 SCI. 2158, 2161 (2002).
[70] B. D. Santer et al., *Forced and Unforced Ocean Temperature Changes in Atlantic and Pacific Tropical Cyclogenesis Regions*, 103 PROC. NAT'L ACAD. SCI. 13905 (2006).
[71] Hansen, *supra* note 54, at 14290.

1.2.3. From Petition to Listing

On March 4, 2004, a U.S. NGO, the Center for Biological Diversity, submitted a petition to NMFS seeking listing of elkhorn and staghorn corals under the ESA.[72] The 111-page petition detailed the decline of the species, projected future threats, and argued that the corals were sufficiently imperiled to warrant the protections of the ESA.[73] Reflective of the scientific literature documenting the species' decline, the petition discussed the various factors negatively affecting the corals, with a particular focus on the current and projected impacts of global warming.[74]

On June 23, 2004, NMFS made a positive ninety-day finding, concluding that the petition had presented "substantial information indicating the petitioned actions may be warranted" and announced the initiation of a formal status review as required by section 4(b)(3)(A) of the ESA.[75] On May 9, 2005, following a public comment period and the completion of a status review, NMFS issued a proposed regulation to list the two species as Threatened.[76] A year later, on May 9, 2006, NMFS finalized the regulations, adding the two species to the official list of Threatened species.[77]

While the final listing of the corals came three months after the statutory deadline, the listing went through the regulatory process with remarkably little opposition or more typical longer delays.[78] Of the more than 1,300 comments submitted by scientists and members of the public during the rulemaking process, not a single comment opposed the listing.[79] And while the listing process explicitly implicated global warming and raised the specter of greenhouse gas regulation, none of the climate deniers or fossil fuels industry associations that have actively opposed virtually all attempts at climate-related regulation in the United States participated in the process. The end result was a final regulation of significant legal effect promulgated by a federal agency under an administration otherwise openly hostile to any greenhouse

---

[72] Center for Biological Diversity, Petition to List Acropora Palmata (Elkhorn Coral), Acropora Cervicornis (Staghorn Coral), and Acropora Prolifera (Fused-Staghorn Coral) as Endangered Species Under the Endangered Species Act (2004), *available at* http://www.biologicaldiversity.org/swcbd/SPECIES/coral/petition.pdf (last visited Dec. 1, 2006). The third species mentioned in the Petition, fused-staghorn coral, was determined by NMFS to be a hybrid of the elkhorn and staghorn corals and therefore not further considered for listing.

[73] *Id.*

[74] *Id.* at 62–67.

[75] Listing Endangered and Threatened Wildlife and Plants and Designating Critical Habitat; 90-Day Finding on a Petition to List Elkhorn Coral, Staghorn Coral, and Fused-Staghorn Coral, 69 FED. REG. 34,995 (June 23, 2004).

[76] Endangered and Threatened Species; Proposed Threatened Status for Elkhorn Coral and Staghorn Coral, 70 FED. REG. 24,359 (May 9, 2005).

[77] Endangered and Threatened Species: Final Listing Determinations for Elkhorn Coral and Staghorn Coral, 71 FED. REG. 26,852 (May 9, 2006); 50 C.F.R. § 223.102.

[78] NMFS, in processing listing petitions, has generally shown less disregard for statutory deadlines than FWS. *See* discussion of polar bear listing process, *infra*. Still, the agency's listing of the corals in just over two years was surprising given the agency's recent history of initially denying almost all listing petitions. *See* Brendan R. Cummings, *Unfulfilled Promise: Using the ESA to Protect Imperiled Marine Wildlife*, 12 WILD EARTH 62 (2002).

[79] 71 FED. REG. 26,852, 26,853–55 (response to comments).

gas regulation and unwilling, to that point, to even acknowledge of the reality of anthropogenic global warming.

The listing of the corals transpired without significant controversy for several reasons. First, given the catastrophic declines in the abundance and distribution of elkhorn and staghorn corals, there was no real scientific dispute as to their endangerment. Second, the decline in the species was multicausal; NMFS did not need to invoke or endorse the science of global warming or rely on predictions of future warming to find the species imperiled. Third, the listing simply went unnoticed; in contrast to the high-profile polar bear listing process, the listing of the corals never attracted significant media attention, and hence entities that may have otherwise mobilized to oppose the rule never became aware of it.

The language in the listing rule itself is a rich example of the actions of scientists operating within the constraints of an agency not allowed to acknowledge the existence of global warming. The phrase "global warming" appears nowhere in the 10,000-word final listing rule. Nor does mention of "greenhouse gases" occur. The phrase "climate change" appears only twice, both times in reference to literature submitted by the public rather than as an actual phenomenon relevant to the rulemaking.[80] Instead, the phrase "elevated sea surface temperature" is sprinkled throughout the document, and is used a total of eleven times with no ascription of causal mechanism(s). One example speaks volumes regarding NMFS's conflicted position:

> The major threats to these species' persistence (i.e., disease, elevated sea surface temperature, and hurricanes) are severe, unpredictable, have increased over the past 3 decades, and, at current levels of knowledge, the threats are unmanageable.[81]

Thus, while not acknowledging any anthropogenic role in these processes, NMFS essentially declares that the impacts of global warming are "unmanageable." In other words, the agency seems to be admitting that current climate policy is nowhere near adequate to address the problem. While NMFS said "unmanageable," "unmanaged" would be a more appropriate term, as the failure to address global warming is a result of policy decisions rather than because the problem itself is (at least as of yet) impossible to manage.

The closest NMFS comes in the rule to acknowledging the existence of global warming as an anthropogenic phenomenon, is a sentence mentioning carbon dioxide "levels" (not "emissions"):

> Along with elevated sea surface temperature, atmospheric carbon dioxide levels have increased in the last century, and there is no apparent evidence the trend will not continue.[82]

---

[80] 71 FED. REG. at 26,855 ("Several comments and journal articles addressing climate change and coral bleaching were received" and "In addition to the comments relating to the proposed listing, the following were also received: (1) peer-reviewed journal articles regarding climate change.")
[81] 71 FED. REG. at 26,858.
[82] Id.

Again, the voice is passive, and there is no mention of a human role in the process, but there is an implicit acknowledgment that existing national and international policies are not sufficiently addressing the problem.

Ultimately, what matters from the listing of the corals is not the language choices by NMFS, but the fact that the species are now listed under the ESA and one of the strongest of U.S. environmental laws can now be turned toward the problem of greenhouse gas emissions and global warming. The consequences of the coral listing are explored further in Section 1.4.

1.3. Case Study II: Polar Bear

1.3.1. The Arctic as a Global Warming Hot Spot

Global warming is already having pronounced impacts in the Arctic. In November 2004, a comprehensive scientific report commissioned by the Arctic Council, the Arctic Climate Impact Assessment (ACIA), painted a stark picture for the future of the region. In Alaska and western Canada, winter temperatures have already increased by as much as 3–4°C in the past fifty years.[83] Over the next 100 years, under a moderate emissions scenario, annual average temperatures are projected to rise an additional 3–5°C over land and up to 7°C over the oceans. Winter temperatures are projected to rise by an additional 4–7°C over land and 7–10°C over the oceans.[84]

In the years subsequent to the ACIA report, both the observed and projected impacts to the Arctic, and particularly to sea ice, have been even more pronounced. The record minimum summer sea-ice extent set in September 2005[85] was smashed in September 2007 when sea-ice extent fell to 1.63 million square miles, about one million square miles below the average minimum sea-ice extent between 1979 and 2000,[86] and 50% lower than conditions in the 1950s to the 1970s.[87] The 2007 minimum was lower than the sea-ice extent most climate models predict would not be reached until 2050 or later.[88]

---

[83] Susan Joy Hassol, IMPACTS OF A WARMING ARCTIC: ARCTIC CLIMATE WARMING ASSESSMENT (2004), at 22, *available at* http://amap.no/acia/ (last visited Dec. 1, 2006).

[84] *Id.* at 26.

[85] National Snow and Ice Data Center, *Arctic Sea Ice Shrinks as Temperatures Rise*, Oct. 3, 2006, *available at* http://www.nsidc.org/news/press/2006_seaiceminimum/20061003_pressrelease.html (last visited Dec. 1, 2006).

[86] National Snow and Ice Data Center (NSIDC), *Overview of Current Sea Ice Conditions*, Sept. 7, 2007, *available at* http://www.nsidc.org/news/press/2007_seaiceminimum/20070810_index.html (last visited May 26, 2008).

[87] Julienne Stroeve et al., *Arctic Sea Ice Extent Plummets in 2007*, 89 EOS 13–20 (2008).

[88] *Id.*; Julienne Stroeve et al., *Arctic Sea Ice Decline: Faster than Forecast*, 34 GEOPHYSICAL RES. LETTERS L09501, doi:10.1029/2007GL029703 (2007); Endangered and Threatened Wildlife and Plants, Determination of Threatened Status for the Polar Bear (Ursus maritimus) throughout Its Range, *supra* note 1, at 28,233.

Both the extent and thickness of winter sea ice are also declining, with a record minimum-low winter sea ice extent in March 2006.[89] Relatively thin, first-year ice covered 72 percent of the Arctic Basin, including the region around the North Pole in March 2008, considerably exceeding the first-year ice cover of March 2007.[90] Since very little first year ice survives the summer melt season (in 2007 only 13 percent of first year ice survived), more first-year winter ice results in lower sea-ice cover in the following summer.[91] The September minimum sea-ice extent in 2008 was the second lowest year on record.[92] Some leading sea-ice researchers now believe that the Arctic could be completely ice free in the summer as early as 2012.[93] Polar bears face a grim future even under relatively optimistic scenarios of sea-ice decline, and as complete loss of summer sea ice within a decade or so becomes increasingly likely, their future is tenuous indeed.

1.3.2. The Polar Bear in a Warming Arctic

The polar bear (*Ursus maritimus*) is completely dependent upon Arctic sea-ice habitat for survival. Polar bears need sea ice as a platform from which to hunt ringed seals and other prey, to make seasonal migrations between the sea ice and their terrestrial denning areas, and for other essential behaviors such as mating.[94] Unfortunately, the polar bear's sea-ice habitat is quite literally melting away.

Canada's Western Hudson Bay population, at the southern edge of the species' range, has been the first to show the impacts of global warming.[95] Breakup of the annual ice in Western Hudson Bay is now occurring on average 2.5 weeks earlier than it did thirty years ago.[96] Earlier ice breakup is resulting in polar bears having less time on the ice to hunt seals. Polar bears must maximize the time they spend on the ice feeding before they come ashore, as they must live off built-up fat reserves for up to eight months before ice conditions allow a return to hunting on the ice. The reduced hunting season has translated into thinner bears, lower female

---

[89] National Snow and Ice Data Center, *Arctic Sea Ice Extent at Maximum Below Average, Thin*, available at http://www.nsidc.org/arcticseaicenews/2008/040708.html (last visited May 26, 2008).
[90] *Id.*
[91] *Id.*
[92] National Snow and Ice Data Center, *Arctic Sea Ice Down to Second-Lowest Extent; Likely Record-Low Volume*, available at http://www.nsidc.org/news/press/20081002_seaice_pressrelease.html (last visited Mar. 11, 2009).
[93] Seth Borenstein, *Arctic Sea Ice Gone in Summer within Five Years?*, Associated Press, Dec. 12, 2007, available at http://news.nationalgeographic.com/news/pf/33860636.html (last visited May 26, 2008); Jonathan Amos, *Scientists in the US Have Presented One of the Most Dramatic Forecasts Yet for the Disappearance of Arctic Sea Ice*, BBC News, Dec. 12, 2007, available at http://news.bbc.co.uk/go/pr/fr/-/2/hi/science/nature/7139797.stm (last visited May 26, 2008).
[94] Andrew E. Derocher, Nicholas J. Lunn & Ian Stirling, *Polar Bears in a Warming Climate*, 44 Integrated Comp. Biology 163 (2004).
[95] *Id.*; Jon Aars, Nicholas J. Lunn & Andrew E. Derocher, Polar Bears: Proceedings of the 14th Working Meeting of the IUCN/SSC Polar Bear Specialist Group, 20–24 June 2005, Seattle, Washington, USA 44–45 (2006).
[96] *Id.*

reproductive rates, and lower juvenile survival rates.[97] At the time of the ACIA report, population declines were not yet reported for Hudson Bay. However, polar bear scientists predicted that if sea-ice trends continue, most female polar bears in the Western Hudson Bay population will be unable to reproduce by the end of the century, and possibly as early as 2012.[98]

While Hudson Bay is showing the earliest signs of global warming's impacts on polar bears, the consequences of future sea-ice reductions for polar bears elsewhere will also be severe. A 2004 peer-reviewed analysis looking at all aspects of global warming's impacts on the polar bear by three of the world's foremost experts on the species concluded that "it is unlikely that polar bears will survive as a species if the sea ice disappears completely as has been predicted by some."[99]

The ACIA comes to a similar conclusion: "polar bears are unlikely to survive as a species if there is an almost complete loss of summer sea-ice cover, which is projected to occur before the end of this century by some climate models."[100]

Even short of complete disappearance of sea ice, projected impacts to polar bears from global warming will affect virtually every aspect of the species' existence:

- The timing of ice formation and breakup will determine how long and how efficiently polar bears can hunt seals. A reduction in the hunting season caused by delayed ice formation and earlier breakup will mean reduced fat stores, reduced body condition, and therefore reduced survival and reproduction.
- Reductions in sea ice will in some areas result in increased distances between the ice edge and land. This will make it more difficult for female bears that den on land to reach their preferred denning areas. Bears will face the energetic trade-off of either leaving the sea ice earlier when it is closer to land or traveling farther to reach denning areas. In either case, the result is reduced fat stores and likely reduced survival and reproduction.
- Reductions in sea-ice thickness and concentration will likely increase the energetic costs of traveling as moving through fragmented sea ice and open water is more energy intensive than walking across consolidated sea ice.
- Reduced sea-ice extent will likely result in reductions in the availability of ice-dependent prey such as ringed seals, as prey numbers decrease or are concentrated on ice too far from land for polar bears to reach.
- Global warming will likely increase the rates of human-bear interactions, as greater portions of the Arctic become more accessible to people and as polar bears are forced to spend more time on land waiting for ice formation. Increased human-bear interactions will almost certainly lead to increased polar bear mortality.

[97] Id.
[98] Derocher, *supra* note 94, at 165.
[99] Id. at 163.
[100] Hassol, *supra* note 83, at 58.

- The combined effects of these impacts of global warming on individual bears' reproduction and survival are likely to ultimately translate into impacts on polar bear populations. Impacts will be most severe on female reproductive rates and juvenile survival. In time, reduction in these key demographic factors will translate into population declines and extirpations.[101]

In sum, changes in sea-ice extent, thickness, movement, fragmentation, location, duration, and timing will have significant and adverse impacts on polar bear feeding, breeding, and movement. Such impacts will likely result in reduced reproductive success and higher juvenile mortality, and in some cases increased adult mortality. If global warming continues unabated, these impacts will ultimately lead to global extinction of the species.

In response to the increasingly recognized threat of global warming on the polar bear, in 2004 the Center for Biological Diversity began preparation of a petition seeking listing of the species under the ESA. Since the petition was filed, new reports of polar bear drownings,[102] cannibalism,[103] starvation,[104] and population decline[105] have been published. The status of the polar bear has grown more dire, and with it, the need for protection all the more compelling.

1.3.3. From Petition to Listing

On February 16, 2005, the Center for Biological Diversity submitted a Petition to FWS to list polar bears as a Threatened or Endangered species under the ESA.[106] The 170-page Petition discussed the status of the species, the science of global warming, and the observed and projected impacts of global warming on the polar bear's sea-ice habitat.[107] The Petition argued that the polar bear was endangered or likely to become so in the foreseeable future given global warming trends and the inadequacy of U.S. and international measures to combat greenhouse gas emissions.[108]

In contrast with the coral petition, where the case for protected status could be made on already documented declines, the decline of the polar bear, at least at the time of the Petition, was something projected for the future. Other than for the Western Hudson Bay population, studies documenting impacts to polar bears

---

[101] Derocher, *supra* note 94; Hassol, *supra* note 83, at 58.
[102] Charles Monnett & Jeffrey S. Gleason, *Observations of Mortality Associated with Extended Open-water Swimming by Polar Bears in the Alaskan Beaufort Sea*, 29 POLAR BIOLOGY 681 (2006).
[103] Steven C. Amstrup et al., *Recent Observations of Intraspecific Predation and Cannibalism among Polar Bears in the Southern Beaufort Sea*, 29 POLAR BIOLOGY 997 (2006).
[104] Eric V. Regehr, S.C. Amstrup & Ian Stirling, Polar Bear Population Status in the Southern Beaufort Sea, U.S. Geological Survey Open-File Report 1337 (2006), at 13–14.
[105] Aars, *supra* note 95, at 41, 44.
[106] Kassie Siegel & Brendan Cummings, *Petition to List the Polar Bear as a Threatened Species under the Endangered Species Act*, Feb. 16, 2006, available at http://www.biologicaldiversity.org/swcbd/SPECIES/polarbear/petition.pdf (last visited Dec. 1, 2006).
[107] *Id.*
[108] *Id.*

from global warming had not yet been published. As such, the Petition was heavily dependent on the forecasts of climate scientists about what conditions for polar bears would be in the coming decades. So while NMFS could skirt the issue of the causal mechanisms of warming oceans and consequent coral decline, acceptance or rejection of the polar bear Petition would require FWS to squarely address the science of global warming. Whatever action FWS took in response to the polar bear Petition would then represent either an explicit agency acceptance of anthropogenic global warming, something the Bush administration had been loath to do, or a rejection of the consensus on the science of global warming, in which case the science of global warming would end up in court under the "best available science" standard of the ESA.[109]

When FWS failed to make a ninety-day finding on the Petition, on October 11, 2005, the Center, now joined by two additional NGOs, Greenpeace and the Natural Resources Defense Council, filed a formal notice of intent to sue as required by the citizen suit provision of the ESA.[110] On December 15, 2005, the organizations filed suit in federal district court in San Francisco, California, to compel FWS to make the overdue ninety-day finding.[111] In response, on February 9, 2006, FWS published its finding in the Federal Register.[112]

The ninety-day finding made by FWS recites the statutory boilerplate that "the petition presents substantial scientific or commercial information indicating that the petitioned action of listing the polar bear may be warranted," but is otherwise devoid, however, of any information or statement that could be interpreted as an acknowledgment of the existence of global warming. The closest the agency comes to acknowledging the primary threat to the species is to solicit information, inter alia, "on the effects of climate change and sea ice change on the distribution and abundance of polar bears and their principal prey over the short- and long-term."[113] So while the polar bear cleared this important first hurdle on the path toward listing, FWS managed to avoid directly confronting the issue of global warming in the finding.

Within a week after the ninety-day finding was made, the one-year deadline from the date of the Petition for FWS to make its now-required twelve-month finding passed.[114] The parties ultimately negotiated a settlement, and filed a stipulation setting forth a date certain for FWS to make the twelve-month finding. On July 5, 2006, the Court issued an order approving the stipulation and setting December 27, 2006, as the judicially enforceable deadline for FWS to make the finding as to whether listing the polar bear under the ESA is or is not warranted.

On December 27, 2006, FWS announced that listing of the polar bear was in fact warranted and that the agency would be publishing a proposed listing rule. The

---

[109] 16 U.S.C. § 1533(b)(1)(A).
[110] 16 U.S.C. § 1540(g).
[111] Ctr. for Biological Diversity v. Norton, No. 05-05191-JSW (N.D. Cal. 2005), complaint *available at* http://www.biologicaldiversity.org/swcbd/SPECIES/polarbear/Complaint12-15-05.pdf.
[112] Endangered and Threatened Wildlife and Plants; Petition to List the Polar Bear as Threatened, 71 FED. REG. 6745 (Feb. 9, 2006).
[113] *Id.*
[114] 16 U.S.C. § 1533(b)(3)(B).

proposed rule was published in the Federal Register on January 9, 2007.[115] As with the ninety-day finding, the proposed rule avoids mention of the terms "global warming," and "greenhouse gases." The rule, however, does go into great depth about the polar bear's dependence on sea ice and ultimately concludes that "polar bear populations throughout their distribution in the circumpolar Arctic are threatened by ongoing and projected changes in their sea ice habitat."[116] Just as NMFS found the now-listed coral species threatened by "elevated sea surface temperatures" without ascribing any causal mechanism to the rising temperatures, FWS never explicitly acknowledges why the sea ice is retreating. Nevertheless, the proposal, to a much greater degree than the coral listing rule, acknowledges that the best available science indicates that temperatures will continue to rise and sea-ice extent will continue to decline.

The Petition filing, sending of the formal notice of intent to sue, subsequent lawsuit, positive ninety-day finding, and eventual proposed listing rule for the polar bear were all accompanied by press releases and all garnered significant media attention.[117] The announcement of the proposed rule generated more than 1,000 news articles, several hundred television reports, and more than 200 editorials, virtually all of which discussed the decision as an important recognition of the reality of global warming by the Bush administration.[118] The FWS received about 670,000 comments on the proposed rule, far more than had been received on any previous ESA proposal.[119] The press attention and public interest triggered by the listing process helped elevate the polar bear to an international symbol of the very real impacts of global warming.[120]

While the coral listing process occurred with relatively little fanfare and virtually no opposition, the listing process for the polar bear has been far more contentious. The State of Alaska, the Alaska Oil and Gas Association, various other fossil fuel industry associations, as well as sport-hunting groups all came out publicly in opposition to listing, and in response the Bush administration instituted a policy prohibiting agency employees from discussing polar bears or global warming while traveling abroad.[121] These entities and others continue to work strenuously against listing.[122]

---

[115] Endangered and Threatened Wildlife and Plants; 12-Month Petition Finding and Proposed Rule to List the Polar Bear (Ursus maritimus) as Threatened throughout Its Range, 72 FED. REG. 1063 (Jan. 9, 2007).
[116] Id. at 1081.
[117] See http://www.biologicaldiversity.org/swcbd/SPECIES/polarbear/index.html for press releases and examples of and links to various media articles and television coverage of the listing process.
[118] Id.
[119] 73 FED. REG. at 28, 235.
[120] The polar bear was featured on the cover of the Apr. 3, 2006, issue of Time magazine accompanying a cover story about global warming (see supra note 5). Similarly, an animation showing a drowning polar bear appears in Al Gore's documentary An Inconvenient Truth.
[121] Dan Joling, Threat of Polar Bear Listing Stirs Politicians, ANCHORAGE DAILY NEWS, May 6, 2007; Dan Joling, 'Threatened' Polar Bear Listing Debated, ANCHORAGE DAILY NEWS, Mar. 2, 2007; Andrew Revkin, Memos Tell Officials How to Discuss Climate, N.Y. TIMES, Mar. 8, 2007.
[122] As discussed supra, listing decisions are to be based solely on science; nevertheless, significant political pressure is often brought to bear on Interior Department decision-makers regarding controversial listings. See supra note 46.

Despite the vociferous opposition from powerful industry groups, the relentless and accelerating warming of the Arctic and increased resources directed toward understanding the future status of polar bears made it increasingly untenable for the Bush administration to refuse to protect the bear. Between the listing proposal and the January 9, 2008, deadline for a final listing decision, both the scientific evidence and public interest in the decision continued to mount. Scientific articles, and subsequent popular press reports, on polar bears drowning from lack of sea ice, starving from lack of access to food, and engaging in cannibalism presumably triggered by food stress, all phenomena without precedent, have appeared with alarming frequency.[123] The Southern Beaufort Sea population in Alaska and Canada, considered "stable or increasing" at the time of the Petition, is now considered to be both "declining" and "reduced."[124]

Most significantly, in 2007, FWS requested that the Department of Interior's U.S. Geological Survey (USGS) address a series of research questions relating to the status of the polar bear. The USGS produced nine administrative reports addressing these questions and in doing so significantly advanced the understanding of sea-ice loss and its implications for polar bears. The USGS conducted polar bear population modeling based on ten general circulation models that most accurately simulate future ice conditions.[125] The USGS used the Intergovernmental Panel on Climate Change ("IPCC") A1B "business as usual" scenario of future emissions to run the climate models.[126] In the A1B scenario, atmospheric carbon dioxide concentrations reach 717 parts per million by 2100.

The USGS divided the world's polar bear populations into four ecological regions: The (1) Seasonal Ice Ecoregion, which includes Hudson Bay, and occurs mainly at the southern extreme of the polar bear range, (2) the Archipelago Ecoregion of the Canadian Arctic, (3) the Polar Basin Divergent Ecoregion, where ice is formed and then advected away from near-shore areas, and (4) the Polar Basin Convergent Ecoregion, where sea ice formed elsewhere tends to collect against the shore.[127]

The USGS projected the future range-wide status of polar bears using two different modeling techniques, and the results are profoundly disturbing. The USGS projects that polar bears will be extinct in the Seasonal Ice and Divergent Ice ecoregions by the middle of this century.[128] These two ecoregions account for two thirds of the world's polar bears, including all of the bears in Alaska. The "good news" is that polar bears may survive in the high Canadian Archipelago and portions of the Convergent Ice Ecoregion through the end of this century. However, their extinction risk is still extremely high: over 40 percent in the archipelago and over 70 percent in Northwest

---

[123] See supra notes 94–105.
[124] Aars, supra note 95, at 34.
[125] Steven C. Amstrup et al., *Forecasting the Range-wide Status of Polar Bears at Selected Times in the 21st Century*, U.S. GEOLOGICAL SURVEY ADMINISTRATIVE REPORT (U.S. Geological Survey, Reston, VA, 2007).
[126] Id.
[127] Id. at 1.
[128] Id.

Greenland.¹²⁹ Moreover, the most likely outcome for each of these ecoregions by the end of this century is also extinction.¹³⁰

In addition, the USGS emphasized that because all of the available climate models have to date underestimated the actual observed sea-ice loss, the assessment of risk to the polar bear is conservative.¹³¹ Perhaps most worrisome is the observation that part of an area in the Canadian Archipelago expected to provide an icy refuge for the polar bear in 2100 lost its ice in the summer of 2007.¹³²

As the January 9, 2008, deadline for a final listing decision approached, the Bush administration thus found itself faced with both irrefutable scientific evidence of the threat to polar bears and mounting media and public interest in the decision. In response, the administration again sought refuge in delay. On January 7, 2008, FWS Director Dale Hall held a press conference and stated that the agency would not meet the deadline, but intended to issue the decision within thirty days. The conservation organizations filed a 60-Day Notice Letter of Intent to Sue for failure to publish a final listing determination, and, when the decision had still not been issued within sixty days, filed suit on March 10, 2008.¹³³ Plaintiffs moved for summary judgment on April 2, 2008, the first day allowed under the Local Rules of Court. On April 28, 2008, the District Court issued an order granting Plaintiffs' Motion for Summary Judgment, finding Defendants in violation of the ESA for failing to publish a final listing decision for the polar bear by January 9, 2008, and directed Defendants to publish a final decision by May 15, 2008, and to make any final regulation effective upon publication pursuant to 5 U.S.C. § 553(d)(3).¹³⁴

On May 14, 2008, Secretary of Interior Dirk Kempthorne announced the Bush administration would list the polar bear as a Threatened species throughout its range.¹³⁵ He accompanied his statement with images showing the rapid melting of the Arctic sea ice. The announcement was remarkable, given both the fact that the administration had spent nearly eight years denying and downplaying the science of global warming and the fact that the administration had shown unprecedented hostility to endangered species and listed far fewer species under the Act than any other. The polar bear was, in fact, the first species listed by Secretary Kempthorne in the United States in over two years.

The decision was a watershed moment in the Bush administration's approach to the science of global warming. The final listing decision clearly and unambiguously adopts the consensus view of the world's scientists on global warming and Arctic melting, and rejects arguments propounded by a tiny number of industry funded spokespeople.¹³⁶ The portions of the rule dealing with climate science and polar bear

---

¹²⁹ *Id.* at 66–67 (Table 8).
¹³⁰ *Id.*
¹³¹ *See, e.g.*, Amstrup et al., *supra* note 125, at 34, 36.
¹³² *Id.* at 35, 96.
¹³³ Ctr. for Biological Diversity v. Kempthorne, No. 08–1339 (CW) (N.D. Cal. filed Mar. 10, 2008).
¹³⁴ *See* Ctr. for Biological Diversity, No. 08–1339, 2008 WL 1902703, at *5, *10 (N.D. Cal. Apr. 28, 2008).
¹³⁵ http://www.doi.gov/secretary/speeches/081405_speech.html.
¹³⁶ *See, e.g.*, 73 FED. REG. at 28,219–28,234 (discussion of Arctic sea ice and climate change); 73 FED. REG. at 28,246 ("We have consistently relied on synthesis documents [such as the IPCC's Fourth

biology are well written, and the importance of these conclusions being included in the binding and precedential final regulation listing the polar bear cannot be overstated.[137] Not surprisingly, however, the decision did not include everything that the law requires.

First, the administration listed the polar bear as "Threatened," rather than as "Endangered." While Threatened status would have been appropriate as of February 2005, when the Petition was first filed, by the time of the listing decision, the science clearly mandated Endangered status. A species that is expected to decline by two-thirds in number, disappear from half of its range, and for which the most likely status by the end of the century is global extinction must be considered "in danger of extinction throughout all or a significant portion of its range."[138] Second, having listed the polar bear as Threatened rather than Endangered, the administration attempted to reduce protections to the polar bear through an Interim Final Section 4(d) ("4(d) Rule"), which authorizes activities that would otherwise be prohibited by the ESA and its implementing regulations.[139] The 4(d) Rule also purports to exempt all greenhouse gas–emitting projects from the ambit of Section 7 of the ESA. Finally, the administration failed to designate critical habitat for the polar bear, stating, nonsensically, that it was currently impossible to determine what habitat is essential to the species.[140] Plaintiffs are challenging these and other shortcomings of the final rulemaking in the ongoing litigation.[141] Global warming has clearly arrived in the Arctic and, if the polar bear is to survive, requires an immediate response. The ESA is a critically important part of the U.S. regulatory response to global warming.

1.4. The ESA and Global Warming

While the ESA, passed by Congress in 1973, was enacted well before global warming was widely recognized as a threat to biodiversity, the statute was written with sufficient breadth, and with understanding of the ever-evolving nature of scientific knowledge, that the law needs no amendment to operate effectively to protect species in a

---

Assessment Report and the Arctic Climate Impact Assessment] that present the consensus view of a very large number of experts on climate change from around the world. We have found that these synthesis reports, as well as the scientific papers used in those reports or resulting from those reports, represent the best available scientific information we can use to inform our decision and have relied upon them and provided citation within our analysis."

[137] That the polar bear decision marked a turning point was reinforced by the administration's release, just two weeks later, of the scientific assessment of climate change impacts in the United States. required by the Global Change Research Act of 1990. The release of this report was also required by a court order in Ctr. for Biological Diversity v. Brennan, No. 06–7062, 2007 WL 2408901 (N.D. Cal. 2007). The scientific assessment, released on May 29, 2008, comprehensively affirms the best available science on climate change that the administration had long sought to question. http://www.climatescience.gov/Library/scientific-assessment/.

[138] 15 U.S.C. § 1531(6).

[139] 73 FED. REG. at 28,306–28,318 (Endangered and Threatened Wildlife and Plants, Special Rule for the Polar Bear) (May 15, 2008) ("4(d) Rule"). 16 U.S.C. § 1538(a); 50 C.F.R. § 17.31.

[140] 73 FED. REG. at 28,297–28,299.

[141] Ctr. for Biological Diversity v. Kempthorne, Civ. No. 08–1339 (CW) (N.D. Cal. filed Mar. 10, 2008).

greenhouse world. What the law need, however, is agency decision-makers willing to heed the "best available science," to accept the reality of global warming, and its causes and solutions, and then take actions consistent with their statutory mandates to mitigate or adapt in the face of this "inconvenient truth." With both the corals and the polar bear now listed, the interplay of the ESA with global warming is no longer theoretical. A brief exploration of the application of the various provisions of the ESA to the problem of global warming, from the listing process through the prohibitions against "jeopardy" and "take" follows.

1.4.1. The Listing Process

As discussed previously, none of the protections of the ESA apply to a species unless it is formally listed under the act. For some species, the impacts of global warming have already been felt and the species are already clearly facing extinction. In such case, they should be listed as Endangered. The polar bear and the elkhorn and staghorn corals fall into this category.[142] In these instances, the listing process can be completed without FWS or NMFS having to squarely address the science of global warming or explicitly acknowledge the inadequate domestic and international climate policy positions leading to the continued growth of greenhouse gas emissions. Again, this is how the coral listing played out.

For many species, however, the worst effects of global warming have yet to be felt; these species are not yet facing extinction. Yet given the amount of additional warming that will occur in the coming decades even under the best scenarios, they clearly will face severe threats to their continued existence in the future. Given the projections for upwards of a third of Earth's species to be committed to extinction as a result of global warming by midcentury,[143] hundreds of thousands, if not millions of species potentially fall into this category. These species would properly warrant listing as Threatened.[144] A species is Threatened, however, only if the extinction threat is in the "foreseeable future."[145] The phrase "foreseeable future" is not defined by the ESA. Climate models regularly extrapolate results out to 2100 and beyond. The IUCN calculates certain types of extinction risk based on 100-year time frames.[146]

---

[142] Judicial review of the listing of the polar bear as Threatened rather than Endangered is ongoing as of the time of this writing, and the corals' listing status will likely be the subject of a future challenge as well.

[143] Chris D. Thomas et al., *Extinction Risk From Climate Change*, 427 NATURE 145 (2004); Jay R. Malcolm et al., *Global Warming and Extinctions of Endemic Species from Biodiversity Hotspots*, 20 CONSERVATION BIOLOGY 538 (2006).

[144] While the Thomas study conveys the magnitude of the impending climate-driven extinction crisis, it does not contain the species-specific information necessary for a listing regulation under the ESA. As a practical matter, therefore, while thousands of climate-imperiled species may warrant listing as Threatened under the ESA, for only a small fraction will there likely be sufficient information to prepare and process a listing petition.

[145] 16 U.S.C. § 1532(20).

[146] H. Resit Akçakaya et al., *Use and Misuse of the IUCN Red List Criteria in Projecting Climate Change Impacts on Biodiversity*, 12 GLOBAL CHANGE BIOLOGY 2037 (2006).

If climate models can predict warming and the IUCN can calculate extinction risk on 100-year time horizons, then it would seem that at least this amount of time is "foreseeable." In the ESA context, however, FWS and NMFS have often taken an unreasonably narrow view of the foreseeable future. In the coral listing rule, NMFS defined the "foreseeable future" as thirty years.[147] Given the already severe declines the coral have undergone, this truncation of what is foreseeable ultimately had little legal effect and the corals were still listed. Similarly, for the polar bear, FWS treated the "foreseeable future" as forty-five years.[148] Again, the decline of the polar bears' sea-ice habitat has been so rapid that even under this rather short time frame the species is clearly imperiled. However, for many species, habitat loss, and hence extinction, may not occur for another 50 to 100 years; limiting the horizon for analysis to the next 30 years or less could lead to a determination that the species is not in fact Threatened. Courts faced with unreasonably short agency treatments of the "foreseeable future" provision have set aside decisions not to list species ranging from salmon[149] to rare plants.[150] Legal wrangling over the "foreseeable future" is likely to be a major element of most efforts to list species imperiled by global warming.[151]

Any decision of FWS or NMFS to not list a species threatened by global warming is subject to judicial review.[152] While court review of most agency decisions is limited by the highly deferential "arbitrary and capricious" standards set out by the Administrative Procedure Act,[153] listing decisions under the ESA must still utilize only the "best available science,"[154] a standard that prohibits reliance on political and economic arguments. Moreover, the "best available science" standard places an emphasis on peer-reviewed science,[155] something the climate skeptics can rarely point toward in their attempts to refute the reality of global warming. This "best available science" standard for decision-making therefore provides an ideal framework to bring the science of global warming into the federal courtroom.

While the goal of filing a petition to list a species threatened by global warming is to see the species listed and the protections of the ESA applied, the "best available science" standard of the statute creates a win-win dynamic for petitioners. It was this standard that forced the Bush administration to list the polar bear and acknowledge

---

[147] 71 FED. REG. at 26,854.
[148] 73 FED. REG. at 28,253.
[149] Or. Natural Resources Council v. Daley, 6 F. Supp. 2d 1139, 1150–52 (D. Or. 1998).
[150] Western Watersheds Project v. Foss, No. 04–168-MHW (D. Idaho 2005) (Aug. 19, 2005, Order on Summary Judgment).
[151] The Supreme Court's decision in Massachusetts v. U.S. EPA, 549 U.S. 497 (2007), may be instructive on this issue as the majority recognized Massachusetts' assertions of damages that would occur over the course of a century.
[152] 16 U.S.C. § 1533(b)(3)(C)(ii).
[153] 5 U.S.C. § 706(2)(A); Chevron U.S.A. Inc. v. Natural Res. Def. Council, 467 U.S. 837, 842–43 (1984).
[154] *See supra* note 48.
[155] FWS/NMFS 1994. Interagency Cooperative Policy for Peer Review in Endangered Species Act Activities, 59 FED. REG. 34,270, and Interagency Cooperative Policy on Information Standards under the Endangered Species Act, 59 FED. REG. 34,271.

the findings of the IPCC, ACIA, and other mainstream literature as the "best available science." High-level officials must have known that a refusal to list the polar bear and continued reference to the opinions of a handful of discredited climate deniers would have led to litigation over what constitutes the "best available" climate science that the administration was sure to lose.

In addition to the actual decision as to whether listing a given species under the ESA is or is not warranted, global warming is likely to factor into several other related provisions of the ESA. At the twelve-month finding stage of processing a listing petition, FWS and NMFS can in limited circumstance make a "warranted but precluded" finding.[156] The agencies can only make such a finding if they are making expeditious progress to list other species facing more immediate threats.[157] As a practical matter, NMFS receives far fewer listing petitions than FWS, and therefore has no listing backlog, and does not and cannot invoke this exception.[158] For FWS, however, more than 250 species are on this waiting list for protection.[159] For many species threatened by global warming, impacts that may not manifest for a decade or more may seem less pressing than those facing species suffering direct and immediate loss of their habitat from other human actions such as logging or development. In such cases, protection of a climate-imperiled species may be deemed "precluded" long enough that the global warming impacts become irreversible before the species ever receives the protections of the ESA.

The Kittlitz's murrelet (*Brachyramphus brevirostris*) is currently in this "warranted but precluded" purgatory. The Kittlitz's murrelet is a glacial relict species occurring in Alaska and Russia.[160] It is strongly associated with tidewater glaciers for foraging. It has one of the smallest populations of any North Pacific seabird and, as its tidewater glacier habitat recedes, is rapidly declining. In May 2001, the Center petitioned to list the Kittlitz's murrelet under the ESA.[161] The murrelet was the first species for which the Center petitioned for ESA listing based upon the threat of global warming. However, as with the corals, the murrelet's decline is likely multicausal and a case for ESA listing could be made even absent the climate threat. FWS ultimately found that listing was "warranted but precluded" and has "recycled" this finding for more than five years.[162] In making the "warranted" part of this finding, FWS never mentions global warming, while for the "precluded" portion of the finding, FWS

---

[156] 16 U.S.C. § 1533(b)(3)(B)(iii). A "warranted but precluded" finding is subject to judicial review, 16 U.S.C. § 1533(b)(3)(C)(ii), and FWS's failure to make expeditious progress on its listing backlog is the subject of substantial litigation. *See, e.g.*, Ctr. for Biological Diversity v. Kempthorne, 466 F.3d 1098 (2006).

[157] *Id.*

[158] *See* discussion *supra* note 78.

[159] Endangered and Threatened Wildlife and Plants; Review of Native Species That Are Candidates or Proposed for Listing as Endangered or Threatened; Annual Notice of Findings on Resubmitted Petitions; Annual Description of Progress on Listing Actions, 71 FED. REG. 53,756 (Sept. 12, 2006).

[160] Information on the Kittlit'z murrelet is summarized in the listing petition available at http://www.biologicaldiversity.org/swcbd/SPECIES/murrelet/index.html.

[161] *Id.*

[162] 71 FED. REG. at 53,780.

finds the threats (whatever they may be) to be "nonimminent."[163] The "warranted but precluded" finding for the Kittlitz's murrelet and 250-plus other species similarly situated is being challenged on the grounds that FWS is not making the required "expeditious progress" on listings.[164] Regardless of the outcome of that particular case, whether global warming threats to a species are "nonimminent," and therefore justify delay in listing will likely be another contested subject as global warming gets inserted into agency decision-making under the ESA.

The ESA requires the designation of "critical habitat" for a species concurrently with listing, or in limited circumstances, within a year of listing.[165] Critical habitat is defined as:

(i) the specific areas within the geographical area occupied by the species, at the time it is listed in accordance with the provisions of section 1533 of this title, on which are found those physical or biological features
(ii) essential to the conservation of the species and
(iii) which may require special management considerations or protection; and
(iv) specific areas outside the geographical area occupied by the species at the time it is listed in accordance with the provisions of section 1533 of this title, upon a determination by the Secretary that such areas are essential for the conservation of the species.[166]

The FWS and NMFS have been slow to take into account changes in species' distribution and habitat that will result from global warming when designating critical habitat, but this is beginning to change. Proposals for species including the Quino checkerspot butterfly[167] and elkhorn and staghorn corals[168] include reference to climate change, though it remains to be seen whether the agencies will correct deficiencies in these proposals. As is evident from the definition of critical habitat, the ESA explicitly grants the authority to designate areas outside of a species' current range as critical habitat if those areas are "essential for the conservation of the species."[169] For many species undergoing rapid range shifts, protection of such areas as critical habitat will be one of the most important regulatory actions that will allow them to persist in a changing climate.

The ESA also requires the preparation, periodic update, and implementation of recovery plans to outline and carry out the steps necessary to conserve each listed

---

[163] Id.
[164] Ctr. for Biological Diversity v. Norton, No. 04-CV-02026 (GK) (D.D.C., filed Nov. 18, 2004).
[165] 16 U.S.C. §§ 1533(a)(3)(A) & (b)(6)(C). As a practical matter, critical habitat designation rarely happens in the absence of litigation to compel such designation.
[166] 16 U.S.C. § 1532(5).
[167] Endangered and Threatened Wildlife and Plants; Revised Designation of Critical Habitat for the Quino Checkerspot Butterfly (Euphydryas editha quino), 73 FED. REG. 3328–3373 (January 17, 2008).
[168] Endangered and Threatened Species; Critical Habitat for Threatened Elkhorn and Staghorn Corals, 73 FED. REG. 6895 (February 6, 2008).
[169] Id.

species.[170] The first recovery plan to mention global climate change was published in 1990. Recovery plans with such references were sporadic, but issued in most years between 1991 and 1999. Between 2000 and May 2008, the percentage of recovery plans with global climate change references grew dramatically. At least 48 present of plans issued in all years between 2004 and May 2008 referenced global climate change. Of those that address global climate change, most call for monitoring and/or mitigation. Few call for reductions in global warming or greenhouse gasses. None specify how reductions should be accomplished, either generally or in the context of the ESA (i.e., through Section 7 consultation, etc.). As recovery plans for the corals and polar bear are developed, the role of recovery planning in addressing the threat of global warming will likely be the center of significant contention. Similarly, as FWS and NMFS revise recovery plans for already listed species, these plans must, if they are to withstand legal scrutiny, analyze the effects of global warming on any such species likely to be harmed by such warming, and lay out a plan for both mitigation of and adaptation to those threats.[171]

Finally, the ESA also requires that every five years the status of all listed species be assessed to determine if they still warrant the protections of the Act, or if a change from Threatened to Endangered status (or the reverse) is warranted.[172] Such reviews open the door to the consideration of global warming's effects on all currently listed species, an essential step if they are to survive under even the most optimistic future scenarios for warming.

1.4.2. The Consultation Process and the Obligation to Avoid Jeopardy

As noted previously, the section 7 consultation process is the heart of the ESA. Section 7 directs all federal agencies to "insure through consultation" with FWS or NMFS, that all actions authorized, funded, or carried out by such agencies are "not likely to jeopardize the continued existence" or "result in the destruction or adverse modification" of "critical habitat" of any listed species."[173] The result of the consultation process is a biological opinion produced by FWS or NMFS concluding whether the action can go forward and suggesting alternatives to the action as necessary to avoid jeopardy to the species or adverse modification of critical habitat.[174]

For section 7 to protect species from global warming, global warming needs to be considered with regard to the action subject to consultation in two key respects: (1) the relevant agencies must take into account, and reduce or eliminate, the greenhouse gas emissions inevitably resulting from the action (mitigation); and (2) the relevant agencies must take into account the observed and projected effects of global warming on the species otherwise affected by the action (adaptation).

[170] 16 U.S.C. § 1533(f).
[171] Id.
[172] 16 U.S.C. § 1533(c)(2).
[173] 16 U.S.C. § 1536(a)(2).
[174] 16 U.S.C. § 1536(b).

Section 7 consultation is required for "any action [that] may affect listed species or critical habitat."[175] Agency "action" is defined in the ESA's implementing regulations to include "all activities or programs of any kind authorized, funded, or carried out, in whole or in part, by Federal agencies in the United States or upon the high seas. Examples include, but are not limited to: (a) actions intended to conserve listed species or their habitat; (b) the promulgation of regulations; (c) the granting of licenses, contracts, leases, easements, rights-of-way, permits, or grants-in-aid; or (d) *actions directly or indirectly causing modifications to the land, water, or air.*"[176]

This regulatory definition of "action" should be sufficiently broad to encompass actions that result in greenhouse gas emissions, as it would be hard to argue that such emissions are not "causing modification to the land, water, or air." The remaining question with respect to the triggering of these requirements for an action resulting in greenhouse gas emissions is whether that action "may affect" the listed species. While it is clear that global warming affects listed species, attributing an individual action's contribution to global warming is more difficult.

Because the goal of section 7 consultation is to avoid jeopardizing any listed species, the regulatory definition of "jeopardy" offers some guidance as to how the consultation requirement for a greenhouse gas–emitting action may be interpreted. To "jeopardize" a species means "to engage in an action that reasonably would be expected, directly or indirectly, to reduce *appreciably* the likelihood of both the survival and recovery of a listed species in the wild by reducing the reproduction, numbers, or distribution of that species."[177] If an action "appreciably" contributed to global warming, that action could then be found to jeopardize a listed species.[178] "Appreciably" has been defined as being "to the degree that can be estimated," while something is "appreciable" if it is "large or important enough to be noticed."[179] So if an action contributes an appreciable amount of greenhouse gas emissions to the atmosphere, that action should undergo the consultation process.

While many federal actions may not contribute appreciable amounts of greenhouse gases to the atmosphere, many clearly do so. For example, the corporate average fuel economy (CAFE) standards for cars and light trucks are set via regulation by the National Highway Transportation Safety Administration. Since the transportation sector represents a large component of U.S. greenhouse gas emissions, the volume of greenhouse gases represented by this single act of rulemaking are certainly "appreciable." Similarly, every five years the Minerals Management Service approves a program for all offshore oil and gas leasing for the entire United States. Again, the greenhouse gases generated through the life cycle of the

---

[175] 50 C.F.R. § 402.14.
[176] 50 C.F.R. § 402.02 (emphasis added).
[177] 50 C.F.R. § 402.02 (emphasis added).
[178] This analysis assumes the validity of the current consultation regulations. An argument can be made that the regulations improperly narrow the reach of the consultation requirements of section 7. However, such a critique of the regulation is beyond the scope of this chapter.
[179] Oxford English Dictionary online, *available at* http://www.askoxford.com.

production and use of these billions of barrels of oil are very "appreciable." The greenhouse gas emissions from numerous other actions, ranging from the approval of new coal-fired power plants, oil shale leasing programs, or limestone mines for cement manufacturing, and dozens, perhaps hundreds, of other projects are individually and cumulatively having an appreciable effect on the atmosphere. These are all agency "actions" as defined by the ESA, which "may affect" listed species, and therefore trigger the consultation requirements of section 7.[180]

While federal agencies have also been slow in consulting on the impacts of greenhouse gas emissions and global warming on ESA-listed species, this is also changing. In May 2007 a biological opinion analyzing the impact of water withdrawals on the delta smelt, a fish that occurs in California's San Joaquin Delta and is impacted by massive water pumping for agricultural and urban purposes, was overturned for failure to consider the impact of global warming on water levels for the fish.[181] Also in 2007, the FWS required consideration of greenhouse gas emissions and global warming in a Section 7 consultation for a new coal fired power plant in New Mexico.[182] With the listing of the polar bear, this trend should accelerate.[183]

Arguments against the applicability of section 7 to greenhouse emissions are premised on a claimed lack of demonstrable connection between greenhouse emissions and harm to listed species. However, the connection between greenhouse gas emissions and sea-ice reductions – and the effect that sea-ice decline has on polar bears – is supported by voluminous scientific literature and, indeed, is the central reason for the decision to place the polar bear on the list of Threatened and Endangered species. Just as there is no requirement to link the thinning of any particular bald eagle egg to any particular molecule of DDT to demonstrate that authorization of the use of DDT may result in a taking of bald eagles, there is no requirement to link any particular molecule of carbon dioxide or other greenhouse pollutant to global warming and the Arctic melt. The Supreme Court stated in *TVA* that Section 7 "admits of no exception," and affords endangered species "the highest of priorities."[184] There is no reason greenhouse gas emissions that jeopardize polar

---

[180] Many of these actions are also "major federal actions" under NEPA, 42 U.S.C. §§ 4321–4347, and the impacts of their emissions should be analyzed under that statute as well. *See, e.g.,* Found. on Econ. Trends v. Watkins, 794 F. Supp. 395 (D.D.C. 1992) (case attempting to force analysis of global warming impacts under NEPA).

[181] Natural Res. Def. Council v. Kempthorne, 506 F.Supp.2d 322 (E.D. Cal. 2007).

[182] Memorandum from Fish and Wildlife Service, Supervisor of New Mexico Ecological Services, to Bureau of Indian Affairs, Regional Director of Navajo Regional office (July 2, 2007) (on file with author).

[183] The Bush administration attempted to exempt gas emissions from Section 7 through a "midnight regulation," which would have substantially changed the nationwide Section 7 regulations in a number of ways, including creating an exemption for greenhouse gas emissions. Interagency Consultation Under the Endangered Species Act, 73 Fed. Reg. 76272–76287 (Dec. 16, 2008). However, these regulations were revoked by the Obama administration. Interagency Consultation Under the Endangered Species Act, 74 Fed. Reg. 20421–20423 (May 4, 2009).

[184] TVA, 437 U.S. at 173–74.

bears should be treated any differently than pesticides that harm salmon or logging that harms owls.

Beyond consultations in which the primary threat is global warming, the issue will likely emerge with respect to species subject to consultation for other reasons. For example, a finding that allowing the destruction of certain coastal wetlands relied upon by a listed species will not equate to jeopardy because sufficient other wetlands still exist in a nearby preserve utterly fails to protect the species if the preserve will no longer exist in fifty years following another half meter or more of sea level rise. Incorporating the changing conditions caused by global warming into agency decision-making is essential if already imperiled species are to survive given the amount of warming we are already committed to even under the best scenarios.

In sum, section 7 of the ESA carries with it the mandate to force actual reductions in greenhouse gas emissions because the federal agencies approving the actions responsible for such emissions have a duty to ensure against jeopardizing all listed species. Moreover, even in instances where section 7 ultimately does not result in actual emissions mitigation, it certainly holds the promise of forcing climate-informed decision-making on actions affecting listed species, such that these species have greater hope of surviving in a greenhouse world.

1.4.2. The Take Prohibition

While section 7 only applies to federal actions and agencies, the prohibitions of section 9 apply far more broadly, reaching the actions of private entities and corporations. Section 9 prohibits the "take" of Endangered species, which includes "harming" and "harassing" listed species in addition to simply killing them directly.[185] Both the legislative history and case law support "the broadest possible" reading of "take."[186] That reading should certainly be broad enough to encompass greenhouse gas emissions.

The impacts polar bears are already experiencing from global warming clearly meet the definition of "harm" and "harass."[187] Moreover, the temperature-induced bleaching of elkhorn and staghorn corals often results in mortality, and therefore also fits within the definition of "take."[188] The problem is one of causation. While it is clear that global warming is causing prohibited take of listed species, current warming is the product of past emissions. Under the citizen suit provision of the ESA, one can only enjoin ongoing or future take.[189] However, since past take is often the best evidence of the likelihood of future take, an appropriate defendant

---

[185] 16 U.S.C. § 1538(a).
[186] Babbitt v. Sweet Home Chapter of Cmtys. for a Great Or., 515 U.S. 687, 704–05 (1995).
[187] *See* Sections 1.1 & 1.3.2 *supra*. Courts have generally treated "harm" and "harass" as complimentary provisions, with an action found to "harm" a species usually also found to "harass" the species as well.
[188] *See* Section 1.2.2 *supra*.
[189] 16 U.S.C. § 1540(g); Nat'l Wildlife Fed'n v. Burlington N. R.R., 23 F.3d 1508, 1511 (9th Cir. 1994).

(or group of defendants) in a section 9 climate case would be an entity that has already contributed measurably to anthropogenic greenhouse gas emissions and, absent regulation, is likely to do so in the future. The utility company defendants in *Connecticut v. American Electric Power Co.*,[190] or any of the major oil companies, are responsible for enough greenhouse gas emissions that they individually, and certainly collectively, reasonably can be considered the proximate cause of the take of listed species.[191]

The section 9 take prohibition, for the most part, applies only to species listed as Endangered.[192] However, section 4(d) of the ESA requires FWS or NMFS to promulgate regulations applying any of the prohibitions of section 9 to Threatened species if such regulations are "necessary and advisable" for the conservation of the species.[193] FWS has promulgated a blanket 4(d) rule applying section 9 to all Threatened species.[194] On rare occasions, such as with the polar bear, FWS has issued species-specific 4(d) rules that alter the terms of the take prohibition.

In the 4(d) rule for the polar bear,[195] FWS has also attempted to exempt greenhouse gas emissions from regulation pursuant to section 9. This regulation is being challenged,[196] and, we believe, highly unlikely to survive judicial scrutiny. The first litigation regarding the role of section 9 in addressing greenhouse gas emissions thus will likely not be brought against a major corporate emitter of greenhouse gases, but rather against FWS to determine whether such regulations are necessary for the conservation of the polar bear.

2. CONCLUSION: WILDLIFE LAW AS A SURROGATE FOR OR SUPPLEMENT TO A NATIONAL POLICY ON GREENHOUSE GAS REGULATION

While we believe that the ESA has a significant role to play in the protection of U.S. and perhaps global biodiversity in the face of global warming, we are under no delusions that the statute represents the "silver bullet" that will result in the United States substantially reducing its greenhouse gas emissions so as to stave off the worst effects of global warming. Without massive policy changes, global warming threatens as many as a third to a half of the species currently living on Earth with

---

[190] Connecticut v. Am. Elec. Power Co., 406 F.Supp.2d 265 (S.D.N.Y. 2004), *appeal docketed*, No. 05-5104 (2nd Cir. Sept. 22, 2005).

[191] For a discussion of the application of section 9 to takings resulting from the combined impacts of multiple independent actors, *see, e.g.*, Federico Cheever & Michael Balster, *The Take Prohibition in Section 9 of the Endangered Species Act: Contradictions, Ugly Ducklings, and Conservation of Species*, 34 ENVTL. L. 363 (2004).

[192] 16 U.S.C. § 1538(a).

[193] 16 U.S.C. § 1533(d).

[194] 50 C.F.R. § 17.31(a).

[195] 73 FED. REG. at 28,306–28,318 (Endangered and Threatened Wildlife and Plants, Special Rule for the Polar Bear) (May 15, 2008) ("4(d) Rule"). 16 U.S.C. § 1538(a); 50 C.F.R. § 17.31.

[196] *In re* Polar Bear Endangered Species Act Listing and 4(d) Rule Litigation, Misc. Action 08-0768 (EGS) (Dist. D.C. 2008).

extinction.[197] Nevertheless, the ESA's breadth and force mean that it stands as one of the most promising mechanisms to force government and corporate entities to disclose, analyze, and mitigate the impacts of their greenhouse gas emissions.

Efforts to address global warming through the ESA have received so much attention in large part because the Bush administration successfully blocked regulation of greenhouse emissions under the Clean Air Act, which should be the first-tier regulator of all major sources of greenhouse emissions in the United States. It is important to recognize that section 7 consultations for greenhouse gas emissions should not occur in a regulatory vacuum, but rather as one part of a comprehensive and somewhat overlapping set of U.S. domestic laws that already address greenhouse emissions. Resistance to addressing greenhouse emissions through section 7 consultations often springs from a lack of understanding of how regulation under the ESA would fit into the comprehensive, complimentary, and in some ways overlapping regulatory system for greenhouse emissions that is already provided by the Clean Air Act, Clean Water Act, ESA, National Environmental Policy Act, Global Change Research Act, and other laws. The irony of the U.S. foot-dragging on global warming is that the United States has the strongest domestic environmental laws in the world, capable of effectively and efficiently reducing greenhouse gas emissions immediately, and each with a proven track record of success in protecting the air we breathe, the water we drink, and the diversity of life on Earth. In contrast to the Bush administration, which actively sought to block use of any U.S. laws to address greenhouse gas emissions, the Obama administration appears to be taking steps, at long last, towards actually implementing existing law. Whether President Obama will act with the speed and decisiveness required to solve the climate crisis remains to be seen.

While new, science-based federal climate legislation that would mandate the deep greenhouse gas reductions from the U.S. economy necessary to address the climate crisis would certainly be welcome, it is critically important that new legislation build upon the successful regulatory regime already in place. For over four decades our flagship environmental laws have saved lives, saved money, and protected the environment, and there is no reason to weaken or eliminate any provision of existing law when adding new climate focused mandates. We have irreversibly altered the natural world; only by acknowledging and urgently responding to those changes will most species still have a chance of persisting in the very different world we have created for them. The ESA, still strong and relevant thirty years after it was passed into law, is essential to that effort.

[197] Thomas (2004), *supra* note 143; Hansen (2006), *supra* note 54, at 14, 292.

8

An Emerging Human Right to Security from Climate Change: The Case Against Gas Flaring in Nigeria

Amy Sinden*

INTRODUCTION

Is there a human right to security from climate change? A recent ruling by the Federal High Court of Nigeria suggests that there is. Royal Dutch/Shell Group (Shell) and the other companies that produce oil in Nigeria have engaged for decades in a practice called "gas flaring," in which natural gas released during oil extraction is burned off, discharging large clouds of greenhouse gases and other pollutants into the atmosphere.[1] Citing the climatic and other environmental impacts of gas flaring on their community, Nigerians living near the flares filed a lawsuit charging that the practice violates their fundamental rights to life and dignity guaranteed under the Nigerian constitution.[2] In a ruling on November 14, 2005, the Federal High Court of Nigeria agreed and ordered Shell and the Nigerian National Petroleum Corporation "to take immediate steps to stop the further flaring of gas" in the plaintiffs' community.[3] Although the court's ruling thus far has had little practical

---

* Associate Professor, Temple University Beasley School of Law, 1719 N. Broad St., Philadelphia, PA 19122, Amy.Sinden@temple.edu, (215) 204-4969.
[1] See The Climate Justice Programme & Environmental Rights Action/Friends of the Earth Nigeria, *Gas Flaring in Nigeria: A Human Rights, Environmental and Economic Monstrosity* (2005) [hereinafter Gas Flaring Report], *available at* http://www.climatelaw.org/media/gas.flaring/report/gas.flaring.in.nigeria.html; Asume (Isaac) Osuoka, *The Shell Report: Continuing Abuses in Nigeria – Ten Years after Ken Saro Wiwa* 23–26 (Environmental Rights Action/Friends of the Earth Nigeria 2005) (on file with author) [hereinafter Shell Report]. Gas flaring produces both of the primary greenhouse gases, carbon dioxide and methane. Gas Flaring Report, at 20; Intergovernmental Panel on Climate Change, *Climate Change 2007: The Physical Science Basis, Summary for Policymakers* 4–5 (Feb. 2007) [hereinafter IPCC 2007 Summary], *available at* http://www.ipcc.ch/SPM2feb07.pdf. Methane has twenty-one times the global warming impact of carbon dioxide. Most of the gas emitted in a flare is burned, producing carbon dioxide from the combustion process. But the combustion process is never 100 percent efficient. As a result, some methane gas – as much as 10 percent – is released directly into the atmosphere without combusting. See Gas Flaring Report, at 20.
[2] *See* Gbemre v. Shell Petroleum Development Co., Suit No. FHC/CS/B/153/2005, Motion Ex Parte under Section 46(1) of the Constitution of the Federal Republic of Nigeria, Statement Pursuant to Order 1, Rule 2(3) of the Fundamental Rights (Enforcement Procedure) Rules, and Verifying Affidavit (July 11, 2005) [hereinafter Gbemre Pleadings], *available at* http://www.climatelaw.org/cases.
[3] Gbemre v. Shell Petroleum Development Co., Suit No. FHC/CS/B/153/2005, Order (Nov. 14, 2005) [hereinafter Gbemre Order], *available at* http://www.climatelaw.org/cases.

effect – the oil companies have yet to comply and are appealing the order – it opens up intriguing possibilities for crafting legal approaches to the problem of climate change.

Little analysis of the plaintiffs' climate change claim accompanied the Nigerian court's ruling, but the notion that actions that contribute substantially to climate change may violate fundamental constitutional or human rights is intuitively appealing. Constitutional rights and international human rights – I will refer to these collectively as "human rights"[4] – invoke a sense of profundity and moral weight that comports with the enormity and gravity of the climate change problem. In my view, this intuition is correct. A right to security from climate change actually fits comfortably within the principles and values that underlie some of the oldest and most venerated rights in the civil and political rights tradition. Even though that tradition was born more than 200 years ago, long before anyone could have conceived of the idea that human activities might one day reach a magnitude and scope sufficient to alter the environment on a global scale, the problem of climate change – at least in its political aspects – is exactly the kind of problem that civil and political rights are aimed at combating. It is a problem that arises fundamentally from the distortion of government decision making by power.

Human rights, of course, have traditionally been defined as imposing duties primarily on States.[5] And yet in many instances the affirmative acts that contribute most dramatically to climate change are committed by private actors, as Shell's involvement in the Nigerian gas flaring illustrates. While the *Gbemre* court did not address this issue, even under traditional doctrine, the close relationship between Shell and the Nigerian government in the operation of the Nigerian National Petroleum Corporation may well warrant a finding of liability against Shell for acting in concert with the State. Moreover, the recognition of rights in the context of climate change may help build momentum toward a reconceptualization of human rights law that broadens the set of duty holders to include not just States, but another set of important actors on the international stage that often wield even more wealth and power than States themselves – multinational corporations.

This chapter proceeds in two parts. The first part provides some background on oil development in the Niger Delta and the gas flaring litigation. The second part explores possible theories under which a right to security from climate change might be grounded in traditional civil and political human rights, as well as how such rights

---

Another action alleging human rights violations in connection with climate change has been filed with the Inter-American Commission on Human Rights by Inuit Circumpolar Conference against the United States. *See* Petition to the Inter-American Commission on Human Rights Seeking Relief from Violations Resulting from Global Warming Caused by Acts and Omissions of the United States (Dec. 7, 2005), *available at* http://www.ciel.org/Publications/ICC_Petition_7Dec05.pdf. For a discussion of that action, see Hari M. Osofsky, *The Inuit Petition as a Bridge? Beyond Dialectics of Climate Change and Indigenous Peoples' Rights*, in this volume.

[4] *See infra* notes 57 to 60 and accompanying text.
[5] *See infra* note 99.

might be either traditionally applied or reconceptualized to impose liability on a private multinational corporation like Shell.

1. BACKGROUND

1.1. Decades of Environmental Devastation and Military Repression in the Niger Delta

Where the Niger River reaches the coastal plain on the south coast of West Africa, it divides into hundreds of small streams and rivulets branching out across the flat landscape. Covering 70,000 square kilometers, the Niger Delta is one of the largest wetlands on Earth.[6] For millennia, it has supported a rich and diverse ecosystem, with one of the highest concentrations of biodiversity on the planet.[7] It is also home to over 10 million people from dozens of different ethnic groups, most of whom depend on subsistence fishing and small-scale farming.[8]

However, the rich ecology of this region and its inhabitants' way of life are now in peril. In the 1950s, Shell and British Petroleum discovered oil in the Delta.[9] Currently, 2.5 billion barrels of crude oil are pumped out of the Niger Delta each day.[10] While a number of big multinational oil companies now have partial stakes in Nigeria's oil industry, Shell remains by far the biggest player. Nearly half of Nigeria's oil is produced by the Shell Petroleum Development Company, a joint venture operated by Shell and owned primarily by Shell and the Nigerian National Petroleum Corporation (NNPC).[11] Oil dominates the Nigerian economy, providing more than 80 percent of government revenues, 90 percent of foreign exchange earnings, and 40 percent of GDP.[12] Three-quarters of Nigeria's oil comes from the Niger Delta.[13]

---

[6] Ibibia Lucky Worika, *Deprivation, Despoliation and Destitution: Whither Environment and Human Rights in Nigeria's Niger Delta?*, 8 ILSA J. INT'L & COMP. L. 1, 4 (2001).

[7] *See* Kaniye S. A. Ebeku, *Biodiversity Conservation in Nigeria: An Appraisal of the Legal Regime in Relation to the Niger Delta Area of the Country*, 16 J. ENVTL. L. 361, 362–65 (2004); ANDY ROWELL, JAMES MARRIOTT & LORNE STOCKMAN, THE NEXT GULF: LONDON, WASHINGTON & OIL CONFLICT IN NIGERIA 8 (2005); IKE OKONTA & ORONTO DOUGLAS, WHERE VULTURES FEAST: SHELL, HUMAN RIGHTS AND OIL IN THE NIGER DELTA 61–63 (2003).

[8] *Id.*; Worika, *supra* note 6, at 4.

[9] Gas Flaring Report, *supra* note 1, at 6; Rowell, *supra* note 7, at 58.

[10] Gas Flaring Report, *supra* note 1, at 5.

[11] *See* ROWELL, *supra* note 7, at 9–10. Shell owns a 30 percent share, NNPC a 55 percent share, the French company ELF-Aquitaine a 10 percent share, and the Italian company AGIP a 5 percent share. *See* OKONTA, *supra* note 7, at 49. Under the Nigerian Constitution, all oil is the property of the federal government, so all oil companies operate in Nigeria as joint ventures with the NNPC. *See* Constitution of the Federal Republic of Nigeria (1999), Art. 44(3); Alison Shinsato, *Increasing the Accountability of Transnational Corporations for Environmental Harms: The Petroleum Industry in Nigeria*, 4 Nw. U. J. INT'L HUM. RIGHTS 186, 191 (2005).

[12] *See* ROWELL, *supra* note 7, at 9.

[13] *See* OKONTA, *supra* note 7, at 18.

Since Shell first struck oil there five decades ago, 159 oil fields and 275 flow stations have been carved out of the fragile Delta ecosystem.[14] Seven thousand kilometers of rusty pipeline – some of it forty years old – snakes across the landscape.[15] The pipeline frequently ruptures, spewing crude oil across the land and water. According to the Nigerian government, 6,817 oil spills occurred in the Niger Delta between 1976 and 2001 (about one a day for twenty-five years). But other experts estimate that the actual amount may be ten times higher.[16] Even as of a decade ago, a report by the CIA estimated that the amount of oil spilled in the Niger Delta was already ten times the amount of the Alaskan *Exxon Valdez* spill.[17]

The spilled oil and the outdated waste disposal techniques practiced by Shell and the other oil companies operating in the Delta have wreaked havoc on the health of the people there and on the ecosystems upon which they depend. But spilled oil and waste is only part of the environmental devastation oil development has brought to the Delta. Oil deposits are often accompanied by natural gas that escapes from the ground when the oil is pumped. Although it is possible to capture this escaping gas and either reinject it into the ground or collect it for sale, the oil companies operating in Nigeria choose instead to simply burn it off. Indeed, gas flaring has been standard practice since oil production first began in Nigeria in the 1950s.[18] Thus, most of the oil wells in Nigeria are accompanied by a raging flame that burns twenty-four hours a day, reaching hundreds of feet into the sky, killing the surrounding vegetation with searing heat, emitting a deafening roar, and belching a cocktail of smoke, soot, and toxic chemicals into the air along with a potent mixture of greenhouse gases.[19] The devastating environmental effects of this wasteful practice have led other countries to reduce gas flaring to a bare minimum. In the United States, less than half of 1 percent of extracted natural gas is flared;[20] in Western Europe, the rate is less than 1 percent.[21] But the same multinational corporations that have virtually stopped the practice in other parts of the world continue to flare 75 percent of the natural gas produced in Nigeria.[22] Because the amount of oil drilling conducted in the Niger Delta is enormous – Nigeria is OPEC's fifth-largest producer of oil and most of it comes from the Delta[23] – the absolute amount of gas being wasted by flaring and the magnitude of the accompanying environmental destruction is staggering. The Niger Delta produces 2.5 billion barrels of crude oil every day, and

[14] See Tom O'Neill, *Curse of the Black Gold: Hope and Betrayal in the Niger Delta*, NAT'L GEOGRAPHIC, Feb. 2007, at 88, *available at* http://www7.nationalgeographic.com/ngm/0702/feature3/index.html.
[15] See Shinsato *supra* note 11; *see also* OKONTA, *supra* note 7, at 77–78.
[16] See O'Neill, *supra* note 14.
[17] See Douglas Farah, *Nigeria's Oil Exploitation Leaves Delta Poor, Poisoned*, WASH. POST, Mar. 18, 2001, at A22.
[18] Shell Report, *supra* note 2, at 23–24; OKONTA, *supra* note 7, at 66.
[19] ROWELL, *supra* note 7, at 67, 245; OKONTA, *supra* note 7, at 73, 78–79, 84–85.
[20] Shinsato, *supra* note 15.
[21] See Peter Roderick, *Environmental Justice, Climate Change, and Environmental Racism in the Niger Delta*, *available at* http://www.foei.org/en/publications/link/env-rights/54.html.
[22] *See* Gas Flaring Report, *supra* note 1, at 11.
[23] *See* Shinsato, *supra* note 15, at 191.

most of the associated 2.5 billion cubic feet of natural gas is burned off into the atmosphere.[24] In 2001, 40 percent of all the natural gas burned throughout Africa was attributable to gas flaring in Nigeria.[25] It is estimated that Nigeria's gas flaring has contributed more greenhouse gases to the atmosphere than all of sub-Saharan Africa combined.[26]

Ironically, the global climate change to which gas flaring in the Niger Delta makes such a significant contribution is likely to produce particularly devastating impacts in Nigeria itself. The Intergovernmental Panel on Climate Change (IPCC) has concluded that "Africa is one of the most vulnerable continents to climate variability and change."[27] The Niger Delta itself is particularly vulnerable to sea level rise. The low-lying marshy lands of the Delta have been gradually subsiding in recent years, a process that has been significantly exacerbated by oil and gas extraction.[28] This subsidence in combination with the sea level rise predicted to occur as a consequence of climate change is likely to cause widespread inundation and dislocation. Studies have estimated that a forty-kilometer-wide strip of the Delta could be inundated within decades and that 80 percent of the Delta's population will have to move.[29] The IPCC reported in 2001 that a one-meter rise in sea level could put 600 square kilometers of land and more than 3 million people at risk in Nigeria.[30] Additionally, increasingly frequent and severe storms triggered by climate change could have devastating impacts. One study concludes that rising sea surface temperatures off Nigeria's coast have the capacity to trigger tornado-type storms in the Niger Delta. Such storms are likely to cause "huge storm surges and catastrophic flooding that will result in unprecedented deaths and collapse or destruction of coastal infrastructure."[31]

Although the dangers of climate change have only recently come to be widely recognized, the people of the Niger Delta have long understood the devastating impacts of the oil industry on their lands and waters. Beginning in the late 1980s, protests against the oil industry in the communities of the Niger Delta met with

---

[24] Gas Flaring Report, *supra* note 1, at 4–5.
[25] *Id.*
[26] *Id.*
[27] IPCC, Climate Change 2007: Climate Change Impacts, Adaptation and Vulnerability, Summary for Policymakers 10 (Apr. 6, 2007) [hereinafter IPCC 2007 Impacts Summary], *available at* http://www.ipcc.ch/SPM6avr07.pdf. The enumerated impacts include the risk of regional conflict over dwindling water resources, declines in agricultural production, potentially irreversible losses in natural resource productivity and biodiversity, risks of increased vector- and water-borne disease, increased desertification, and flooding of coastal areas due to sea level rise. *Id.*
[28] *See* ROWELL, *supra* note 7, at 235.
[29] *Id.*
[30] *See* Intergovernmental Panel on Climate Change, IPCC Special Report on the Regional Impacts of Climate Change: An Assessment of Vulnerability, ch. 2.3.4.1.2, (1997) [hereinafter IPCC Special Report on Regional Impacts] *available at* http://www.grida.no/climate/ipcc/regional/index.htm.
[31] D. O. Adefolalu & J. F. Adeyemi, *Climate Change: Potential Impact on the Niger Delta – the Economic Nerve-Center of Nigeria*, International Conference on Energy, Environment and Disasters, Charlotte, N.C., July 24–30, 2005.

increasingly brutal and violent repression by the Nigerian military.[32] In the early 1990s in the Ogoniland region of the Niger Delta, Nigerian security forces killed 2,000 people and razed thirty villages in an attempt to quell mounting protests by the Ogoni people against oil development in their region.[33] There is evidence that Shell played a major role in instigating and supporting this violence, supplying helicopters and boats to transport the troops for these operations and paying bonuses to the military personnel who participated.[34] There is also evidence that Shell was involved or complicit in the Nigerian government's 1994 arrest and subsequent execution of Ogoni political leader Ken Saro Wiwa in what was widely condemned as a sham proceeding.[35]

1.2. The Gas Flaring Litigation

The events described previously received considerable publicity and triggered two lawsuits alleging human rights violations in connection with the repression of the Ogoni and the environmental destruction caused by oil development in Ogoniland.[36] But a series of lawsuits filed in 2005 in the Nigerian courts are the first to focus specifically on the practice of gas flaring in the Niger Delta. Seven cases have been filed in various local divisions of the federal court system in Nigeria.[37] In each

---

[32] See ROWELL, supra note 7, at 83–84. In 1987, Nigeria's Military Police Force came to the small fishing village of Ito at the request of Shell to quell protests, arriving in speedboats belonging to Shell. By the end of the ensuing confrontation, the police had killed two demonstrators and destroyed forty houses, leaving 350 villagers homeless. Id.

[33] See Paul Lewis, Blood and Oil: A Special Report: After Nigeria Represses, Shell Defends Its Record, N.Y. TIMES, Feb. 13, 1996, at A1.

[34] See Douglass Cassel, Corporate Initiatives: A Second Human Rights Revolution?, 19 FORDHAM INT'L L.J. 1963, 1965–66 (1996).

[35] See Paul Lewis, Blood and Oil: A Special Report: After Nigeria Represses, Shell Defends Its Record, N.Y. TIMES, Feb. 13, 1996, at A1; Douglass Cassel, Corporate Initiatives: A Second Human Rights Revolution?, 19 FORDHAM INT'L L.J. 1963, 1966–67 (1996); Rowell, supra note 7, at 1–7. Two prosecution witnesses recanted their testimony during the trial, claiming they had been bribed to provide testimony implicating Saro Wiwa in the murders. In a filmed statement and sworn affidavit, one said that he had been promised money and contracts with Shell to testify against Saro Wiwa and that Shell representatives were present when the offer was made. Shell Denies Foul Play in Nigerian Murder Trial, GUARDIAN, Sept. 29, 1995, at 13.

[36] One, filed in U.S. federal court in 1996 under the Alien Tort Claims Act against Shell, is still proceeding in the trial court. See Wiwa v. Royal Dutch Petroleum Co., 226 F.3d 88 (2d Cir. 2000). The other, filed against the Nigerian government before the African Commission on Human and Peoples' Rights, resulted in a ruling that the environmental destruction of Ogoniland and the military repression of the Ogoni people violated their rights to life, to nondiscrimination, to property and family, to health and a satisfactory environment, and to the free disposal of wealth and natural resources under the African Charter on Human and Peoples' Rights. See Social & Econ. Rights Action Ctr. for Econ. and Social Rights v. Nigeria, African Commission on Human and People's Rights, Comm. No. 155/96 (2001), available at http://www1.umn.edu/humanrts/africa/comcases/155-96b.html; see also Justice C. Nwobike, The African Commission on Human and People's Rights and the Demystification of Second and Third Generation Rights under the African Charter: Social and Economic Rights Action Center (SERAC) and the Center for Economic and Social Rights (CESR) v. Nigeria, 1 AFRICAN J. LEGAL STUD. 129 (2005) (analyzing decision).

[37] Conversation with Peter Roderick, Climate Justice Programme, July 28, 2006.

one, the members of the local community as a class have sued the oil companies engaged in gas flaring in their locality. Four of the cases name Shell as a defendant. Chevron, the French company ELF-Aquitaine, and the Italian company AGIP are also each named in one of the other suits. Each suit also names as defendants the Attorney General of Nigeria and the Nigerian National Petroleum Corporation.

In each case, the plaintiffs allege that the defendants' practice of gas flaring violates their fundamental rights to life and dignity guaranteed under the Nigerian constitution as well as their rights to life, integrity of the person, health, and a satisfactory environment guaranteed under the African Charter on Human and Peoples' Rights.[38] The basis for these claims is the damage to the environment and the health of the local people caused by the air pollution and noise emitted by the flares: Plaintiffs point to an "increased risk of premature death, respiratory illnesses, asthma and cancer," acid rain, and reduced crop production.[39] But perhaps most interestingly, plaintiffs also cite the contribution of gas flaring to climate change as a basis for their constitutional and human rights claims.[40]

The first case was filed in the federal court in Benin City on July 11, 2005, by Jonah Gbemre on behalf of himself and the Iwherekan Community in Delta State (the "applicants").[41] It named Shell, the NNPC, and the Attorney General of Nigeria as respondents.[42] In their "counter-affidavits" to Gbemre's claims, Shell and NNPC denied that they engaged in gas flaring in the Iwherekan Community and denied any causal connection between their activities and the adverse environmental impacts cited by the applicants.[43] Shell and NNPC then engaged in a series of procedural maneuvers apparently designed to delay a decision on the merits, but that instead seemed only to irritate the trial judge.

When their attempt to convince the trial judge to delay a ruling on the merits in order to dispose of certain procedural motions failed, the lawyers representing Shell and NNPC initiated multiple proceedings in the court of appeals. They then made repeated motions before the trial court for stays pending appeal. The trial judge denied these motions and repeatedly directed defense counsel to present argument in response to the merits of plaintiffs' claims. At each successive hearing, however, defense counsel avoided arguing the merits by making additional motions for stays, continuances, and even recusal, all of which the judge denied. At one hearing, relations between defense counsel and the judge became so strained that defense

---

[38] Gbemre Pleadings, *supra* note 2, Statement at B.2, C.1, C.2.
[39] *Id.*, Verifying Affidavit at 11(b), (f), (g).
[40] *Id.* at 11(a), (c).
[41] *See* Gbemre Pleadings, *supra* note 2. The court granted Mr. Gbemre permission to bring his case on July 21, 2005. *See* Gbemre v. Shell Petroleum Dev. Co., Suit No. FHC/CS/B/153/2005, Judgment 1 (Nov. 14, 2005) [hereinafter Gbemre Judgment], *available at* http://www.climatelaw.org/cases.
[42] Gbemre Pleadings, *supra* note 2.
[43] *See* Gbemre v. Shell Petroleum Dev. Co., Suit No. FHC/CS/B/153/2005, Counter Affidavit of Mrs. Enobong Ozor (Aug. 30, 2005); Gbemre v. Shell Petroleum Dev. Co., Suit No. FHC/CS/B/153/2005, Counter Affidavit for 2nd Defendant by Mary Akujobi (Sept. 15, 2005) [on file with author].

counsel abruptly stood up and, with all his junior associates in tow, walked out of court "without," in the words of the court, "the usual courtesy of bowing to the bench."[44]

The trial judge's growing irritation with these maneuverings comes through in the final judgment, in which he accuses the lawyers for Shell and NNPC of acting "in bad faith" and calls their repeated motions for stays "an abuse of the process of this Court."[45] Ultimately, the court ruled that Shell and NNPC were "hereby foreclosed from presenting any further Reply" to the applicants' claims,[46] and on November 14, 2005, the court issued its final judgment.

The court ruled that the constitutional rights cited by plaintiffs "inevitably include[] the rights to [a] clean poison-free healthy environment," and that the defendants' gas flaring constitutes "a gross violation of [the plaintiffs'] fundamental [constitutional] right to life and dignity."[47] While the court's judgment referenced the plaintiffs' assertions in their affidavit that gas flaring leads to the emission of greenhouse gases and "contributes to adverse climate change,"[48] the court made no specific findings with respect to climate change and offered no analysis of whether the climate change impacts of gas flaring in particular formed part of the basis for its holding that a violation of constitutional rights had occurred. The court's order "restrained [Shell and NNPC] from further flaring of gas in Applicant's Community" and ordered those defendants "to take immediate steps to stop the further flaring of gas."[49]

After the court's order, the flares continued to burn. The companies took no steps to stop gas flaring as they had been directed to do by the court and provided no indication that they intended to do so. In December 2005, the applicants went back to the trial court with a motion seeking to have certain officials of Shell and NNPC held in contempt.[50] When the court ruled on this motion it softened its original order somewhat, giving the defendants an additional year, until April 2007, to stop the flaring. But the judge also ordered officials of Shell and NNPC to personally appear before the court on May 31, 2006, to present a quarterly program for ending the flaring by the April 2007 deadline.[51] That hearing never occurred, however. Shortly after the trial court's ruling, Shell and NNPC's repeated attempts to get the appeals court involved in the case finally bore fruit. The appeals courts actually issued an order restraining the trial court from sitting on May 31, 2006, thus halting the contempt proceedings.[52] Subsequently, the trial judge was

[44] Gbemre Judgment, *supra* note 4, at 27.
[45] *Id.* at 24.
[46] *Id.* at 28.
[47] *Id.* at 29.
[48] *Id.* at 4–5.
[49] Gbemre Order, *supra* note 3, at 4.
[50] *See* Climate Justice Programme, *Contempt of Court Proceedings Against Shell*, http://www.climatelaw.org/media/nigeria.shell.contempt.dec05 (posted Dec. 16, 2005).
[51] *See id.*
[52] Conversation with Peter Roderick, July 28, 2006.

removed from the case and transferred to a different district in the north of the country.[53]

The April 2007 deadline has now come and gone, but Shell and NNPC have not stopped the flaring, nor have they submitted a plan for doing so.[54] Meanwhile, the procedural issues as well as Shell's and NNPC's appeal of the trial court's ruling on the merits are in front of the court of appeal and a ruling is expected some time this year. The case will then likely make its way to the Nigerian Supreme Court. Thus, as a legal matter, the trial court's ruling stands, though it has had little practical effect.

While the trial court's judgment in *Gbemre* contains broad and definitive language reading an environmental right into the Nigerian Constitution's right to life, it does not explicitly analyze the question of whether and under what circumstances climate change impacts can provide a basis for finding a violation of that right. The sparseness of the court's analysis may reflect the fact that the respondents defended the case almost entirely on procedural grounds and presented little to the court in the way of substantive opposition to the applicants' claims. Nonetheless, the court's judgment does make reference to the evidence submitted by the applicants linking gas flaring to greenhouse gas emissions and climate change.[55] On that basis, one could read it as implicitly recognizing a right to security from climate change. Under any reading, though, the court's opinion leaves a lot of questions about the justifications for and scope of such a right unaddressed. The following analysis begins to explore these questions.

2. LOCATING A RIGHT TO SECURITY FROM CLIMATE CHANGE IN THE HUMAN RIGHTS TRADITION

2.1. *The Human Rights Tradition*

While there are clearly important distinctions between domestic constitutional rights and international human rights, for purposes of this analysis, the similarities are more important than the differences.[56] Both constitutional rights and international human rights are traditionally understood to encompass a particular set of rights that individuals enjoy against the State.[57] As such, they are fundamentally different

---

[53] *See* Friends of the Earth, *Shell Fails to Obey Gas Flaring Court Order* (May 2, 2007), *available at* http://www.foe.co.uk/resource/press_releases/shell_fails_to_obey_gas_fl_02052007.html.
[54] *See id.*
[55] *See supra* note 48.
[56] Constitutional rights and international human rights obviously differ in important respects. *See generally* Gerald L. Neuman, *Human Rights and Constitutional Rights: Harmony and Dissonance*, 55 STAN. L. REV. 1863 (2003). International human rights, for example, stand even more firmly outside the State than do constitutional rights, since they are defined by international law and have the capacity to be enforced by international rather than domestic tribunals.
[57] *See* Rex Martin, *Human Rights and Civil Rights*, *in* THE PHILOSOPHY OF HUMAN RIGHTS 75, 79–81 (Morton E. Winston ed., 1989).

from the private rights of tort and contract that government enables individuals to enforce against each other.[58] Constitutional and international human rights come into play when for some reason we cannot trust the political system to protect certain individual interests through the usual forms of private law. Usually that occurs when there is some reason to worry about abuse of power by the State itself. In those instances, because we cannot trust the State to police itself, we need some higher source of authority to act as a check on State power. Within domestic legal systems, that higher source of authority is the constitution.[59] In international law, it is international human rights norms.[60] Accordingly, both constitutional rights and international human rights share this key characteristic of standing outside and above the State in order to constrain abuses of State power.[61] I use the term "human rights" to refer to both kinds of rights.

Human rights have evolved roughly in three waves. Civil and political human rights arose during the Enlightenment and form the basis for the U.S. Bill of Rights and the French Declaration of the Rights of Man. They are rooted in a conception of the person as an autonomous individual, and they stress the protection of individual dignity and autonomy from government interference. Economic and social rights, by contrast, arose in the mid-twentieth century and are grounded in the notion that government has affirmative obligations to protect individuals from deprivation of the basic material necessities of life.[62] Finally, in the last several decades, a "third generation" of human rights has begun to emerge. These rights attach to groups rather than individuals and are aimed at the preservation of cultural identity and self-determination.[63]

Second and third generation rights may seem at first blush more amenable to the accommodation of a climate change right. Economic and social rights typically include a right to health and sometimes even an explicit right to a healthy environment,[64] and third generation rights often include a right to the free use of natural resources. But second and third generation rights are generally less enforceable

---

[58] See LOCKE: TWO TREATISES OF GOVERNMENT 271–72, 357–63 (Peter Laslett ed., 1988) (government's purpose to protect rights of individuals against invasion by each other, but in serving that function, government necessarily accrues power, which it has duty to citizens not to abuse).

[59] See, e.g., Stanley v. Illinois, 405 U.S. 645, 656 (1972) (observing that the Bill of Rights was "designed to protect the fragile values of a vulnerable citizenry" from overbearing government officials); Wolff v. McDonnell, 418 U.S. 539, 558 (1974) ("The touchstone of due process is protection of the individual against arbitrary action of government.").

[60] See Steven Ratner, Corporations and Human Rights: A Theory of Legal Responsibility, 111 YALE L.J. 443, 466 (2001); Louis Henkin, International Rights as Human Rights, in THE PHILOSOPHY OF HUMAN RIGHTS 129, 131 (Morton E. Winston ed., 1989).

[61] See TOM CAMPBELL, RIGHTS: A CRITICAL INTRODUCTION 37–39 (2006).

[62] See THE PHILOSOPHY OF HUMAN RIGHTS 4–5, 18–19 (Morton E. Winston ed., 1989); CAMPBELL, supra note 61, at 5–10.

[63] See generally Philip Alston, A Third Generation of Solidarity Rights: Progressive Development or Obfuscation of International Human Rights Law? 29 NETH. INT'L L. REV. 307 (1982).

[64] See 1988 Protocol to the American Convention on Human Rights, Art. 11 ("Everyone shall have the right to live in a healthy environment").

than civil and political rights.⁶⁵ First, they are typically expressed in less binding terms. For example, the International Covenant on Economic, Social, and Cultural Rights only calls on States to "take steps" to achieve the enumerated rights "up to the maximum of available resources."⁶⁶ The International Covenant on Civil and Political Rights, by contrast, directs each State to "undertake to respect and to ensure [the enumerated rights] to every individual within its territory."⁶⁷ Moreover, second and third generation rights are often framed in explicitly nonjusticiable terms. Many constitutions, for example, include them in a separate section designated for nonjusticiable rights.⁶⁸ Indeed, the Nigerian constitution includes an environmental rights provision, but it is in a separate section of the constitution that the courts have interpreted as nonjusticiable.⁶⁹ For this reason, the applicants in *Gbemre* did not even cite the environmental rights clause in their pleadings.⁷⁰ They did cite certain second and third generation rights from the African Charter on Human and Peoples' Rights, but rather than relying on these rights as the basis for a free-standing claim, they simply referenced them as "reinforce[ing]" the civil and political rights in the Nigerian constitution.

In sum, civil and political rights, with their centuries-old pedigree, enjoy far more acceptance and are far more likely to be viewed as enforceable by the courts than second and third generation rights. Accordingly, a climate change right is likely to be far more effective both rhetorically and legally if it is grounded in traditional civil and political rights.

Additionally, the values and concerns that underlie our civil and political rights tradition are of particular salience in the context of climate change. As I have argued elsewhere, civil and political rights are grounded fundamentally in concerns about power imbalance and its distorting effect on government decision making. Thus, many of the rights that we consider central to our civil and political rights tradition aim at counteracting the disparity of power between the State and the individual in

---

⁶⁵ See Robin R. Churchill, *Environmental Rights in Existing Human Rights Treaties*, in HUMAN RIGHTS APPROACHES TO ENVIRONMENTAL PROTECTION 89, 100 (Alan E. Boyle & Michael R. Anderson eds. 1996) (calling the environmental rights provision in Article 11 of the 1988 Protocol to the American Convention on Human Rights "rather weak, since it requires party States essentially to do no more than what they feel able to do, in the light of their available resources").

⁶⁶ International Covenant on Economic, Social and Cultural Rights, Art. 2(1).

⁶⁷ International Covenant on Civil and Political Rights, Art. 2(1).

⁶⁸ Barry E. Hill, Steve Wolfson & Nicholas Targ, *Human Rights and the Environment: A Synopsis and Some Predictions*, 16 GEO. INT'L ENVTL. L. REV. 359, 381–82 (2004); Michael R. Anderson, *Individual Rights to Environmental Protection in India*, in HUMAN RIGHTS APPROACHES TO ENVIRONMENTAL PROTECTION 199, 213–14 (Alan E. Boyle & Michael R. Anderson eds., 1996) (discussing constitution of India).

⁶⁹ The environmental provision states: "The state shall protect and improve the environment and safeguard the water, air and land, forest and wildlife of Nigeria." Constitution of the Federal Republic of Nigeria, Art. 20. It is contained in a chapter entitled "Fundamental Objectives and Directive Principles of State Policy" rather than the chapter entitled "Fundamental Rights," which contains civil and political rights including those on which the *Gbemre* applicants relied.

⁷⁰ Conversation with Peter Roderick, Climate Justice Programme, July 28, 2006.

criminal proceedings.[71] Similarly, the right to free speech is often justified on the ground that by allowing public criticism of government officials, it provides a crucial check on government power.[72] And the equal protection guarantee was added to the U.S. Bill of Rights after the Civil War in response to what is perhaps the most extreme example of power imbalance in society – the institution of slavery.[73] Indeed, Cass Sunstein argues that much of modern constitutional doctrine reflects "a single perception of the underlying evil: the distribution of resources or opportunities to one group rather than another solely because those benefited have exercised the raw power to obtain government assistance."[74] By acting as "trumps," civil and political rights aim to counteract that underlying evil by effectively putting a thumb on the scale in favor of the weaker party.[75]

In the context of climate change, there is an enormous power imbalance between the interests that stand to gain from climate change regulation and those that stand – in the short run at least – to lose. Those who stand to lose are those who profit from the extraction and combustion of fossil fuels. These are some of the wealthiest and most powerful corporations in the world. Multinational oil companies and car manufacturers dominate the list of the top revenue-producing corporations in the world. Shell, for example, earned more than $25 billion in profits in 2006, second only to Exxon Mobil.[76] The influence that these corporate giants wield over government decision making is undeniable even in the developed world.[77] But the power that a company like Shell exerts over a poor cash-strapped government like Nigeria, that derives more than 80 percent of its revenues from oil production, is monumental.

On the other side, those who stand to gain from climate change regulation are primarily individual people, like the people of the Niger Delta who will be inundated by rising seas and battered by increasingly severe storms as the Earth warms. These "gainers" from climate change regulation are large in number, disproportionately

---

[71] See United States v. Gouveia, 467 U.S. 180, 189 (1984) (right to counsel aimed at correcting the imbalance of power between the government and the accused); Miranda v. Arizona, 384 U.S. 436, 460 (1966) (right against self incrimination aimed at ensuring "the proper scope of governmental power over the citizen ... and maintaining a fair state-individual balance"); Susan Bandes, "We the People" and Our Enduring Values, 96 MICH. L. REV. 1376, 1389, 1391 (1998) (arguing that the criminal procedure amendments "serve to address the inequality of power between the government and the individual and the need to curtail abuse of that power.").

[72] See Vincent Blasi, The Checking Value in First Amendment Theory, AM. B. FOUND. RES. J. 521 (1977).

[73] Since then, the Equal Protection Clause has been interpreted to address the subordination of other stigmatized groups as well. See Ruth Colker, Anti-Subordination Above All: Sex, Race, and Equal Protection, 61 N.Y.U. L. REV. 1003, 1007 (1986); Owen M. Fiss, Groups and the Equal Protection Clause, 5 PHIL. & PUB. AFF. 107, 154–55 (1976).

[74] Cass R. Sunstein, Interest Groups in American Public Law, 38 STAN. L. REV. 29, 50–51 (1985).

[75] See RONALD DWORKIN, TAKING RIGHTS SERIOUSLY 234–35, 184–205 (1977).

[76] Terry Macalister, Exxon and Shell See Profits Rocket, GUARDIAN UNLIMITED, Feb. 1, 2007, available at http://business.guardian.co.uk/story/0,2003392,00.html.

[77] See Richard B. Stewart, Pyramids of Sacrifice? Problems of Federalism in Mandating State Implementation of National Environmental Policy, 86 YALE L.J. 1196, 1213 (1977).

poor,[78] widely dispersed, and have interests that are often hard to measure in precise economic terms and not likely to be felt until well into the future. In contrast to the oil companies, this is just the kind of group that has a particularly hard time organizing politically.[79]

This kind of power imbalance has the capacity to grossly distort government decision making. Arguably, this is exactly what has happened in Nigeria, where the government has been unable to effectively regulate the widespread practice of gas flaring despite a long-standing recognition of its devastating environmental and health effects. This is precisely the kind of situation that human rights are intended to address. Indeed, the trial court's ruling is a classic, triumphal human rights story, in which the politically powerless communities of the Niger Delta use human rights to beat back the Goliath of corporate-backed government power. In the end, of course, fighting vast and well-entrenched power disparities is a difficult business, and the human rights tool is only as strong as the judiciary that enforces it. The impunity with which Shell and the NNPC have ignored the trial court's orders and their apparent ability to inspire the court system to take extraordinary measures to prevent the trial judge from issuing further rulings in the case is a testament to just how enormous and intractable this particular power imbalance has become. Even if the lawsuit does not ultimately result in an enforceable order ending gas flaring, however, framing this conflict as a human rights issue still serves an important rhetorical purpose by bringing into stark relief the power imbalance at its root.

2.2. Security from Climate Change as a Civil and Political Right

A number of decisions from international and domestic tribunals have already begun to find a basis for environmental rights in certain well-established civil and political rights, like the right to life, the right to privacy and family life, and the right to information. These precedents may also provide support for a right to security from climate change.

The right to life, dignity, and personal security (or some variant thereof) appears in every human rights document. It is, perhaps, the most fundamental of all human rights. A number of domestic and international tribunals have found this right implicated in the context of environmental harms.[80] In a case brought by the Ogoni people against the Nigerian government, for example, the African Commission on

---

[78] See Intergovernmental Panel on Climate Change, Climate Change 2001: Synthesis Report, Summary for Policymakers 12 (2001), available at http://www.grida.no/climate/ipcc_tar/vol4/english/005.htm.
[79] See MANCUR OLSON, THE LOGIC OF COLLECTIVE ACTION: PUBLIC GOODS AND THE THEORY OF GROUPS 16–23 (1965).
[80] See generally Hari M. Osofsky, Learning from Environmental Justice: A New Model for International Environmental Rights, 24 STAN. ENVTL. L.J. 71 (2005). In 1972, the Stockholm Declaration was the first international instrument to draw an explicit connection between environmental protection and the right to life: "Both aspects of man's environment, the natural and the man-made, are essential to his wellbeing and to the enjoyment of basic human rights – even the right to life itself." Declaration of the United Nations Conference on the Human Environment, 11 I.L.M. 1416 (1972).

Human and Peoples' Rights held that the pollution and environmental degradation caused by oil production in the Niger Delta constituted a violation of the Ogoni's right to life under Article 4 of the Charter.[81] Similarly, in a study on the human rights situation in Ecuador, the Inter-American Commission on Human Rights found that environmental degradation connected with oil development activities in that country violated the residents' right to life under the American Convention on Human Rights.[82] In a case brought by Canadian citizens challenging a radioactive waste facility near their homes, the United Nations Human Rights Committee found that the case raised "serious issues" regarding the right to life under Article 6(1) of the International Covenant on Civil and Political Rights, even though it ultimately dismissed the case for failure to exhaust domestic remedies.[83] Additionally, domestic courts in India,[84] Columbia, and now Nigeria, have found enforceable rights to a clean environment under constitutional guarantees of the right to life.[85]

Thus, where plaintiffs can show they will suffer some risk of death or personal injury from the impacts of climate change, they may be able to claim a violation of the core civil and political rights to life, dignity, and personal security.[86] These are the rights the plaintiffs relied on in *Gbemre* and in which the Nigerian court found a generic "right to a clean poison-free, pollution-free and healthy environment."[87] The pleadings do not specifically describe any particular climate change impacts that would cause personal injury to the residents of the Niger Delta and thereby potentially violate this right. But, as discussed earlier, rising sea surface temperatures associated with climate change are expected to trigger increasingly severe storms in the Niger Delta. Especially in combination with the Delta's increasing vulnerability to flooding due to climate change–induced sea level rise, such storms could well result in personal injuries and loss of life.[88]

---

[81] Social and Econ. Rights Action Ctr. for Econ. and Social Rights v. Nigeria, African Commission on Human and People's Rights, Comm. No. 155/96, ¶ 70 (2001), *available at* http://www1.umn.edu/humanrts/africa/comcases/155-96b.html.

[82] Inter-Am. C.H.R., Report on the Situation of Human Rights in Ecuador, OEA/Ser.L/V/II.96, doc. 10 rev. 1 (1997).

[83] EHP v. Canada, Communication No. 67/1980, CCPR/C/17/D/67/1980 (U.N. Human Rights Comm. Oct. 27, 1982).

[84] *See* Charan Lal Sahu v. Union of India, AIR 1990 SC 1480, 717; Michael R. Anderson, *Individual Rights to Environmental Protection in India*, *in* HUMAN RIGHTS APPROACHES TO ENVIRONMENTAL PROTECTION 199, 215–16 (Alan E. Boyle & Michael R. Anderson eds., 1996).

[85] Barry E. Hill, Steve Wolfson & Nicholas Targ, *Human Rights and the Environment: A Synopsis and Some Predictions*, 16 GEO. INT'L ENVTL. L. REV. 359, 382–87 (2004).

[86] *See* PRUE TAYLOR, AN ECOLOGICAL APPROACH TO INTERNATIONAL LAW: RESPONDING TO THE CHALLENGES OF CLIMATE CHANGE 197–200 (1998).

[87] Gbemre Judgment, *supra* note 41, at 29. *See id.* at 19 (noting applicant's argument that right to life should be broadly construed as "not just [a right not] to have one's head cut or guillotined, but also . . . [as] the right of a human being to have his organs function properly and to the enjoyment of all his faculties").

[88] *See supra* notes 27 to 31 and accompanying text.

Even where the injuries associated with climate change are not life threatening, they may violate the right to privacy and family life.[89] The European Court of Human Rights (ECHR) has found this right violated where pollution prevents people from living in their homes. In *Lopez-Ostra v. Spain*, for example, pollution and fumes from a tannery waste treatment plant that the government allowed to operate without a license forced the plaintiffs to move from their homes.[90] The court held that "severe environmental pollution may affect individuals' well-being and prevent them from enjoying their homes in such a way as to affect their private and family life adversely, without, however, seriously endangering their health."[91] Similarly, in *Guerra v. Italy*, the ECHR found the Italian government violated the right to privacy and family life of residents living near a chemical factory by failing to provide them with information on the risks posed by the factory.[92] Thus, under these precedents, if rising sea levels caused by climate change displace people from their homes even without causing them physical injury, the right to privacy and family life might well be violated.

Activities contributing to climate change may also implicate a right to information. The right to information is often contained in statutes requiring the preparation of environmental impact assessments, but it sometimes also appears in human rights instruments[93] and is increasingly viewed as derivative of the long-standing and fundamental civil and political right to freedom of expression.[94] While no explicit right to information appears in the Nigerian constitution, in *Gbemre*, the applicants made an effort to derive such a right from the right to life. They alleged that Shell's and NNPC's failure to prepare an environmental impact assessment violated Nigeria's Environmental Impact Assessment Act and "contributed to the

---

[89] International Covenant on Civil and Political Rights, Art. 8(1).
[90] 20 EHRR 277, ECHR 16798/90 (1994).
[91] *Id.* at 51.
[92] 26 EHRR 357, ECHR 14967/89 (1998). The African Commission on Human and Peoples' Rights also found a violation of the rights to property and family in connection with the destruction of Ogoni homes and villages and forced evictions perpetrated by the Nigerian military in retaliation for protests against the environmental harms caused by oil development. *See* Social and Econ. Rights Action Ctr. for Econ. and Social Rights v. Nigeria, African Commission on Human and People's Rights, Comm. No. 155/96, 61–66 (2001).
[93] *See* International Covenant on Civil and Political Rights, Art. 19(2) (freedom of expression includes "freedom to seek, receive and impart information and ideas of all kinds"). The yet-to-be-ratified Charter of Fundamental Rights of the European Union also contains a right of access to European Parliament, Council, and Commission documents at Article 42. *See* Charter of Fundamental Rights of the European Union, Art. 42, 2000/C 364/01, 2000 O.J., (C 364), *available at* http://www.europarl.europa.eu/charter/pdf/text_en.pdf.
[94] *See* Patrick Birkinshaw, *Freedom of Information and Openness: Fundamental Human Rights?*, 58 ADMIN. L. REV. 177 (2006). In its Resolution of the General Assembly of December 14, 1946, the United Nations declared that "[f]reedom of information is a fundamental human right and is the touchstone for all freedoms to which the United Nations is consecrated." G.A. Res. 59(I), at 95, U.N. Doc. A64 (Dec. 14, 1946). Many international environmental treaties and declarations also contain explicit provisions requiring governments to provide access to environmental information. *See* Alexandra Kiss, *The Right to Conservation of the Environment*, in LINKING HUMAN RIGHTS & THE ENVIRONMENT 31, 33–36 (Romina Picolotti & Jorge Daniel Taillant eds., 2003).

violation of the Applicant's...fundamental rights to life and dignity" under the Nigerian Constitution.[95]

An explicit right to information appears in Article 10 of the European Convention on Human Rights and Fundamental Freedoms, but the European Court of Human Rights has construed this provision narrowly, as simply imposing a duty on the State not to interfere with efforts to obtain information from public or private entities willing to share it.[96] This crabbed reading has been widely condemned and many commentators have argued that a broader interpretation of the right to information is more in keeping with foundational principles of democracy and open government. The right can easily be interpreted, for example, as creating an obligation on the part of government to release information about its own projects. An even broader but still reasonable interpretation would impose a duty on the government to both obtain and disseminate information on public and private projects that may impact the environment.[97]

While it has the potential to be broadly applicable in a variety of contexts, the right to information is particularly important with respect to environmental harms, the causes of which are often not superficially apparent.[98] Understanding such causes frequently requires access to sophisticated scientific and technical information that may often be in the control of government or corporate officials. Ensuring public access to such information is thus crucial to the proper functioning of democratic processes. These concerns are particularly salient in the context of climate change, where the causal chain between the activities triggering the harm and the harm itself is extremely complex and nonintuitive.

In sum, there is a significant potential for existing civil and political rights to form the basis for a claim arising from climate change–induced harms. In particular, there is precedent finding the right to life, dignity, and personal security, the right to privacy and family life, the right to information, and a number of the other core rights implicated in the context of environmental harms. Such precedent may be persuasive in the context of a climate change claim.

2.3. The State Action Problem

Much of the activity around the world that is contributing most significantly to climate change is conducted by private actors, often multinational corporations. Indeed, gas flaring in Nigeria provides a case in point. Yet human rights have traditionally been understood as rights that individuals enjoy against governments,

---

[95] Gbemre Pleadings, *supra* note 2, Statement at B.3.
[96] *See* Leander v. Sweden, ECHR (1987); Guerra v. Italy, 26 EHRR 357, ECHR 14967/89 (1998).
[97] *See* Kiss, *supra* note 94, at 33–36.
[98] *See* Claudia Saladin, *Public Participation in the Era of Globalization*, *in* LINKING HUMAN RIGHTS & THE ENVIRONMENT 57 (Romina Picolotti & Jorge Daniel Taillant eds., 2003).

not private actors.[99] Certainly, the climate change–inducing activities c parties can be constrained to some extent by the private law of tort and cont where the governments that define such private rights themselves face incentives to encourage the very activities that drive climate change, such private rights may be ineffective. Where that is so, can human rights of the sort alleged in *Gbemre* be invoked to constrain the actions of private parties directly?

The court in *Gbemre* did not explicitly address this issue but, depending on the circumstances, such direct liability against private corporate actors for their contributions to climate change may be possible. First, under existing doctrine in some jurisdictions, private actors can be held liable for human rights violations where they act in concert with State actors. Moreover, there has in recent years been an increasing chorus of voices in the academic literature calling for an extension of existing doctrine in order to impose human rights duties directly on multinational corporations even in the absence of concerted action.[100]

Under existing U.S. constitutional law[101] and international human rights law as interpreted by U.S. courts,[102] for example, a private actor can be held liable for violations of constitutional or international human rights where it acts in concert with State agents. The plaintiff must show that the private actor is a "willful participant in joint action with the State or its agents" in violating such rights,[103] or that "there is a substantial degree of cooperative action between the State and the private actors

---

[99] *See* Tom Campbell, *Moral Dimensions of Human Rights*, in HUMAN RIGHTS AND THE MORAL RESPONSIBILITIES OF CORPORATE AND PUBLIC SECTOR ORGANISATIONS 14 (Tom Campbell & Seamas Miller eds., 2004); Ratner, *supra* note 60, at 465–66. There are a few exceptions. Human rights against genocide, war crimes, and crimes against humanity have, since the aftermath of World War II, been enforceable against private individuals. *See id.* at 466–68; Kadic v. Karadzic, 70 F.3d 232, 239–44 (2d Cir. 1995); Convention on the Prevention and Punishment of the Crime of Genocide, Dec. 9, 1948, art. 4, S. Exec. Doc. O, 81–1 (1949) ("[P]ersons committing genocide shall be punished whether they are constitutionally responsible rulers, public officials, or private individuals.").

[100] *See* Ratner, *supra* note 60; Campbell, *supra* note 99, at 11; Menno T. Kamminga, *Holding Multinational Corporations Accountable for Human Rights Abuses: A Challenge for the EU*, in THE EU AND HUMAN RIGHTS 553 (Philip Alston ed., 1999); NICOLA JAGERS, CORPORATE HUMAN RIGHTS OBLIGATIONS: IN SEARCH OF ACCOUNTABILITY (2002); Rebecca M. Bratspies, *"Organs of Society": A Plea for Human Rights Accountability for Transnational Business Enterprises and Other Business Entities*, 13 MICH. ST. J. INT'L L. 9 (2005); Amy Sinden, *Power and Responsibility: Why Human Rights Should Address Corporate Wrongs*, in THE NEW CORPORATE ACCOUNTABILITY: CORPORATE SOCIAL RESPONSIBILITY & THE LAW (Doreen McBarnet, Aurora Voiculescu & Tom Campbell eds.) [forthcoming, Cambridge Univ. Press].

[101] Dennis v. Sparks, 449 U.S. 24, 27 (1980) (articulating standard for establishing violation of U.S. constitutional rights under 42 U.S.C. § 1983).

[102] *See* Kadic, 70 F.3d at 245 (adopting § 1983 "under color of law" test for establishing violation of international human rights law by private actor under Alien Tort Claims Act); *see also* Doe v. Unocal, 963 F. Supp. 880, 891 (C.D. Cal. 1997); Wiwa v. Royal Dutch Petroleum Co., No. 96 Civ. 8386, 2002 WL 319887, at *10 (S.D.N.Y. 2002) (quoting *Dennis* and *Unocal*). *See generally* Hari M. Osofsky, *Environmental Human Rights under the Alien Tort Statute: Redress for Indigenous Victims of Multinational Corporations*, 20 SUFFOLK TRANSNAT'L L. REV. 335 (1997).

[103] Dennis, 449 U.S. at 27.

in effecting the deprivation of rights."[104] Such tests have not generally been applied by international human rights tribunals because the jurisdictional rules of those forums only permit suits against States, but Steven Ratner has argued that a similar test should be applied to hold corporations accountable where they act in concert with government agents to commit human rights violations.[105]

The plaintiffs in *Gbemre* should be able to make a strong showing that this test is met with respect to Shell. There is certainly good reason to believe that there has been a close relationship between Shell and the Nigerian government, at least in the past. Indeed, significant evidence exists that joint action between them has led to human rights violations. Although the allegations have yet to be proved in court, in the Alien Tort Claims Act suit against Shell for human rights violations connected with Ken Saro Wiwa's death, petitioners have successfully defeated a motion to dismiss in a U.S. district court in New York based on the "joint action" theory.[106] While the acts that form the basis for that suit were committed by the Nigerian government itself, evidence indicates that Shell assisted in those efforts by helping to plan attacks against the Ogoni, providing financial and logistical support to the Nigerian military, and participating in bribing witnesses. The climate change claims alleged in *Gbemre* would arguably present an even clearer case of joint action. There, the culpable act is the flaring of gas, which like all oil development activities in Nigeria, is conducted as part of a joint venture between Shell and the NNPC. Since the legal entity that is conducting the gas flaring – the Shell Petroleum Development Company – is actually jointly owned by Shell and the Nigerian government, it is hard to imagine a clearer case of "joint action."[107]

Not all climate change activities involve such joint ventures between government and private entities. But in a world in which a number of multinational corporations wield more wealth than many countries and the power of multinationals to affect the conditions of daily existence for individuals often rivals that of government, the notion that human rights norms impose duties only on State actors may be gradually losing traction.[108] Indeed, multinationals often exercise considerable power over States themselves, particularly in the developing world, where cash-strapped, debt-ridden governments are desperate for the foreign investment that multinationals can bring. In this environment, the ability of domestic governments to regulate the activities of multinational corporations is significantly compromised, a situation that is frequently exacerbated by the fact that a multinational may be incorporated in a different country from the one in which it is conducting business.[109] Thus, as stories of environmental atrocities committed by powerful multinational corporations unchecked by domestic regulation continue to emerge from various corners of

---

[104] Unocal, 963 F. Supp. at 891.
[105] See Ratner, *supra* note 60, at 498.
[106] See Wiwa v. Royal Dutch Petroleum Co., No. 96 Civ. 8386, 2002 WL 319887, at *11 (S.D.N.Y. 2002).
[107] See *supra* note 11 and accompanying text.
[108] See Kamminga, *supra* note 100, at 553.
[109] See Ratner, *supra* note 60, at 463.

the globe, there have been increasing calls for the imposition of human rights duties directly on such corporations.

I have argued elsewhere that the same concerns that animated the conceptualization of civil and political rights in the eighteenth century as rights against government warrant the imposition of such rights directly against multinational corporations in the twenty-first century.[110] Civil and political rights were grounded largely in concerns about power imbalances, and during the Enlightenment, when States were the largest aggregations of power in society, such rights were crafted to protect individuals from abuses of State power. But in today's world, where the power of multinational corporations rivals that of States, civil and political rights should protect against certain abuses of corporate power as well. Thus, I have argued that at least in situations in which multinationals are not checked by any domestic government and thus wield final unappealable power of the type that States traditionally wield, human rights duties should be imposed directly on those corporations.[111] Indeed, gas flaring in Nigeria presents a classic example of a weak, cash-strapped government unable or unwilling to rein in corporate power.[112] A series of feeble attempts by the Nigerian government to regulate gas flaring over the past several decades has been entirely ineffectual.[113]

In sum, while many of the actors contributing most significantly to climate change are private rather than State actors, even under traditional doctrine, such private actors may face human rights liability where they participate in joint action with the State. And even where such joint action cannot be shown, emerging theories of human rights eventually may justify the imposition of liability directly on multinational corporations, at least in situations in which they exercise State-like power.

CONCLUSION

Climate change may well be the most profound moral issue ever to confront the human species. While humans have altered their environment on a local scale probably for as long as they have walked the Earth, the impact of anthropogenic greenhouse gas emissions on the fundamental forces that drive the global climate system marks the first time that human activity can literally be said to have altered every spot on Earth. In the words of Bill McKibben, global climate change may indeed signal "the end of nature."[114] Such a profound moral issue demands a profound response from law. Human rights, with all the gravity and moral weight they

---

[110] See Sinden, *supra* note 100.
[111] Arguably, this justification for the imposition of human rights duties on multinational corporations is even more salient in the context of climate change, where governments are disabled from regulating corporate behavior not only because of the inordinate power wielded by corporations themselves but because of the inescapable logic of the tragedy of the commons.
[112] See ROWELL, *supra* note 7, at 96–112.
[113] The Associated Gas Reinjection Decree, enacted by the Nigerian government in 1979 was supposed to stop all flaring by 1984 but has had little or no effect. See OKONTA *supra* note 7, at 73–74.
[114] BILL MCKIBBEN, THE END OF NATURE (1989).

have come to express, may well be an appropriate part of that response. The *Gbemre* case holds out hope that the recognition of a human right to security from climate change may provide a vehicle for courts to issue orders that begin to nudge those actors responsible for substantial greenhouse gas emissions toward more responsible behavior. And, just as importantly, *Gbemre* suggests how treating climate change as a human rights issue may serve to imbue it with a sense of gravity and moral urgency that has been too often missing from the public debate.

9

Tort-Based Climate Litigation

David A. Grossman*

INTRODUCTION

Discussions about how to address climate change usually focus on politics, policies, and programs. Until recently, the potential role of climate change litigation had been virtually ignored. But in the past few years, the idea of using litigation as a tool to address the causes and impacts of climate change has picked up steam, as illustrated in many chapters of this book. Perceiving a lack of meaningful political action – and given the increasing scientific evidence that "[m]ost of the observed increase in global average temperatures since the mid-20th century is *very likely* [greater than 90% likelihood] due to the observed increase in anthropogenic greenhouse gas concentrations"[1] – lawyers around the world have begun exploring litigation strategies and, in some cases, initiating actions.[2] This chapter evaluates the viability of one type of climate change litigation – what some see as the most novel or radical idea – namely, applying tort law to hold companies emitting substantial amounts of greenhouse gases liable for at least some of the harms caused by climate change.

---

* President and founder of Green Light Group, a consulting practice that provides research, writing, and strategic advice on climate and energy projects related to policy, politics, law, and international development. More information is available at http://www.GreenLightGroup.org. This chapter is based on an article first published in 2003, before any tort-based climate change suits had been filed. The research has been updated and the argument refined for this book. The original article is David A. Grossman, *Warming Up to a Not-So-Radical Idea: Tort-Based Climate Change Litigation*, 28 COLUM. J. ENVTL. L. 1 (2003).

[1] WORKING GROUP I, INTERGOVERNMENTAL PANEL ON CLIMATE CHANGE (IPCC), CLIMATE CHANGE 2007: THE PHYSICAL SCIENCE BASIS, SUMMARY FOR POLICYMAKERS, FOURTH ASSESSMENT REPORT 10 (2007) (emphasis original).

[2] *See generally* Climate Justice Programme, http://www.climatelaw.org, for a description of various climate change–related legal efforts around the world. For another account of climate change–related legal efforts, see Kristin Choo, *Feeling the Heat: The Growing Debate over Global Warming Takes on Legal Overtones*, ABA J., July 2006, at 29–35. For an account of relatively early explorations of climate change litigation, see Katharine Q. Seelye, *Global Warming May Bring New Variety of Class Action*, N.Y. TIMES, Sept. 6, 2001, at A14.

There are those who argue that it is not useful to pursue such climate change claims in the courts.³ But harm caused by human activity is a central concern of tort law,⁴ and many of climate change's costs are harms produced at least partially as a result of human actions. Further, because of the uneven nature and distribution of the effects of climate change, some localized groups (e.g., those living in coastal areas or at high latitudes) are bearing, and will continue to bear, the brunt of climate change's harms and costs. This existing allocation raises the question of whether we should continue to ask the victims of climate change to bear these costs or transfer them to those who have most substantially contributed to creating the harm. Allocation of the costs of harms is another central tort concern.⁵

There may be several areas of tort law that could be relevant in the climate change context. This chapter focuses on two that may seem most applicable at first blush. Section 1 examines the applicability of public nuisance to climate change, looking at both pending and potential cases. The section explains some of the public rights that defendants have arguably unreasonably infringed upon and the importance of defendants retaining control of the mechanisms of harm.

Section 2 explores the applicability of products liability. Although at least some products liability climate claims are probably viable, the defenses available to defendants and the need to extend the manufacturers' duty of care ultimately make products liability a weaker tort claim than public nuisance, which could explain why no plaintiffs have filed a products liability climate case yet.

The chapter then turns to some of the general issues underlying all climate tort suits. For instance, although a tort framework might be applicable, some would dispute the propriety of such litigation, contending that climate change requires a political solution. Ultimately, it is surely correct that litigation alone will not solve the problems posed by climate change. But the point of tort-based climate change litigation is to provide redress for harms caused or to provide injunctive relief to prevent further harms – tasks that courts are well equipped to address. Section 3 analyzes this question of justiciability, as well as other jurisdictional hurdles such as standing and preemption that climate change plaintiffs must overcome to reach a hearing on the merits. It is possible that some climate tort claims could overcome these hurdles.

Plaintiffs will face other challenges when dealing with the merits of the claims. Section 4 explores how a plaintiff in climate change litigation might establish generic, specific, and proximate causation. The section also explains the basis for naming certain types of companies as defendants. Section 5 then describes the

---

3   See, e.g., Associated Press, *To Curb Global Warming, Eight States and New York City Vow to Sue Nation's Largest Power Companies*, July 21, 2004 (quoting AEP spokesman: "A lawsuit is not a constructive way to deal with climate change.").
4   See Eduardo M. Penalver, *Acts of God or Toxic Torts? Applying Tort Principles to the Problem of Climate Change*, 38 NAT. RESOURCES J. 563, 569 (1998).
5   See id.

standards for injunctive relief or damages, the types of damages that plaintiffs could allege in their tort claims, some rules that might restrict damage recovery, and the extent to which defendants could be liable for the total costs of climate plaintiffs' harms.

This chapter therefore lays out several of the key elements involved in climate tort suits. Courts have already encountered a few of these cases, and there probably will be more. The cases are not as radical as some may think, and they are part of the new reality of climate change in the courts.

1. PUBLIC NUISANCE

Public nuisance claims focus on "unreasonable injury" – in other words, such claims are generally more concerned with the harm caused than with defendants' conduct or intentions.[6] Accordingly, plaintiffs have used public nuisance suits for decades to address pollution.[7] The application of nuisance law to the problem of climate change does not appear to be that novel an extension.[8]

1.1. Existing Climate Change Public Nuisance Cases

Plaintiffs have already filed climate suits under a public nuisance theory, and this section focuses on the three most prominent examples.

In *Connecticut v. American Electric Power Co.*,[9] eight state attorneys general and the City of New York, plus three private land trusts, brought suit against the five largest electric utilities in the United States[10] seeking injunctive relief in the form of an order "(i) holding each of the Defendants jointly and severally liable for contributing to an ongoing public nuisance, global warming, and (ii) enjoining each of the Defendants to abate its contribution to the nuisance by capping its emissions of carbon dioxide and then reducing those emissions by a specified percentage each year for at least a decade."[11] The District Court dismissed the case, holding that the "actions present non-justiciable political questions that are consigned to the political

---

[6] *See, e.g.*, Wood v. Picillo, 443 A.2d 1244, 1247 (R.I. 1982) ("Distinguished from negligence liability, liability in nuisance is predicated upon unreasonable injury rather than upon unreasonable conduct.").
[7] *See, e.g.*, Georgia v. Tenn. Copper Co., 206 U.S. 230 (1907) (interstate air pollution); Illinois v. Milwaukee, 406 U.S. 91 (1972) (*Milwaukee I*) (interstate water pollution).
[8] *See* Cox v. City of Dallas, Tex., 256 F.3d 281, 291 (5th Cir. 2001) ("The theory of nuisance lends itself naturally to combating the harms created by environmental problems.").
[9] 406 F. Supp. 2d 265 (S.D.N.Y. 2005), *appeal docketed*, No. 05-5104-cv (2d Cir. 2006).
[10] The plaintiffs are the states of Connecticut, New York, California, Iowa, New Jersey, Rhode Island, Vermont, and Wisconsin, the City of New York, and the Open Space Institute, Open Space Conservancy, and Audubon Society of New Hampshire. The defendants are AEP, Southern Company, Tennessee Valley Authority, Xcel Energy, and Cinergy.
[11] 406 F. Supp. 2d at 270 (quoting Complaint, California *ex rel.* Bill Lockyer, Attorney Gen. v. Gen. Motors Corp., Case No. C06-05755 ¶ 6 (N.D. Cal. Sept. 20, 2006), *available at* http://ag.ca.gov/newsalerts/cms06/06-082_oa.pdf).

branches, not the Judiciary."[12] As of the writing of this chapter, the case is on appeal with the Second Circuit.

California filed another public nuisance climate change suit, *California v. General Motors Corp.*, in September 2006 against six automakers.[13] Unlike the relief sought in *Connecticut v. AEP*, California "seeks a judgment holding each Defendant jointly and severally liable for contributing to a public nuisance" and requests monetary damages, attorneys' fees, and declaratory judgment for future monetary expenses and damages "incurred by California in connection with the nuisance of global warming."[14] As of the writing of this chapter, the district court has dismissed the case (also because of "non-justiciable political questions"), and the case is on appeal with the Ninth Circuit.[15]

In February 2008, the Native Village of Kivalina and the City of Kivalina, located in northwest Alaska, filed a public nuisance suit against twenty-four oil, gas, and power companies.[16] As in *California*, the plaintiffs seek a judgment that holds "each defendant jointly and severally liable for creating, contributing to, and maintaining a public nuisance," attorneys' fees, and "declaratory judgment for such future monetary expenses and damages as may be incurred by Plaintiffs in connection with the nuisance of global warming."[17] Plaintiffs also allege civil conspiracy and concert of action.[18] As of the writing of this chapter, the *Kivalina* case has not yet been heard.

1.2. Basics of Public Nuisance

The basic elements of a public nuisance claim are quite uniform throughout the country, since most states follow the approach embodied in the Restatement (Second) of Torts. To be liable, defendants must carry on, or participate to a substantial

---

[12] *Id.* at 274. For more on justiciability, *see infra* Section 3.3.
[13] No. C06-05755, 2007 U.S. Dist. LEXIS 68547 (N.D. Cal. Sept. 17, 2007). The six defendants are General Motors, Ford, Chrysler, Toyota North America, Honda North America, and Nissan North America. The insurance implications of this case are explored in depth by Jeffrey W. Stempel in his chapter *Insurance and Climate Change Litigation*, this volume.
[14] *Id.* at *4.
[15] *Id.* at *17; James Boles, *Appeals Pending for Public Nuisance Climate Change Litigation*, GLOBAL CLIMATE L. BLOG, Jan. 28, 2009, *available at* http://www.globalclimatelaw.com/tags/california-v-general-motors-co/. Hurricane Katrina victims in Mississippi also filed a class action in U.S. District Court in April 2006 against oil and coal companies for contributing to global warming, which plaintiffs assert contributed to the severity of the hurricane; claims include unjust enrichment, civil conspiracy (against the American Petroleum Institute), public and private nuisance, trespass, negligence, and fraudulent misrepresentation. The plaintiffs also sued chemical companies for contributions of halocarbons to climate change. The case was dismissed in August 2007. *Comer v. Murphy Oil*, Case No. 1:05-cv-00436-LG-RHW (S.D. Miss. 2007). For more on the challenges such a claim faces, *see infra* note 158.
[16] Native Vill. of Kivalina v. ExxonMobil Corp., No. CV-08-1138 (N.D. Cal. filed Feb. 26, 2008).
[17] Complaint, Native Vill. of Kivalina v. ExxonMobil Corp. at 67, *available at* http://www.globalclimatelaw.com/uploads/file/Kivalina Complaint.pdf .
[18] *Id.*

extent in carrying on, activities that create "an unreasonable interference with a right common to the general public."[19]

The first critical element of the definition of public nuisance is "a right common to the general public." Such a right is collective; if, for instance, pollution prevents the use of a public beach or kills the fish in a navigable stream and thus potentially affects all members of the community, it impinges on a public right and can be characterized as a public nuisance.[20] Pollution, in fact, often impinges on public rights. In one of the early public nuisance cases, the Supreme Court recognized the right of "a sovereign that the air over its territory should not be polluted on a great scale by sulphurous acid gas, that the forests on its mountains... should not be further destroyed or threatened by the act of persons beyond its control, that the crops and orchards on its hills should not be endangered from the same source."[21] Similarly, courts have recognized "the right of the public in the waters of Lake Champlain to have those waters preserved from oil-spill pollution,"[22] the right of the public against "great harm, annoyance and discomfort" caused by "continuing and unreasonable discharges of malodors,"[23] and other such rights. In *Connecticut v. AEP*, the plaintiffs claimed interference with "the right to public comfort and safety, the right to protection of vital natural resources and public property, and the right to use, enjoy, and preserve the aesthetic and ecological values of the natural world."[24] California asserted interference with the same rights in its case against the automakers.[25] Kivalina alleged "substantial and unreasonable interference with public rights, including, inter alia, the rights to use and enjoy public and private property in Kivalina."[26]

The second critical element of a public nuisance claim is that the defendants' interference with the public right is unreasonable. The Restatement recognizes three independent and sufficient grounds for establishing unreasonableness: (1) defendants' conduct significantly interferes with the public safety, health, peace, comfort, or convenience; (2) it is continuing conduct, or has produced a permanent or long-lasting effect, and defendants know or have reason to know that it has a significant effect upon the public right; or (3) defendants' conduct is unlawful.[27]

---

[19] RESTATEMENT (SECOND) OF TORTS §§ 821B(1), 834 (1979); see also infra Section 4.2 for more on substantiality.
[20] RESTATEMENT (SECOND) OF TORTS § 821B cmt. g (1979). Some states have statutes defining a public nuisance to be an interference with "any considerable number of persons," under which no public right as such need be involved. *Id.*
[21] Georgia v. Tenn. Copper Co., 206 U.S. 230, 238 (1907).
[22] United States v. Ira S. Bushey & Sons, Inc., 363 F. Supp. 110, 120 (D. Vt. 1973).
[23] Concerned Citizens of Bridesburg v. Philadelphia, 643 F. Supp. 713, 722 (E.D. Penn. 1986).
[24] Complaint, Connecticut v. Am. Elec. Power Co. ¶ 154, *available at* http://www.ct.gov/ag/lib/ag/press_releases/2004/enviss/global%20warming%20lawsuit.pdf.
[25] Complaint, *supra* note 11, ¶ 59.
[26] Complaint, *supra* note 17, at 62.
[27] RESTATEMENT (SECOND) OF TORTS § 821B(2) (1979). Liability for a public nuisance may arise even though a party complies in good faith with laws and regulations. *See* City of Boston v. Smith & Wesson Corp., No. 1999-02590, 2000 Mass. Super. LEXIS 352, at *60 (Mass. Super. July 13, 2000).

The first and second grounds just listed seem readily applicable to the climate change context. Defendants' greenhouse gas emissions substantially contribute to climate change and its resulting effects, thereby threatening public safety, health, comfort, and convenience. Climate change is also a "permanent or long-lasting effect" that defendants could have foreseen would interfere with these public rights.[28] In *California v. GM*, for example, California claimed that the automakers "knew or should have known, and know or should know, that their emissions of carbon dioxide and other greenhouse gases contribute to global warming and to the resulting injuries and threatened injuries to California, its citizens and residents, environment, and economy,"[29] and the plaintiffs in *Connecticut v. AEP* made similar claims, adding that the electric utilities "are knowingly, intentionally or negligently creating, maintaining or contributing to a public nuisance – global warming – injurious to the plaintiffs and their citizens and residents."[30] Kivalina made comparable claims as well.[31]

Courts also sometimes consider a third element, namely that the defendants failed "to take reasonable actions within their control that would eliminate, ameliorate, or minimize the harm."[32] It is not clear whether this element is required, so long as defendants' conduct creates or contributes to the nuisance.[33] Nevertheless, in many instances, it is clear that defendants failed to take meaningful mitigating action, and that some in fact acted to prevent public pressure for such mitigation.[34]

---

[28] For instance, the first IPCC assessment report came out in 1990. Courts appear to be in agreement that manufacturers are held to the knowledge and skill of an expert, at a minimum keeping abreast of scientific knowledge, discoveries, and advances. *See, e.g.*, Borel v. Fibreboard Paper Prods. Co., 493 F.2d 1076, 1089 (5th Cir. 1973), *cert. denied*, 419 U.S. 869 (1974).

[29] Complaint, *supra* note 11, ¶ 61.

[30] Complaint, *supra* note 24, ¶ 153.

[31] Complaint, *supra* note 17, at 63.

[32] David Kairys, *The Governmental Handgun Cases and the Elements and Underlying Policies of Public Nuisance Law*, 32 CONN. L. REV. 1175, 1177 (2000).

[33] *Id.* at 1177 n.7.

[34] *See, e.g.*, Ross Gelbspan, *Beyond Kyoto*, AMICUS J. 22, 24 (Winter 1998) ("To date, fossil fuel interests, with few exceptions, have been devoting enormous resources to confounding the public with an appalling public relations campaign of deception and disinformation"). For a detailed account of fossil fuel companies' early efforts to shape public debate, see generally ROSS GELBSPAN, THE HEAT IS ON: THE HIGH STAKES BATTLE OVER EARTH'S THREATENED CLIMATE (1997). Misinformation efforts have declined in recent years as many companies have started taking actions to address climate change, but misinformation and obfuscation continue. The Competitive Enterprise Institute – a conservative think tank funded at times partly by ExxonMobil, Ford, and other business interests – released TV ads in May 2006 questioning the existence of global warming. *See Exxon Blinks in the Global Warming Debate: Oil Giant Gives No Money to Group That Denies Global Climate Change – For Now*, CNNMONEY.COM, Sept. 20, 2006 *available at* http://money.cnn.com/2006/09/20/news/companies/exxon_funding/index.htm; David Adam, *Royal Society Tells Exxon: Stop Funding Climate Change Denial*, GUARDIAN, Sept. 20, 2006, *available at* http://environment.guardian.co.uk/climatechange/story/0,,1876538,00.html; Andrew Leonard, *How the World Works: Is That Climate Change Egg All over Ford's Face?*, SALON.COM, May 18, 2006, *available at* http://www.salon.com/tech/htww/2006/05/18/ford/index.html.

Some courts, particularly in handgun decisions, have held that public nuisance is inapplicable in the context of products, whether they are defective or not.[35] They have contended that nuisance law is, at its heart, not about products but rather about wrongful use of property.[36] (Other courts addressing handgun actions, however, have rejected the idea that public nuisances must arise from activities on or related to property and have allowed public nuisance claims to proceed.)[37] If climate plaintiffs pursue claims focused on the ramifications of the use of products such as motor vehicles, these holdings could be relevant. However, courts rejecting the applicability of nuisance law to products have seemed primarily concerned about the issue of control.[38] As explained further in Section 4, with respect to proximate causation, climate plaintiffs likely could establish that defendants retained control of the mechanisms of harm at all steps of the causal chain (i.e., there are no intervening third parties using products such as automobiles in some unintended way).[39] Accordingly, the "products" issue might not pose an obstacle to some climate nuisance suits – although it does raise the question of whether a doctrine designed specifically for harmful products could apply in the climate context.

2. PRODUCTS LIABILITY

Products liability is another tort theory potentially applicable to climate change, although it seems to be a significantly weaker claim than public nuisance, which may be why no plaintiffs have filed climate change products liability suits to date. The basic elements of a products liability claim are: (1) a product has a defect that makes it unreasonably dangerous; (2) this defect existed when the product left the defendant's control; and (3) the defect proximately caused plaintiff's injuries.[40]

Under either a strict liability or a negligence theory, three types of defects can result in an unreasonably dangerous product.[41] A warning defect occurs when there is reason to anticipate that danger may result from a product, but the manufacturer

---

[35] See, e.g., Camden County Bd. of Chosen Freeholders v. Beretta, U.S.A. Corp., 273 F.3d 536, 540–41 (3d Cir. 2001); Tioga Pub. Sch. Dist. v. U.S. Gypsum Co., 984 F.2d 915, 920 (8th Cir. 1993); City of Philadelphia v. Beretta U.S.A., Corp., 126 F. Supp. 2d 882, 909–10 (E.D. Pa. 2000). See also Kairys, supra note 32, at 1182. For more on defective products, see infra Section 2.
[36] City of Philadelphia, 126 F. Supp. 2d at 910 (citing Detroit Bd. of Educ. v. Celotex Corp., 493 N.W.2d 513, 521 (Mich. Ct. App. 1992)).
[37] See City of Boston v. Smith & Wesson Corp., No. 1999-02590, 2000 Mass. Super. LEXIS 352, at *61 (Mass. Super. July 13, 2000).
[38] See Camden County Bd. of Chosen Freeholders, 273 F.3d at 541 ("[T]he limited ability of a defendant to exercise control beyond its sphere of immediate activity may explain why public nuisance law has traditionally been confined to real property and violations of public rights."); City of Philadelphia, 126 F. Supp. 2d at 910–11 (noting that harms are from intervening third-party criminals over whom defendants exercise no control).
[39] See infra Section 4.2.
[40] See, e.g., Gebhardt v. Mentor Corp., 191 F.R.D. 180, 184 (D. Ariz. 1999) (citing Gosewisch v. Am. Honda Motor Co., 737 P.2d 376, 379 (Ariz. 1987)).
[41] See id.

fails to warn users of that danger. The warning defect inquiry thus focuses more on how the manufacturer acted than on the physical state of the product.[42] A manufacturing defect occurs when a manufacturer makes a product in a way that does not accord with its intended design.[43] A design defect occurs when the harm arises from the design of the product itself.[44] In both manufacturing and design defect cases, the focus of the inquiry is more product oriented than conduct oriented. Manufacturing defects do not appear to be relevant here, since the harms caused by products that have contributed to climate change (e.g., cars) do not stem from shoddy manufacture. As elaborated subsequently, and using motor vehicles as an example, warning and design defect claims do not seem like a particularly good fit either.

2.1. Warning Defects

Generally, a product has a warning defect "when the foreseeable risks of harm posed by the product could have been reduced or avoided by the provision of reasonable instructions or warnings by the seller [or manufacturer] and the omission of the instructions or warnings renders the product not reasonably safe."[45] Manufacturers warn the user about the risk so that he or she can avoid harm either by appropriate conduct during use or by choosing not to use the product.[46] Climate plaintiffs might thus seek to bring a warning defect claim against defendants for failure to warn users of the climate-changing dangers associated with their products' carbon dioxide emissions. They could argue, for instance, that if car manufacturers had advertised fuel efficiency standards as early as they could have,[47] consumers could have chosen more fuel-efficient cars or other transportation alternatives.

But climate change plaintiffs are unlikely to prevail on a warning defect theory for at least three reasons. First, some state statutes provide that liability for a warning defect attaches only if the absence of the warning makes the product "not reasonably fit, suitable or safe for its intended purpose."[48] Failure to warn about climate-changing impacts in no way makes products such as cars unfit for their intended purposes.

Second, other states have determined that warning defect liability attaches only if the manufacturer knew or should have known about the risk and failed to provide a warning that a manufacturer exercising reasonable care would have provided, in

---

[42] JAMES T. O'REILLY & NANCY C. CODY, THE PRODUCTS LIABILITY RESOURCE MANUAL 5 (General Practice Section, American Bar Association 1993).
[43] Id.
[44] Id. at 6.
[45] RESTATEMENT (THIRD) OF PRODUCTS LIABILITY § 2(c) (1998).
[46] See id. § 2 cmt. i; see also Borel v. Fibreboard Paper Prods. Co., 493 F.2d 1076, 1089 (5th Cir. 1973), cert. denied, 419 U.S. 869 (1974).
[47] See generally JACK DOYLE, TAKEN FOR A RIDE: DETROIT'S BIG THREE AND THE POLITICS OF POLLUTION (2000) (explaining how the automobile industry failed to do so).
[48] See, e.g., Dennis v. Pertec Computer Corp., 1996 U.S. Dist. LEXIS 18906, at *22 (D.N.J. Nov. 18, 1996) (N.J. law) (citing N.J. Stat. Ann. § 2A:58C-2).

light of the likelihood that the product would cause harm of the plaintiff's type and in light of the likely severity of that harm.[49] Potential climate change defendants probably have been aware of the climate-changing risks posed by their products for at least several years,[50] and the likelihood and severity of harm is fairly high.[51] However, even reasonable manufacturers may not have seen the need to provide warnings, under the belief that they would not make a significant difference in consumers' practices with regard to the purchase and use of products such as cars.

This ties into the third weakness in a climate change warning defect claim, namely that a plaintiff must show that the failure to provide adequate warning was a proximate cause of the harm.[52] Even if manufacturers provided warnings about the climate-changing emissions of their products, most consumers' behavior probably would not have changed meaningfully. There would still be few viable alternatives to these products available to consumers. Given these considerations, warning defect claims do not seem readily applicable in the context of climate change.

2.2. Design Defects

As a general rule, a product is defective in design "when the foreseeable risks of harm posed by the product could have been reduced or avoided by the adoption of a reasonable alternative design and the omission of the alternative design renders the product not reasonably safe."[53] Inherent features of a product, such as a knife's sharp edge, are not design defects.[54] This fact would seem to rule out design defect suits against oil or coal companies, since there is no feasible way to burn their products without producing carbon dioxide.[55] For a product such as an automobile, however, greenhouse gas emissions are not an "inherent" feature, since manufacturers can design cars and engines in ways to reduce or eliminate carbon dioxide emissions. Accordingly, climate plaintiffs might be able to bring a design defect claim against car manufacturers, arguing that the "defect" of the automotive designs is the unnecessary production of significant amounts of greenhouse gases, which substantially contribute to plaintiffs' harms from global climate change.[56]

---

[49] See, e.g., Hisrich v. Volvo Cars of N. Am., Inc., 226 F.3d 445, 450 (6th Cir. 2000) (Ohio law).
[50] See supra note 28.
[51] See William C. G. Burns & Hari M. Osofsky, Overview: The Exigencies That Drive Potential Causes of Action for Climate Change, this volume.
[52] See Port Auth. v. Arcadian Corp., 189 F.3d 305, 320 (3d Cir. 1999); Gebhardt v. Mentor Corp., 191 F.R.D. 180, 185 (D. Ariz. 1999).
[53] RESTATEMENT (THIRD) OF PRODUCTS LIABILITY § 2(b) (1998).
[54] O'REILLY & CODY, supra note 42, at 7 (citing RESTATEMENT (SECOND) OF TORTS § 402A (1965)); McCarthy v. Olin Corp., 119 F.3d 148, 155 (2d Cir. 1997).
[55] See, e.g., DOYLE, supra note 47, at 238 (describing how reformulating gasoline might help air pollution but would have no effect on global warming, since any form of gasoline contains the same amount of carbon as another).
[56] The standard for legal causation is substantiality, and a court can find all actors that are substantial causes jointly and severally liable for the harm, subject to apportionment if feasible. See infra Sections 4.2 and 5.4.

Most jurisdictions seem to use some variant of a risk-utility or risk-benefit test in design defect cases,[57] balancing the severity and the likelihood of the potential harm against the product's benefits and the burden that effective precautions would impose.[58] If the risk outweighs the utility, a court can consider the product to have a design defect.[59] Climate change's present and projected impacts are quite severe, involving loss of land, buildings, infrastructure, species, ecosystems, and communities, and the likelihood of these harms occurring is fairly high.[60] The "foreseeable risk" is thus substantial.[61] Undeniably, the benefits of motor vehicles are also high, but the existence of potential alternatives might detract from the weight of these benefits.

To better evaluate the "benefit" side of the equation, most courts require plaintiffs to prove the existence of an alternative design that is feasible and that could have avoided the injury in question.[62] Courts often look to whether the alternative design is safer, is technologically and economically feasible, does not impair the usefulness of the product, and does not create other equal or greater risks.[63] For automobiles, such alternative designs probably do exist; manufacturers can design cars to use cleaner energy sources and to use fossil fuels more efficiently.[64] A court cannot judge past and alternative designs, however, by contemporary expectations; it must measure a design defect against standards as of the time of marketing.[65] Some alternative

---

[57] Some states apply the "consumer expectation test" in design defect cases, assessing whether the risk of harm from the product is greater than the ordinary consumer would have expected. See, e.g., Kelley v. Rival Mfg. Co., 704 F. Supp. 1039, 1042 (W.D. Okla. 1989). Some other states look to whether a reasonably prudent manufacturer who knew the product's risks would have placed the product on the market. See, e.g., Nichols v. Union Underwear Co., 602 S.W.2d 429, 433 (Ky. 1980).

[58] O'REILLY & CODY, supra note 42, at 64–66; see also Andrew J. McClurg, The Tortious Marketing of Handguns, 19 SETON HALL LEGIS. J. 777, 779 (1995) (citing Prentis v. Yale Mfg. Co., 365 N.W.2d 176, 183 (Mich. 1984)). Benefits of the product that a court may consider include its cost, effectiveness for an intended function, utility for multiple uses, durability and strength, convenience of use, collateral safety (protecting against some other risks), and appearance and aesthetics. O'REILLY & CODY, supra note 42, at 66. The burden includes engineering costs to change the current design. See Caterpillar Tractor Co. v. Beck, 593 P.2d 871, 885–86 (Alaska 1979).

[59] O'REILLY & CODY, supra note 42, at 64.

[60] See Burns & Osofsky, supra note 51.

[61] See supra note 28; see also Putman v. Gulf States Utils., 588 So. 2d 1223, 1228–29 (La. Ct. App. 1991) (noting that the standard of knowledge, skill, and care in design defect cases is that of expert).

[62] O'REILLY & CODY, supra note 42, at 67. See, e.g., West v. Searle & Co., 806 S.W.2d 608, 612 (Ark. 1991).

[63] See, e.g., Barker v. Lull Eng'g Co., 573 P.2d 443, 455 (Cal. 1978); Wilson v. Piper Aircraft Corp., 577 P.2d 1322, 1326 (Or. 1978).

[64] In 2000, for instance, Toyota and Honda introduced gas-electric hybrids with greatly improved fuel efficiency. Since then, choices have expanded from two models to at least eleven. See Tara Baukus Mello, Hybrid Popularity Skyrockets, EDMUNDS.COM, Oct. 17, 2007, available at http://www.edmunds.com/advice/hybridcars/articles/101677/article.html.

[65] See Quintana-Ruiz v. Hyundai Motor Corp., 303 F.3d 62, 72 (1st Cir. 2002); Cover v. Cohen, 461 N.E.2d 864, 866 (N.Y. 1984); O'REILLY & CODY, supra note 42, at 69. Some state courts have even held that alternative designs that "are feasible but not demanded or expected by consumers or external standards" at the time "are not retrospectively held to be necessary in the context of a later design defect trial." O'REILLY & CODY, supra note 42, at 72.

vehicle designs, such as the gas-electric hybrid cars currently on the market, are recent developments. That these designs may not have been technologically or economically feasible until recently may ultimately defeat a claim of defectiveness.

Two points are worth noting here, however. First, other designs, such as electric cars, multivalve engines, and lighter automotive components, have been around for decades, were known by consumers, and were put into commercial production, if at all, later than they could have been.[66] Second, the fact that companies continue to make and market products that do not employ alternative designs – car manufacturers are still producing fuel-inefficient vehicles such as SUVs – may facilitate design defect suits targeting recent products.

2.3. Negligence, Breach of Duty, and Defenses

In products liability, plaintiffs can sue under either a strict liability or negligence theory. A manufacturer will be held strictly liable in tort when it places a product on the market, knowing that it is to be used without inspection for defects, and the product proves to have a defect that causes injury to a person.[67] If the risks of a design outweigh its utility, pure strict liability would impose liability without regard to whether the manufacturer knew or should have known about those risks.[68] Most courts have eschewed this approach, however, often looking at reasonableness even in what are ostensibly strict liability cases.[69]

To establish a traditional negligence case, plaintiffs must prove (1) a duty of care owed to plaintiffs by the defendants; (2) breach of that duty by the defendants; (3) defendants' breach as a proximate cause of plaintiffs' damages; and (4) cognizable injury or harm to the plaintiffs.[70] That the harms alleged by plaintiffs such as the states in *California v. GM* and *Connecticut v. AEP* are cognizable is addressed in Section 3,[71] so only the first three elements are discussed here.

Reasonable foresight and knowledge of a product's potential risks usually define the scope of a manufacturer's duty in product design.[72] At the level of expert

---

[66] See generally DOYLE, supra note 47.
[67] Greenman v. Yuba Power Prods., Inc., 377 P.2d 897, 900 (Cal. 1963); see also Kennedy v. S. Cal. Edison Co., 268 F.3d 763, 771 (9th Cir. 2001).
[68] O'REILLY & CODY, supra note 42, at 137.
[69] See McClurg, supra note 58, at 800–01. See also Port Auth. v. Arcadian Corp., 189 F.3d 305, 313 (3d Cir. 1999) ("[U]nder New York law, theories of negligence and strict liability for design and warning defects are functionally equivalent.") (citations omitted).
[70] O'REILLY & CODY, supra note 42, at 135. Judge Learned Hand considered a party to be negligent if the expected costs of accidents, discounted by the likelihood that the accident will occur, are greater than the costs of avoiding those accidents. See Penalver, supra note 4, at 576–77. This is essentially a risk-balancing test. The costs of climate change are and will be enormous, and the likelihood of destructive effects is fairly high. See Chapter 1. The costs to defendants of avoiding or minimizing these harms, while potentially large, are likely to be less than the costs of climate change. Under Learned Hand's approach, therefore, courts would likely deem climate change defendants negligent.
[71] See infra Section 3.1.
[72] O'REILLY & CODY, supra note 42, at 30; see also McClurg, supra note 58, at 796. In some states, the ultimate determination of the existence of a duty is more "a question of fairness and public policy"

knowledge, potential climate change defendants likely have known of the climate-changing risks of their products for quite some time.[73] Manufacturers' duties are usually restricted to those who foreseeably would consume or use their products.[74] When the products in question are something like motor vehicles, it seems fair to say that virtually everyone is a foreseeable user.[75] Climate change plaintiffs, however, are not harmed in their capacity as users or consumers of vehicles.

Nevertheless, two cases indicate that plaintiffs might still be able to demonstrate a duty on the part of defendants. In a case involving the contamination of plaintiffs' wells by MTBE in gasoline, the court found that the defendant oil companies could owe the plaintiffs a duty to warn.[76] The court acknowledged that some courts have extended the duty to "third persons exposed to a foreseeable and unreasonable risk of harm by the failure to warn."[77] The court then found that despite the fact that the contamination "was [not] the direct result of [plaintiffs'] own use of gasoline containing MTBE, [plaintiffs'] allegations are sufficient to show that the harm suffered by the plaintiffs was a foreseeable result of defendants' placement of gasoline containing MTBE in the marketplace."[78] Although climate plaintiffs probably will not pursue a duty-to-warn claim, the logic of extending defendants' duty to all those foreseeably exposed to risk seems equally applicable to design defects.

A 2000 handgun case supports the extension of this duty-to-warn logic. In that case, the court noted that defendants owed a duty of care to all people "to whom injury may reasonably be anticipated as a probable result of manufacturing, marketing, and distributing a product with an alleged negligent design."[79] Since climate plaintiffs' harms are arguably a foreseeable result of placement of defendants' products in the marketplace, defendants might owe plaintiffs a duty of care.

Plaintiffs' second requirement in establishing a negligence claim is to prove a breach of the relevant duty of care. In the products liability context, breach occurs when a product is defective; so the risk-benefit test described earlier is also a test for breach.[80] To preclude a finding of breach, defendants in a negligence suit can assert that their actions were reasonable. The most important such defense in products liability is that the defendants took due care by meeting the "state of the art."[81] To proffer the "state of the art" defense, manufacturers do not have to operate at

---

than of foreseeability, though foreseeability is still important. See *Arcadian Corp.*, 189 F.3d 305 at 315–16 (citing Kuzmicz v. Ivy Hill Park Apartments, Inc., 688 A.2d 1018, 1020 (N.J. 1997)).

[73] See *supra* note 28.
[74] See, e.g., Morris v. Chrysler Corp., 303 N.W.2d 500, 502–03 (Neb. 1981).
[75] Cf. *In re* Methyl Tertiary Butyl Ether ("MTBE") Prod. Liab. Litig., 175 F. Supp. 2d 593, 625–26 (S.D.N.Y. 2001).
[76] *Id.*
[77] *Id.* at 625 (citing McLaughlin v. Mine Safety Appliances Co., 181 N.E.2d 430, 433 (N.Y. 1962)).
[78] *Id.*
[79] White v. Smith & Wesson, 97 F. Supp. 2d 816, 828–29 (N.D. Ohio 2000) (citing Gedeon v. E. Ohio Gas Co., 190 N.E. 924, 926 (Ohio 1934)).
[80] See *supra* Section 2.2.
[81] Some states have established statutory presumptions that a product is not defective if its design conforms to the "state of the art." O'REILLY & CODY, *supra* note 42, at 72.

the forefront of technology; the term sometimes refers to economic feasibility, the existence of generally recognized industry practices, or the existence of industry or government design standards.[82]

As previously noted, climate change plaintiffs might have trouble showing that all alternative designs were economically feasible.[83] The government standards in Title II of the Clean Air Act (CAA)[84] might also bolster industry claims of meeting the state of the art in automotive emissions and fuel efficiency, though foreign manufacturers adopted some basic technologies well before U.S. manufacturers, resulting in marked differences in fuel efficiency.[85] This fact might help defeat any claims of an "industry practice." Furthermore, even if no manufacturer adopts or considers alternative designs, a plaintiff still can introduce expert testimony to show that, as a practical matter, manufacturers could have adopted a reasonable alternative design.[86] For instance, climate change plaintiffs might be able to show the existence of such an alternative design for the current production of fuel-inefficient SUVs.

The third step in climate plaintiffs' negligence claim is to establish proximate causation. Section 4 addresses this issue.[87] Defendants in a negligence suit, though, can offer evidence of plaintiffs' conduct to defeat or mitigate a finding that defendants were the proximate cause of plaintiffs' injuries. These defenses fall into two general categories: (1) contributory negligence or comparative fault; and (2) assumption of risk.[88] Contributory negligence means that plaintiffs were negligent in a way that contributed to their injuries. Historically, and still in a few jurisdictions, contributory negligence defeats any liability for defendants.[89] Most jurisdictions, however, utilize comparative fault, in which courts reduce the defendants' liability proportionate to the plaintiffs' degree of fault.[90] "Assumption of risk" means that courts bar plaintiffs from recovery because plaintiffs knew of the product's danger but nevertheless unreasonably proceeded to use it.[91]

---

[82] *Id.* at 154. Courts may reject a scientifically sound alternative design because its expense would prevent it from being commercially viable or because government or formal private standards could be said to express the state of the art of safe design. *Id.*; *see also* RESTATEMENT (THIRD) OF TORTS: PRODUCTS LIABILITY § 2 cmt. d (1998).

[83] *See supra* notes 62–66 and accompanying text.

[84] 42 U.S.C. §§ 7521–7590 (2001).

[85] *See, e.g.,* DOYLE, *supra* note 47, at 253 (noting difference in Toyota and Ford fuel efficiencies in 1988, largely due to Toyota's adoption of multivalve engines); *id.* at 255, 261 (noting high miles per gallon (MPG) achieved by French Citroens AX-10 and by Honda Civics sold in America in 1990, compared to declining MPGs of U.S.-manufactured cars); *see also supra* note 64.

[86] *See* RESTATEMENT (THIRD) OF TORTS: PRODUCTS LIABILITY § 2 cmt. d (1998).

[87] *See infra* Section 4.2.

[88] Defendants can also claim contributory negligence, comparative fault, and assumption of risk in strict liability cases, but only if plaintiff's conduct is voluntary and unreasonable. *See* Borel v. Fibreboard Paper Prods. Corp., 493 F.2d 1076, 1097–98 (5th Cir. 1973), *cert. denied,* 419 U.S. 869 (1974); *see also* O'REILLY & CODY, *supra* note 42, at 164.

[89] O'REILLY & CODY, *supra* note 42, at 28.

[90] *Id.* About two-thirds of states have comparative fault legislation or decisions. *Id.* at 164.

[91] RESTATEMENT (SECOND) OF TORTS § 402A cmt. n (1965). Cigarette manufacturers often used this defense in tobacco lawsuits brought by smokers.

At first glance, defendants' defenses appear to have merit in a climate suit. Defendants could argue that plaintiffs have been well aware that products like cars produce emissions that aggravate climate change, yet plaintiffs, their agents, and their citizens have continued to use those products with that knowledge. Climate change plaintiffs have strong rebuttals to these defenses, however. First, citizens' awareness of the risks posed by use of fossil fuels is debatable, although such awareness has recently been on the rise.[92] Second, even assuming that citizens are aware of the risks, it would be difficult for defendants to show that plaintiffs acted unreasonably, especially given the few practical alternatives to using these manufacturers' products.

All things considered, therefore, climate change plaintiffs' strongest products liability claim would appear to be a design defect suit. However, recognition of manufacturers' duties to climate change victims outside of their capacity as users or consumers of products that emit carbon dioxide is by no means certain, and potential defendants might be able to present strong "state of the art" defenses. While a products liability claim might be viable, therefore, these caveats suggest that it is a much weaker claim than public nuisance.

3. JURISDICTIONAL HURDLES

Climate plaintiffs seeking to press a tort claim – whether public nuisance, products liability, or some other tort – cannot, of course, go right to the merits of the case. They must first clear various jurisdictional hurdles. The principal ones in this context are standing, preemption, and justiciability.

3.1. *Standing*

As stated by the Supreme Court, "to satisfy Article III's standing requirements, a plaintiff must show (1) it has suffered an injury in fact that is (a) concrete and particularized and (b) actual or imminent, not conjectural or hypothetical; (2) the injury is fairly traceable to the challenged action of the defendant; and (3) it is likely, as opposed to merely speculative, that the injury will be redressed by a favorable decision."[93]

---

[92] *See supra* note 34; *see also* Penalver, *supra* note 4, at 577 n.71 ("The public campaigns carried out by fossil fuel companies have made it very difficult for the average consumer to accurately weigh the risks involved in continued use of fossil fuels."). For information on recently increasing awareness, see, e.g., Zogby International/National Wildlife Federation Survey, Aug. 11–15, 2006, *at* http://www.zogby.com/wildlife/NWFfinalreport8–17-06.htm ("Three-fourths of likely voters (74%) are more convinced from events over the past two years that global warming is happening, with two in five (40%) saying they are much more convinced.").

[93] Friends of the Earth v. Laidlaw Envtl. Servs. (TOC), 528 U.S. 167, 180–81 (2000) (citing Lujan v. Defenders of Wildlife, 504 U.S. 555, 560–61 (1992)). The general rule is that a plaintiff must show a particularized harm; if all citizens are affected in the same way, the assumption is that they should go to their legislature. *See, e.g.*, Fla. Audubon Soc'y v. Bentsen, 94 F.3d 658, 667 n.4 (D.C. Cir. 1996) ("[T]he plaintiff must show that he is not simply injured as is everyone else, lest the injury be too general for court action, and suited instead for political redress.").

The Supreme Court recently addressed many of these issues in *Massachusetts v. EPA*, in which a group of states, local governments, and nongovernmental organizations sued the Environmental Protection Agency over its rejection of a petition to regulate greenhouse gas emissions from new motor vehicles under the Clean Air Act.[94] Ruling on the standing of state petitioner Massachusetts, and recognizing both the state's "quasi-sovereign ... interest independent of and behind the titles of its citizens" and the fact that the state itself "owns a great deal of the 'territory alleged to be affected,'" the Court first noted that the state is "entitled to special solicitude in our standing analysis."[95] This "special solicitude" might in fact be even greater in a public nuisance case. The underlying basis for public nuisance is "to protect the public from lawful and even productive activities that are substantially incompatible with the public's common rights. Public nuisance is the only tort designed and equipped to protect the public from activities or conduct that is incompatible with public health, safety, or peace."[96] Given this underlying public basis, the typical plaintiff in a public nuisance action is a governmental entity or official seeking to protect the public, such as mayors and other city executive officials, county executive officials, governors, and state attorneys general.[97]

In determining the first requirement of standing, the Court in *Massachusetts v. EPA* noted that "[t]he harms associated with climate change are serious and well recognized," including "reduction in snow-cover extent" and "the accelerated rate of rise of sea levels," and that the fact that "these climate-change risks are 'widely shared' does not minimize Massachusetts' interest in the outcome of this litigation."[98] The

---

[94] 549 U.S. 497 (2007).
[95] *Id.* at 518–20 (first quotation quoting Georgia v. Tenn. Copper, 206 U.S. 230, 237 (1907)).
[96] Kairys, *supra* note 32, at 1178.
[97] *See id.* at 1175, 1177 n.9, 1181; *see also id.* at 1176 ("A public nuisance claim is the vehicle provided by civil law for executive-branch officials to seek immediate relief to stop and remedy conduct that is endangering the public."); State v. Lead Indus. Ass'n, No. 99-5226, 2001 R.I. Super. LEXIS 37, at **6–7 (R.I. Super. Apr. 2, 2001). *But cf.* Ganim v. Smith & Wesson Corp., 780 A.2d 98, 131–33 (Conn. 2001) (finding city of Bridgeport and its mayor not to have standing in public nuisance handgun suit because harms alleged were derivative and remote). Citizens can also bring public nuisance actions, although the Restatement limited the class of private plaintiffs who could recover damages to those who had "suffered harm of a kind different from that suffered by other members of the public exercising the right common to the general public that was the subject of the interference." RESTATEMENT (SECOND) OF TORTS § 821C(1) (1979). Without such a particularized injury, victims generally must seek a remedy through the public authorities. *See* Connerty v. Metro. Dist. Comm'n, 495 N.E.2d 840, 845 (Mass. 1986). *But see* Akau v. Olohana Corp., 652 P.2d 1130, 1134 (Haw. 1982) (holding that member of the public without special injury has standing to sue to enforce rights of public if he or she can show injury-in-fact and satisfy the court that concerns of multiplicity of suits will be satisfied by any means, including class action). Even if they could establish such particularized injury, however, citizen plaintiffs in a climate change suit would face great difficulties in showing causation. *See infra* Section 4.1.
[98] 549 U.S. at 521–22. *See also* Lujan v. Defenders of Wildlife, 504 U.S. 555, 581 (1992) (Kennedy, J., concurring in part and concurring in the judgment) ("While it does not matter how many persons have been injured by the challenged action, the party bringing suit must show that the action injures him in a concrete and personal way."); Warth v. Seldin, 422 U.S. 490, 501 (1975) (holding that plaintiff may be able to satisfy Article III standing requirements "even if it is an injury shared by a large class of other possible litigants.").

Court also found particularized injury in the fact that "rising seas have already begun to swallow Massachusetts' coastal land," which affects the state "in its capacity as a landowner."[99] Other states and governmental parties suing for similar climate harms should likewise be able to establish injury in fact.

The second standing requirement is traceability. In *Massachusetts v. EPA*, the Court noted that because EPA did not dispute the causal connection between man-made greenhouse gas emissions and global warming, the agency's "refusal to regulate such emissions 'contributes' to Massachusetts' injuries."[100] Despite the fact that EPA claimed its decision was an insignificant contributor to the state's injuries, the Court noted that "U.S. motor-vehicle emissions make a meaningful contribution to greenhouse gas concentrations and hence, according to petitioners, to global warming."[101] Similarly, as explained in more detail in Section 4, plaintiffs in a climate change tort suit could likely trace their harms in part to the emissions contributed by defendants.[102]

The third prong requires plaintiffs to show that a favorable judicial decision would redress their harms. Clearly, an award of damages could compensate climate plaintiffs for present harms and expenses incurred. If plaintiffs seek injunctive relief to enjoin defendants' emissions,[103] the issue seems more complicated, since defendants' reductions would still leave numerous other emitters and a large amount of greenhouse gases already in the atmosphere. Defendants could argue, therefore, that plaintiffs' harms will occur regardless of whether they reduce their emissions. The Supreme Court essentially rejected this argument, however, in *Massachusetts v. EPA*. The Court acknowledged that "regulating motor-vehicle emissions will not by itself *reverse* global warming," but noted that this did not mean that it "lack[ed] jurisdiction to decide whether EPA has a duty to take steps to *slow* or *reduce* it."[104] Despite the fact that increases from other emission sources would dwarf the amount of reductions achieved, a "reduction in domestic emissions would slow the pace of global emissions increases, no matter what happens elsewhere."[105] In sum, the Court held that petitioners had standing because "[t]he risk of catastrophic harm, though remote, is nevertheless real. That risk would be reduced to some extent if petitioners received the relief they seek."[106]

After *Massachusetts v. EPA*, the ability of plaintiffs in a climate tort case to establish standing therefore appears greatly enhanced. This is particularly so for sovereign climate plaintiffs, who are entitled to "special solicitude," but even climate plaintiffs

---

[99] 549 U.S. at 522.
[100] *Id.* at 523.
[101] *Id.* at 525.
[102] *See infra* Section 4.
[103] In *Laidlaw*, the Supreme Court explained that the plaintiff must demonstrate standing for each form of relief sought. 528 U.S. at 185.
[104] 549 U.S. at 525 (emphasis in original).
[105] *Id.* at 526.
[106] *Id.*

that are not sovereigns seem to be on stronger footing given the Court's Article III standing analysis.

3.2. Preemption

Climate plaintiffs seeking to press a federal or state common law tort claim such as public nuisance may have to address the issue of preemption.[107] The preemption standards and analyses for federal common law and state common law are different, since "[f]ederal courts, unlike state courts, are not general common-law courts and do not possess a general power to develop and apply their own rules of decision."[108]

3.2.1. Preemption of Federal Common Law Claims

The Supreme Court has recognized essentially two limited instances in which federal common law may exist: (1) where Congress has given the courts power to develop substantive law, and (2) where a federal rule of decision is needed to protect "uniquely federal interests."[109] If either instance applies, the Court will allow plaintiffs to invoke federal common law, unless displaced by a federal statute. The first instance does not seem to apply in the climate change context, as Congress clearly has not given courts explicit authority to develop substantive law in the area of climate change harms. The second instance, however, could be relevant.

"Uniquely federal interests" exist only in particular narrow areas, such as disputes concerning "the rights and obligations of the United States" and "interstate and international disputes implicating the conflicting rights of States or our relations with foreign nations."[110] The Court recognizes federal common law in such disputes because "our federal system does not permit the controversy to be resolved under state law, either because the authority and duties of the United States as sovereign are intimately involved or because the interstate or international nature of the controversy makes it inappropriate for state law to control."[111] In situations in which a state (as a state or under parens patriae) is suing sources outside of its own territory because they are causing air pollution within the state, the Court thus has been willing to recognize a federal common law tort claim.[112] Analogizing

---

[107] Although many of the cases below deal with nuisance claims, the same ordinary preemption principles and analysis apply to products liability suits. See, e.g., Geier v. Am. Honda Motor Co., 529 U.S. 861 (2000); Nathan Kimmel, Inc. v. DowElanco, 275 F.3d 1199 (9th Cir. 2002); Choate v. Champion Home Builders Co., 222 F.3d 788 (10th Cir. 2000).

[108] City of Milwaukee v. Illinois, 451 U.S. 304, 312 (1981) (*Milwaukee II*).

[109] Texas Indus. v. Radcliff Materials, Inc., 451 U.S. 630, 640 (1981).

[110] *Id.* at 641.

[111] *Id.*

[112] See Nat'l Audubon Soc'y v. Dep't of Water, 869 F.2d 1196, 1205 (9th Cir. 1988) (holding dispute at issue to be solely domestic and thus not properly asserted under the federal common law developed under *Illinois v. Milwaukee*, 406 U.S. 91, 107 n.9 (1972) (*Milwaukee I*)); *Georgia v. Tenn. Copper Co.*, 206 U.S. 230, 237 (1907); and *Missouri v. Illinois*, 200 U.S. 496, 520–21 (1906)); see also Ouellette v. Int'l Paper

this idea to the climate context, states could bring suits based on federal common law if they would be suing sources outside their territory for internal harms incurred.

Federal statutes or regulations will preempt federal common law if they "fully authorized" defendants' behavior, established a "comprehensive set of legislative acts or administrative regulations governing the details of a particular kind of conduct," or "spoke directly to a question" at issue in the dispute.[113] The question of whether a federal statute preempts federal common law "involves an assessment of the scope of the legislation and whether the scheme established by Congress addresses the problem formerly governed by federal common law."[114]

The Court's 2007 decision in *Massachusetts v. EPA* raises the possibility that the Clean Air Act (CAA) might preempt a federal common law tort claim on climate change, since the Court has now held that the EPA has authority to regulate greenhouse gases from new motor vehicles as air pollutants under the CAA,[115] and it is possible that this reasoning could extend to stationary sources of emissions (e.g., power plants) as well.[116] Although President Obama's EPA is likely to develop regulations shortly to limit greenhouse gas emissions from new mobile sources, there are still no actual regulations yet, which means federal common law likely still remains available.

In *Illinois v. City of Milwaukee* (*Milwaukee I*), the Court allowed a federal common law nuisance suit on interstate water pollution to proceed, finding that the water quality legislation in existence at the time did not contain the remedy sought by Illinois and noting that "[u]ntil the field has been made the subject of comprehensive legislation or authorized administrative standards, only a federal common law basis can provide an adequate means for dealing with such claims as alleged federal rights."[117] The Court observed that "[i]t may happen that new federal laws and new federal regulations may in time pre-empt the field of federal common law of nuisance. But until that comes to pass, federal courts will be empowered to appraise the equities of the suits alleging creation of a public nuisance by water

---

Co., 666 F. Supp. 58, 61 (D. Vt. 1987) ("The *Milwaukee I*, *Wyandotte*, and *Milwaukee II* decisions are noncontrolling in this case because those decisions involved states which, when acting as states, filed actions under the Supreme Court's original jurisdiction to resort to the 'necessary expedient' of federal common law to obtain relief from interstate pollution. Because federalism concerns precluded the state sovereigns from resorting to state law claims, the Court applied federal common law in the *Milwaukee* dispute because it was 'concerned in that case that Illinois did not have any forum in which to protect its interests unless federal common law were created.'").

[113] RESTATEMENT (SECOND) OF TORTS § 821B cmt. f (1979); *Milwaukee II*, 451 U.S. at 315, 319 n.14.
[114] *Milwaukee II*, 451 U.S. at 315 n.8.
[115] 127 S. Ct. at 1459–62.
[116] *See, e.g.*, Robert Meltz, Legislative Attorney, American Law Division, Congressional Research Service, *The Supreme Court's Climate Change Decision: Massachusetts v. EPA*, CRS Report for Congress RS22665, May 18, 2007, 6 ("The stationary-source provisions of the CAA [42 U.S.C. § 7408(a)(1)-(2)] use terms similar to that of Section 202 – in particular, 'air pollutant,' 'in his judgment,' and 'may reasonably be anticipated to endanger public health and welfare.'").
[117] 406 U.S. 91, 103, 107 n.9 (1972) (citation omitted).

pollution."[118] Such new laws came to pass with amendments creating the comprehensive Clean Water Act, so when the Court revisited the issue in *Milwaukee v. Illinois (Milwaukee II)*, the Court found that Congress had supplanted federal common law when it "occupied the field through the establishment of a comprehensive regulatory program supervised by an expert administrative agency," representing "an all-encompassing program of water pollution regulation."[119] It therefore seems that federal common law is displaced only when there is a comprehensive set of regulations in place that cover the particular issue.

The Court also emphasized in *County of Oneida v. Oneida Indian Nation of New York State* that "federal common law is used as a 'necessary expedient' when Congress has not 'spoken to a *particular* issue.'"[120] In *Oneida*, the Court found that the Oneida Indian Tribes had a federal common law cause of action for the occupation and use by counties of aboriginal tribal land, since the Nonintercourse Act of 1793 "does not speak directly to the question of remedies for unlawful conveyances of Indian land."[121] In other words, because the law did not provide a remedy for the particular claim advanced, federal common law remained available.

In the climate tort context, there remains no comprehensive federal regulatory scheme akin to the Clean Water Act that governs greenhouse gas emissions,[122] nor is any regulation in place that provides a remedy to states and other plaintiffs harmed by greenhouse gas emissions. *Massachusetts v. EPA* says the EPA has authority to regulate such emissions, but it does not require such regulation,[123] no regulation currently exists, and no regulations may exist for a while. Although the existence of EPA authority to regulate greenhouse gases may lead some to suggest that federal

---

[118] *Id.* at 107.
[119] 451 U.S. 304, 317–19 (1981).
[120] 470 U.S. 226, 237 (1985) (emphasis added in original to quote from *Milwaukee II*, 451 U.S. at 313–14).
[121] *Id.*
[122] The CAA is not as comprehensive as the Clean Water Act (CWA). While the CWA prohibits every point source discharge into navigable water without a permit, the CAA prohibits only unpermitted emissions of certain listed pollutants that have been found to threaten the air-quality standards promulgated by the EPA. New England Legal Found. v. Costle, 666 F.2d 30, 32 n.2 (2d Cir. 1981); Nat'l Audubon Soc'y v. Dep't of Water, 869 F.2d 1196, 1212–14 (9th Cir. 1988) (Reinhardt, J., dissenting). Two district courts have found the CAA to preempt federal common law in air pollution cases. In United States v. Kin-Buc, Inc., 532 F. Supp. 699, 701–02 (D.N.J. 1982), the court acknowledged the difference in comprehensiveness between the CWA and CAA but nonetheless found that Congress had occupied the field. In Reeger v. Mill Serv., Inc., 593 F. Supp. 360, 363 (W.D. Pa. 1984), the court, without comparing comprehensiveness, also found that the CAA's regulatory scheme was similar to the CWA's and thus applied "the same principle of preemption." Higher courts have explicitly not reached the issue – Nat'l Audubon Soc'y v. Dep't. of Water, 869 F.2d 1196, 1205 (9th Cir. 1988); New England Legal Found., 666 F.2d at 32, 32 n.2 – and at least one Ninth Circuit judge would have ruled differently (Nat'l Audubon Soc'y, 869 F.2d at 1212–14 (Reinhardt, J., dissenting)).
[123] 127 S. Ct. at 1463. *See also* Robert Meltz, Legislative Attorney, American Law Division, Congressional Research Service, *The Supreme Court's Climate Change Decision: Massachusetts v. EPA*, CRS Report for Congress RS22665, May 18, 2007, 1 ("The decision does not compel EPA to regulate greenhouse gas (GHG) emissions from new motor vehicles, but it does limit the range of options available to the agency that would justify not doing so.").

common law is preempted,[124] it seems likely that climate plaintiffs will be able to pursue federal common law tort claims until regulations exist that comprehensively govern the field or that provide a remedy for states harmed by greenhouse gas emissions.

3.2.2. Preemption of State Common Law Claims

Unlike federal courts, state courts are general common law courts.[125] As such, federal law preempts state law (including state common law) only when (1) it is the "clear and manifest purpose of Congress,"[126] (2) the federal law is "sufficiently comprehensive to make reasonable the inference that Congress 'left no room' for supplementary state regulation,"[127] or (3) a state law "actually conflicts with a valid federal statute"[128] in that it "stands as an obstacle to the accomplishment and execution of the full purposes and objectives of Congress."[129]

The Supreme Court found in *International Paper Co. v. Ouellette* that the Clean Water Act preempted the nuisance law of a state affected by water pollution, since it stood as an obstacle to the Act's comprehensive scheme regulating every point source discharge.[130] The Court, however, found that "nothing in the Act bars aggrieved individuals from bringing a nuisance claim pursuant to the law of the *source* State."[131] On remand, the district court found that "the same concerns that led the *Ouellette* Court to require application of the source state's law in interstate water disputes are equally applicable to [the CAA and to private party] plaintiffs' air claims."[132] Similarly, it is likely that the CAA would not preempt a climate change tort claim based on the common law of a source state – for instance, one in which many coal-fired electric utilities reside.

It is possible, however, that the CAA might preempt state common law claims against automobile or gasoline manufacturers, since Congress did "speak directly to" the issue of automobile emissions and fuels. Under section 209 of the CAA, "[n]o State or any political subdivision thereof shall adopt or attempt to enforce any standard relating to the control of emissions from new motor vehicles or new

---

[124] *Cf.* Mattoon v. City of Pittsfield, 980 F.2d 1, 5 (1st Cir. 1992) ("The comprehensiveness of the legislative grant is not diminished, nor is the congressional intent to occupy the field rendered unclear, merely by reason of the regulatory agency's discretionary decision to exercise less than the total spectrum of regulatory power with which it was invested."). *Mattoon* seems contrary to the Supreme Court's emphasis in *Oneida* that a regulatory scheme must speak to the "particular" issue in order to displace federal common law.
[125] *Milwaukee II*, 451 U.S. at 312.
[126] Rice v. Santa Fe Elevator Corp., 331 U.S. 218, 230 (1947).
[127] Hillsborough County v. Automated Med. Lab., Inc., 471 U.S. 707, 713 (1985).
[128] Ray v. Atlantic Richfield Co., 435 U.S. 151, 158 (1978).
[129] *Hillsborough County*, 471 U.S. at 713 (quoting Hines v. Davidowitz, 312 U.S. 52, 67 (1941)).
[130] 479 U.S. 481, 494–97 (1987).
[131] *Id.* at 497 (emphasis in original).
[132] Ouellette v. Int'l. Paper Co., 666 F. Supp. 58, 62 (D. Vt. 1987).

motor vehicle engines."[133] The Supreme Court has noted that "Congress has largely pre-empted the field with regard to 'emissions from new motor vehicles,' and motor vehicle fuels and fuel additives."[134]

In sum, it seems likely that the CAA would not preempt federal common law claims, at least until comprehensive regulations are in place (and perhaps even then, if the regulations do not provide a remedy for harms), and would not preempt claims based on a source state's common law, so long as the emissions at issue in the state common law claim are not from motor vehicles.

3.3. Justiciability

One final jurisdictional hurdle to consider is justiciability, which was the basis of dismissal in both *Connecticut v. AEP* and *California v. GM*. In *Connecticut*, the district court dismissed the case "[b]ecause resolution of the issues presented here requires identification and balancing of economic, environmental, foreign policy, and national security interests, [so] 'an initial policy determination of a kind clearly for non-judicial discretion' is required."[135] The court accordingly concluded that "these actions present non-justiciable political questions that are consigned to the political branches, not the Judiciary."[136] Similarly, in *California*, the court wrote that "[j]ust as in *AEP*, the adjudication of Plaintiff's claim would require the Court to balance the competing interests of reducing global warming emissions and the interests of advancing and preserving economic and industrial development. The balancing of those competing interests is the type of initial policy determination to be made by the political branches, and not this Court. . . . [T]he Court finds that the claim presents a non-justiciable political question."[137]

The political question doctrine "is designed to restrain the Judiciary from inappropriate interference in the business of the other branches of Government."[138] In *Baker v. Carr*, the Supreme Court explained that:

> Prominent on the surface of any case held to involve a political question is found a textually demonstrable constitutional commitment of the issue to a coordinate political department; or a lack of judicially discoverable and manageable standards for resolving it; or the impossibility of deciding without an initial policy determination of a kind clearly for nonjudicial discretion; or the impossibility of a court's

---

[133] 42 U.S.C. § 7543(a) (2001); *see also id.* § 7545(c)(4)(A) (fuels). Section 209(b) of the Act contains an exception for California, *id.* § 7543(b), and other states have the option of adopting the California emission standards, *id.* § 7507.

[134] Washington v. Gen. Motors Corp., 406 U.S. 109, 114 (1972) (citations omitted); *see also* Am. Auto. Mfrs. Ass'n v. Cahill, 152 F.3d 196, 198 (2d Cir. 1998) (explaining exception for California and for states opting in to California standards).

[135] 406 F. Supp. 2d 265, 274 (2005).

[136] *Id.*

[137] No. C06–05755, 2007 U.S. Dist. LEXIS 68547, at **23–24, *48 (N.D. Cal. Sept. 17, 2007) (citation omitted).

[138] United States v. Munoz-Flores, 495 U.S. 385, 394 (1990).

undertaking independent resolution without expressing lack of the respect due coordinate branches of government; or an unusual need for unquestioning adherence to a political decision already made; or the potentiality of embarrassment from multifarious pronouncements by various departments on one question. Unless one of these formulations is inextricable from the case at bar, there should be no dismissal for non-justiciability on the ground of a political question's presence.[139]

Most of the *Baker* factors do not seem particularly relevant to the climate tort context. For instance, there appears to be no "textually demonstrable constitutional commitment" of climate change abatement or damages to Congress or the Executive.[140] Similarly, it seems that courts have extensive experience with nuisance cases seeking damages from and/or abatement of pollution by the defendants before them,[141] suggesting both that there could be standards for resolving climate tort suits and that courts would not be showing disrespect to other branches by resolving an interstate nuisance dispute.[142] Additionally, there does not appear to be a coherent national political decision already made about greenhouse gas abatement and damages,[143] and court action concerning particular defendants and plaintiffs would not "inappropriate[ly] interfere[]" with the other branches of government continuing their own climate change efforts (Congress can, in fact, override through legislation any result from a federal common law court decision).

As noted, the district courts in *Connecticut v. AEP* and *California v. GM* both cited the third factor as the most relevant, concerning "an initial policy determination."[144] On the one hand, asking a court to mandate greenhouse gas emission reductions from power plant defendants could seem like a court dictating aspects of energy or climate policy, but on the other hand, courts are well equipped to determine whether pollution from out-of-state actors is harming a state and so must be abated.[145] For

---

[139] 369 U.S. 186, 217 (1962).

[140] *But see* California v. Gen. Motors Corp., 2007 U.S. Dist. LEXIS 68547, at *43 ("[T]he Court finds that Plaintiff's federal common law global warming nuisance tort would have an inextricable effect on interstate commerce and foreign policy – issues constitutionally committed to the political branches of government.").

[141] *See, e.g.*, Georgia v. Tenn. Copper, 206 U.S. 230 (1907). *See also* Weinberger v. Romero-Barcelo, 456 U.S. 305, 314 n.7 (noting that the objective of the Clean Water Act is "in some respects similar to that sought in nuisance suits, where courts have fully exercised their equitable discretion and ingenuity in ordering remedies"); Reserve Mining Co. v. EPA, 514 F.2d 492, 535–42 (8th Cir. 1975) (describing remedy).

[142] *But see* California v. Gen. Motors Corp., 2007 U.S. Dist. LEXIS 68547, at *46 ("[T]he cases cited by Plaintiff do not provide the Court with legal framework or applicable standards upon which to allocate fault or damages, if any, in this case. The Court is left without guidance in determining what is an unreasonable contribution to the sum of carbon dioxide in the Earth's atmosphere, or in determining who should bear the costs associated with the global climate change that admittedly result from multiple sources around the globe.").

[143] This is particularly evident given the current debates in Congress about whether and how to enact a greenhouse gas abatement scheme.

[144] California v. Gen. Motors Corp., 2007 U.S. Dist. LEXIS 68547, at *17; Connecticut v. Am. Elec. Power Co., 406 F. Supp. 2d 265, 272 (S.D.N.Y. 2005).

[145] *Massachusetts v. EPA* seems to have settled that even incremental abatements in greenhouse gas emissions would partially redress plaintiff harms. *See supra* notes 104–106 and accompanying text.

instance, the Court has previously affirmed the justiciability of interstate nuisance actions, noting that "[w]hile we have refused to entertain, for example, original actions that seek to embroil this tribunal in 'political questions,' this Court has often adjudicated controversies between States and between a State and citizens of another State seeking to abate a nuisance that exists in one State yet produces noxious consequences in another."[146] Furthermore, this issue seems somewhat clearer when suits involve damages claims instead of injunctive relief, as a court need make no policy judgments about issues such as abatement in order to determine that a state has been harmed by climate change, that defendants are substantial contributors to climate change, and that the state is entitled to damages.[147]

In sum, it seems that the political question doctrine would be inapplicable to many climate tort claims, although both district courts to have ruled on climate nuisance cases have found otherwise. However, it is worth remembering that the mere fact that "these cases present issues that arise in a politically charged context does not transform them into cases involving nonjusticiable political questions."[148]

4. CAUSATION AND SUBSTANTIALITY

Causation in any climate change tort suit will be a complicated issue, as plaintiffs must show that their harms are traceable to defendants' actions. This section analyzes the issue of causation in the climate change context and explains why causation is easier to establish for certain types of defendants.

4.1. *Generic and Specific Causation*

In many toxic tort cases, as in a climate case, simple causal chains do not exist. Instead, plaintiffs must rely on more statistical or probabilistic means. In mass exposure cases such as Agent Orange, for instance, plaintiffs often had to rely on epidemiological studies to demonstrate the association between exposure to a substance and deleterious health effects.[149] Such studies attempt to establish *generic causation* – whether it can be said that the substance, as a general proposition, causes the sort

---

[146] Ohio v. Wyandotte Chems. Corp., 401 U.S. 493, 496 (1971) (citations omitted).
[147] *Cf. Massachusetts v. EPA*, 127 S. Ct. 1438, 1462–63 (2007) (noting that although the Court has "neither the expertise nor the authority to evaluate the[] policy judgments" offered as justifications for EPA inaction, such as that "a number of voluntary executive branch programs already provide an effective response to the threat of global warming" or that "regulating greenhouse gases might impair the President's ability to negotiate with 'key developing nations' to reduce emissions," these issues "have nothing to do with whether greenhouse gas emissions contribute to climate change."). *But see* California v. Gen. Motors Corp., 2007 U.S. Dist. LEXIS 68547, at *23 ("Regardless of the type of relief sought, the Court must still make an initial policy decision in deciding whether there has been an 'unreasonable interference with a right common to the general public.'") (citation omitted).
[148] Kadic v. Karadzic, 70 F.3d 232, 249 (2d Cir. 1995).
[149] *See* Paul Sherman, *Agent Orange and the Problem of the Indeterminate Plaintiff*, 52 BROOK. L. REV. 369, 383 (1986) (citing *In re* "Agent Orange" Prod. Liab. Litig., 597 F. Supp. 740 (E.D.N.Y. 1984)).

of injuries afflicting the plaintiffs.[150] In the climate context, scientists use computer models to project the past and future course of Earth's climate and to demonstrate the probabilistic association between increased greenhouse gas emissions and climatic effects.[151] The studies and models such as those the IPCC relied upon – along with the studies that scientists are continually publishing in peer-reviewed journals – soundly establish beyond the "more likely than not" standard used in the legal arena the general causal link between greenhouse gas emissions, climate change, and effects such as higher temperatures and sea-level rise.[152]

Generally, courts have not considered statistical associations like those that epidemiological studies produce to be adequate proof of *specific causation* – whether it can be said that the substance caused plaintiffs' particular injuries.[153] This individual causation is often the most problematic for toxic tort plaintiffs, who have to grapple with the existence of background levels of the injuries and other risk factors that may contribute to the injuries.[154] These complications mean that even where plaintiffs can show that defendants are responsible for a significant proportion of the cases of a harm, no single plaintiff can prove that he or she is one of those cases.[155] Given these difficulties, plaintiffs in toxic tort cases have had to supplement epidemiological evidence with supporting scientific evidence, statistical evidence, expert testimony, or further epidemiological evidence that shows that, more probably than not, the risk factor in question caused their individual injuries, as opposed to any other cause.[156]

---

[150] See JAMES HENDERSON & AARON TWERSKI, PRODUCTS LIABILITY: PROBLEMS AND PROCESS 143 (3d ed. 1997).

[151] The basic idea of climate models is that parameters (such as temperature) numerically describing the dynamics of the climate are represented on a grid covering the planet, dividing the globe into little boxes. More boxes means a finer resolution for the model, but it also means more data, more calculations, and more time, so climate models are usually averaged over relatively large geographical areas. Climate models face uncertainty due to the complexity and interdependence of the climate system, feedback loops, vegetation changes, ocean circulation, clouds, and many other factors. *Modelling the Climate*, CLIMATEPREDICTION.NET, *available at* http://www.climateprediction.net/science/model-intro.php (visited Jan. 2, 2008). Confidence in climate models has improved due to advances in their performance on a range of space and time scales. WORKING GROUP I, IPCC, *supra* note 1, at 10, 13.

[152] See WORKING GROUP I, IPCC, *supra* note 1. See also Chapter 1.

[153] HENDERSON & TWERSKI, *supra* note 150, at 143.

[154] Penalver, *supra* note 4, at 580.

[155] Tom Christoffel & Stephen P. Teret, *Epidemiology and the Law: Courts and Confidence Intervals*, 81 AM. J. PUB. HEALTH 1661, 166–63 (1991).

[156] See Penalver, *supra* note 4, at 580–81; *see also* O'REILLY & CODY, *supra* note 42, at 32, and *In re* "Agent Orange" Prod. Liab. Litig., 611 F. Supp. 1223, 1253 (E.D.N.Y. 1985) (finding inadmissible expert opinions that do not show that Agent Orange was more likely than anything else to be the cause of plaintiffs' harms). *But see* Heckman v. Fed. Press Co., 587 F.2d 612, 617 (3d Cir. 1978) ("Expectancy or statistical data about a group do not establish concrete facts about an individual."). Courts sometimes translate this requirement to mean a "relative risk" of at least two. Relative risk is the difference in risk of acquiring a given condition between exposed and unexposed populations. If a given action has doubled the risk of a harm occurring, then one can say that it is more probable than not that a particular incidence of that harm was caused by that action. See Penalver, *supra* note 4, at 580–81.

Showing specific causation in the climate change context could be similarly difficult in some cases. The complexity of the climate system means that several factors are involved in producing shifts in climatic activity, such as more intense storms or higher temperatures, which are also subject to natural fluctuations.[157] These multiple causes and background levels of climatic effects make it difficult to show that defendants' contributions to anthropogenic climate change caused any particular incidence of a phenomenon.[158] Harms caused by one particularly intense hurricane or heat wave, for instance, are difficult to tie to global climate change, as such intense phenomena do sometimes occur naturally.

For some harms, however, there is no one particular "incident" for plaintiffs to attribute. For example, the erosion that is damaging Alaskan coastal villages is very directly connected to the general effect of retreating and thinning sea ice, which is clearly tied to the rapidly warming Arctic.[159] In other words, some harms are caused more by trends exacerbated by climate change than by particular events.

In addition, the obstacle that specific causation poses is mitigated when governments as opposed to individuals are the plaintiffs. When states bring tort claims, the plaintiffs have almost infinite "lifespans" and cover large amounts of territory, allowing for an aggregation of effects over both space and time. The aggregation of these harms makes it easier to rule out confounding factors; for instance, it is easier to attribute one sinkhole in an Alaskan road to factors other than climate change than it is to do so for a state full of roads and infrastructure damaged by thawing permafrost. Natural fluxes and confounding factors still exist, since climate change may not cause some portion of the harms within the aggregation, but aggregation allows plaintiffs to better establish that *some* present harms from climate change exist in the broader geographic and temporal range.[160]

---

[157] Penalver, *supra* note 4, at 581.

[158] Although the scientific consensus is that a clear anthropogenic signal can be detected despite these natural variations and confounding factors, see WORKING GROUP I, IPCC, *supra* note 1, and WORKING GROUP II, INTERGOVERNMENTAL PANEL ON CLIMATE CHANGE (IPCC), CLIMATE CHANGE 2007: IMPACTS, ADAPTATION AND VULNERABILITY, SUMMARY FOR POLICYMAKERS, FOURTH ASSESSMENT REPORT (2007), that reaffirms only generic causation. The variations and confounding factors mean that it is difficult to attribute to climate change any one manifestation of a harm generally linked to climate change, since natural fluxes or other factors could be the cause in that particular case. See David R. Hodas, Standing and Climate Change: Can Anyone Complain About the Weather?, 15 J. LAND USE & ENVTL. L. 451, 456 (2000). This will be a serious challenge to claims like those in *Comer v. Murphy*, *supra* note 15.

[159] See Ola M. Johannessen et al., *Arctic Climate Change: Observed and Modeled Temperature and Sea-Ice Variability*, 56A TELLUS 328, 330, 337 (2004) ("[A]nthropogenic forcing is the dominant cause of the recent pronounced warming in the Arctic.... [T]here are strong indications that neither the warming trend nor the decrease of ice extent and volume over the last two decades can be explained by natural processes alone.").

[160] *Cf.* Recent Legislation: Torts – Products Liability – Florida Enacts Market Share Liability for Smoking-Related Medicaid Expenditures, 108 HARV. L. REV. 525, 528 (1994) (describing Florida law that allows State to use statistical evidence to prove causation and damages by allowing State to aggregate harms that large population suffers: "Although a statistical snapshot of excess death and disease may be an

Once plaintiffs establish these harms, the question is no longer whether defendants have caused harms. Rather, the pertinent question becomes whether the amount of their contributions is sufficient to find liability for damages.

4.2. Proximate Causation and the Substantiality Requirement

One could say that the entire global community is responsible for climate change to some degree; no group of defendants could be entirely responsible for global climate change. Nevertheless, the law of torts does not predicate defendant liability on causing all of the plaintiffs' harms.

An actor's tortious conduct can be a legal cause of another's harm if the conduct is a "substantial factor" in bringing it about.[161] "Substantial" means that the defendant's conduct "has such an effect in producing the harm as to lead responsible men to regard it as a cause, using that word in the popular sense, in which there always lurks the idea of responsibility."[162] As such, "substantial cause" is something of a fuzzy concept akin to "proximate cause," of which substantiality is a critical element;[163] inquiries into both concepts focus on similar issues of defendants' involvement in and control over plaintiffs' harms.[164]

---

inexact measure of harm to a particular individual, it accurately measures the harm to a State suing for treatment of hundreds of thousands of Medicaid recipients.").

[161] RESTATEMENT (SECOND) OF TORTS § 431 (1965); see also id. § 834 ("One is subject to liability for a nuisance caused by an activity, not only when he carries on the activity but also when he participates to a substantial extent in carrying it on."); RESTATEMENT (THIRD) OF TORTS: PRODUCTS LIABILITY § 16(a) (1998) ("When a product is defective at the time of commercial sale or other distribution and the defect is a substantial factor in increasing the plaintiff's harm beyond that which would have resulted from other causes, the product seller is subject to liability for the increased harm."); Shetterly v. Raymark Indus., 117 F.3d 776, 780 (4th Cir. 1997) ("In order to sustain an action against Raymark for asbestos related injuries, Plaintiffs must prove that Raymark products were a substantial causative factor in their injuries.") (internal quotation marks and citations omitted).

[162] RESTATEMENT (SECOND) OF TORTS § 431 cmt. a (1965); see also id. § 433. The considerations in section 433 are relevant only to the degree they dilute or make insignificant the actor's conduct in a particular case. Id. at § 433 cmt. d. That a third party is also a substantial factor does not in itself protect the actor from liability. Id. at § 439.

[163] See Laborers Local 17 Health & Benefit Fund v. Philip Morris, Inc., 191 F.3d 229, 235–36 (2d Cir. 1999), cert. denied, 528 U.S. 1080 (2000) (noting that critical elements of proximate cause are direct injury, defendant's acts being substantial cause of injury, and plaintiff's injury being reasonably foreseeable). See also Young v. Bryco Arms, 821 N.E.2d 1078, 1086 (Ill. 2004) ("The proper inquiry regarding legal cause involves an assessment of foreseeability, in which we ask whether the injury is of a type that a reasonable person would see as a likely result of his conduct.").

[164] See City of Bloomington, Ind. v. Westinghouse Elec. Corp., 891 F.2d 611, 614 (7th Cir. 1989) (holding Monsanto not to be liable for nuisance because "Westinghouse was in control of the product purchased and was solely responsible for the nuisance it created."). Others outside of defendant's control might contribute to the harm, but courts can still find liability if the defendant also has some element of control by means of its tortious conduct and participation. See In re Methyl Tertiary Butyl Ether ("MTBE") Prod. Liab. Litig., 175 F. Supp. 2d 593, 628–29 (S.D.N.Y. 2001).

In several cases dealing with municipal claims against handgun manufacturers and distributors, courts grappled with proximate causation and the degree of defendants' control.[165] In general, the plaintiffs in these cases alleged that the defendants should be held liable for creating and fostering illegal markets for handguns, which in turn allow guns to get into the hands of criminals, who then use them for illegal and often deadly purposes.[166] In one such case, the court found the causal chain involved to be "simply too attenuated to attribute sufficient control to the manufacturers" of handguns.[167] The main element that made the chain "attenuated" was that the manufacturers did not have an adequate degree of control over criminals and those who diverted guns to them, and thus were not in a position to prevent the wrongs caused by handguns diverted to unauthorized owners and criminal use.[168]

The causal chain in climate change tort suits would likely look something like the following: (1) companies produce fuel, power, vehicles, etc.; (2) consumer use of these items generates greenhouse gas emissions, which rise into the atmosphere; (3) the emissions combine with other greenhouse gas emissions to warm the Earth; (4) this warming causes sea levels to rise, snowpack to melt, etc.; and (5) these effects cause damage to plaintiffs' property. Arguably, this end result has been foreseeable for several years.[169] Further, the relevant companies' control does not appear to be lacking. The only intervening parties are consumers, whose intervention is quite foreseeable. Moreover, customers are not misusing the goods but rather are the intended owners using the goods in the intended way;[170] in fact, no misuse of power, fuels, etc., seems possible. Given this uninterrupted causal chain, climate change plaintiffs might thus be able to establish proximate causation.[171]

---

[165] See Camden County Bd. of Chosen Freeholders v. Beretta, U.S.A. Corp., 273 F.3d 536, 541 (3d Cir. 2001), and City of Philadelphia v. Beretta U.S.A., Corp., 126 F. Supp. 2d 882, 910–11 (E.D. Pa. 2000); Young v. Bryco Arms, 821 N.E.2d at 1089–90 (all finding lack of sufficient control).

[166] See, e.g., Camden County Bd. of Chosen Freeholders, 273 F.3d at 538–39.

[167] Id. at 539, 541 ("(1) the manufacturers produce firearms at their places of business; (2) they sell the firearms to federally licensed distributors; (3) those distributors sell them to federally licensed dealers; (4) some of the firearms are later diverted by unnamed third parties into an illegal gun market, which spills into Camden County; (5) the diverted firearms are obtained by unnamed third parties who are not entitled to own or possess them; (6) these firearms are then used in criminal acts that kill and wound County residents; and (7) this harm causes the County to expend resources to prevent or respond to those crimes.").

[168] Id. at 541. Where there are intervening third parties, the issue of legal causation is whether the intervening cause is of the type that a reasonable person would foresee as a likely result of his conduct. RESTATEMENT (SECOND) OF TORTS § 442 (1965).

[169] The first IPCC report, for instance, was in 1990. See supra note 28.

[170] See RESTATEMENT (SECOND) OF TORTS § 442 (1965).

[171] It should be noted that at least one circuit held that proximate causation is not needed in public nuisance claims, though this does not appear to be the majority approach. Allegheny Gen. Hosp. v. Philip Morris, Inc., 228 F.3d 429, 446 (3d Cir. 2000) ("The Hospitals' remaining claims of public nuisance, aiding and abetting and civil conspiracy, restitution, unjust enrichment, quantum meruit, and indemnity do not require proximate cause.") (applying Pennsylvania law). But cf. Camden County Bd. of Chosen Freeholders, 273 F.3d at 541 ("The County argues that proximate cause, remoteness, and control are not essential to a public nuisance claim. . . . But the relevant case law shows that, even

Of course, being a "substantial factor" not only involves control but also implies something about the size of defendants' contributions. It is clearly possible to identify defendants who have contributed substantially to climate change and its resulting effects. The plaintiffs in *California v. GM* and *Connecticut v. AEP*, for instance, targeted automakers and electric utilities, respectively. In 2004, 98% of total U.S. carbon dioxide emissions and more than 82% of total U.S. greenhouse gas emissions were from fossil fuel combustion.[172] Electricity generators were responsible for consuming 34% of U.S. energy from fossil fuels and emitted 40% of the CO_2 emissions from fossil fuel combustion; 82% of these emissions came from coal.[173] U.S. energy-related CO_2 emissions represent about 24% of the world total.[174] Overall, the generation of electricity resulted in a larger portion (33%) of total U.S. greenhouse gas emissions in 2004 than any other activity.[175] Transportation activities were not far behind; transportation accounted for roughly 28% of greenhouse gas emissions in 2004.[176] Transportation activities accounted for 33% of CO_2 emissions from fossil fuel combustion; more than 60% of these emissions resulted from gasoline consumption for personal vehicle use.[177]

There are limited numbers of relevant companies in these sectors. The big three American automakers (GM, Ford, and Chrysler) accounted for 58.7% of the U.S. market in 2004, while the big three Japanese automakers (Toyota, Honda, and Nissan) accounted for another 31%.[178] More than 5,000 power plants generate electricity in the United States, but the 100 largest power producers in the United States own nearly 2,000 of them and accounted for 88% of the electric power generated (including 93% of all coal-fired power) and 89% of the industry's reported emissions in 2004.[179] Seven electric power producers contributed 25% of the industry's carbon dioxide emissions; nineteen producers accounted for half.[180]

There are also limited numbers of companies in related sectors. In 1997, twenty of the world's petroleum and coal companies collectively accounted for roughly half of the world's carbon emissions.[181] In 2003, the largest five oil companies operating

---

if the requisite element is not always termed 'control,' the New Jersey courts in fact require a degree of control by the defendant over the source of the interference that is absent here.").

[172] ENERGY INFORMATION ADMINISTRATION (EIA), EMISSIONS OF GREENHOUSE GASES IN THE UNITED STATES 2004, x, xii (Dec. 2005).

[173] ENVIRONMENTAL PROTECTION AGENCY (EPA), INVENTORY OF U.S. GREENHOUSE GAS EMISSIONS AND SINKS: 1990–2004, EXECUTIVE SUMMARY 7–8 (2006); EIA, *supra* note 172, at 22.

[174] EIA, *supra* note 172, at 3.

[175] EPA, *supra* note 173, at 13.

[176] *Id.* at 13–14.

[177] *Id.* at 7.

[178] Christine Tierney, *Big 3 Market Share Dips to All-Time Low*, DETROIT NEWS, Jan. 5, 2005, *available at* http://www.detnews.com/2005/autosinsider/0501/06/A01-50668.htm (visited Jan. 2, 2008).

[179] CERES, NATURAL RESOURCES DEFENSE COUNCIL, & PUBLIC SERVICE ENTERPRISE GROUP, BENCHMARKING AIR EMISSIONS OF THE 100 LARGEST ELECTRIC POWER PRODUCERS IN THE UNITED STATES – 2004 at 1, 3, 7 (5th ed. Apr. 2006).

[180] *Id.* at 3.

[181] NATURAL RESOURCES DEFENSE COUNCIL (NRDC) ET AL., KINGPINS OF CARBON: HOW FOSSIL FUEL PRODUCERS CONTRIBUTE TO GLOBAL WARMING, PART 1 (1999).

in the United States (Exxon Mobil, Chevron Texaco, ConocoPhillips, BP, and Royal Dutch Shell) controlled more than 14% of global oil production, 48% of domestic oil production, more than half of domestic refining capacity, more than 61% of the retail gasoline market, and more than 21% of domestic natural gas production.[182] ExxonMobil alone is responsible for some 5% of global anthropogenic CO_2 emissions since 1882.[183] The top five coal producers in the United States (Peabody, Rio Tinto, Arch, CONSOL, and Foundation) accounted for more than half of total coal production in 2005.[184]

These companies could all, it seems, be defendants in a climate change suit. Some might argue, however, that the true cause of greenhouse gas emissions is consumption of these products rather than production of them.[185] After all, it is not car manufacturers per se, but rather the millions of individual drivers who use their products that emit greenhouse gases. Several law and policy considerations, however, support holding producers, but not individual consumers, liable for the harms of climate change. First, individual consumers such as drivers and users of electricity do not contribute "substantially" to climate change; as such, their small individual contributions would not meet the standards for legal causation.[186] Second, the degree to which individual consumers maintain real "control" over the harms is debatable. Individual consumers have few meaningful alternatives to fossil fuels and the products that rely on them. Moreover, some fossil fuel companies' efforts to encourage public uncertainty about global climate change have compromised the level of consumer knowledge about the risks posed by fossil fuel use.[187] Finally, tort law's goal of reducing the cost of "accidents" would not be furthered by placing the costs of climate change on individual consumers, but rather by holding liable producers who can incorporate the various costs of climate change into the prices of their products (or produce different products).[188]

5. RELIEF IN A CLIMATE CHANGE TORT SUIT

Climate change plaintiffs seeking either of the two basic tort remedies – damages or injunctive relief – must consider certain issues. This section explores the standards

---

[182] PUBLIC CITIZEN, MERGERS, MANIPULATION AND MIRAGES: HOW OIL COMPANIES KEEP GASOLINE PRICES HIGH, AND WHY THE ENERGY BILL DOESN'T HELP (Mar. 2004), *available at* http://www.citizen.org/documents/oilmergers.pdf (visited Jan. 2, 2008).

[183] FRIENDS OF THE EARTH INTERNATIONAL, EXXON'S CLIMATE FOOTPRINT: THE CONTRIBUTION OF EXXONMOBIL TO CLIMATE CHANGE SINCE 1882 (Jan. 2004).

[184] ENERGY INFORMATION ADMINISTRATION, 2005 ANNUAL COAL REPORT, tbl.10.

[185] NRDC ET AL., KINGPINS OF CARBON, *supra* note 181, EXECUTIVE SUMMARY (comparing this portrayal to the war on drugs focusing primarily on users rather than suppliers).

[186] *See* RESTATEMENT (SECOND) OF TORTS § 834 cmt. d (1979) ("Thus if the operation of a dance hall unreasonably interferes with the comfortable enjoyment of a neighboring residence, the proprietor is liable, but a patron normally does not participate in the objectionable activity to such an extent as to justify imposing liability upon him for the invasion.").

[187] *See supra* note 34.

[188] *See* Penalver, *supra* note 4, at 591.

for injunctive relief and damages (particularly in nuisance suits), the potential damages claims for which plaintiffs could seek recovery in tort-based litigation, possible restrictions on damage recovery, and potential ways of apportioning liability among defendants.

5.1. Standards for Damages and Injunctions

The climate nuisance suits described earlier involve two different types of relief.[189] In *California v. GM*, for instance, the State of California is seeking damages. In determining whether to award damages in a public nuisance suit, "the court's task is to decide whether it is unreasonable [for the defendants] to engage in the conduct without paying for the harm done."[190] To be compensated with a damage award, climate change plaintiffs must have actually incurred significant harm,[191] which California asserts it has in the form of property damage and expenses for preventative measures.[192] The question then becomes the unreasonableness of allowing defendants to continue their behavior without providing compensatory damages to the plaintiffs. Climate change plaintiffs could contend that even if defendants' activities are of great utility to society, as one could reasonably argue, it could still be unreasonable to inflict the harm on plaintiffs without compensating them.[193]

In a public nuisance action for injunctive relief, the question is whether the defendants' activity itself is so unreasonable that the court must stop or reduce it. Plaintiffs need only be threatened with harm and need not actually have incurred harm yet.[194] The plaintiffs in *Connecticut v. AEP* allege a range of threatened harms, including:

> increased heat deaths due to intensified and prolonged heat waves; increased ground-level smog with concomitant increases in respiratory problems like asthma; beach erosion, inundation of coastal land, and salinization of water supplies from accelerated sea level rise; reduction of the mountain snow pack in California that provides a critical source of water for the State; lowered Great Lakes water levels, which impairs commercial shipping, recreational harbors and marinas, and hydropower generation; more droughts and floods, resulting in property damage and hazard to human safety; and widespread loss of species and biodiversity, including the disappearance of hardwood forests from the northern United States.[195]

---

[189] *See supra* Section 1.1.
[190] RESTATEMENT (SECOND) OF TORTS § 821B cmt. i (1979).
[191] *Id.*
[192] "Human-induced global warming has, among other things, reduced California's snow pack (a vital source of fresh water), caused an earlier melting of the snow pack, raised sea levels along California's coastline, increased ozone pollution in urban areas, increased the threat of wildfires, and cost the State millions of dollars in assessing those impacts and preparing for the inevitable increase in those impacts and for additional impacts." Complaint, *supra* note 11, ¶ 1. For more on types of damages that plaintiffs can claim, *see infra* Section 5.2.
[193] *See* RESTATEMENT (SECOND) OF TORTS § 821B cmt. i (1979).
[194] *Id.*
[195] Complaint, *supra* note 24, ¶ 3.

Courts often grant injunctions "when damages are inadequate, such as with ongoing nuisances in which numerous suits or future damage awards would be required."[196] Future suits and damage awards will be likely during the continued progress of global climate change. It would be unreasonable and unwise, however, for a court to enjoin all emissions from defendants, since that would destroy those companies and imperil the economy. Instead, should a court determine that defendants' substantial carbon dioxide or other greenhouse gas emissions are unreasonable, it could arguably focus injunctive relief directly on the source of plaintiffs' harms by enjoining defendants from continuing greenhouse gas emissions at their current levels, which is in fact the injunctive relief that the plaintiffs in *Connecticut* seek.[197] Injunctions also might theoretically mandate conservation and efficiency measures or improvement and updating of technology and equipment, if a court thought such drastic intervention through its equity powers was warranted.

It is important to realize some of the potential implications of seeking damages versus injunctive relief. If climate plaintiffs seek damages, they might get compensation for the harms caused to them; at the same time, however, courts may be hard pressed to deal with the many possible suits by the numerous governments harmed by climate change, and defendants seeking to prevent such suits might push for passage in Congress of a liability shield.[198] Plaintiffs seeking damages might also be able to recover from defendants only a relatively small percentage of the actual damages incurred, reflecting defendants' proportionate contribution to the harm.[199] On the other hand, climate plaintiffs seeking injunctive relief can more directly affect the activities that are causing the present harms and risks, but the costs of existing harms remain on the plaintiffs. It is also possible that a court could view such relief as dictating national energy or climate policy and therefore as prohibited by separation of powers or other justiciability concerns, as discussed earlier.[200]

5.2. *Potential Types of Damages Claims*

Some of plaintiffs' claims for damages would involve present harms. For instance, in *California v. GM*, the State of California has asserted present damages including reduced snow pack (a source of fresh water), higher sea levels along the coastline, and increased ozone pollution in urban areas.[201] Many of plaintiffs' damages, however, might be less tied to property already lost than to efforts that have

---

[196] Cox v. City of Dallas, 256 F.3d 281, 291 (5th Cir. 2001) (citing *Development in the Law – Injunction*, 78 HARV. L. REV. 997, 1001 (1965)).
[197] 406 F. Supp. 2d at 270 (quoting Complaint, *supra* note 24, 6).
[198] For descriptions of discussions in the 109th Congress concerning a liability shield, *see Backers of CO2 Curbs Eye Liability Relief to Bolster Industry Support*, 27 INSIDEEPA, (No. 36), Sept. 8, 2006.
[199] *See infra* Section 5.4.
[200] *See supra* Section 3.3.
[201] Complaint, *supra* note 11, 1.

been or need to be taken to prevent future harm.²⁰² California, for instance, asserts that:

> The State is spending millions of dollars on planning, monitoring, and infrastructure changes to address a large spectrum of current and anticipated impacts, including reduced snow pack, coastal and beach erosion, increased ozone pollution, sea water intrusion into Sacramento Bay-Delta drinking water supplies, and to respond to impacts on wildlife, including endangered species and fish, wildfire risks, and the long-term need to monitor on-going and inevitable impacts.²⁰³

The general tort rule is that plaintiffs harmed by defendants are entitled to recover their reasonable expenditures needed to abate, mitigate, or prevent future recurrences of those harms.²⁰⁴ For more than a century, for example, courts have held that plaintiffs can recover from defendants their reasonable expenditures for erecting walls to keep water off of their property.²⁰⁵ These precedents seem directly applicable to efforts to prevent or mitigate harms from sea-level rise (e.g., building levees, elevating houses and infrastructure). Given the high levels of confidence with which the IPCC and other scientists have established these present and projected impacts, preventative measures seem reasonable.²⁰⁶ Plaintiffs thus should be able to properly include the expenses for these preventative measures in the scope of damages.

In addition, one could view damages sought for monitoring as analogous to a medical monitoring claim, often asserted in cases involving exposure to a substance such as asbestos.²⁰⁷ The basic claim in those cases is that defendants' negligence in exposing plaintiffs to hazardous substances so increased the risk of adverse health consequences that the defendants should be liable for the present, quantifiable costs to plaintiffs of being tested periodically for signs of the illness.²⁰⁸ Some courts require

---

²⁰² Although such claims involve future harms, they should not be hindered by courts' general reluctance to award damages for future or latent injuries. In cases involving asbestos, for example, courts have hesitated to award damages for fear that the injuries may never actually occur. See, e.g., Lavelle v. Owens Corning Fiberglass Corp., 507 N.E.2d 476, 479 (Ohio Ct. Com. Pl., Cuyahoga Cty. 1987); Mauro v. Raymark Indus., 561 A.2d 257 (N.J. 1989). These climate change claims, in contrast, would seek compensation not for future harms, but rather for the present actions needed to prevent them.

²⁰³ Complaint, supra note 11, 4.

²⁰⁴ See generally M.O. Regensteiner, Annotation, *Expense Incurred by Injured Party in Remedying Temporary Nuisance or in Preventing Injury as Element of Damages Recoverable*, 41 A.L.R. 2d 1064 (2001). See also RESTATEMENT (SECOND) OF TORTS § 930(3)(b) (1979) (allowing damages for past and prospective invasions of land to include compensation for reasonable cost to plaintiff of avoiding future invasions).

²⁰⁵ See Regensteiner, supra note 204, at § 3(c) (citing Comstock v. New York C. & H.R.R. Co., 48 Hun. 225 (N.Y. 1888) (holding building owner who constructed concrete wall to prevent water from broken pipe on defendant's premises from flowing into his cellar entitled to recover expense of constructing wall) and Piedmont Cotton Mills, Inc. v. Gen. Warehouse, Inc., 149 S.E.2d 72 (Ga. 1966) (holding that where defendant's diversion of stream into artificial watercourse resulted in flooding to abutting property of plaintiff, cost of plaintiff's protective measures were properly element of damage)).

²⁰⁶ See WORKING GROUP II, IPCC, supra note 158. See also Chapter 1.

²⁰⁷ See, e.g., Marine Asbestos Cases v. Am. Hawaiian Cruises, Inc., 265 F.3d 861 (9th Cir. 2001).

²⁰⁸ See id. at 866; see also David Rosenberg, *The Causal Connection in Mass Exposure Cases: A "Public Law" Vision of the Tort System*, 97 HARV. L. REV. 851, 886 (1984). Even some courts that reject enhanced-risk claims accept the less-speculative medical monitoring claims. See, e.g., Mauro v.

present physical impacts or symptoms in order to make such a claim,[209] but many focus solely on whether the monitoring is reasonably necessary and will produce a real benefit.[210] Courts recognizing monitoring claims often do so to encourage early detection and mitigation of the harm.[211] In the climate context, a system to monitor water supply levels, for instance, could have real benefit by allowing early mitigation action.[212] Plaintiffs such as California thus might be able to recover reasonable monitoring expenses as well.

5.3. Restrictions on Damage Recovery

Although the damages may be cognizable, plaintiffs still might be unable to recover for some of their harms. If the property damaged or for which plaintiffs took preventative actions is private property, the remoteness doctrine might preclude governmental plaintiffs from recovering their own costs. The doctrine of remoteness bars recovery in tort for indirect harm suffered as a result of injuries directly sustained by another person.[213] Courts sometimes deem expenditures by a state that are inescapably contingent on direct or speculative harm to state residents to be too derivative or remote to support a tort claim.[214] The damages to homes from sea-level rise, for instance, might prove too derivative for a state to assert.

Nonderivative harms, however, could sustain a tort claim. For instance, if the damaged property was public, as are many beaches, roads, and other infrastructure, then the government itself is harmed. Diminished property tax revenues and lower property values also can harm states apart from any harm to individuals.[215] Further, even if some of the states' injuries arise from harm to others, states and their citizens have a relationship that can sometimes overcome the remoteness doctrine.[216]

Raymark Indus., Inc., 561 A.2d 257, 263 (N.J. 1989); *In re* Paoli R.R. Yard PCB Litig., 916 F.2d 829, 850 (3d Cir. 1990).

[209] *See, e.g.,* Mergenthaler v. Asbestos Corp. of Am., 480 A.2d 647, 651 (Del. 1984); Villari v. Terminix Int'l, Inc., 677 F. Supp. 330, 338 (E.D. Pa. 1987). Climate "symptoms" were described in Chapter 1.

[210] *See, e.g., Paoli,* 916 F.2d at 851 ("[T]he appropriate inquiry is not whether it is reasonably probable that plaintiffs will suffer harm in the future, but rather whether medical monitoring is, to a reasonable degree of medical certainty, necessary in order to diagnose properly the warning signs of disease."); *see also Marine Asbestos Cases,* 265 F.3d at 866; *Mauro,* 561 A.2d at 263.

[211] *See Paoli,* 916 F.2d at 852 (describing these as "conventional goals of the tort system"); *see also* Potter v. Firestone Tire & Rubber Co., 863 P.2d 795, 824 (Cal. 1993).

[212] An imperfection in the analogy is that for medical monitoring claims, the damage is to the human body, whereas the harms from climate change are largely to property or other interests. Courts might view these damages to be less urgent and therefore less in need of legal innovation. Monitoring systems to detect something like the northward spread of tropical diseases as the climate warms, however, might fit more squarely into the usual human health paradigm.

[213] *See* State v. Lead Indus. Ass'n, No. 99-5226, 2001 R.I. Super. LEXIS 37, at *44 (R.I. Super. Apr. 2, 2001) (unpublished opinion).

[214] *See, e.g., id.* at **44–46.

[215] *See* City of Boston v. Smith & Wesson Corp., No. 1999-02590, 2000 Mass. Super. LEXIS 352, at *24 (Mass. Super. July 13, 2000) (memorandum of decision and order on defendants' motion to dismiss).

[216] *See* Massachusetts v. EPA, 127 S. Ct. 1438, 1454 (2007) (quoting Georgia v. Tenn. Copper, 206 U.S. 230, 237 (1907): "This is a suit by a State for an injury to it in its capacity of *quasi*-sovereign. In that

The municipal cost recovery rule could pose another obstacle to recovery of some damages by climate change plaintiffs. Generally, a municipality may not recover the costs of providing public services.[217] "The cost of public services for protection from a safety hazard is to be borne by the public as a whole, not assessed against a tortfeasor whose negligence creates the need for the service."[218] What the cases barring recovery under the municipal cost recovery rule have in common is that the acts causing the damage were of the sort that the municipality reasonably could have expected to occur.[219] Courts have recognized, however, that governments can recover their expenses for abatement of public nuisances (such as cleanup of toxic wastes discharged into drinking water supplies) and for protection of the government's own property.[220] Which climate-related expenses are truly unexpected will largely be a question for the courts.[221]

5.4. Liability for and Apportionment of Damages

When multiple actors cause a harm, which appears to be the case with climate change plaintiffs' injuries, the critical question is the amount of damages for which courts should hold the defendants collectively and individually liable. If there is a reasonable basis for dividing the harm according to each defendant's contribution, each is liable only for that portion of the total harm that each has caused.[222] If the harm is an indivisible harm, all parties that are legal causes of the harm are jointly and severally liable for the entire harm.[223]

On their face, the harms from climate change appear to be indivisible. In *Michie v. Great Lakes Steel Division*,[224] several people residing near Ontario, Canada, sued three corporations operating seven plants immediately across the Detroit River in

---

capacity the State has an interest independent of and behind the titles of its citizens, in all the earth and air within its domain. It has the last word as to whether its mountains shall be stripped of their forests and its inhabitants shall breathe pure air."). *See also* City of Boston, 2000 Mass. Super. LEXIS 352 at **26–27.

[217] City of Philadelphia v. Beretta U.S.A., Corp., 126 F. Supp. 2d 882, 894–95 (E.D. Pa. 2000); *City of Boston*, 2000 Mass. Super. LEXIS 352, at **30–32.

[218] *City of Philadelphia*, 126 F. Supp. 2d at 894 (quoting City of Pittsburgh v. Equitable Gas Co., 512 A.2d 83, 84 (Pa. Commw. Ct. 1986)).

[219] *City of Boston*, 2000 Mass. Super. LEXIS 352, at **33–34.

[220] *See* City of Flagstaff v. Atchison, Topeka & Santa Fe Ry. Co., 719 F.2d 322, 324 (9th Cir. 1983); Town of E. Troy v. Soo Line R.R. Co., 653 F.2d 1123, 1132 (7th Cir. 1980), *cert. denied*, 450 U.S. 922 (1981).

[221] Courts can disagree on what are reasonably expected municipal costs. *Compare City of Philadelphia*, 126 F. Supp. 2d at 894–95 (noting that at least three courts – in Ohio, Florida, and Connecticut – have held that municipal cost recovery rule bars cities' suits against gun industry for recovery of expenses of policing cities), with *City of Boston*, 2000 Mass. Super. LEXIS 352, at *34 ("Plaintiffs allege wrongful acts [by the gun industry] which are ... [not] of the sort a municipality can reasonably expect.").

[222] RESTATEMENT (SECOND) OF TORTS § 881 (1979); *see also* RESTATEMENT (THIRD) OF TORTS: PRODUCTS LIABILITY §§ 16(a), (b) (1998).

[223] RESTATEMENT (SECOND) OF TORTS §§ 875, 879 (1979); *see also* RESTATEMENT (THIRD) OF TORTS: PRODUCTS LIABILITY § 16(c) (1998).

[224] 495 F.2d 213 (6th Cir. 1974).

the United States, claiming that pollutants emitted by defendants' plants were a nuisance. Each plaintiff sought damages from all three corporate defendants jointly and severally. The Sixth Circuit held that "although there is no concert of action between tort-feasors, if the cumulative effects of their acts is a single indivisible injury which it cannot certainly be said would have resulted but for the concurrence of such acts, the actors are to be held liable as joint tort-feasors."[225] This joint liability exists and operates regardless of the existence of "other corporations, persons and instrumentalities" that contributed to the air pollution "so as to make it impossible to prove whose emissions did what damage to plaintiffs' persons or homes."[226] If the judge or jury determines that it is not practicable to apportion the harm among the tortfeasors, "the entire liability may be imposed upon one (or several) tortfeasors subject, of course, to subsequent right of contribution among the joint offenders."[227] The *Michie* court therefore shifted from the injured party to the defendants the burden of proof as to which defendant was responsible for the relevant harms, and to what degree.[228]

One understandably could view holding defendants jointly and severally liable for the entirety of plaintiffs' harms from climate change as unfair. Because greenhouse gases have long lifespans in the atmosphere, past emissions are contributors to climate change. Accordingly, if courts assign all damages to current companies, those companies would be liable for past emissions to which they have no connection.[229] Furthermore, although those companies are "substantial" contributors, there are still other parties who have contributed somewhat to climate change as well. To avoid such inequity, courts may require apportionment even where harms seem indivisible if some means of fair and rational apportionment is possible without causing injustice to any of the parties.[230] In pollution cases, for instance, courts can treat a seemingly

---

[225] *Id.* at 216 (quoting Watts v. Smith, 134 N.W.2d 194 (Mich. 1965)).

[226] *Id.* at 218.

[227] *Id.* at 217; *see also* Martin v. Owens-Corning Fiberglass Corp., 528 A.2d 947 (Pa. 1987).

[228] *Michie*, 495 F.2d at 218. Courts have replicated the *Michie* holding in other pollution contexts. *See*, *e.g.*, Landers v. E. Tex. Salt Water Disposal Co., 248 S.W.2d 731, 734 (Tex. 1952); Velsicol Chem. Corp. v. Rowe, 543 S.W.2d 337, 342–43 (Tenn. 1976); Commonwealth v. PBS Coals, Inc., 534 A.2d 1130, 1139 (Pa. 1987). The Restatement incorporated the holding as well. *See* RESTATEMENT (SECOND) OF TORTS §§ 433A, 433B, 875, 879 (1965). A similar rule exists in products liability, in which a manufacturer is liable for the increased harm caused to plaintiffs by his product (beyond the harm that would have otherwise occurred). If the manufacturer cannot show what harm would have occurred absent the product defect, he can be liable for all of plaintiffs' harms. *See* RESTATEMENT (THIRD) OF TORTS: PRODUCTS LIABILITY § 16 (1998).

[229] *See* Christopher D. Stone, *Beyond Rio: "Insuring" Against Global Warming*, 86 AM. J. INT'L L. 445, 468 (1992). However, global CO_2 emissions skyrocketed in the twentieth century, particularly in the latter half, and have continued to do so at the start of the twenty-first century, thus playing a much larger role in altering the atmospheric concentrations of carbon dioxide. *See* UNEP/GRID-Arendal, *Global atmospheric concentration of CO2* (2005) (graph showing the increase in CO_2 levels in the atmosphere from 1870 to 2004 and predicted levels to the year 2100), *available at* http://maps.grida.no/go/graphic/global_atmospheric_concentration_of_co2 (visited Jan. 2, 2008).

[230] *See* RESTATEMENT (SECOND) OF TORTS § 433A cmt. d (1965); *see also id.* § 433B cmt. e.

indivisible harm as divisible and apportion it among defendants on the basis of evidence of their respective quantities of pollution discharged.[231] In the climate context, this division could involve apportioning damages (appropriately reduced to account for past emissions) based on the global warming potential of their greenhouse gas emissions, to correspond as much as possible to each defendant's contributions to climate change.[232] On the other hand, apportioning damages in accordance with each defendant's contribution to climate change might provide plaintiffs with only a small percentage of the damages they are suffering. Courts will have to find a balance between equitable apportionment and adequately compensatory damages.

6. CONCLUSION

Tort-based climate change litigation strikes many people as a strange idea at first. Basic tort principles, however, combined with the overwhelming scientific consensus that global climate change is occurring and is having present detrimental effects, may provide a basis for liability claims against major corporate emitters for some of climate change's effects.

Public nuisance seems to be the strongest of the climate tort claims, although it is possible that a products liability suit or some other tort claim could prove viable. Particularly in light of the recent Supreme Court decision in *Massachusetts v. EPA*, climate plaintiffs should be able to establish standing; the other jurisdictional issues of preemption and justiciability, while much less clear cut than standing, also may not present insurmountable obstacles. Plaintiffs should be able to establish generic and specific causation, particularly state plaintiffs who can aggregate harms over time and space, and there are several potential defendants who could be said to have proximately caused the plaintiffs' harms by means of their substantial contributions to global climate change. Should climate plaintiffs clear the jurisdictional hurdles and succeed on the merits of their cases, they could seek injunctive relief or damages for a range of costs incurred, likely apportioned among defendants in some way that recognizes both their relative contributions of greenhouse gas emissions and the compensatory needs of plaintiffs.

---

[231] *See id.* § 433A cmt. d (1965).
[232] *See* Penalver, *supra* note 4, at 592. Global warming potentials provide a quantified measure of the impact of a particular greenhouse gas on climate change. Carbon dioxide is the biggest contributor to climate change because of its volume, but other greenhouse gases are much more potent. *See* EPA, *supra* note 173, at 3. The Agent Orange settlement employed a similar system for apportioning damages, with reference to both market-share and dioxin content. *See* Ellen Tannenbaum, Note, *The Pratt-Weinstein Approach to Mass Tort Litigation*, 52 BROOK. L. REV. 455, 486 n.190 (1986) (citing *The Bargaining Behind the Agent Orange Deal*, BUS. WK., May 21, 1984, at 39). *But see* Allen Rostron, *Beyond Market Share Liability: A Theory of Proportional Share Liability for Nonfungible Products*, 52 UCLA L. REV. 151, 170–73, 215 (2004) (describing skepticism of courts, scholars, and Restatement drafters to applying "risk-adjusted market share liability," though arguing that courts should recognize that "fungibility is not essential if liability can be allocated in a way that reasonably accounts for the differing levels of risk created by each defendant.").

In sum, although there are some interesting issues involved and courts have dismissed the early cases, it seems that at least some tort-based climate change suits have strong legal merits and may be capable of succeeding. Like sea level and temperatures, the number of such cases likely will continue to rise over the next several years.

10

Insurance and Climate Change Litigation

Jeffrey W. Stempel*

INTRODUCTION

To date, most discussion of insurance in relation to climate change has concentrated on the problem of insuring against damage wrought by natural disasters such as Hurricane Katrina, and the particularly destructive 2004 hurricane season.[1] Significant media attention has addressed the posited impact of climate change upon the severity of what insurers term "environmental risks" – property losses occasioned by hurricane, flood, storms, or Earth movement.[2] Most reports focus on property insurance held by persons or entities suffering loss. These losses involve so-called

---

* Doris S. & Theodore B. Lee Professor of Law, William S. Boyd School of Law, University of Nevada Las Vegas.
[1] See, e.g., Robert J. Rhee, *Catatrophic Risk and Governance After Hurricane Katrina: A Postscript to Terrorism Risk in a Post-9/11 Economy*, 39 ARIZ. ST. L.J. 582, 591–602 (2007); Lavonne Kuykendall, *Chubb to Offer Flood Insurance for Some Upscale Customers*, WALL ST. J., Dec. 21, 2006, at D2, col. 1; Liam Pleven, *Hurricane Losses Prompt Allstate to Pursue New Path*, WALL ST. J., Nov. 24, 2006, at A1, col. 1.
[2] See, e.g., *Killer Hurricanes: No End in Sight*, NAT'L GEOGRAPHIC (Aug. 2006). See generally Jeffrey W. Stempel, *From Johnstown to New Orleans: The Law and Practice of Insurance of Environmental Risks in the USA*, in DIE VERSICHERUNG VON UMWELTRISIKEN ("Insurance of Environmental Risks") (Alexander Bruns & Zdenko Grobenski eds.) (Kompentenzzunturm Versicherungswissenschaft 2007) (describing major categories of environmental risks and the structure of United States risk management and legal environment regarding risks); *2006 Catastrophes Cost World's Insurers $15 Billion, Lightest Hit in Years, Swiss Re Estimates*, INS. J., Dec. 20, 2006, available at http://www.insurancejournal.com/news/international/2006/12/20/75293.htm.

In this context, "environmental risk" means risk of loss from widespread natural phenomenon, as distinguished from pollution-related losses that may also be labeled "environmental" losses. Environmental risks are viewed as particularly difficult by insurers because they involve substantial destruction across a wide expanse and thus present "correlated" risk (i.e., a single hurricane or flood affects an entire neighborhood or town) rather than the "uncorrelated" risk insurers prefer (e.g., fire, theft, and even most weather conditions such as hail or lightning, which tend not to affect every house on the block) since under these conditions the insurer is unlikely to be faced with a large number of claims at the same time. Gradually rolling uncorrelated losses permit insurers to earn more investment income. See MARK S. DORFMAN, INTRODUCTION TO RISK MANAGEMENT AND INSURANCE 19–23 (8th ed. 2004); EMMETT J. VAUGHAN & THERESE VAUGHAN, FUNDAMENTALS OF RISK AND INSURANCE 652 (8th ed. 1999); Stempel, *From Johnstown to New Orleans*, supra, at 5–10. By contrast, a large correlated property loss such as Hurricane Katrina or the September 11, 2001, destruction of the World Trade Center towers, produces a "capital shock" to the insurance markets due to the bunching of so many claims.

"first-party" insurance in which the policyholder (the first party) makes a claim for repayment of its loss by the insurer (the second party), who in return for a fee (the premium) agreed to indemnify the policyholder should it incur a covered loss.[3] As Hurricane Katrina made all too clear, many Americans are underinsured, particularly with respect to flood insurance.[4]

The public policy debate over effective national risk management policy for property promises to be interesting, even though the realities of political gridlock give little optimism for progress. But however interesting or important, these debates about national insurance and disaster policies tend not to include much analysis of the type of insurance most relevant to climate change litigation, liability insurance, which is potentially at issue when third parties sue policyholders who allegedly inflict injury.

This chapter uses *California v. General Motors*[5] as a starting point for exploring the liability insurance issues potentially raised by climate change litigation. In *California v. General Motors*, the State of California brought an action under the tort liability theory of "public nuisance." The State alleged that automobile manufacture and sale resulted in substantial production of carbon gases, in turn leading to adverse climate change causing discernible harm to State property. This chapter introduces the case, puts it in the broader context of liability insurance, considers potential insurer defenses to providing coverage, and explores the political and economic implications of insurance company payments for climate change litigation.

1. IMPLICATIONS FOR LIABILITY INSURANCE OF PENDING CLIMATE CHANGE LITIGATION

Most litigation in the United States alleging climate change–based causes of action against defendants has focused on either regulatory compliance or mandating reduced carbon emissions. *California v. General Motors* is the most prominent case thus far seeking damages from private entities rather than regulatory change by government.[6] Although this case, like other climate nuisance cases to date, was

---

[3] *See* Emeric Fischer, Peter Nash Swisher & Jeffrey W. Stempel, Principles of Insurance Law § 1.01 (3d ed. 2004) (addressing first-party/third-party distinction and differences between property and liability insurance); Robert H. Jerry, II, Understanding Insurance Law § 13 (3d ed. 2002) (addressing same).

[4] *See, e.g.*, Leonard v. Nationwide Ins. Co., 2006 U.S. Dist. LEXIS 60079 (S.D. Miss. Aug. 15, 2006) (homeowner unsuccessfully seeks compensation under windstorm coverage provided by standard homeowner's policy, which excludes flood-related losses; court holds exclusion to encompass losses caused by storm surge from hurricanes).

[5] *See* Complaint, California *ex rel.* Bill Lockyer, Attorney Gen. v. Gen. Motors Corp., Case No. C06-05755 (N.D. Cal. Sept. 20, 2006), available at http://ag.ca.gov/newsalerts/cms06/06-082_oa.pdf.

[6] For example, in the Introduction to this book, coeditors Wil Burns and Hari Osofsky note that U.S. climate change cases have ranged from 1. *Massachusetts v. EPA*; 2. a suit against two U.S. government agencies – the Export Import Bank and the Overseas Private Investment Corporation – for allegedly funding transnational fossil fuel-based energy projects that generate substantial greenhouse gas emissions without adequate environmental assessment under the National Environmental

dismissed at the district court level on political question grounds, the possibility of appellate reversal and future actions make this construct important to explore.[7]

In regulatory litigation, the identity of the defendants and the regulatory nature of the case effectively preclude involvement by liability insurers. Liability insurance ordinarily is triggered only when the policyholder is sued in an action seeking to hold the policyholder liable for causing bodily injury or property damage in which

---

Policy Act (NEPA); 3. a public nuisance suit against several major power companies; and 4. a petition to the U.S Fish and Wildlife Service to list the polar bear as an endangered species under the Endangered Species Act on the grounds that climate change is imperiling the [polar bear's] future. In a German case which recently settled, advocates demanded information from an agency providing risk insurance about the greenhouse gas emissions produced by its overseas projects. A Nigerian federal court action characterized gas flaring as a constitutional rights violation not only because of the direct small-scale impacts, but also because of those localities' vulnerability to the effects of climate change.

See *Overview: The Exigencies That Drive Potential Causes of Action for Climate Change*, this volume; see also Hari M. Osofsky, *The Geography of Climate Change Litigation: Implications for Transnational Regulatory Governance*, 83 WASH. U. L.Q. 1789, 1819–50 (2006) (discussing these and other cases in more detail); Hari M. Osofsky, *Climate Change Litigation as Pluralist Legal Dialogue?*, 26 STANFORD ENVTL. L.J. & 43 STANFORD J. INT'L L. 181 (2007) (discussing these and additional cases).

With the exception of example No. 3, *Connecticut v. Am. Elec. Power Co.*, 406 F. Supp. 265 (S.D.N.Y. 2005), a public nuisance suit against power companies, all of these actions are so primarily regulatory that they would appear to fall outside the scope of the liability insurance policies held by the defendants irrespective of the manner in which plaintiffs' claims were pled. Had the suit sought monetary damages, it could have triggered duties on the part of the power companies' liability insurers. See TAN 17–19, 26–30, *infra*.

[7] For the dismissal, see Order Granting Defendants' Motion to Dismiss, People v. Gen. Motors, 2007 WL 2726871 (Sept. 17). For another example, in *Comer v. Nationwide Mutual Insurance Co.*, a putative class of plaintiffs who incurred property damage due to Hurricane Katrina sought damages from a number of commercial entities (insurers, mortgage lenders, chemical companies, oil companies, and an industry trade association) engaged in activity alleged to have contributed to adverse climatic changes that increased the frequency and severity of hurricanes. The Court refused class action treatment on the ground that the plaintiffs' claims and injuries were too individuated and later dismissed the case on the grounds that it presented a nonjusticiable "political question" and the plaintiffs lacked legal "standing." See 2006 U.S. Dist. LEXIS 33123 (S.D. Miss. Feb. 23, 2006) (*Comer* action alleges injury from climate change and seeking monetary damages); Order of Aug. 30, 2007, in Comer v. Murphy Oil USA, Inc., Civil Action No. 1:05-CV-436 (S.D. Miss.) (dismissing case); *see also* JOHN E. NOWAK & RONALD D. ROTUNDA, CONSTITUTIONAL LAW §§ 2.12(f), 2.15 (6th ed. 2000) (reviewing legal doctrines of standing and political question). In *Northwest Environmental Defense Center v. Owens Corning Corp.*, 434 F. Supp. 2d 9557 (D. Or. 2006), plaintiffs alleged violations of the Clean Air Act and contribution to global warming, seeking civil penalties as relief. Many insurers would take the position that a request for civil penalties is not covered under a standard liability insurance policy. However, as discussed herein, the standard commercial general liability ("CGL") policy agrees to provide coverage for actions seeking to impose liability upon the policyholder because of property damage to a third party. Depending on the nature of the claim and the underlying facts, a request for civil penalties may or may not qualify as an attempt to impose liability for property damage. If, for example, the civil penalties are imposed merely for violating permitting processes, the insurer position that the policyholder is not being sued for damages may prevail. If, however, the civil penalties are assessed because of property damage inflicted by the policyholder's failure to observe proper permitting process, the policyholder may prevail in obtaining defense of the action and coverage; *see also In re* Methyl Tertiary Butyl Ether ("MTBE") Product Liability Litigation, 438 F. Supp. 2d 291 (S.D.N.Y. 2006); Barasich v. Columbia Gulf Transmission Co., 2006 U.S. Dist. LEXIS 86062 (E.D. La. Sept. 28, 2006), at *8–10.

the plaintiff seeks monetary damages. Actions aimed at regulatory compliance are generally outside the scope of general liability insurance coverage irrespective of whether the defendant is a public or private entity. In addition, government entities frequently self-insure or "go bare" without liability insurance in place. To date, the most prominent example of what I call "regulatory" climate change litigation – lawsuits directed toward prompting improved regulatory responses to climate change – is *Massachusetts v. EPA*,[8] in which several states successfully challenged the EPA's view that it was not authorized under the Clean Air Act to regulate greenhouse gas emissions from new motor vehicles and that the stated reasons for its refusal were "consistent with the statute."[9]

Actions that seek judicially mandated emissions reductions directly from private parties are also unlikely to implicate third-party liability insurance unless the plaintiff seeks money damages from a defendant. Standard general liability policies agree to pay the amounts for which a policyholder is liable "as damages," a requirement that generally has been held to preclude coverage for actions against the policyholder that seek only declaratory or injunctive relief.[10] The plaintiffs in the bulk of climate change cases to date seek a change in government policy or a change in defendant behavior rather than a monetary award.[11] For example, in *Connecticut v. American Electric Power Co.*,[12] the plaintiffs sought an order forcing the defendant utilities to abate the "public nuisance" aspects of their business activity but did not request the type of "damages" that are traditionally required to invoke liability insurance coverage.[13]

Despite the inapplicability of insurance questions to many of the lawsuits to date, there exists potential for substantial liability insurance involvement in climate change litigation seeking damages from private parties through formulations like the one in California's recent lawsuit against automobile manufacturers. In *California v.*

---

[8] 549 U.S. 497, 504 (Apr. 2, 2007).

[9] *See id.* The *Massachusetts v. EPA* majority (Justices Stevens, Kennedy, Souter, Ginsburg, and Breyer) found that the plaintiff states had standing to challenge the agency action, that the statute clearly provided the EPA with the requisite authority to regulate, and that the agency's preferred reasons for refraining to consider action were meritless. *See* 549 U.S. at 504, 514–35. Chief Justice Roberts and three other Justices (Scalia, Thomas, and Alito) dissented.

[10] *See* BARRY R. OSTRAGER & THOMAS R. NEWMAN, HANDBOOK OF INSURANCE COVERAGE DISPUTES §§ 5.01, 5.02 (12th ed. 2006); JEFFREY W. STEMPEL, STEMPEL ON INSURANCE CONTRACTS §§ 9.03; 14.01–14.04, 14.09 (3d ed. 2006); EUGENE R. ANDERSON, JORDAN S. STANZLER & LORELIE S. MASTERS, INSURANCE COVERAGE LITIGATION §§ 3.01–3.05 (2d ed. 2000 & Supp. 2006), Jerry, *supra* note 4, § 111; ALLAN WINDT, INSURANCE CLAIMS & DISPUTES ch. 4 (3d ed. 1995).

[11] *See, e.g.*, Nw. Envt'l Def. Ctr. v. Owens Corning Corp., 434 F. Supp. 2d 957 (D. Or. 2006); Korinsky v. EPA, 2005 U.S. Dist. LEXIS 21778 (S.D.N.Y. Sept. 29, 2005); Friends of the Earth, Inc. v. Watson, 2005 U.S. Dist. LEXIS 42335 (N.D. Cal. Aug. 23, 2005); *see also* Cent. Valley Chrysler-Jeep v. Witherspoon, 2006 U.S. Dist. LEXIS 67933 (E.D. Cal. Sept. 11, 2006); *In re* Quantification of Envtl. Costs, 578 N.W.2d 4794 (Minn. 1998).

[12] 406 F. Supp. 2d 265 (S.D.N.Y. 2006).

[13] *See* Complaint, Connecticut v. Am. Elec. Power Co., *available at* http://www.ct.gov/ag/lib/ag/press_releases/2004/enviss/global%20warming%20lawsuit.pdf.

General Motors, the State of California notes that auto emissions account for more than 30 percent of carbon dioxide emissions in California and contends that the defendants, by promoting carbon emissions, contribute substantially to global warming, producing adverse impacts on the State.[14] The California complaint is styled as one sounding in federal common law and California public nuisance law and seeks declaratory relief. More important for insurance purposes, the complaint also specifically requests an "[a]ward [of] monetary damages" and the declaratory relief request is "for such future monetary expenses and damages as may be incurred by California in connection with the nuisance of global warming."[15]

The California action seeks compensatory damages against the automakers because the State has suffered: "reduced . . . snow pack (a vital source of fresh water)"; "an earlier melting of the snow pack"; "raised sea levels along California's coastline"; "increased ozone pollution in urban areas"; and "increased . . . threat of wildfires." In addition, the complaint alleges that as a result of climate change caused in part by auto emissions, the State has expended substantial funds "assessing impacts and preparing for the inevitable increase in those impacts and for additional impacts," including expenditures already made "to address the declining snow pack and earlier melting of the snow pack in order to avert future water shortages and flooding."[16] These contentions of property damage and financial consequence due to global warming are made throughout the complaint.[17] The complaint arguably alleges bodily injury to California residents as well, contending that climate change is "having severe impacts on the health and well-being of California's residents" and on the State's health-care system because adverse climate change increases "the frequency, duration, and intensity of extreme heat events, conditions that are favorable to the formation of smog," which increases "the risk of injury or death caused by dehydration, heatstroke, heart attack, and respiratory problems."[18]

Due to these allegations of property damage and bodily injury, *California v. General Motors* appears to trigger a duty of the defendant's insurers to defend against the litigation and raises the prospect that commercial liability insurers may be forced to become involved in climate change litigation. Depending on how appellate courts rule in this and other pending nuisance actions, similar cases could perhaps

---

[14] *See* Complaint, California v. Gen. Motors Corp., *supra* note 6, at ¶ 3.
[15] *See id.* at 14 ("Relief Requested" 2, 3).
[16] *See id.* 1, 4.
[17] *See id.* 21 (alleging "significant damage" to the state's natural resources); 43 (injured land held in public trust by state); 45 (injury to "natural resources, including water, snow pack, rivers, streams, wildlife, coastline, and air quality"); 46 (disruption of water storage systems); 47 (significant increase in average temperatures in Sierra Nevada mountain region); 48 (detailing further damage to rivers, streams, and wildlife due to shrinkage of snow pack); 49 ("greater risk of flooding"); 51 (increased flooding); 52 (beach erosion, beach closures, costs to rectify beach problems); 53 (disappearance of sandy beaches, shoreline erosion leading to increased flooding); 54 (increased salt infiltration into fresh water); 56 ("increased risk and intensity of wildfires, risk of prolonged heat waves, loss of moisture . . . and related impacts on forests and other ecosystems, and a change in ocean ecology as water warms.").
[18] *See id.* 55.

proceed and trigger insurer defense duties. The possibility of this scenario makes an exploration of insurance coverage for climate change litigation critical.

2. APPLICABLE INSURANCE COVERAGE FOR CLIMATE CHANGE LITIGATION

2.1. *The History, Nature, and Structure of General Liability Insurance*

Large automakers typically purchase comprehensive or commercial general liability ("CGL") insurance. The CGL policy was created by the insurance industry during the 1940s by fusing together a number of other liability insurance products, principally owner's public liability insurance ("PL") with owner's, landlord's, and tenant's ("OTL") insurance.[19] The CGL policy firmly institutionalized what had been a growing practice among liability insurers since the 1920s: the "duty to defend" in which, as part of the liability insurance product, the insurer agreed not only to pay tort judgments against the policyholder (provided there was no applicable exclusion to coverage) but also to defend the lawsuits brought by plaintiffs against the policyholder even if the suit was "groundless, false or fraudulent."[20] Liability insurance in effect became "litigation insurance."

Over the years, courts and insurers have adopted the view that a liability insurer's duty to defend is "broader" than its duty to indemnify because the insurer must defend any suit that raises a "potential for coverage" while the insurer need only pay judgments or settlements that in fact fall within coverage.[21] If there are facts known to the insurer (or made known by the policyholder) that create a potential for coverage, most states require the insurer to defend even if these facts are not set forth on the face of the plaintiff's complaint.[22] Over time, insurers amended the CGL (substantially in 1955, 1966, 1973, and 1986)[23] but retained its breadth. CGL insurers defended hundreds of thousands of lawsuits and paid billions of dollars in settlements or judgments. In many cases, the plaintiffs' claims in these lawsuits were initially viewed as exotic or quixotic. For example, asbestos liability, pollution liability, strict product liability, and liability for inadequate security are

---

[19] See STEMPEL, *supra* note 10, § 14.04[A]; ELMER W. SAWYER, COMPREHENSIVE LIABILITY INSURANCE 12 (1943); S.S. HUEBNER, KENNETH BLACK, JR. & ROBERT S. CLINE, PROPERTY AND LIABILITY INSURANCE ch. 6 (2d ed. 1976).

[20] See STEMPEL, *supra* note 10, § 14.01[B].

[21] See WINDT, *supra* note 10, §§ 4.01–4.03; OSTRAGER & NEWMAN, *supra* note 10, § 5.02; ANDERSON, STANZLER & MASTERS, *supra* note 10, § 3.01–3.05.

[22] See ANDERSON, STANZLER & MASTERS, *supra* note 10, § 3.03; JERRY, *supra* note 3, § 111. See, e.g., Fitzpatrick v. Am. Honda Motor Corp., 78 N.Y. 61, 575 N.E.2d 90, 571 N.Y.S.2d 672 (1991).

[23] See DONALD S. MALECKI & ARTHUR L. FLITNER, COMMERCIAL GENERAL LIABILITY INSURANCE (8th ed. 2005); ANDERSON, STANZLER & MASTERS, *supra* note 10, § 1.02 and Appendix A; *see also* George C. Tinker, *Comprehensive General Liability Insurance – Perspective and Overview*, 25 FED'N INS. COUNS. Q. 217 (1975) (extensive discussion of 1973 revision); John J. Tarpey, *The New Comprehensive Policy: Some of the Changes*, 33 INS. COUNS. J. 223 (1966) (describing 1966 revision).

claims for relief that initially met with skepticism by commercial defendants and their insurers but ultimately proved viable causes of action.

Where certain types of litigation became unduly problematic or expensive, insurers ceased covering them. For example, the 1986 revisions to the CGL policy included a broad exclusion of all asbestos-related claims as well as what is now known as the "absolute" pollution exclusion.[24] During the past twenty years, the CGL policy has also been narrowed in other ways that may be pertinent to insurance coverage for climate change defendant policyholders. For example, CGL policies since 1986 are more likely than earlier policies to be written on a "claims-made" basis, in which a third party's action against the policyholder (rather than injury itself) is the trigger of coverage, a change that can limit the applicability of earlier years of insurance coverage. Some modern CGL policies are also written on a "defense costs within limits" basis in which funds spent defending litigation are applied to reduce the amount of remaining policy limits available to pay claims, effectively shrinking the overall insurance protection enjoyed by the policyholder. Nonetheless, claims like those of the *California v. General Motors* case may trigger application of older policies that lack such restrictive provisions because defendant conduct contributing to climate change arguably extends several decades into the past.

2.2. Liability Insurance Coverage for Injuries Inflicted by Business Policyholders and the Liability Insurer's "Duty to Defend" Climate Change Lawsuits

2.2.1. The Breadth of the CGL Insurer's Duty to Defend

Climate change lawsuits seeking monetary damages, such as *California v. General Motors*, pose a significant possibility of insurers being required to defend.[25] As noted earlier, the California Attorney General's complaint contains several allegations that the defendant automakers caused "property damage" to the State within the meaning of the CGL policy. Under the standard form CGL policy, covered property damage is defined as "physical injury to tangible property."[26] The California complaint alleges erosion (which is physical) of beaches and coastline (which are tangible);

---

[24] The 1986 revisions to the CGL made changes in terminology that were designed to make it less likely that courts would seize on the breadth of some of the policy language to require coverage in cases beyond the contemplation of insurers subscribing to the basic CGL form.

[25] See TAN 14–19, *supra*.

[26] See, e.g., Insurance Services Office, Commercial General Liability Form CG 00 01 10 01 (2000). The Insuring Agreement in Section 1, Coverage A provides: "We will pay those sums that the insured becomes legally obligated to pay as damages because of 'bodily injury' or 'property damage' to which this insurance applies. We will have the right and duty to defend the insured against any "suit" seeking those damages.... We may, at your discretion, investigate any 'occurrence' and settle any claim or 'suit' that may result." In the Definitions section of the CGL policy, "property damage" is defined as "physical injury to tangible property."

flooding (physical injury) to land (tangible property); shrinkage (physical) to snow-pack (tangible); and warming or pollution (physical) of the atmosphere (tangible, although some insurers may dispute that air is tangible).[27]

The complaint's allegations of "increased threat" of flood, wildfires, or other calamities probably fail to satisfy the CGL definition of property damage because these are not allegations of physical injury to tangible property. Similarly, the complaint's allegations that the state must engage in preventative measures in response to global warming[28] are not allegations of actual physical injury today and thus may not be covered unless the need for future state expenditures is linked to property loss that has already befallen California. Preventive measures designed to avoid future loss are often considered outside of the scope of liability insurance coverage. Conversely, the defendant policyholders may successfully argue that if they are required to fund California's preventive measures, necessitated by defendants' creation of a public nuisance, this constitutes currently cognizable injury and covered "property damage" within the meaning of the CGL policy. The presence of any uncovered allegations does not vitiate a liability insurer's duty to defend. The ironclad legal rule in all states is that the presence of a single allegation creating a potential for coverage requires the insurer to defend the entire lawsuit against the policyholder.[29]

2.2.2. The Requirement that the Policyholder Submit the Claim to the Insurer for Defense

Under the rules of insurance, the policyholder must make a timely "tender" of the claim to its insurer in order to trigger this duty.[30] Presumably, the automaker defendants will do so unless there are business factors militating against submitting the case to the respective insurers. The automaker defendants presumably have a strong incentive to notify their insurers and to seek a defense so that they cannot be accused of giving late notice that prejudices the insurer's opportunity to defend, and therefore excuses the insurers from coverage otherwise provided under the CGL.[31]

---

[27] See TAN 14–19, *supra*.
[28] See, *e.g.*, Complaint, California v. Gen. Motors Corp., *supra* note 5, 49, 56, 63.
[29] See STEMPEL, *supra* note 10, § 9.03[C]; JERRY, *supra* note 3, § 111. See, *e.g.*, Aerojet-Gen. Corp. v. Transport Indem. Co., 948 P.2d 909, (Cal. 1997); Buss v. Sup. Ct., 939 P.2d 766, (Cal. 1997).
[30] See STEMPEL, *supra* note 10, §§ 9.01, 9.03; JERRY, *supra* note 3, §§ 81[a], 111. See, *e.g.*, Country Mut. Ins. Co. v. Livorsi Marine, 856 N.E.2d 333 (Ill. 2006); Gazis v. Nat. Catholic Risk Retention Group, Inc., 892 A.2d 1277 (N.J. 2006).
[31] However, it is possible that some of the automaker defendants will attempt to defend the California claim without insurer assistance in order to maintain peaceable relations with insurers, who might be less inclined to sell coverage in future years should they become embroiled in protracted test case climate change litigation. See, *e.g.*, W. Bay Exploration Co. v. AIG Spec. Ag. of Tex., 915 F.2d 1030 (6th Cir. 1990) (applying Michigan law) (policyholder and agent delay in giving notice out

2.3. The Distinction between General Liability and Auto Liability Insurance and the Likely Inapplicability of the Auto Exclusion to the CGL Policy

As noted previously, the CGL policy is just that – a "general" liability policy designed to cover the basic tort liability risks faced by the average business. By intent and policy text, the CGL historically, and currently specifically, excludes a number of significant liability risks faced by some businesses. Auto risks (which are dependent upon area population, road conditions, alcohol consumption, crime, law enforcement, car safety devices, and other factors) differ from general liability risks (which depend more heavily upon the nature of the commercial policyholder's work, its potential for causing injury to others, and evolving tort law doctrine), requiring different analyses and premium setting. The instinctive notion that the automobile exclusion might well apply in a case involving automakers, however, falters upon closer examination. The auto liability exclusion found in general liability policies is designed to prevent the CGL policy from becoming an unintended automobile policy that provides coverage for a commercial fleet of vehicles as well as the general operations of the policyholder.[32] The automobile use exclusion is not designed to strip the policyholder of core CGL coverage.

The policyholder has the burden of showing that a claim falls within the insuring agreement. Once the policyholder has established that a claim falls within coverage, the burden shifts to the insurer when it attempts to escape coverage on the basis of an exclusion. A standard axiom of insurance policy interpretation is that exclusions in a policy are strictly construed and that insurers have the burden to establish clearly the applicability of an exclusion in order to deny coverage successfully.[33]

This burden cannot be met regarding the automobile exclusion. The California complaint does not allege that GM's "use" of automobiles caused climate change injury to the state. Rather, California alleges that the automaker's design, manufacture, sale, and distribution of inefficient fossil fuel engines caused destructive climate change.[34] Liability of a business for its design, manufacture, or distribution

---

of concern for cancellation or premium increases, resulting in loss of coverage when notice was finally given).

[32] See MALECKI & FLITNER, supra note 23, at 55–57.

[33] See Gustafson v. Cent. Iowa Mut. Ins. Ass'n, 277 N.W.2d 609, 614 (Iowa 1979); Merced Mut. Ins. Co. v. Mendez, 261 Cal. Rptr. 273 (Cal. App. 1989); STEMPEL, supra note 11, § 2.03; OSTRAGER & NEWMAN, supra note 11, § 1.01; see also Stonewall Ins. Co. v. Asbestos Claims Mgmt. Corp., 73 F.3d 1178, 1205 (2d Cir. 1995).

[34] In summarizing the complaint in this manner, I am being a bit charitable to California counsel, who could have, in my view, done a better job of framing the complaint to make it clear that the wrongdoing of the automakers was putting so many engines (literally and figuratively) of climate change into the field where their carbon emissions would wreak destructive climate change. For example, the First Cause of Action in the complaint (public nuisance under federal common law) alleges:

> Defendants have engaged in and are engaging in activities that have caused and continue to cause injury to the State of California. Defendants, by their emissions of carbon dioxide and other greenhouse gases from the combustion of fossil fuels in passenger vehicles and trucks, have knowingly created or contributed to and are knowingly creating or contributing

of products or services lies at the core of general liability as understood by insurers, policyholders, and insurance intermediaries.

General Motors and the other automakers are not accused of bad driving. They are being sued for alleged corporate irresponsibility in the design of their products, which in turn caused tortious injury to State property. Seen in this light, it seems quite clear that CGL insurers will not be able to use the automobile use exclusion as a basis for refusing to defend against climate change lawsuits and covering successful climate change claims.

3. ANTICIPATING INSURER DEFENSES TO COVERAGE OF CLIMATE CHANGE CASES

In the face of requests for defense or coverage, insurers likely will interpose a number of defenses based on policy language, exclusions, or basic concepts of insurance. The most likely battlegrounds are discussed below.

3.1. *The Problem of When an Insurer's Responsibility Is "Triggered"*

An insurance policy is issued for a particular "policy period," usually one year. A policy is said to be "triggered" when a covered event takes place during the policy period.[35] There are two primary triggers – occurrence basis and claims made – that may be used in liability policies. By far the most common is the occurrence-based policy, which provides that it is triggered by a claim (for duty to defend purposes) of bodily injury or property damage taking place during the policy period as a result of

> to a public nuisance – global warming – injurious to the State of California, its citizens and residents.... [This] constitutes a substantial and unreasonable interference with public rights in [California] including, among other things, public comfort and safety, natural resources and public property, and aesthetic and ecological values.

See Complaint, California v. Gen. Motors Corp., *supra* note 6, 58–59. The Second Cause of Action (public nuisance under California law) restates these allegations as violations of Calif. Civ. Code sections 3479 and 3480. See *id.* 66–70.

Although these allegations are in my view sufficient to make it clear that California is aggrieved because of what the automakers have made and sold (and not what they drove), CGL insurers may argue that allegations that defendants have engaged in "emissions of carbon dioxide and other greenhouse gases from the combustion of fossil fuels in passenger vehicles and trucks" suggest that the automaker defendants caused harm through driving the vehicles rather than making them available for others to drive. Insurers may also seize on this portion of the complaint to argue that the automakers are being sued over the direct release of carbon gases rather than for their role in making and distributing the engines used by others to release carbon gases, thus making the CGL policy's "pollution exclusion" applicable. See TAN 51–55, *infra*.

Because insurers bear the burden to show clear applicability of an exclusion in a CGL policy, I am relatively confident that this sort of insurer argument would not persuade courts to excuse CGL carriers from their duties to defend the automaker defendants. However, the complaint could have more clearly alleged harm resulting from defendant corporate activities unrelated to actual use of motor vehicles by the defendants or their agents.

[35] See OSTRAGER & NEWMAN, *supra* note 10, ch. 9; STEMPEL, *supra* note 10, § 14.09[A].

an "occurrence."³⁶ The triggering event under an occurrence policy is the date (or dates) of injury to a third person or damage to the property of a third person. The time of actual wrongdoing or negligence is not relevant.³⁷

The public nuisance action commenced by California would appear not to require the state to prove any particular knowledge of the harm being wrought in order for plaintiffs to prevail. However, once discovery ensues in litigation, very damning facts about defendant knowledge could be unearthed. Insurers could try to use this type of information to argue that the injuries inflicted upon plaintiff by the policyholder were not "accidental" and therefore need not be defended under the CGL policy.³⁸ For example, in the asbestos litigation, the evidence unearthed during discovery showed that some manufacturers were aware of the medical evidence of asbestos danger as early as the 1930s, some three or four decades prior to the initial plaintiff claims.³⁹ A similar scenario could unfold in cases such as the California climate change litigation (assuming that courts do not foreclose the litigation entirely on the grounds that it is too speculative or novel). Because its claim for relief sounds in nuisance, California may be able to succeed without even needing to prove that the automakers had any reason to know of the harmful consequences of the use of their products.⁴⁰

To the extent that defendant knowledge or state of mind matters in a nuisance action, automakers likely have been aware of auto pollution dangers since at least 1970, when the dangers were of sufficient concern to Congress to prompt passage of the Clean Air Act.⁴¹ Although climate change may only have recently achieved a high public profile and potential saliency as a political issue, it is not unreasonable to argue that as a scientific matter, the posited linkage between carbon fuels and environmental harm has been well established for approximately forty to fifty years, with the linkage to climate change seriously discussed for more than twenty years. Even in the absence of a successful public nuisance claim, if automakers are charged with actual or constructive knowledge on these points, it may not be far fetched to impose liability upon them for continuing to manufacture and market vehicles that produced very high levels of carbon dioxide emissions into the marketplace.

The prevailing view is that an occurrence basis policy is triggered by "actual injury" but that the injury or property damage need not be visible, palpable, or manifest. So long as property damage can be shown to have occurred at least in part in a policy

---

³⁶ See OSTRAGER & NEWMAN, supra note 10, ch. 9; STEMPEL, supra note 10, § 14.09[A].
³⁷ See OSTRAGER & NEWMAN, supra note 10, ch. 9; STEMPEL, supra note 10, § 14.09[A].
³⁸ See TAN 44–47, infra.
³⁹ See PAUL BRODEUR, OUTRAGEOUS MISCONDUCT (1985).
⁴⁰ See, RESTATEMENT (SECOND) OF TORTS § 821E (1979) (discussing elements of private nuisance). Public nuisance is defined as "an unreasonable interference with a right common to the general public." See RESTATEMENT § 821B. Accord DAN B. DOBBS & PAUL T. HAYDEN, TORTS AND COMPENSATION 670–76 (5th ed. 2005); DAN B. DOBBS, THE LAW OF TORTS §§ 465–470 (2000).
⁴¹ See 42 U.S.C. § 7521(a)(1).

year, the CGL policy for that year is triggered.[42] Particularly at the duty to defend stage of the litigation, it would appear impossible to use the face of the California complaint to preclude the triggering of many of the automakers' CGL policies. If this is correct, there may be decades of CGL policies at risk in the California climate change litigation and in similar lawsuits in progress or on the horizon. It is unlikely that insurers will be able to limit coverage to only a few policies or to only the more modern policies that contain additional defenses to coverage.

3.2. Lack of an Occurrence/Expected or Intended Injury/Lack of Fortuity

Whatever policies are at issue, CGL insurers are sure to argue that climate change claims are outside the scope of CGL coverage because the claimed damage results from the volitional conduct of the defendants and therefore does not satisfy the definition or concept of an "occurrence" (accidentally caused injury) set forth in the policy. In a similar fashion, insurers probably will argue that the climate change damage alleged by plaintiffs results from an intentional act or was "expected or intended" from the standpoint of the policyholder and that the claims are therefore excluded. Although insurance is available only for fortuitous, accident-like losses and not for losses willfully brought about by the policyholder,[43] it is unlikely that insurers can defeat policyholder requests for climate change coverage on this basis.

The California complaint, for example, is styled as a public nuisance action, which may succeed in imposing liability on the automaker defendants almost without regard to whether the fault or state of mind of the defendants because nuisance liability appears to stem more from the consequences of the defendants' activity than from the reasonableness of the activity. However, as noted here, insurers may argue that because nuisance requires an "intentional" invasion of property, any successful nuisance suit would require a finding of nonfortuitous, intentional conduct.[44] Although perhaps initially beguiling, such arguments should fail.

Notwithstanding that fortuity-based defenses to CGL coverage travel under names such as the "intentional act" exclusion, the exclusion does not bar coverage merely because a defendant's act was volitional. Rather, coverage is barred only when the policyholder as a matter of subjective state of mind intended to inflict injury or actually knew that the claimed injury was practically certain to take place as a result of the policyholder's conduct.[45] Climate change defendants may have fully intended to conduct business, but there is at this juncture nothing to indicate that they intended to contribute to destructive climate change or that they were practically certain that their activities would bring about deleterious climate change.

---

[42] See STEMPEL, supra note 10, §§ 14.09[A][1], 14.05[b][2].
[43] See id. § 1.06[A]; JERRY, supra note 4, §§ 63–63C.
[44] See TAN 44, supra.
[45] See STEMPEL, supra note 10, § 1.06.

For example, the *California v. General Motors* complaint, although styled as a public nuisance action, suggests negligence and perhaps even callous corporate disregard. However, nothing on the face of the complaint or inherent in the nature of the action suggests that the automaker defendants intended to create a public nuisance or to inflict harm on any third parties or the public generally. They may have been sadly oblivious to the impact of their products and too shortsighted to see the wisdom of more fuel-efficient, environmentally sensitive vehicles. But these corporate shortcomings do not make them intentional despoilers. Negligent, reckless, or even irresponsible conduct is not the same as conduct done with the intent to inflict injury upon others. Although California sues the automakers for doing what they intended to do (make cars with fossil fuel engines), the State does not allege that automakers specifically intended to effect adverse climate change.[46] Mere intent to conduct activity that leads to liability claims does not remove those claims from the broad scope of CGL coverage.

3.3. *The Number of Occurrences and Available Policy Limits*

CGL policies provide that the insurer's obligations end when the total policy limits are exhausted from the payment of judgments or settlements, so long as the expenditures have been reasonably made in good faith.[47] Thus, if an insurer is not recklessly or intentionally paying out its limits in order to terminate any defense obligations, it is possible for both the insurer's duty to pay claims and duty to defend to come to an end. Large auto manufacturers facing product liability and other substantial claims may have already exhausted much of their insurance that would otherwise be available to pay a successful global warming claim. However, so long as any single insurance policy retains even a dollar of available, unexhausted coverage, the CGL insurer must defend a potentially covered claim.

The California climate change litigation presents an interesting problem for causal analysis in determining the number of occurrences. On one hand, the manifestations of climate change contained in the State of California's complaint can be seen as having been caused by a single corporate policy of building vehicles that produce carbon emissions. At the other end of the spectrum, each relatively discrete instance of injury from climate change (e.g., erosion of this beach or of that coastline, melting of snowpack on Mountain A or on Mountain B, etc.) could be construed as a separate occurrence. In between, a court could find separate causes based on differentiating among corporate decisions about engine design, fuel efficiency, whether to warn, or even on the creation of each and every individual car model for the respective years of development.

---

[46] See TAN 14–19, *supra* (discussing allegations of California complaint).
[47] See, e.g., ISO CGL Policy Form No. CG 00 01 10 01, *supra* note 26.

As a general matter, it appears that courts tend to determine a number of occurrences in a manner that maximizes coverage for the policyholder.[48] Consequently, a realistic prospect is that courts will identify a set of discrete defendant decisions or actions as key events by automakers that have substantially contributed to climate change. For example, a court might regard each engine type or each line of products (e.g., heavy trucks, consumer trucks, SUVs, passenger automobiles) as separate occurrences causing climate change injury. Whatever course taken by the courts on the issues of occurrence and policy limits (presuming that courts are willing to let California go forward with its effort to achieve a pathbreaking imposition of liability), it is unlikely that insurers will be able to dramatically limit their coverage exposure on the basis of arguments concerning the number of occurrences.

3.4. The Pollution Exclusion

As previously mentioned, modern insurance policies (issued after 1986, and to some extent, from the 1970s onward) often contain a pollution exclusion that if clearly applicable may preclude even the potential for coverage necessary to invoke the CGL insurer's duty to defend. Insurers have enjoyed considerable success invoking the absolute/total pollution exclusion to bar coverage whenever a plaintiff's claim against a policyholder involves chemically related injury, even when the liability does not involve classic instances of pollution such as smokestack fumes, effluent runoff, or wetlands contamination. Many courts also have prevented the absolute pollution exclusion from being construed so broadly as to gut core general liability coverage,[49] but other courts have been very willing to read the exclusion broadly, notwithstanding the axiom that exclusions are to be narrowly construed against insurers.[50]

Against this jurisprudential backdrop, insurers would have considerable ground for arguing that climate change claims are pollution-related claims. But even if the insurer view prevails on this point concerning construction of the absolute pollution exclusion, policyholders may still have substantial CGL coverage for climate change claims. Prior to the 1970s, many CGL policies contained no pollution exclusion at all. Furthermore, before the mid-1980s, numerous policies contained only the qualified

---

[48] See JERRY, supra note 3, § 65.

[49] See, e.g., Stoney Run Co. v. Prudential-LMI Commercial Ins. Co., 47 F.3d 34, 37 (2d Cir. 1995) (applying New York law) (coverage for claims arising out of carbon monoxide poisoning from defectively installed furnace); Associated Wholesale Grocers, Inc. v. Americold Corp., 934 P.2d 65 (Kan. 1997) (coverage for damage to inventory due to smoke from nearby fire); W. Am. Ins. Co. v. Tufco Flooring E., Inc., 409 S.E.2d 692 (N.C. Ct. App. 1991) (coverage for injury stemming from inadequately ventilated fumes released during carpet installation).

[50] See, e.g., Deni Assocs. of Fla., Inc. v. State Farm Fire & Casualty Ins. Co., 711 So. 2d 1135 (Fla. 1998) (no coverage when blueprint machine tipped over during relocation, spilling ammonia); E.C. Fogg III v. Fla. Farm Bureau Mut. Ins. Co., 711 So. 2d 1135 (Fla. 1998) (no coverage for crop duster that inadvertently sprayed bystanders).

pollution exclusion that in many states was construed to provide coverage so long as the policyholder did not specifically intend to discharge harmful material. In the states that interpret this version of the exclusion to bar coverage only for liability resulting from intentional pollution-related injury, policyholders would appear to have coverage, at least with respect to the duty to defend. In states requiring that a pollution-related liability result from an abrupt or swift discharge of pollutants, insurers probably would prevail, at least in cases such as *California v. General Motors*, which asserts ongoing automaker misfeasance rather than any abrupt release of pollutants.

More threatening to insurers and the efficacy of the pollution exclusion is the possibility that climate change defendants and CGL policyholders will succeed in convincing courts that they are not being accused of actually releasing pollutants but of having engaged in wrongful conduct that fostered the greater release of carbon dioxide contributing to climate change. Under this characterization of climate change litigation, such as in *California v. General Motors*, neither the qualified nor the absolute pollution exclusion would seem to be applicable.

However, the decision in *Massachusetts v. EPA* will make such a characterization harder. The complaint in that case[51] posited that the Agency wrongfully refused to characterize carbon dioxide as a "pollutant." The plaintiff states' consequent victory in the litigation, which established that carbon dioxide can be considered a "pollutant" subject to EPA regulation, could provide commercial insurers with a stronger argument for imposition of the CGL pollution exclusion to bar coverage.

CGL insurers covering the post-1986 time period and D&O insurers[52] with policies containing broad pollution exclusions have significant defenses to coverage, but pollution exclusion arguments are hardly certain to succeed even though noxious or unwanted gases or emissions are widely regarded as "pollution." Add this to the significant number of potentially applicable occurrence-basis CGL policies that may be triggered in climate change cases that could be old enough to contain no pollution exclusion, or only the qualified, "sudden and accidental" version of the pollution exclusion, and insurers are at considerable risk of at least being compelled to defend against climate change claims.

Exploring potential application of the pollution exclusion requires considerable focus on the nuances of plaintiff theories of liability. For the *California v. General Motors* action, even the modern, absolute pollution exclusion would appear not to bar coverage. By both its text and its structure, the modern pollution exclusion provision is designed to bar general liability insurance coverage for liability arising

---

[51] *See* TAN 8–9, *supra*.
[52] Directors and Officers liability insurance or "D&O" insurance is written on a claims-made basis. The D&O policies that will be "in the field" in the event that predicted shareholder suits emerge against policyholders for failure to adequately anticipate and deal with the impact of climate change on company fortunes will be D&O forms with modern, post-1986 pollution exclusion language favorable to insurers. In short, D&O insurers may have a strong argument against coverage that many CGL insurers lack.

from a policyholder's own polluting activities at its places of business and closely linked operations (e.g., transport or waste disposal). Despite its breadth and broad definition of what constitutes a "pollutant,"[53] the exclusion is not designed to block coverage merely because a policyholder is sued in connection with a consumer's "downstream" use of a product that involves the release of chemicals or other pollution activity.

Applied to the *California v. General Motors* litigation, this analysis strongly suggests that even the absolute pollution exclusion is not a bar to coverage. Use of automobile engines results in discharge of a number of "pollutants," including carbon dioxide, which exacerbates deleterious climate change. But (with the exception of product testing and test drives around a dealer's parking lot), none of these discharges are perpetrated by the policyholder. Moreover, the emissions do not take place on policyholder premises, policyholder work sites, or in the course of policyholder business operations. Couple this with the fact that the pollution exclusion is (to perhaps state the obvious) an exclusion (on which the CGL insurer bears the burden of persuasion in the face of strict construction favoring the policyholder) and it appears that a pollution exclusion defense will be a loser for the automaker insurers.[54]

By contrast, if the plaintiffs in *Connecticut v. American Electric Power Co.*, in which California is one of the parties, were to obtain reversal of the political question dismissal currently on appeal and amend their complaint against the defendant utilities to seek monetary relief, utility insurers would have a near-ironclad defense to coverage under the modern pollution exclusion. In *Connecticut v. American Electric Power*, the plaintiffs allege that the defendants' regular, ongoing "smokestack pollution" at their plant sites contributed to adverse climate change. However, as discussed previously, climate change cases arguably involve insurance policy periods during which the relevant CGL policies contained no pollution exclusion, or only the qualified pollution exclusion.

3.5. *Seeking Recoupment of Defense Costs*

A number of states, including California, permit an insurer that has defended a lawsuit involving both potentially covered claims and uncovered claims to seek reimbursement for defense expenditures related to the uncovered claims. The

---

[53] The modern CGL typically defines pollutants as meaning "any solid, liquid, gaseous or thermal irritant or contaminant, including smoke, vapor, soot, fumes, acids, alkalis, chemicals and waste [which includes] material to be recycled, reconditioned or reclaimed." *See* CGL Policy Form, *supra* note 28, § V (Definitions), No. 15, *reprinted in* FISCHER, SWISHER & STEMPEL, *supra* note 3, at app. E-17.

[54] *But see* Complaint at ¶ 3 and *supra* note 44 (a portion of plaintiff's allegations can be interpreted as suggesting that automaker defendants were actually involved in the direct release of pollutants but Complaint as a whole is more consistent with the view that defendants are accused of selling polluting products and not the direct release of pollutants).

leading pro-insurer decision in this context was rendered by a California court. In *Buss v. Superior Court*,[55] liability insurers for Los Angeles Lakers owner Jerry Buss successfully argued that they were entitled to seek recoupment of defense costs in connection with defending a nineteen-count complaint (involving mostly uncovered breach of contract allegations) against Buss that involved only one potentially covered claim (defamation, which is covered under the "personal injury" provisions of the CGL policy rather than the more commonly invoked bodily injury and property damage aspects of the policy).[56]

Under *Buss* and similar decisions, the insurer must defend any lawsuit in which the face of the complaint seeks monetary damages and alleges at least one potentially covered claim for relief against the policyholder. However, if the insurer can adequately separate the costs of defending the covered claims and the not-even-potentially-covered claims in a lawsuit, the insurer may seek reimbursement from the policyholder for defense expenditures regarding the uncovered claims. In large or protracted litigation, the amounts at stake can be substantial. For example, in *Buss*, the insurer asserted that of the approximately $1,000,000 it spent on defense costs, only about $50,000 involved the potentially covered defamation claim.[57]

States are divided over the propriety of insurer efforts to recoup fees. In two important post-*Buss* decisions, Illinois and Wyoming have rejected the *Buss* approach and refused to permit insurer actions to recoup fees.[58] The fee recoupment issue could affect climate change litigation in the following manner. Lawsuits seeking to impose liability for climate change–related damage will likely contain a mix of covered claims (e.g., for damages due to negligent infliction of property damage, or perhaps even bodily injury, due to the defendant's contribution to global warming and attendant problems) and uncovered claims (e.g., for injunctive relief, for a mere declaration of responsibility, or for a regulatory determination). Depending on the nature of the plaintiff's theory of the case against a climate change defendant, the resulting litigation could center more or less on covered claims.

4. THE ECONOMIC AND POLITICAL IMPLICATIONS OF INSURER PARTICIPATION IN CLIMATE CHANGE LITIGATION

Climate change litigation is controversial. Even many environmental progressives are skeptical about the prospect of holding automobile manufacturers, electric utilities, and other commercial actors liable for the consequences of the worldwide

---

[55] 939 P.2d 766 (Cal. 1997).
[56] *See* CGL Form, *supra* note 28, § I, Coverage B (Personal and Advertising Injury Coverage); STEMPEL, *supra* note 11, §§ 14.05–06.
[57] *See*, 939 P.2d 766, 771 (Cal. 1997).
[58] *See* Gen. Agents Ins. Co. of Am., Inc. v. Midwest Sporting Goods Co., 828 N.E.2d 1092 (Ill. 2005); Shoshone First Bank v. Pac. Employers Ins. Co., 2 P.3d 510 (Wyo. 1998).

problem of global warming.⁵⁹ Some of the reluctance undoubtedly stems from a sense that it seems unfair to impose liability upon commercial actors who were acting legally in a manner designed to make and sell products at a profit. More reluctance probably stems from concerns that climate change litigation presents significant problems of proof of liability, causation, calculation of damages, and the limits of tort, evidence, and civil procedure doctrine.

Even if such conduct is considered sufficient to ascribe liability, it remains difficult to say how much of each defendant's respective conduct produced a recognizable quantum of injury. Thereafter, it remains difficult to place a dollar figure on the damage stemming from an individual defendant's activity. Although a "market share" theory of liability⁶⁰ might solve some of the damage calculation difficulties, other problems would remain. While one can be certain that automobile use contributes significantly to climate change, it is harder to say that auto manufacturers breached an established tort duty to California by engaging in commercial enterprise that was not only legal but often actively encouraged by local governments eager to attract auto manufacturing business that was perceived as job creating and tax-revenue enhancing.

There is even concern that in the face of these difficulties, California may not continue to pursue its claim against the automakers.⁶¹ Climate change litigation, like many emerging causes of action, presents litigants and courts with a host of difficult issues. In addition to the issue of whether the claim has sufficient chance

---

[59] At least this is how I interpret extensive discussion on an environmental law professors' listserve that ensued during fall 2006 in the wake of the filing of the *California v. General Motors* complaint. The litigation prompted at least thirty back-and-forth messages among a number of well-known law professors, with several expressing reservations over the utility of the lawsuit.

The continuing controversial nature of climate change litigation (despite the increasing consensus that global warming is taking place and presents a problem) is reflected in the 5–4 *Massachusetts v. EPA* decision. See 127 S. Ct. 1438 (2007). Compare 127 S. Ct. at 1446 (majority opinion of Justice Stevens joined by Justices Kennedy, Souter, Ginsburg, and Breyer) with 127 S. Ct. at 1463 (dissent of Chief Justice Roberts, joined by Justices Scalia, Thomas, and Alito) (expressing a dramatically different view of whether Massachusetts and other plaintiffs have legal "standing" to challenge EPA action) and at 1471 (Justice Scalia's separate dissent joined by the other dissenters) (expressing a dramatically different view of the statute in question and EPA prerogatives). To perhaps state the obvious, a slight change in Court composition could overrule the *Massachusetts v. EPA* holding. This division extends far outside the Court. Ten states actively opposed the climate change regulation efforts of the plaintiff states. See 127 S. Ct. at 1446, n.5.

[60] See, e.g., Sindell v. Abbott Labs., 607 P.2d 924 (Cal. 1980) (assessing liability of drugmakers according to their share of sales); Hall v. E.I. Du Pont De Nemours, 345 F. Supp. 353 (E.D.N.Y. 1972) (same for unidentifiable blasting caps). See generally DOBBS, supra note 40, § 176 (discussing market share liability).

[61] See Amanda Bronstad, *Some of AG's Signal Suits May Not Survive: Two Candidates Vying for AG's Office Doubt Merits of Global Warming Suit*, NAT'L L.J., Oct. 30, 2006, at 4. But see Edmund G. Brown, Jr., *Letter to the Editor*, WALL ST. J., Feb. 12, 2007, at B6, col. 1 (new Attorney General "would prefer to reach solutions outside of the courthouse" but defends *California v. General Motors* as "well grounded in precedent"). The federal trial court dismissed California's federal common law nuisance claims (see supra note 6). However, this does not automatically terminate the litigation. First, California may successfully appeal the federal trial ruling. Second, the trial court's order specifically permits California to refile its state law–based nuisance claims against the automakers in state court.

for success to merit continued prosecution, there are many questions regarding the ultimate fate of inventive legal claims. As with other novel theories of liability, one can expect the early years of climate change litigation to focus on whether such claims can be brought at all.

A parallel perhaps lies in asbestos bodily injury litigation. During the late 1960s and early 1970s, asbestos claims met with mixed success, but plaintifs' counsel pressed on during what Deborah Hensler has referred to as the "heroic" phase of asbestos litigation.[62] With the Fifth Circuit's 1973 *Borel* decision,[63] plaintiff's counsel achieved a great doctrinal victory when a federal appeals court rejected many standard defenses and ushered in the era of the asbestos mass tort. Although subsequent developments have raised issues of excess and greed regarding asbestos claims,[64] almost everyone now acknowledges that asbestos is a dangerous material, that many asbestos defendants long knew of its dangerous properties but failed to warn or protect users, and that victims are entitled to compensation.[65] Climate change litigation may achieve the success of asbestos claims or sexual harassment litigation,[66] both premised on what were initially considered nearly radical theories of liability, or it may be relegated to the category of creative but ultimately unsuccessful claims, in the same vein as comparable worth employment discrimination litigation[67] or efforts to ban imposition of the death penalty based on data showing that the race of the victim is strongly correlated with the jury's decision to execute.[68] Which scenario ensues may hinge on the relative persuasiveness of plaintiff and defense counsel, which in turn may hinge on the effectiveness of climate change defendants, insurers, and counsel in resisting and resolving climate change causes of action. If I am correct in this assessment, both plaintiff and policyholder counsel must engage important practical considerations. Other climate change plaintiffs (such as the Connecticut plaintiffs in the suit against electric utilities) should perhaps take a page from the *California v. General Motors* book and seek not only declaratory and injunctive relief

---

[62] *See* Deborah Hensler, *Asbestos Litigation in the United States: Triumph and Failure of the Civil Justice System*, 12 CONN. INS. L.J. 1, 5–7 (2006).

[63] *See* Borel v. Fibreboard Corp., 493 F.2d 1076 (5th Cir. 1973) (applying Texas law) *see also* Brodeur, *supra* note 39 (describing in detail the background of *Borel* case and trial court action).

[64] *See* WALTER K. OLSON, THE RULE OF LAWYERS 207 (2003) (criticizing high settlement values and counsel fees for marginal or speculative asbestos injuries).

[65] *See* Hensler, *supra* note 63, at 4.

[66] *See* Vinson v. Meritor Savs. Bank, 477 U.S. 57 (1986) (adopting the view that sexual harassment in the workplace constituted a violation of Title VII of 1964 Civil Rights Act).

[67] *See* Alexander v. Chattahoochee Valley Cmty. Coll., 325 F. Supp. 2d 1274, 1294 (M.D. Ala. 2004) ("courts have held that comparable worth claims are not cognizable under either the Equal Pay Act or Title VII"); *see also* Great Am. Savs. & Loan v. Novotny, 442 U.S. 366 (1979) (rejecting as a matter of law a 242 U.S.C. § 1985(3) claim based on alleged conspiracy to deprive plaintiff of her Title VII rights). Thus far, courts appear receptive to defendant arguments that climate change claims are nonjusticiable political questions involving policy choices that must be decided by other branches of government. However, the appellate process remains to run its course, making it premature to pronounce the death of current theories of liability in climate change litigation.

[68] *See* McCleskey v. Kemp, 481 U.S. 279 (1987).

against actors allegedly contributing to climate change damage, but also specifically request monetary damages from these defendants.

In addition, well-drafted plaintiffs' complaints should take pains to allege that the damages complained of flow from physical injury to tangible property or from bodily injury, including sickness or disease. Both of these pleading efforts (which of course are not legitimate unless made in good faith and supported by counsel's factual investigation) will make it much more difficult for any liability insurer to successfully refuse to defend climate change claims against the policyholder. In this way, a plaintiff alleging injury from climate change can both pursue recompense and aid the policyholder (without consultation or collusion) in seeking optimal insurance coverage. Alternatively, climate change plaintiffs may make a knowing and intentional decision to avoid seeking monetary damages or current or past property damage inflicted by the respective defendants in order to attempt to ensure that climate change defendants are unable to access liability insurance coverage and insurer-funded defense of the claims. Similarly, plaintiffs might (where supported by information sufficient to satisfy honesty-in-pleading requirements) specifically allege that defendants intended or expected to cause injury from their climate change activities.

In this manner, plaintiffs can – if this is their goal – force defendants to bear the brunt of the litigation without the aid of their insurers. Clever plaintiffs' counsel might even purposely plead cases in a way that makes the duty to defend question problematic in hopes of forcing a defendant to both pay its own way at the outset and engage in protracted coverage litigation with its insurers, perhaps even sowing the seeds for future attorney fee recoupment disputes between insurers and policyholders.[69]

Ordinarily, this is not the goal of plaintiffs, most of whom wish to maximize the prospects for a larger monetary recovery by ensuring that the defendants' insurers are drawn into the action. However, climate change plaintiffs may rationally wish to force defendants to internalize the costs of global warming litigation in order to inflict more economic pain on the defendants in hopes that this will make defendants more pliable regarding settlement and less likely to discount their environmental decisions in the future.

My own view is that such a strategy would be misplaced. If courts do not smother climate change litigation in the cradle through rejection on political question or other doctrinal grounds, it poses a substantial economic threat, even to defendants with plenty of insurance and the resources and expertise to battle insurers over coverage. The U.S. Supreme Court's willingness to decide the controversy between Massachusetts and its allied states and the federal Environmental Protection Agency

[69] In an adversary system, clever, ulterior motive–laden pleading of this type is generally permissible so long as the complaint does not misstate or fabricate facts. However, collusion between plaintiffs and insurers to plead a matter "out of coverage" would constitute bad faith by the defendant's insurer. *See* Lockwood Int'l, B.V. v. Volm Bag Co., 273 F.3d 741 (7th Cir. 2001) (applying Wisconsin law), *reprinted in* FISCHER, SWISHER & STEMPEL, *supra* note 3, at 962.

suggests that efforts to keep these disputes out of court on technical legal grounds may not succeed.[70] If defendants such as auto manufacturers and oil companies can add insurance proceeds to their already ample coffers, plaintiffs may extract sufficient funds to achieve significant progress in holding back adverse climate change (provided they can mount sufficient proof of liability to bring defendants and insurers to the bargaining table).

In responding to climate change lawsuits, defendants will make choices regarding whether to defend such claims with chosen counsel or whether to tender the cases to their liability insurers. If climate change cases grow, and if the insurance industry becomes heavily engaged in defending such claims, one would reasonably expect insurer involvement to shape the nature and degree of defendant responses. In addition, one would anticipate that insurers, perhaps looking down the barrel of another asbestos-like mass tort (unless courts roundly reject climate change theories, a significant possibility), would consider a holistic approach to resolving such claims and limiting their financial exposure.

Because climate change cases will involve a mixture of covered and uncovered claims, there exists, at least in theory, substantial opportunity for policyholder defendants and liability insurers to collaborate in trying to resolve climate change lawsuits as efficiently and effectively as possible. If the climate change plaintiff is a state like California or Connecticut (rather than an individual or commercial entity paying counsel fees), the plaintiffs also have a substantial incentive to work for resolution rather than to "shoot the moon" in hopes of obtaining large damage awards. Although some may decry this as de facto legislation through novel litigation, my own view is that this sort of outcome may serve as a needed filling of the vacuum created by government inaction, opposition, or denial of the problem.

All of this suggests to me that the prospect of insurer obligations in the emerging field of climate change litigation, such as those raised by *California v. General Motors*, is generally a good thing, perhaps adding an infusion of additional litigation expertise and capital that will enable "private" dispute resolution to accomplish some of what the national government has failed to address. At a minimum, a commercial policyholder's success in getting the CGL carrier involved in defense of a claim means that the insurer money becomes part of the mix, which may prompt more serious insurer attention to climate change claims and may provide an additional source of funds and settlement expertise as well as additional legal resources ensuring that climate change disputes will be thoroughly litigated and (at least in theory) correctly decided.

---

[70] *See* Massachusetts v. EPA, 549 U.S. 497 (2007). *Accord* Noel C. Paul, Student Article, *The Price of Emission: Will Liability Insurance Cover Damages Resulting from Global Warming?*, 19 Loy. Consumer L. Rev. 468 (2007) ("It is not unlikely that emitters of greenhouse gases will ultimately face a public nuisance suit on the part of states' attorneys general, similar to the tobacco litigation brought against cigarette manufacturers in the 1990s. If plaintiffs prevail . . . defendants should receive coverage under their standard CGL policies.") (footnotes omitted).

With luck, plaintiffs, defendants, insurers, and courts will recognize both the contractual and doctrinal exposure of insurers to defense and coverage of climate change litigation, as well as the potential gains from rational resolution of such litigation aided by the insurer involvement and the possible infusion of insurance policy proceeds. Better still, perhaps these parties will "make their own luck" by prosecuting, defending, and resolving climate change litigation with an appreciation of the potential positive role of liability insurance in addressing climate change disputes.

PART III

SUPRANATIONAL CASE STUDIES

11

The World Heritage Convention and Climate Change: The Case for a Climate-Change Mitigation Strategy beyond the Kyoto Protocol

Erica J. Thorson*

INTRODUCTION

Between 2004 and 2006, nongovernmental organizations (NGOs) from several countries submitted four petitions and a report (collectively, the Petitions) to the World Heritage Committee[1] to list certain World Heritage sites on the "List of World Heritage in Danger" (the "in danger" list) because of the deterioration these sites have endured as a result of climate change.[2] These sites include Sagarmatha National Park in Nepal, Huascarán National Park in Peru, the Great Barrier Reef in Australia, Waterton-Glacier International Peace Park in the United States and Canada, and Belize's Barrier Reef Reserve System,[3] which suffer from two of the most dramatic effects of climate change on natural areas – coral bleaching and glacial ice loss.[4] The Petitions argue that pursuant to their obligations under the World Heritage Convention (WHC),[5] State Parties must develop a mitigation strategy that prevents anthropogenic interference with the climate system sufficient to halt further deterioration of World Heritage sites threatened by climate change. At the heart of the Petitions, then, is a call for all State Parties to the WHC to make drastic cuts in their greenhouse gas emissions.

---

* Clinical Professor of Law, International Environmental Law Project, Lewis & Clark Law School, 10015 SW Terwilliger Blvd., Portland, Oregon 97219, 503.768.6715, ejt@lclark.edu. The author greatly appreciates the valuable suggestions of Prof. Chris Wold of Lewis & Clark Law School and Peter Roderick of the Climate Justice Programme during the preparation of this manuscript.

[1] The World Heritage Committee implements the World Heritage Convention. It consists of representatives from twenty-one State Parties, which are elected for terms of up to six years by the General Assembly of the WHC. For further discussion of the World Heritage Committee, see http://whc.unesco.org/en/comittee/ (last visited Aug. 25, 2006).

[2] See http://www.climatelaw.org/media (offering links to press releases regarding the Petitions) (last visited Aug. 25, 2006).

[3] For copies of the Petitions and more information on the NGO's action, see http://www.climatelaw.org/.

[4] For a more concise summary of the relationship between climate change and coral bleaching, see Contribution of Working Group II to the Third Assessment Report of the Intergovernmental Panel on Climate Change, *Climate Change 2001: Impacts, Adaptation, and Vulnerability*, 361 (2001). For a summary of the effects of climate change on glaciers and small ice caps, see *id.* at 208.

[5] See Convention Concerning the Protection of the World Cultural and Natural Heritage, Nov. 16, 1972, 27 U.S.T. 37, 1037 U.N.T.S. 151, (entered into force Dec. 17, 1975) (defining cultural and natural heritage of "outstanding universal value") [hereinafter World Heritage Convention].

The World Heritage Committee first considered the Petitions, except the petition concerning Waterton-Glacier International Peace Park (Waterton-Glacier), at its regular meeting in Durban, South Africa, during July 2005.[6] The Committee adopted a decision recognizing the threat climate change poses to the integrity of World Heritage sites. Moreover, it encouraged State Parties to incorporate responses to these threats in management plans developed for World Heritage sites and requested the creation of a working group of experts to study the effects of climate change on world heritage.[7] The expert working group called for in the decision met in March 2006 at a meeting at the World Heritage Centre[8] in Paris, France.[9]

One month prior to the meeting, a group of NGOs submitted the petition to list Waterton-Glacier on the "in danger" list. This petition, along with the election of the United States to the World Heritage Committee, raised the stakes of the review of the Petitions because of the claim that the WHC requires State Parties to reduce greenhouse gas emissions to protect World Heritage sites.[10] The United States, under the Bush administration, withdrew from the Kyoto Protocol, which would have imposed a binding international obligation to reduce its emissions. At the time of the petitions, Australia had also refused to ratify the Kyoto Protocol (though it has since done so), despite the fact that the Great Barrier Reef had suffered a series of devastating coral bleaching episodes linked to climate change.[11] By suggesting that the WHC requires a climate change mitigation strategy independent of the Kyoto Protocol, the Petitions argue that all State Parties, including the United States and Australia, may have an obligation to cut greenhouse gas emissions that exceeds their obligations under the Kyoto Protocol or, in the case of the United States and Australia at that time, what would have been their Kyoto Protocol obligations.

---

[6] See World Heritage Committee, *General Issues: Threats to World Heritage Properties*, Decision 29 COM 7B.b (July 2005). The Petition to list Waterton-Glacier International Peace Park as a World Heritage site "in danger" was submitted on Feb. 16, 2006.

[7] *Id.* at paras. 5–7.

[8] The World Heritage Centre is tasked with managing the day-to-day affairs of the World Heritage Convention. For a further discussion of the World Heritage Centre, see http://whc.unesco.org/en/134/ (last visited Aug. 25, 2006).

[9] See Decision 29 COM 7B.b, *supra* note 6, at para. 8.

[10] See Wil Burns et al., *International Environmental Law*, 40 INT'L LAW. 197, 199 (Summer 2006) (reporting that the Bush administration continues to refuse to ratify the Kyoto Protocol, emphasizing insead voluntary approaches and funding of technology development and transfer).

[11] See Sydney Centre for International and Global Law, *Global Climate Change and the Great Barrier Reef: Australia's Obligations Under the World Heritage Convention* (Sept. 21, 2004) (describing coral bleaching events and other climate change effects in the Great Barrier Reef), *available at* http://www.law.usyd.edu.au/scigl/SCIGLFinalReport21_09_04.pdf (last visited Aug. 25, 2006). The United States, Australia, and other Asian countries are collaborating on climate change issues within the context of the "Asia-Pacific Partnership for Clean Development and the Climate." The Partnership focuses on nonbinding, voluntary mechanisms, including technology development and transfer, information exchange and increasing national energy security, as means of combating long-term climate change. See Burns, *supra* note 10, at 199 (reporting on recent atmosphere and climate developments); *see also* The White House, Fact Sheet: The Asia-Pacific Partnership on Clean Development and Climate, *available at* http://www.whitehouse.gov/news/releases/2006/01/20060111–8.html (last visited Aug. 16, 2006).

The United States issued a policy and position paper prior to the March 2006 meeting of the expert working group contending that the Petitions are "invalid" for a number of substantive and procedural reasons.[12] The primary premise is that the root cause of climate change is not necessarily anthropogenic. The United States asserts that "[c]limate change is as old as the earth itself" and that "there is not enough data available to distinguish whether climatic changes at the named world heritage sites are the result of human-induced climate change or natural variability."[13] Whether due to U.S. influence or not, the World Heritage Committee's most recent decision on climate change fails to endorse a mitigation strategy that adequately implements the State Parties' WHC obligations.[14]

This chapter examines the relationship between climate change mitigation and the WHC and responds to the views articulated in the position paper of the United States. The chapter concludes that because climate change is threatening world heritage, State Parties are obligated to take mitigation action pursuant to the substantive provisions and the spirit of the WHC. Section 1 of this chapter provides a general overview of the relevant WHC provisions. Section 2 briefly discusses the threat of climate change to World Heritage sites. Section 3 examines the nature and extent of the WHC obligations, concluding that the provisions of the WHC create legally binding duties despite qualifying language; it then applies the WHC obligations to the threats posed by climate change and suggests that the WHC requires that State Parties adopt a "deep cuts" mitigation strategy. Section 3 also explores the failures of the World Heritage Committee's climate change efforts to date. Section 4 concludes that although the Petitions have heightened attention to climate change within the WHC, the protection of world heritage requires that the World Heritage Committee and State Parties to the WHC take more proactive mitigation action sooner rather than later.

1. OVERVIEW OF THE WORLD HERITAGE CONVENTION

The General Assembly of the United Nations Educational, Scientific and Cultural Organization (UNESCO) adopted the World Heritage Convention at its seventeenth session on November 16, 1972.[15] As of April 28, 2006, 182 countries had ratified the WHC, making it one of the most widely adopted international agreements.[16] The WHC's history reflects the global community's growing understanding that

---

[12] United States, "Position of the United State [sic] of America on Climate Change with Respect to the World Heritage Convention and World Heritage Sites," *available at* http://www.elaw.org/assets/word/u.s.climate.US%20position%20paper.doc (last visited Aug. 17, 2006) [hereinafter U.S. Position Paper].
[13] *Id.* at 4–5.
[14] World Heritage Committee, *Issues Relating to the State of Conservation of World Heritage Properties: The Impacts of Climate Change on World Heritage Properties*, Decision 30 COM 7.1 (July 2006).
[15] *See* http://whc.unesco.org/en/169/ (detailing history of the World Heritage Convention) (last visited Aug. 25, 2006).
[16] *See* http://whc.unesco.org/en/statesparties/ (last visited Aug. 9, 2006).

conservation of culture and nature requires international cooperation and commitments.[17] As René Maheu, Director-General of UNESCO during the WHC negotiations, stated in an address to the drafters of the Convention, "[Member States] should be responsible not only for combating deterioration and damage to the cultural and natural heritage, but also for investigating their causes in order that the evil may be attacked at its root."[18] As of March 2007, the State Parties have inscribed 830 sites to the list, including 644 cultural sites, 162 natural areas, and 24 mixed cultural and natural properties in 138 countries.[19]

The "World Heritage List" is the core focus of the WHC. Article 11 of the Convention provides that the World Heritage Committee must compose a list of World Heritage sites based on inventories of world heritage submitted by State Parties.[20] The List serves as a locus for the World Heritage Committee's energies, fund distribution, and international protection. "Outstanding universal value" is the foundational criterion for listing a site as World Heritage under the Convention, and a property may be of "outstanding universal value" based on either its cultural or natural values.[21]

At its core, the WHC recognizes that the "deterioration or disappearance of any item of the cultural or natural heritage constitutes a harmful impoverishment of the heritage of all nations of the world... [they] therefore need to be preserved as part of the world heritage of mankind as a whole."[22] The provisions of the treaty implement the principle that international cooperation is essential to protect world heritage, but they also explicitly respect national sovereignty.

Article 4 of the Convention defines the obligations of State Parties respecting World Heritage sites within their territories. It states that

> [e]ach State Party to this Convention recognizes that the duty of ensuring the identification, protection, conservation, presentation and transmission to future generations of the cultural and natural heritage... situated on its territory, belongs primarily to that State. It will do all it can to this end, to the utmost of its own resources....[23]

Thus, State Parties accept the responsibility to expend resources and take all necessary actions possible to preserve World Heritage sites for future generations. To fulfill

---

[17] The decision to build the Aswan Dam first sparked international interest in safeguarding cultural monuments. The dam was to flood the valley containing the Abu Simbel temples. Subsequent to an appeal from Egypt and Sudan, UNESCO campaigned to safeguard the temples. Its success led to other campaigns, and soon the idea for a convention to protect cultural heritage arose. A few years later, the United States began work to include natural heritage. For further discussion, see *id.*

[18] Address by Mr. René Maheu, Director-General of the United Nations Educational, Scientific and Cultural Organization, 72//DG/72/4, *at* http://whc.unesco.org/archive/1972/dg-72-4e.pdf (last visited Aug. 22, 2006).

[19] *See* http://whc.unesco.org/en/list/ (last visited Aug. 9, 2006).

[20] For a complete list of World Heritage sites, see http://whc.unesco.org/pg.cfm?cid=31 (last visited Aug. 23, 2006).

[21] World Heritage Convention, *supra* note 5, at arts. 1 and 2.

[22] *Id.* at preamble, sixth and seventh recitals.

[23] *Id.* at art. 4.

this obligation, Article 5, among other things, requires that State Parties endeavor to implement operating methods that "will make the State capable of counteracting the dangers that threaten its cultural or natural heritage"[24] and to "take the appropriate legal, scientific, technical, administrative and financial measures necessary for the identification, protection, conservation, presentation and rehabilitation of this heritage[.]"[25]

Although each State Party is first and foremost the protector of World Heritage sites situated in its territory, the Convention, as stated in its Preamble, recognizes that national effort alone is often insufficient to address the threats facing world heritage. Article 6 provides that State Parties recognize "that such heritage constitutes a world heritage for whose protection it is the duty of the international community as a whole to co-operate."[26] State Parties agree "to give their help in the identification, protection, conservation and presentation of the cultural and natural heritage... if the States on whose territory it is situated so request."[27] Finally, "[e]ach State Party... undertakes not to take any deliberate measures which might damage directly or indirectly the cultural and natural heritage... situated on the territory of other States Parties to this Convention."[28] Together, these provisions comprise the responsibility to cooperate in global efforts to protect world heritage and to ensure that actions taken within a national territory do not cause damage or deterioration of the world heritage situated in any other national territory.

2. LOSING WORLD HERITAGE – THE THREAT OF CLIMATE CHANGE

The dramatic nature of the effects of climate change is most readily apparent in ecosystem change, particularly in fragile and vulnerable ecosystems. According to the Millennium Ecosystem Assessment Series, "[s]ome systems – including coral reefs, glaciers, mangroves, boreal and tropical forests, polar and alpine systems, prairie wetlands, and temperate native grasslands – are particularly vulnerable to climate change because of limited adaptive capacity and may undergo significant and irreversible damage."[29] Many examples of these ecosystems are treasured throughout the world as World Heritage sites, and their particular vulnerability to climate change has led to devastating and sometimes catastrophic consequences. A recent survey by the World Heritage Centre of all State Parties of the WHC demonstrates the nature and extent of the consequences of climate change on valuable natural areas. The responses of eighty-three State Parties revealed that climate change threatened

---

[24] *Id.* at art. 5(c). Article 5 also suggests that State Parties endeavor to develop comprehensive planning and protection programs, train and educate protected-area staff, scientists, and community members, and undertake scientific and technical studies and research. *Id.* at art. 5.
[25] *Id.* at art. 5(d).
[26] *Id.* at art. 6(1).
[27] *Id.* at art. 6(2).
[28] *Id.* at art. 6(3).
[29] 1 MILLENNIUM ECOSYSTEM ASSESSMENT, ECOSYSTEMS AND HUMAN WELL-BEING: CURRENT STATE AND TRENDS 379 (Rashid Hassan, Robert Scholes, & Neville Ash eds., 2005).

a reported 125 World Heritage sites.[30] Of these sites, seventy-nine were listed as natural or mixed heritage (of both cultural and natural significance), including sixteen coastal areas (seven of which are coral reefs), fourteen glacier sites (seven of which are glaciated mountain areas), twenty-eight terrestrial biodiversity sites, and fourteen mixed biome and other sites.[31] Among the concerns listed were glacial retreat, sea level rise, loss of biodiversity, species migration and tree-line shifts, coral bleaching, and droughts.[32] The responses to this survey indicate that State Parties are aware of the devastating consequences of climate change on world heritage, and they further highlight how pervasive climate change consequences are in natural areas.

This chapter focuses on natural areas because the Petitions for "in danger" listing due to climate change concern either the imminent loss of mountain glaciers or the dying off of coral reefs. However, the World Heritage Centre survey reveals that many man-made, cultural sites are also under threat from climate change.[33] As such, the following sections of this chapter concerning the legal obligations of State Parties to the WHC should also be read with the protection and preservation of cultural heritage in mind.

3. CLIMATE CHANGE MITIGATION

Although the Petitioners employed the "in danger" listing process to highlight the devastating consequences of climate change and to urge immediate attention for particular areas, the language of the Convention text, which is implicated when a site is listed simply as "world heritage," demands that State Parties engage in effective climate change mitigation even before a site is listed as "in danger." Climate change mitigation is defined as "an anthropogenic intervention to reduce the sources of greenhouse gases or enhance their sinks."[34] Certainly, if climate change is causing deterioration of World Heritage sites, then climate change mitigation is at least one of the "appropriate" legal, scientific, and technical undertakings because mitigation is necessary to prevent total deterioration of many vulnerable World Heritage sites. For example, although adaptation mechanisms could address floods resulting from glacial melt, only mitigation addresses the root cause of current glacial melt trends – namely, an excess of greenhouse gases in the atmosphere. As was recognized by an expert working group on climate change and world heritage, preventive actions,

---

[30] May Cassar et al., *Predicting and Managing the Effects of Climate Change on World Heritage: A Joint Report from the World Heritage Centre, Its Advisory Bodies, and a Broad Group of Experts to the 30th Session of the World Heritage Committee*, para. 42 (2006) [hereinafter *Joint Report*].

[31] *Id.* at para. 43.

[32] *Id.* at para. 44.

[33] *Id.* at para. 45–46. For example, sea level rise due to climate change threatens the World Heritage Sites near the Thames River, including the Palace of Westminster and the Tower of London, because the river's floodplain continues to expand. *See Joint Report, supra* note 30, at box 7–8 (describing climate change effects on cultural sites).

[34] Working Group III to the Third Assessment Report to the Intergovernmental Panel on Climate Change, *Climate Change 2001: Mitigation* (Summary for Policymakers), at 3 (2001); *see also Joint Report, supra* note 30, at para. 10 (restating the definition of mitigation as "reducing the emission and enhancing the sinks of greenhouse gases").

including mitigation, "need to be taken to safeguard heritage."[35] However, the World Heritage Committee has thus far failed to recommend the type of aggressive mitigation strategy required by the WHC.

3.1. The Nature and Extent of State Parties' Obligations under Articles 4, 5, and 6

The nature and extent of how the obligations set forth in Articles 4, 5, and 6 bind State Parties – namely, whether the operative provisions impose mere recommendations entirely left to State Party discretion to implement or whether, in a given context, such as climate change, they impose substantive obligations – is a key interpretive question. Articles 4 and 5 are broad, potentially leaving much room for State Party discretion as to the exact nature of the respective responsibilities. They contain qualifying language such as "as far as possible," employ precatory verbs such as "endeavor," and merely require that State Parties "recognize" certain responsibilities. In fact, some would argue that the language of Articles 4 and 5 is so broad and imparts so much discretion that it eviscerates any binding obligation.[36] The only case to examine the nature of the obligations imposed by Articles 4 and 5 is *Commonwealth v. Tasmania*, a case of the High Court of Australia.[37] Despite the qualifying language of Articles 4 and 5, a majority of the High Court of Australia determined that both Articles impose legally binding obligations, essentially because the qualifying language would be superfluous if, in fact, no obligation existed.[38] Although having

---

[35] World Heritage Committee, *The Impacts of Climate Change on World Heritage Properties*, WHC-06/30.COM/7.1, para. 13 (This document reports on the outcome of the expert working group on climate change and world heritage, which met in Paris during March 2006, and was prepared for presentation at the 30th Session of the World Heritage Committee, which took place in July 2006.) [hereinafter *Strategy*].

[36] *See* Commonwealth of Australia v. Tasmania (Tasmanian Dam Case), 158 CLR1 (1983), para. 69 (C.J. Gibbs) ("It is however impossible to conclude that Arts. 4 and 5 were intended to impose a legal duty... on State Parties to the Convention. If the conduct which those articles purport to prescribe was intended to be legally enforceable, the obligations thereby created would be of the most onerous and far reaching kind.... The very nature of these obligations is such as to indicate that the States Parties did not intend to assume a legal obligation to perform them."); *see also* Michael I. Jeffery, QC, *An International Legal Regime for Protected Areas*, in IUCN ENVIRONMENTAL LAW & POLICY PAPER NO. 49, 23 (John Scanlon & Françoise Burhenne-Guilmin, eds. 2004) (suggesting that the phraseology is so subjective that some argue no legal obligations may exist).

[37] Tasmanian Dam Case, *supra* note 36. Although the case primarily concerned the relationship of Commonwealth and State power, the decision turned, in part, on whether the WHC imposed binding obligations and the nature of these obligations. In the case, Tasmania challenged the Australia Commonwealth's legislation providing for the protection of World Heritage areas. The argument revolved around the division of powers between the Australian federal government and individual state governments, such as Tasmania. *See id.* at paras. 2, 25 (C.J. Gibbs).

[38] *Id.* at para. 31 (J. Mason). Judge Mason's opinion states: "Article 5 cannot be read as a mere statement of intention. It is expressed in the form of a command requiring each party to endeavour to bring about the matters dealt with in the lettered paragraphs. Indeed, there would be little point in adding the qualifications 'in so far as possible' and 'as appropriate for each country' unless the article imposed an obligation." *See also* Jeffery, *supra* note 36, at 23 ("Although terminology such as 'to the utmost of its own resources' and 'in so far as possible' might be seen as adding a subjective mechanism from which States can easily escape responsibility, it still places a legal obligation on each contracting party.").

found that Articles 4 and 5 of the WHC impose binding legal obligations, the Court nonetheless recognized that the duties are so broadly articulated that State Parties have much latitude as to how they implement the Convention. As one judge stated in his opinion: "[T]here may be an element of discretion and value judgment on the part of the State to decide what measures are necessary and appropriate."[39] This discretion, however, is not without bounds. This judge further noted, "There is a distinction between a discretion as to the manner of performance and discretion as to performance or non-performance."[40]

The Australian case clarifies that Articles 4 and 5 impose discretionary obligations, but international law defines the nature of State Parties' discretion. With respect to treaty implementation, the principle of *pacta sunt servanda* guides State Party discretion.[41] This principle provides that States are bound by their international agreements and that they must implement such agreements in good faith.[42] Thus, Articles 4 and 5 of the WHC impose discretionary obligations, but "good faith" is the touchstone for implementation, and the aims of the Convention – namely, the protection and conservation of world heritage – guide operationalization of State Parties' good faith.

Unlike Articles 4 and 5, Article 6 is not qualified with language of limitation.[43] The provisions of Article 6 are less discretionary, stating that State Parties are not to

---

[39] Tasmanian Dam Case, *supra* note 36, at para. 29 (J. Mason).
[40] *Id.* at para. 31 (J. Mason).
[41] *See* I. M. SINCLAIR, THE VIENNA CONVENTION ON THE LAW OF TREATIES 3 (1973) (describing *pacta sunt servanda* as "the most fundamental principle of treaty law"). *See generally* LORD MCNAIR, THE LAW OF TREATIES 493–505 (1961) (explaining extensively the principle of *pacta sunt servanda*); Josef L. Kunz, *The Meaning and the Range of the Norm Pacta Sunt Servanda*, 39 AM. J. INTL. L. 180 (1945).
[42] IAN BROWNLIE, PRINCIPLES OF PUBLIC INTERNATIONAL LAW 620 (5th ed. 1998). The Vienna Convention on the Law of Treaties (Vienna Convention) states the principle in the following manner: "Every treaty in force is binding upon the parties to it and must be performed by them in good faith." Vienna Convention on the Law of Treaties, May 23, 1969, U.N. Doc. A/CONF. 39/27, 1155 U.N.T.S. 331 (entered into force Jan. 27, 1980) [hereinafter Vienna Convention]. The Vienna Convention entered into force in 1980, after the WHC, and therefore might not be applicable retroactively. However, much of the Vienna Convention embodies customary international law and, as such, would be applicable. *See* Case Concerning the Gabcikovo-Nagymaros Project, Hungary/Slovakia, 1997 I.C.J. 3 (stating that although the Vienna Convention may not be directly applicable to an earlier international agreement, those provisions of the Vienna Convention that state customary international law are relevant). The United States has not ratified the Vienna Convention, but the State Department has stated that the Vienna Convention is evidence of the customary law on treaties. It describes the Vienna Convention as "constituting a codification of the customary international law governing international agreements and therefore as foreign relations law of the United States even though the United States has not adhered to the convention." SEN. EXEC. DOC. L., 92nd Cong., 1st Sess. (1971). Further, in the letter of submittal of the Vienna Convention, the Secretary of State described it to be "'generally recognized as the authoritative guide to current treaty law and practice.'" Maria Frankowska, *The Vienna Convention on the Law of Treaties Before United States Courts*, 28 VA. J. INT'L L. 281, 298 (1988) (quoting Secretary of State Rogers' Report to the President, Oct. 18, 1971, 65 Department of State Bulletin 684, 685 (1971)).
[43] The Australian Court suggested that these provisions more clearly impose binding obligations on Parties to the Convention; however, the Court did not directly rule on the issue.

undertake deliberate measures that might damage world heritage.[44] A simple textual analysis of the plain meaning of the provision supports this interpretation. Under fundamental rules of treaty interpretation, as provided by the Vienna Convention, a treaty must "be interpreted in good faith in accordance with the ordinary meaning of the terms of the treaty in their context and in light of its object and purpose."[45] The plain language of Article 6(3) sets forth a nondiscretionary duty to forgo deliberate undertakings that may damage world heritage.

> The *travaux préparatoires* (the negotiating history of the treaty) supports this plain language interpretation.[46] Early drafts of the Convention did contain qualifying language, but the drafters pointedly excluded it from the final version of Article 6. In early drafts, Article 6(3) read: "The States Parties to this Convention undertake to respect the cultural and natural heritage enjoying international protection under this Convention by refraining so far as possible from acts which might damage them."[47]

The adopted language is far less discretionary and imposes a binding, articulable legal obligation on State Parties. In fact, the drafters specifically eliminated "in so

---

[44] Tasmanian Dam Case, *supra* note 36, at para. 32 (J. Mason).
[45] The Vienna Convention is widely understood to codify customary international law regarding interpretation of treaties. SINCLAIR, *supra* note 41, at 153 ("There is no doubt that articles 31 to 33 of the Convention constitute a general expression of the principles of customary international law relating to treaty interpretation."); *see also* BROWNLIE, *supra* note 42, at 608 (stating that "a good number" although not all, of the provisions of the Vienna Convention express general international law, and those that do not "constitute presumptive evidence of emergent rules of general international law"). Indeed, the textual approach to interpretation of treaty provisions codified in Article 31 has attained the status of customary international law. *See* Sir Gerald Fitzmaurice, *The Law and Procedure of the International Court of Justice 1951–1954*, 33 BRIT. Y.B. INT'L. L. 203, 204 (1957) (suggesting that the International Court of Justice favors the textual approach); *and see, e.g.*, Territorial Dispute Case (Libyan Arab Jamahiriya v. Chad), 1994 I.C.J. Reports 6, para. 41; OPPENHEIM'S INTERNATIONAL LAW, 1271–1275 (Jennings & Watts eds., 9th ed. 1992).
[46] The textual approach to treaty interpretation excludes resort to the negotiating history of a treaty to discern the meaning of a term. Typically, recourse to negotiating documents only occurs when, after an analysis of the plain meaning, treaty terms remain ambiguous. BROWNLIE, *supra* note 42, at 635. However, the negotiating work, or the *travaux préparatoires*, may verify or confirm an interpretation emerging from a textual analysis. *Id.* Article 32 of the Vienna Convention states that "[r]ecourse may be had to supplementary means of interpretation, including the preparatory work of the treaty and the circumstances of its conclusion, in order to confirm the meaning resulting from . . . [a textual interpretation], or to determine the meaning when . . . [a textual interpretation] leaves the meaning ambiguous or obscure; or leads to a result which is manifestly absurd or unreasonable." Vienna Convention, *supra* note 42, at art. 32.
[47] Special Committee of Government Experts to Prepare a Draft Convention and a Draft Recommendation to Member States Concerning the Protection of Monuments, Groups of Buildings and Sites, *Draft Convention Concerning the Protection of Cultural and Natural World Heritage*, SHC-72/Conf.37/5 (Apr. 7, 1972), *available at* http://whc.unesco.org/archive/1972/shc-72-conf37–5e.pdf. Another earlier draft read: "Each Party shall respect all areas and sites inscribed in the Register by refraining so far as possible from acts which might damage them." Special Committee of Government Experts to Prepare a Draft Convention and a Draft Recommendation to Member States Concerning the Protection of Monuments, Groups of Buildings and Sites, *Draft Convention Concerning the Protection of Cultural and Natural World Heritage*, SHC-72/Conf.37/4 (Apr. 7, 1972), *available at* http://whc.unesco.org/archive/1972/shc-72-conf37–4e.pdf.

far as possible," indicating that this provision was meant to be implemented in a less discretionary manner than Articles 4 and 5.[48] Article 6, as adopted, codifies the object and purpose of the Convention – international cooperation for the protection of world heritage.

As the Preamble evinces, the WHC's object and purpose is twofold. First, protection of "[world] heritage at the national level often remains incomplete because of the scale of the resources which it requires and of the insufficient economic, scientific, and technological resources of the country where the property" is located.[49] In other words, the State Parties recognized that in many circumstances national-level efforts are insufficient to provide adequate protection. Second, to work toward resolving the inadequacies inherent in national-level protection, the State Parties understand that "it is incumbent on the international community as a whole to participate in the protection of the cultural and natural heritage of outstanding universal value, by the granting of collective assistance which, although not taking the place of action by the State concerned, will serve as an efficient complement thereto."[50] Essentially, the Preamble, while recognizing the primary nature of national effort, makes clear that the State Parties recognize that to ensure protection they must engage in an internationally cooperative effort.[51]

Rules of treaty interpretation, including the Vienna Convention, indicate that the object and purpose of a treaty evinces the ordinary meaning of treaty language.[52] The Preamble to a treaty provides context for the meaning of treaty terms, and often the Preamble elucidates the object and purpose of the treaty.[53] The WHC Preamble supports the interpretation that Articles 4, 5, and 6 impose binding legal obligations. It makes clear that the WHC's object and purpose is to foster international cooperation, coupled with national efforts, to protect world heritage.

---

[48] In fact, according to Robert Meyer, the author of an article entitled *"Travaux Preparatoires* for the UNESCO World Heritage Convention," "The words 'so far as possible'... were considered an overly broad loophole, so the word 'deliberate' was substituted." Robert L. Meyer, *Travaux Preparatoires for the UNESCO World Heritage Convention*, 2 EARTH L.J. 45, 53 (1972). Meyer's article also suggests that the drafters did not intend this provision to subject State Parties to strict liability for unintentional damage caused by pollution. *Id.* The desire not to impose strict liability, however, does not eviscerate the plain meaning of the provision. The word "deliberate" can be construed according to its plain meaning to impart an intent requirement. In other words, State Parties are only obligated not to take deliberate measures that might damage World Heritage sites; they are not obliged to protect sites from their unintended actions.

[49] World Heritage Convention, *supra* note 5, at preamble, third recital.

[50] *Id.*

[51] The Preamble states that the treaty seeks to establish "an effective system of collective protection of the cultural and natural heritage of outstanding universal value, organized on a permanent basis and in accordance with modern scientific methods." *Id.* at preamble, eighth recital.

[52] *See* Vienna Convention, *supra* note 42, at art. 31(1).

[53] *See id.* at art. 31(2); *see also* BROWNLIE, *supra* note 42, at 634 (stating that for purposes of interpretation, the "context" of the treaty includes its preamble); Sir Gerald Fitzmaurice, *The Law of Procedure of the International Court of Justice: Treaty Interpretation and Other Treaty Points*, 28 BRIT. Y.B. INT'L. L. 1, 4 (1951) (indicating that "a preamble does have legal force and effect from the *interpretative* standpoint") (emphasis in original).

3.2. The Mitigation Strategy Required by the World Heritage Convention

The obligations imposed by Articles 4, 5, and 6 of the WHC require that State Parties engage in an aggressive climate-change mitigation strategy because they mandate the protection of World Heritage sites and the "outstanding universal values" therein. Articles 4 and 5 call for State Parties to act aggressively to protect world heritage within their territories, and Article 6 obliges all State Parties to forgo actions that might damage World Heritage sites. Together, these provisions require that all State Parties engage in an aggressive climate change mitigation strategy entailing sharp reductions in greenhouse gas emissions.

The Petitions suggest that the Kyoto Protocol targets could provide useful guidelines for State Party implementation of WHC obligations respecting climate change; however, "appropriate" mitigation measures for many State Parties would necessarily include reductions beyond those called for by the Kyoto Protocol, because the WHC states that State Parties recognize that they must do all they can to the utmost of their resources.[54] In the case of many State Parties to the WHC, this would entail greater reductions than those provided by the Kyoto Protocol. In fact, although the Kyoto Protocol sets greenhouse gas reduction targets with the aim of preventing dangerous anthropogenic interference with the climate system, it calls for developed countries to reduce greenhouse gas emissions by an average of only 5.2% against a 1990 baseline during the period of 2008–2012.[55] Many State Parties to the WHC can, within their resources, reduce greenhouse gas emissions further. National and localized efforts to take action above and beyond Kyoto Protocol requirements make this clear.[56]

Indeed, if State Parties are to protect World Heritage sites from climate change, then all Parties to the WHC may be obligated to implement a regime of so-called "deep cuts" in greenhouse gas emissions. As is commonly understood, the reductions proposed by the Kyoto Protocol will not stabilize concentrations of greenhouse gas emissions in the atmosphere, and they certainly will not reverse current global climate change trends. The Chairman of the Intergovernmental Panel on Climate Change, Dr. Rajendra Pachauri, has warned that the world "'has reached the level

---

[54] World Heritage Convention, *supra* note 5, at art. 6(3); *see also* Scott Barrett, *The Problem of Averting Global Catastrophe*, 20 CHI. J. INT'L L. 527, 549–50 (2006) (describing failure of the Kyoto Protocol to achieve reductions commensurate with Parties' capacity).

[55] Kyoto Protocol to the U.N. Framework Convention on Climate Change, Dec. 10, 1997, 37 I.L.M. 22, *available at* http://UNFCCCc.int/resource/docs/convkp/kpeng.pdf [hereinafter Kyoto Protocol]; *see also* David W. Childs, *The Unresolved Debates that Scorched Kyoto: An Analytical Framework*, 13 U. MIAMI INT'L & COMP. L. REV. 233, 251 (2005) (noting that climatologists estimate that reductions would need to increase 40% to 50% to stabilize greenhouse gas concentrations in the atmosphere).

[56] *See generally* Randall S. Abate, *Kyoto or Not, Here We Come: The Promise and Perils of Piecemeal Approach to Climate Change Regulation in the United States*, 15 CORNELL J. L. & PUB. POL'Y 369 (2006); *see also* Matthew Bramley, *The Case for Deep Reductions: Canada's Role in Preventing Dangerous Climate Change – An Investigation by the David Suzuki Foundation and the Pembina Institute*, sec. 0.3 (2005) (summarizing government commitments to reduce greenhouse gas emissions).

of dangerous concentrations of carbon dioxide in the atmosphere' and has called for immediate and 'very deep' cuts in carbon dioxide emissions."[57]

Thus, the goal of the UNFCCC provides helpful guidance regarding WHC obligations. The UNFCCC's "ultimate objective" is "stabilization of greenhouse gas concentrations in the atmosphere at a level that would prevent dangerous anthropogenic interference with the climate system."[58] The consensus of the scientific community, as well as many governments, suggests that, to avoid "dangerous climate change," the global average surface temperature must not increase beyond 2°C above preindustrial temperatures.[59] To avoid temperature increases beyond 2°C, the global community must limit cumulative greenhouse gas emissions to no more than 15% above 1990 levels by 2020 and reduce emissions to at least 30% to 50% below 1990 levels by 2050.[60] This daunting task requires substantially more reductions in greenhouse gas emissions than the global community can achieve either through implementation of the Kyoto Protocol or through other nonbinding, multilateral measures.

The UNFCCC's goal of preventing dangerous human-induced climate change could provide a basis for implementation of the WHC obligations regarding climate change because it expresses nearly the entire international community's sentiment and would achieve the protection necessary for World Heritage sites that is contemplated by the WHC – namely, that such sites should be preserved for future generations by preventing damaging anthropogenic interference with the climate system.[61] However, the UNFCCC's Kyoto Protocol does not adequately implement the WHC's obligations to prevent climate change effects. State Parties to the

---

[57] Quoted in Geoffrey Lean, "Global Warming Approaching Point of No Return, Warns Leading Climate Expert," (Jan. 23, 2005), *available at* http://www.commondreams.org/headlines05/0123-01.htm (last visited Aug. 16, 2006).

[58] United Nations Framework Convention on Climate Change, May 9, 1992, S. TREATY DOC. NO. 102-38, 1771 U.N.T.S. 107, at art. 2 [hereinafter UNFCCC].

[59] *See* 3 MILLENNIUM ECOSYSTEM ASSESSMENT, ECOSYSTEMS AND HUMAN WELL-BEING: POLICY RESPONSES 375 (Kanchan Chopra et al. eds., 2005) ("The best guidance that can currently be given suggest that efforts be made to limit the increase in global mean surface temperature to less than 2°C above pre-industrial levels[.]"); *see also* Bramley, *supra* note 57, at sec. 0.2 (stating that the European Council first endorsed a 2°C limit and that the Climate Action Network International "has concluded that 'climate action must be driven by the aim of keeping global warming as far below 2°C as possible'").

[60] Bramley, *supra* note 56, at sec. 3.1, Table 1. The table presents a comparative look at data from three climate change studies. *See* Bill Hare & Malte Meinshausen, *How Much Warming Are We Committed To and How Much Can Be Avoided?*, *available at* http://www.pik-potsdam.de/pik_web/publications/pik_reports/reports/pr.93/pr93.pdf; Michael den Elzen & Malte Meinshausen, *Meeting the EU 2°C Climate Target: Global and Regional Emission Implications*, *available at* http://www.gci.org.uk/briefings/rivm.pdf; and Niklas Höhne et al., *Options for the Second Commitment Period of the Kyoto Protocol*, *available at* http://www.umweltbundesamt.de/fpdf-1/2847.pdf, for the climate change studies cited.

[61] RODA VERHEYEN, CLIMATE CHANGE DAMAGE AND INTERNATIONAL LAW 55 (2005) (noting that "stabilisation is linked to the prevention of dangerous interference with the climate system, which implies that the actual objective of the [UN]FCCC is the stabilization of the climate itself at safe levels").

WHC have an obligation independent of the obligations they may have under the UNFCCC and the Kyoto Protocol to prevent dangerous human-induced climate change and eliminate the threat of climate change to world heritage. This obligation arises directly from Article 4's call for State Parties to do all they can and the request in Article 5 that State Parties undertake the appropriate legal, technical, administrative, and scientific measures. In light of current climate change trends, these provisions require that State Parties undertake to make "deep cuts" in their greenhouse gas emissions to protect the world heritage within their territories. Thus, although the UNFCCC provides the same goal State Parties must have when executing their WHC obligations, the current implementation strategies under the UNFCCC, that is, the Kyoto Protocol, have failed to achieve the necessary emissions reductions.

In addition to the obligations State Parties have to protect threatened world heritage within their territories, Article 6 states that all State parties must "not take any deliberate measures which might damage directly or indirectly" World Heritage sites.[62] Thus, whereas Articles 4 and 5 specifically concern State Party obligations to protect and preserve their own world heritage, Article 6 reiterates the recognition that world heritage is, in fact, part of the common heritage of humankind and thus all State Parties must undertake to protect all world heritage. With respect to climate change, this obligation means that all State Parties must act to reduce or limit their greenhouse gas emissions whether or not climate change threatens World Heritage sites within their respective jurisdictions.

In its position paper, the United States mischaracterizes the nature of the obligations in Article 6. The United States reads the Petitions as arguing that State Parties have failed to reduce greenhouse gas emissions and thus have not prevented climate change, leading to a violation of Article 6(3).[63] The United States correctly states the Petitioner's position, but Petitioners do not argue that the failure to reduce greenhouse gas emissions is a violation of Article 6(3), as the United States suggests. The United States argues that "[n]ot taking an action, such as not reducing greenhouse gas emissions, or not signing on to an agreement like the Kyoto Protocol, does not constitute a 'deliberative measure which might damage' a site."[64] Thus, the United States concludes that a violation of Article 6(3) has not occurred. This is a specious, end-run argument based on semantics. Article 6(3) obliges State Parties "not to take deliberate measures" that directly or indirectly damage world heritage. The relevant action is emission of greenhouse gases, not their reduction. This is the central argument of the Petitions. State Parties have an obligation to reduce their greenhouse gas emissions because emitting greenhouse gases is a deliberate measure directly and indirectly damaging World Heritage sites. In other words, the Convention obliges State Parties not to emit greenhouse gases to the extent that they are contributing to anthropogenic interference with the climate system. Unlike

---

[62] World Heritage Convention, *supra* note 5, at art. 6(3).
[63] U.S. Position Paper, *supra* note 12, at 2.
[64] *Id.* (quotation in original).

the targets and timetables of the Kyoto Protocol, which bind only certain nations to specific reductions,[65] the climate change responsibilities under the WHC bind all State Parties similarly, whether affected world heritage lies within a State Party's territory or beyond.[66] However, these obligations must be read with international principles of equity in mind, primarily the concept of common but differentiated responsibilities.[67] Principle 7 of the United Nations' Rio Declaration is the foremost statement of this concept. It states: "States shall cooperate in a spirit of global partnership to conserve, protect and restore the health and integrity of the Earth's ecosystem. In view of the different contributions to global environmental degradation, States have common but differentiated responsibilities."[68] Article 3.1 of the UNFCCC specifically recognizes this principle's application to climate change responsibility, stating that "the Parties should protect the climate system... on the basis of equity and in accordance with their common but differentiated responsibilities and respective capabilities."[69] The relevant provisions of the WHC recognize that responsibilities may vary depending on availability of capacity and resources. Article 4 specifies that a State Party must do all it can "to the utmost of its own resources," and Article 5 indicates that State Parties must endeavor to undertake the specified requirements "in so far as possible." The widely accepted principle of "common but differentiated responsibilities" and the recognition in the text of the WHC of varying degrees of capacity present a conceptual framework for compromise and cooperation in meeting the challenge of reducing global greenhouse gas emissions.

3.3. *The World Heritage Committee's Stance on Mitigation*

In response to the petitions, the World Heritage Committee commissioned a joint report entitled "Predicting and Managing the Effects of Climate Change on World Heritage" (Joint Report); however, thus far the work of the World Heritage Committee does not meet WHC obligations to adequately protect World Heritage sites

---

[65] The Kyoto Protocol obligates Annex I Parties (developed countries) to collectively reduce their greenhouse gas emissions at least 5% below 1990 levels by 2008–2012, but non-Annex I Parties (developing countries) are not subject to binding reduction targets. See generally P. G. Harris, *Common but Differentiated Responsibility: The Kyoto Protocol and United States Policy*, 7 N.Y.U. ENVTL. L.J. 27 (1999).

[66] The United States recognizes this concept in its position paper but argues that because the provisions bind all State Parties equally, it does not confer any climate change obligations. See U.S. Position Paper, *supra* note 12, at 3 (arguing that developed nations have not violated Article 6(3) because "even if this provision applied to not taking particular actions, it would apply equally to all State Parties, not just the developed country Parties").

[67] For background and the history of "common but differentiated responsibilities," see ANITA MARGRETHE HALVORSSEN, EQUALITY AMONG UNEQUALS IN INTERNATIONAL ENVIRONMENTAL LAW: DIFFERENTIAL TREATMENT FOR DEVELOPING COUNTRIES (1999).

[68] Rio Declaration on Environment and Development, adopted June 13, 1992, U.N. Doc. A/CONF.151/26 (1992), 31 I.L.M. 874, Principle 7 (1992).

[69] UNFCCC, *supra* note 58, at art. 3(1). See also Christin Batruck, *'Hot Air' as Precedent for Developing Countries? Equity Considerations*, 17 UCLA J. ENVTL. L. & POL'Y 45, 50–3 (describing rationale for including principle of "common but differentiated responsibilities").

from climate change.[70] The Joint Report recognizes that only mitigation absolutely alleviates the threats caused by climate change; however, it stops far short of recommending that State Parties implement a general mitigation strategy to protect World Heritage sites. Both the Joint Report and a document prepared by the World Heritage Committee based on an expert working group meeting concerning climate change and the WHC (the Strategy)[71] indicate that climate change mitigation initiatives are within the sole province of the UNFCCC and Kyoto Protocol.[72] In fact, both documents provide that climate change mitigation under the auspices of the WHC ought to occur only as site-specific projects. For example, the Joint Report suggests that some World Heritage sites may be involved in sequestering carbon dioxide but concludes that any quantitative effect is negligible. It also indicates that World Heritage site managers could be encouraged to promote "improved technology to reduce emissions throughout the World Heritage network."[73]

Effectively, neither the Joint Report nor the Strategy prescribes clear-cut action on climate change mitigation. Many World Heritage sites will never be preserved for transmission to future generations unless the State Parties, led by the World Heritage Committee, act more proactively than merely supporting site-specific mitigation. For example, Waterton-Glacier International Peace Park was listed as a World Heritage site, in part, because of its unique geophysical landscape, including its iconic glaciers.[74] However, today only 27 glaciers remain in Glacier National Park (the U.S. portion of Waterton-Glacier), less than one-fifth of the approximately 150 glaciers that existed within the park's current boundaries in 1850.[75] In fact, since 1850, the area covered by glaciers in the park has decreased by 73%.[76]

In Waterton-Glacier, climate change is deleteriously affecting nearly all of the "outstanding universal values" of the park. In the 2004 *Report on the State of Conservation of Waterton Glacier International Peace Park*, a regular report submitted to the World Heritage Committee,[77] park managers indicated that:

> Climate change has and will continue to have important impacts to the International Peace Park natural resources. Scientific data collected in Glacier indicates that park glaciers have shrunk dramatically over the past century; that the park's tree

---

[70] *See* World Heritage Committee, *General Issues: Threats to World Heritage Properties*, Decision 29 COM 7B.b, para. 9 (July 2005). For a full cite to the Joint Report, see *supra* note 30.

[71] For a full cite to the Strategy, see *supra* note 35.

[72] *See Joint Report*, *supra* note 30, at para. 7 (providing that mitigation is the mandate of the UNFCCC and Kyoto Protocol); *Strategy*, *supra* note 35, at para. 13 (stating that the "UNFCCC is the instrument through which mitigation strategies at the global and State Parties level is being addressed").

[73] *Joint Report*, *supra* note 30, at paras. 124–25.

[74] *See* World Heritage Committee, WHC-95/CONF.203/16, § VIII(A.1) (describing characteristics for which the parks were listed as a World Heritage site).

[75] U.S. National Parks Service, *Glacier National Park, Environmental Management Plan*, at 5 (August 2004), *available at* http://www2.nature.nps.gov/air/features/docs/GlacFinalEMS200408.pdf.

[76] *Id.*

[77] For more on the required "State of Conservation" reports, see http://whc.unesco.org/en/soc/ (last visited Aug. 27, 2006).

line is creeping higher in elevation; that the alpine tundra zone is shrinking, and that subalpine meadows are filling in with tree species. The ecological significance of losing the park's glaciers is likely affecting stream baseflow in late summer and increasing water temperatures thus influencing the distribution and behavior of aquatic organisms and food webs.[78]

Any climate change mitigation occurring within the park's boundaries, while commendable, is inevitably inadequate to address the devastating consequences of climate change within the park.[79] Even a total ban on greenhouse gas emissions within the park would not slow, and could never reverse, the climate change effects on glacial melt. Yet this type of mitigation is all that the Joint Report and the Strategy suggest should occur — a wholly inadequate response to the threat of climate change because it will not protect the outstanding universal values of the park.

Though the park managers of Waterton-Glacier recognize the need to manage for threats occurring because of climate change, they are incapable of adequately addressing these threats because the cause — high rates of greenhouse gas emissions — occurs almost exclusively outside the park's boundaries.[80] This is true for all World Heritage sites threatened by climate change, and as a result, site-specific mitigation could never ameliorate the climate change threats to these sites in any meaningful way. Thus, the recommendations in the Joint Report and the Strategy neither fully nor adequately implement State Parties' obligations to engage in climate change mitigation because they do not specifically address the cause of the threats to world heritage due to climate change. The World Heritage Committee's weak approach may be politically palatable, especially to State Parties struggling to address their greenhouse gas emissions adequately, but it falls far short of the type of mitigation required to protect World Heritage sites.

4. CONCLUSION

The World Heritage Convention requires State Parties to develop a comprehensive mitigation strategy to protect and preserve World Heritage sites. Although the broad

---

[78] U.S. Department of the Interior and Parks Canada, *Periodic Report on the Application of the World Heritage Convention, Report on the State of Conservation of Waterton-Glacier International Peace Park*, at § 5b (considered by the World Heritage Committee July 2005), *available at* http://www.nps.gov/oia/topics/Waterton-Glacier.pdf.

[79] Because Glacier National Park's managers recognize that the park is experiencing climate change consequences, they have taken steps to reduce greenhouse gas emissions within the park, including using alternative fuel buses as shuttles for employees and increasing energy efficiency in park buildings. National Park Service, *Glacier National Park Environmental Management Plan* 8–10 (2004), *available at* http://www2.nature.nps.gov/air/features/docs/GlacFinal GHGInventory.pdf (last visited Aug. 27, 2006).

[80] A symposium on national park management in the United States noted that "[e]cologically sound management requires active management and a vision which looks beyond artificial boundaries at environmental concerns, whether they originate locally, regionally, nationally, or internationally. [NPS] must have the capacity to respond to threats, whether they come from a dam at the park boundary, air pollution from a facility 100 miles away, or climate change caused by increased greenhouse gas concentrations in the atmosphere." WILLIAM J. BRIGGLE ET AL., NATIONAL PARKS FOR THE 21ST CENTURY: THE VAIL AGENDA 106 (1993).

language of the Convention facilitates flexibility and discretionary approaches to these obligations, it does not mean that State Parties may entirely abdicate any responsibility to remedy the threat to World Heritage sites arising from climate change. For many State Parties, these conclusions may seem like an end run around the Kyoto Protocol, particularly to State Parties struggling or hesitant to meet even those commitments, but the obligations under the WHC are clear. Certainly, the negotiators of the WHC did not foresee the threat of climate change, but they knew that they could not foresee all potential threats to World Heritage sites. As a result, the WHC provides broad protections against all threats, and if the WHC is to remain a meaningful tool to protect natural areas of outstanding universal value, including mountain glaciers and barrier reefs, then the World Heritage Committee must effectively engage State Parties in an aggressive climate change mitigation strategy.

12

The Inuit Petition as a Bridge? Beyond Dialectics of Climate Change and Indigenous Peoples' Rights

Hari M. Osofsky*

INTRODUCTION

The rapid pace of climate change in the Arctic poses serious challenges for the Inuit peoples living there. A petition filed with the Inter-American Commission on Human Rights in December 2005 on behalf of Inuit in the United States and Canada claims that U.S. climate change policy violates their rights. Upon filing the petition, Sheila Watt-Cloutier, Chair of the Inuit Circumpolar Conference, made a statement at the 2005 Conference of Parties of the United Nations Framework Convention on Climate Change. She summarized the severity of the stakes involved as follows:

> What is happening affects virtually every facet of Inuit life – we are a people of the land, ice, snow, and animals. Our hunting culture thrives on the cold. We need it to be cold to maintain our culture and way of life. Climate change has become the ultimate threat to Inuit culture.... How would you respond if an international assessment prepared by more than 300 scientists from 15 countries concluded that your age-old culture and economy was doomed, and that you were to become a footnote to globalization?[1]

---

* Associate Professor, Washington and Lee University School of Law; B.A., J.D., Yale University. The author can be contacted at osofskyh@wlu.edu. This chapter is a republication, with permission and minor editorial changes, of Hari M. Osofsky, *The Inuit Petition as a Bridge? Beyond Dialectics of Climate Change and Indigenous Peoples' Rights*, 31 AM. INDIAN L. REV. 675 (2007). This paper benefited greatly from the interchange at the 2006 University of Idaho College of Law International Law Symposium – which reflected the tremendous organizational skills and insights of Rebecca Bratspies and Russell Miller – as well as at LatCrit XI and the Northwest Tribal Water Rights Third Annual Conference. I also am grateful for the conceptual and editorial input of William Burns, Joshua Gitelson, Donald Goldberg, Stefanie Herrington, Lillian Aponte Miranda, Margie Paris, Radha Pathak, Martin Wagner, and Lua Kamal Yuille. I appreciate the generous financial and collegial support of the University of Oregon School of Law and, in particular, the Dean's Advisory Council Endowment Fund that made my work on this project possible. Finally, I would like to thank Kristina Bell for her patient and helpful stewardship of this piece, and Lindsay Goodner, Keneisha Green, Sheila Southard, and Michael Waters for their assistance with its editing and production in the *American Indian Law Review*.

[1] *See Presentation by Sheila Watt-Cloutier, Chair, Inuit Circumpolar Conference Eleventh Conference of Parties to the UN Framework Convention on Climate Change*, Montreal, Dec. 7, 2005, http://www.inuitcircumpolar.com/index.php?ID=318&Lang=En.

The Inuit Petition as a Bridge?

The Inter-American Commission provided a two-paragraph response to the petition on November 16, 2006, that "the information provided does not enable us to determine whether the alleged facts would tend to characterize a violation of the rights protected by the American Declaration."[2] Watt-Cloutier, in conjunction with Earthjustice and the Center for International Environmental Law, requested additional information on that decision, as well as a hearing on the linkages between climate change and human rights.[3] The Commission agreed to this broader hearing, which took place on March 1, 2007, and the Commission is currently deliberating on the basis of it.[4]

Although a positive decision from the Commission on the specific claims brought by the Inuit appears unlikely at this point and U.S. climate policy has evolved significantly under the Obama administration, the Inuit petition serves as an important example of creative lawyering in both substance and form. It reframes a problem typically treated as an environmental one through a human rights lens, and moves beyond the confines of U.S. law to a supranational forum. In so doing, the petition lies at the intersection of two streams of cases occurring at multiple levels of governance: (1) environmental rights litigation and petitions and (2) climate change litigation and petitions.[5]

In addition, the petition raises critical issues about the mix of advocacy tools needed to address pressing problems. For example, Watt-Cloutier presented the petition as part of a dialogue with the U.S. government and openly acknowledged the difficulties of formal enforcement.[6] An examination of the Inuit petition thus

---

[2] Letter from the Organization of American States to Sheila Watt-Cloutier, et al. regarding Petition No. P-1413-05, Nov. 16, 2006, *available at* http://graphics8.nytimes.com/packages/pdf/science/16commissionletter.pdf.

[3] Letter from Sheila Watt-Cloutier, Martin Wagner, and Daniel Magraw to Santiago Cantón, Executive Secretary, Inter-American Commission on Human Rights, Jan. 15, 2007 (on file with author); *see also* Jane George, *ICC Climate Change Petition Rejected*, NUNATSIAQ NEWS, Dec. 15, 2006, *available at* http://www.nunatsiaq.com/news/nunavut/61215_02.html.

[4] *See* Letter from the Organization of American States to Sheila Watt-Cloutier, et al. regarding Petition No. P-1413–05, Feb. 1, 2007 (on file with author).

[5] I have analyzed both of these streams in my previous scholarship. *See* Hari M. Osofsky, *The Geography of Climate Change Litigation: Implications for Transnational Regulatory Governance*, 83 WASH. U. L.Q. 1789 (2005); Hari M. Osofsky, *Learning from Environmental Justice: A New Model for International Environmental Rights*, 24 STAN. ENVTL. L.J. 71 (2005). For additional scholarship discussing environmental rights litigation, see sources *infra* note 47; HUMAN RIGHTS APPROACHES TO ENVIRONMENTAL PROTECTION (Alan E. Boyle & Michael R. Anderson eds., 1996); Natalie L. Bridgeman, *Human Rights Litigation under the ATCA as a Proxy for Environmental Claims*, 6 YALE HUM. RTS. & DEV. L.J. 1 (2003); Linda A. Malone & Scott Pasternack, *Exercising Environmental Human Rights and Remedies in the United Nations System*, 27 WM. & MARY ENVTL. L. & POL'Y REV. 365 (2002); Mariana T. Acevedo, Student Article, *The Intersection of Human Rights and Environmental Protection in the European Court of Human Rights*, 8 N.Y.U. ENVTL. L.J. 437 (2000). For additional scholarship discussing climate change litigation, see sources *infra* note 23; *see also* JOSEPH SMITH & DAVID SHEARMAN, CLIMATE CHANGE LITIGATION: ANALYSING THE LAW, SCIENTIFIC EVIDENCE & IMPACTS ON THE ENVIRONMENT, HEALTH & PROPERTY (2006); RODA VERHEYEN, CLIMATE CHANGE DAMAGE AND INTERNATIONAL LAW, PREVENTION DUTIES AND STATE RESPONSIBILITY (2005).

[6] *See* Presentation by Sheila Watt-Cloutier, *supra* note 1.

opens broader questions about the best way to address crosscutting environmental problems like climate change.

This chapter will focus on these questions by exploring the intersectional nature of the Inuit petition. The piece will break apart the petition to uncover the relational dynamics imbedded in it. In particular, this chapter will rely upon two conceptual approaches to dissect the petition: (1) a law and geography perspective and (2) an exploration of the limits of dialectical analysis.[7] Through this unpacking process, the piece attempts to engage what might constitute progress on climate change and indigenous peoples' rights.

Section 1 will use geographic analysis – examining the way in which key actors and claims tie to place – to illustrate the many places, individuals, and entities interacting through the filing of this petition.[8] For the purposes of this discussion, the piece uses "place" to refer to physical location, "space" to refer to the sociopolitical and legal structures, and "scale" to refer to the level of governance.[9] This section is thus "geographical" because it locates the actors physically, socially, and politically.

Section 2 builds from that analysis – and Watt-Cloutier's own geographic framing of the petition – to consider how these complex relationships might help build bridges. It engages the extent to which the petition creates links across several types of divisions generally recognized in the law, such as the one between public and private. In particular, the section explores the limitations of dialectical analysis with respect to substantive categories, legal structures, and legal approaches that occur within this petition.[10]

The piece concludes with some reflections on the extent to which this kind of advocacy strategy in general – and the petition in particular – can be part of much-needed progress in protecting the rights of indigenous peoples. It discusses the ability of this petition to address legitimacy problems embedded in interactions between

---

[7] Robert Ahdieh introduces the idea of "dialectical regulation" in his recent work, which he describes as the strongest form of intersystemic regulatory behavior in which institutions are engaged in "an active, iterative, and potentially even institutionalized, pattern of substantive regulatory engagement across jurisdictional lines, between simultaneously competing and coordinating regulators." Robert B. Ahdieh, *Dialectical Regulation*, 38 CONN. L. REV. 863, 870 (2006). Paul Berman builds on this concept in his latest work on global legal pluralism. *See* Paul Schiff Berman, *Global Legal Pluralism*, 80 S. CAL. L. REV. 1155 (2007). This piece's critique of dialectical framing does not focus on the specific institutional dynamics that they discuss but rather the limitations of construing crosscutting legal problems through a Hegelian dance between two categories. I discuss this idea in more depth in Hari M. Osofsky, *The Geography of Climate Change Litigation Part II: Narratives of* Massachusetts v. EPA, 8 CHI. J. INT'L L. 573 (2008).

[8] This analysis builds upon my discussion of the Inuit petition in *The Geography of Climate Change Litigation, supra* note 5.

[9] The definitions I choose represent only one version of how place and space have been defined in the geography literature. For examples of analyses of the concepts of place and space, see R.J. JOHNSTON, A QUESTION OF PLACE: EXPLORING THE PRACTICE OF HUMAN GEOGRAPHY (1991); DOREEN MASSEY, FOR SPACE (2005); THE POWER OF PLACE: BRINGING TOGETHER GEOGRAPHICAL AND SOCIOLOGICAL IMAGINATIONS (John A. Agnew & James S. Duncan eds., 1989); YI-FU TUAN, SPACE AND PLACE: THE PERSPECTIVE OF EXPERIENCE (1977).

[10] As noted *supra* note 7, I explore the need to move beyond dialectics in more depth in Osofsky, *The Geography of Climate Change Litigation Part 2*.

the international legal system and indigenous peoples, and what might constitute a "win" in this context.

1. THE GEOGRAPHY OF THE INUIT PETITION

A geographic analysis of the Inuit petition reflects the complex power dynamics contained within it. Both the actors and claims have multiple, overlapping ties to place which are important to understanding the framing and potential impact of the petition. These ties help to situate the petition within the webs of relationships[11] that underlie the production of greenhouse gases, the impact of human-induced climate change on the Inuit, and possibilities for addressing this problem. Understanding these linkages provides the basis for further analysis into the constructive role of the petition.

1.1. Actors

The primary actors in the case are those on the respective sides of the petition – the petitioners and respondent – and the decision maker – the Inter-American Commission on Human Rights. This section considers how each of these actors connects to place and space, and in so doing, provides a map of dynamic interactions.

1.1.1. Petitioners

The Inuit petitioners have subnational, national, and supranational identities that include layered sociopolitical and legal connections. At a subnational level, they have strong ties to particular local communities, ties that form a part of the human rights claims they are making. Moreover, these communities sometimes form larger regional groupings.[12]

The petitioners are simultaneously citizens of particular nation-states, the United States and Canada, and of states within those nation-states.[13] Moreover, those governmental entities recognize their village and tribal affiliations. In the Alaskan context, part of that recognition includes viewing the Inuit as part of Alaska Native Regional Corporations.[14]

---

[11] I have argued that climate change litigation in general, and the Inuit petition in particular, is simultaneously multiscalar, multibranch, and multiactor. *See* Osofsky, *The Geography of Climate Change Litigation, supra* note 5, at 1813–18, 1843–51.

[12] *See* Petition to the Inter American Commission on Human Rights Seeking Relief from Violations Resulting from Global Warming Caused by Acts and Omissions of the United States (submitted Dec. 7, 2005), at 13–20, *available at* http://www.earthjustice.org/library/reports/ICC_Human_Rights_Petition.pdf [hereinafter, Inuit Petition].

[13] *See id.*

[14] For an analysis of Alaskan Native Economic Corporations and their economic performance, see Stephen Colt, *Alaskan Natives and the "New Harpoon": Economic Performance of the ANSCA Regional Corporations*, 25 J. LAND, RESOURCES, & ENVTL. L. 155 (2005).

Dual identities also exist at a supranational level. The Inuit Circumpolar Conference unites the Inuit across the national borders that artificially subdivide them. It represents approximately 150,000 Inuit from three countries – the United States, Canada, and Russia – and an administratively self-governing entity, Greenland, which is a division of Demark.[15] However, the petition includes only the Inuit from the United States and Canada because Greenland and Russia are not part of the Inter-American regional grouping.[16]

These affiliations together construct the Inuit petitioners' complex ties to place and space. Each of these ties helps to shape the identity of the "Inuit petitioners." The narrative of the petition interweaves these connections to form a coherent representation that conforms to the requirements of an adjudicative body.

1.1.2. Respondent

The petition was filed against the United States, a nation-state. However, the United States, especially with respect to climate change policy, is far from a monolithic entity.

The executive branch sets climate change policy and negotiates international agreements but also evolves in its approach to both over time, particularly as administrations change. That branch, then headed by President George W. Bush, decided to withdraw from the Kyoto Protocol, a decision that the petition argues forms a key part of the U.S. failure to control its greenhouse gas emissions adequately.[17] In February 2007, just before the Commission's hearing on climate change and human rights, President Bush presented his first State of the Union address that discussed the need "to confront the serious challenge of global climate change."[18] Now, under President Obama, the executive branch has committed to addressing climate change more seriously in a myriad of ways.[19]

---

[15] Inuit Circumpolar Conference, *available at* http://www.inuitcircumpolar.com/index.php?ID=16&Lang=En (last visited Mar. 1, 2006).

[16] For a list of OAS members, see Organization of American States website, Member States and Permanent Missions, http://www.oas.org/main/main.asp?sLang=E&sLink=http://www.oas.org/documents/eng/memberstates.asp.

[17] President George W. Bush, Speech Discussing Global Climate Change (June 11, 2001), *available at* http://www.whitehouse.gov/news/releases/2001/06/20010611-2.html.

[18] *President Bush Delivers State of the Union Address*, Jan. 23, 2007, http://www.whitehouse.gov/news/releases/2007/01/20070123-2.html (full transcript); *The State-of-the-Union Message: Bush Loses the Upper Hand*, ECONOMIST, Jan. 27, 2007.

[19] *See* President Barack Obama, Address to Joint Session of Congress (Feb. 24, 2009), *available at* http://www.whitehouse.gov/the_press_office/Remarks-of-President-Barack-Obama-Address-to-Joint-Session-of-Congress/; Remarks by the President on Jobs, Energy Independence, and Climate Change, East Room of the White House, Jan. 26, 2009, *available at* http://www.whitehouse.gov/blog_post/Fromperiltoprogress/; Obama for America, Barack Obama and Joe Biden: New Energy for America, *available at* http://www.barackobama.com/pdf/factsheet_energy_speech_080308.pdf (last visited Dec. 22, 2008); *see also* Ceci Connolly & R. Jeffrey Smith, *Obama Positioned to Quickly Reverse Bush Actions*, WASH. POST, Nov. 9, 2008, *available at* http://www.washingtonpost.com/wp-dyn/content/article/2008/11/08/AR2008110801856_pf.html (last visited Nov. 11, 2008); John M. Broder & Andrew C.

The legislative branch creates statutory law that underlies U.S. energy policy and other decision making relevant to climate change. This legislative role has become even more relevant since the petition was filed. In the wake of the 2006 midterm elections, Barbara Boxer indicated that "[a]s the new chair of the [U.S. Senate] Committee [on Environment and Public Works], [she was] already planning for vigorous oversight and legislation to make sure that the U.S. Senate is once again an environmental leader in protecting the health of our families and our children and addressing pressing concerns like global warming."[20] Nancy Pelosi, as Speaker of the House, has been part of a similar push for new legislation.[21] With President Obama's commitment to major cap-and-trade legislation, Congress will likely play an even greater role in addressing greenhouse gas emissions than it has before, assuming it can overcome political hurdles.[22]

The resolved and pending U.S. court cases, which rely on a variety of legal theories regarding the approaches of different entities to global climate change, may ultimately influence U.S. policy in a range of ways.[23] For the first time, on November 29, 2006, the U.S. Supreme Court heard one of the cases challenging federal regulatory decisions on climate change;[24] on April 2, 2007, it issued a landmark

---

Revkin, *Hard Task for New Team on Energy and Climate*, N.Y. TIMES, Dec. 16, 2008, at A24, *available at* http://www.nytimes.com/2008/12/16/us/politics/16energy.html?_r=1&scp=3&sq=Salazar&st=cse; John Vidal, *Obama Victory Signals Rebirth of US Environmental Policy*, GUARDIAN, Nov. 5, 2008, *at* http://www.guardian.co.uk/environment/2008/nov/05/climatechange-carbonemissions. For examples of President Obama's efforts during his first week in office, see Memorandum on The Energy Independence and Security Act of 2007 (Jan. 26, 2009), *available at* http://www.whitehouse.gov/the_press_office/Presidential_Memorandum_fuel_economy/; Memorandum on State of California Request for Waiver under 42 U.S.C. 7543(b), the Clean Air Act (Jan. 26, 2009), *available at* http://www.whitehouse.gov/the_press_office/Presidential_Memorandum_EPA_Waiver/; Memorandum from Lisa P. Jackson, Adm'r, Envtl. Prot. Agency, to All EPA Employees (Jan. 23, 2009), *available at* http://www.epa.gov/administrator/memotoemployees.html; *see also* John M. Broder, *E.P.A. Expected to Regulate Carbon Dioxide*, N.Y. TIMES, Feb. 18, 2009, at A15, *available at* http://www.nytimes.com/2009/02/19/science/earth/19epa.html.

[20] *See* Greenwire, Nov. 9, 2006, *available at* http://www.eenews.net/gw/.

[21] *See, e.g.*, Press Release, *Pelosi and Reed: We Should Work Together to Take American in a New Direction*, Jan. 27, 2007, *available at* http://www.house.gov/pelosi/press/releases/Jan07/SOTU.html; *Is U.S. Energy Independence a Pipe Dream?*, NPR Talk of the Nation, Jan. 24, 2007 ("Today Speaker of the House Nancy Pelosi upped the ante and called for energy independence within 10 years).''

[22] For an analysis of the complexities of energy legislation, see Jackie Calmes & Carl Hulse, *Obama's Budget Faces Test Among Party Barons*, N.Y. TIMES, Mar. 10, 2009, at A1, *available at* http://www.nytimes.com/2009/03/10/us/politics/10chairmen.html?_r=1&scp=3&sq=obama%20climate%20change&st=cse.

[23] *See* William C. G. Burns, *The Exigencies That Drive Potential Causes of Action for Climate Change Damages at the International Level*, 98 AM. SOC'Y INT'L L. PROC. 223 (2004); Bradford C. Mank, *Standing and Global Warming: Is Injury to All Injury to None?*, 35 ENVTL. L. 1 (2005); Osofsky, *The Geography of Climate Change Litigation*, *supra* note 5; Richard W. Thackeray, Jr., Note, *Struggling for Air: The Kyoto Protocol, Citizens' Suits Under the Clean Air Act, and the United States' Options for Addressing Global Climate Change*, 14 IND. INT'L & COMP. L. REV. 855 (2004).

[24] *See* Massachusetts v. EPA, 415 F.3d 50 (D.C. Cir. 2005), *cert. granted*, 2006 WL 1725113 (U.S. Dist. Col. June 26, 2006) (No. 05-1120). For a transcript of the oral argument, see http://www.supremecourtus.gov/oral_arguments/argument_transcripts/05-1120.pdf.

decision in *Massachusetts v. EPA* against the Bush administration.[25] Although the EPA under President Bush delayed taking action to implement the decision, the Obama administration's EPA is moving rapidly to respond.[26]

Moreover, a focus on federal governmental actors reveals only a piece of the construction of "U.S." climate change policy. Numerous other actors create the backdrop against which national efforts to regulate greenhouse gases occur and help to shape the nature of national policy.

In particular, state and local government play an important role in determining the overall national level of emissions.[27] Some of them increasingly make policy decisions to regulate greenhouse gases more aggressively than the federal government and file lawsuits targeted at influencing the behavior of the federal government and corporations. California's landmark fall 2006 legislation,[28] which aimed at dramatically reducing its emissions, and its involvement – as well as the involvement of its localities – as petitioners and respondents in climate change litigation in federal courts[29] serve as just one instance of this phenomenon. Another indicator of the

---

[25] Massachusetts v. EPA, 549 U.S. 497 (2007).
[26] Proposed Endangerment and Cause or Contribute Findings for Greenhouse Gases Under Section 202(a) of the Clean Air Act; Proposed Rule, 74 Fed. Reg. 18885 (proposed Apr. 24, 2009) (to be codified at 40 C.F.R. ch. 1); John M. Broder, *E.P.A. Clears Way for Greenhouse Gas Rules*, N.Y. TIMES, Apr. 17, 2009, at A15, *available at* http://www.nytimes.com/2009/04/18/science/earth/18endanger.html.
[27] *See* BARRY G. RABE, STATEHOUSE AND GREENHOUSE: THE EMERGING POLITICS OF AMERICAN CLIMATE CHANGE POLICY (2004); Donald A. Brown, *Thinking Globally and Acting Locally: The Emergence of Global Environmental Problems and the Critical Need to Develop Sustainable Development Programs at State and Local Levels in the United States*, 5 DICK. J. ENVTL. L & POL'Y 175 (1996); Ann E. Carlson, *Federalism, Preemption, and Greenhouse Gas Emissions*, 37 U.C. DAVIS L. REV. 281 (2003); David R. Hodas, *State Law Responses to Global Warming: Is It Constitutional to Think Globally and Act Locally?*, 21 PACE ENVTL. L. REV. 53 (2003); Laura Kosloff & Mark Trexler, *State Climate Change Initiatives: Think Locally, Act Globally*, 18 WTR NAT. RESOURCES & ENV'T 46 (2004); Robert B. McKinstry, Jr., *Laboratories for Local Solutions for Global Problems: State, Local and Private Leadership in Developing Strategies to Mitigate the Causes and Effects of Climate Change*, 12 PENN ST. ENVTL. L. REV. 15 (2004); Hari M. Osofsky, *Local Approaches to Transnational Corporate Responsibility: Mapping the Role of Subnational Climate Change Litigation*, 20 PAC. MCGEORGE GLOBAL BUS. & DEV. L.J. 143 (2007) (Conference Proceedings Issue); Barry G. Rabe, *North American Federalism and Climate Change Policy: American State and Canadian Provincial Policy Development*, 14 WIDENER L.J. 121 (2004); Judith Resnik, *Law's Migration: American Exceptionalism, Silent Dialogues, and Federalism's Multiple Ports of Entry*, 115 YALE L.J. 1564, 1643–47 (2006).
[28] California Global Warming Solutions Act of 2006 (AB 32), Cal. Health & Safety Code §§ 38,500 *et seq.*; *see also* Press Release, Office of the Governor, Governor Schwarzenegger Signs Landmark Legislation to Reduce Greenhouse Gas Emissions, Sept. 27, 2006, *available at* http://www.climatechange.ca.gov/documents/2006-09-27_AB32_GOV_NEWS_RELEASE.PDF.
[29] *See* http://www.climatechange.ca.gov/; *see also* Complaint, Connecticut v. Am. Elec. Power Co., 406 F. Supp. 2d 265 (S.D.N.Y. 2005) (No. 04 Civ 5669), *available at* http://caag.state.ca.us/newsalerts/2004/04-076.pdf; *Massachusetts*, 127 S. Ct. at 1438; Complaint for Declaratory and Injunctive Relief (Second Amended), Friends of the Earth, Inc., v. Watson, No. 02–4106 (N.D. Cal. Sept. 3, 2002), *available at* http://www.climatelawsuit.org/documents/Complaint_2Amended_Declr_Inj_Relief.pdf.; Complaint, State of California v. Gen. Motors Corp., *available at* http://ag.ca.gov/newsalerts/cms06/06-082_oa.pdf; Complaint, Central Valley Chrysler-Jeep v. Witherspoon, *available at* 2004 WL 5001055 (E. D. Cal.); Non-Binding Statement of Issues of Petitioners, Coke Oven Envtl. Task Force v. U.S. EPA, Case No. 06–1131 (Sept. 3, 2003); Los Angeles v. Nat'l Highway Traffic Safety Admin.,

breadth of subnational involvement is that the number of U.S. cities and counties participating in the International Council for Local Environmental Initiatives Climate Protection Campaign grew from 40 in 2003 to 152 in 2005.[30]

Similarly, many nongovernmental actors who are not directly included in the petition influence the dialogue over climate change policy greatly. Corporate actors are major contributors of the greenhouse gases that the United States is allegedly underregulating.[31] However, corporations and nongovernmental organizations not only advocate on both sides of the climate debate, but have joined mixed public-private initiatives to achieve greenhouse gas emissions reductions.[32] Although none of those actors is a respondent in any formal sense[33] – and, in fact, some of them help to reduce to overall level of emissions – an examination of their roles sheds additional light on the dynamics framing U.S. policy.

1.1.3. Adjudicative Authority

The Inter-American Commission on Human Rights, where the petition was filed, is a regional human rights body established by the Organization of American States (OAS).[34] All of the North and South American nation-states belong to the OAS.[35] The Commission's headquarters are in Washington, D.C., and its members are elected by the OAS General Assembly.[36] The adjudicative body thus represents a region that partially overlaps with the Arctic.

Together, the Inuit petitioners, United States, and Inter-American Commission represent a geography that spans place and space. They have connections to many

---

912 F.2d 478 (D.C. Cir. 1990), *overruled in part by* Fla. Audubon Soc'y v. Bentsen, 94 F.3d 658 (D.C. Cir. Aug. 20, 1996).

[30] *See* International Council for Local Environmental Initiatives, Cities for Climate Protection, *available at* http://www.iclei.org/co2/index.htm.

[31] *See* Complaint, State of California v. Gen. Motors Corp., *available at* http://ag.ca.gov/newsalerts/cms06/06-082_oa.pdf; Complaint, Connecticut v. Am. Elec. Power Co., 406 F. Supp. 2d 265 (S.D.N.Y. 2005) (No. 04 Civ 5669), *available at* http://caag.state.ca.us/newsalerts/2004/04-076.pdf.

[32] *See* http://www.pewclimate.org/ (Pew Center on Climate Change); http://www.clintonglobalinitiative.org/home.nsf/pt_cmt_topic?open&cat=climate (Clinton Global Initiative).

[33] Petitions can be brought only against OAS members, and thus potential respondents are limited to nation-states in this forum. *See infra* note 34 and accompanying text. I have previously analyzed the ways in which forum constraints shape who are parties to environmental rights suits. *See* Osofsky, *Learning from Environmental Justice, supra* note 5, at 120–21.

[34] *See* Statute of the Inter-American Commission on Human Rights arts. 2–3, Oct. 31, 1979, O.A.S. G.A. Res. 447 (IX-0/79), *available at* http://www.iachr.org/Basicos/basic15.htm; Inter-American Commission on Human Rights, What Is the IACHR?, http://www.iachr.org/what.htm (last visited Apr. 7, 2006).

[35] *See* Organization of American States website, Member States and Permanent Missions, *available at* http://www.oas.org/main/main.asp?sLang=E&sLink=http://www.oas.org/documents/eng/memberstates.asp. Cuba, however, has not been allowed to participate since 1962. *See id.*

[36] Inter-American Commission on Human Rights Website, What Is the IACHR?, *available at* http://www.oas.org/main/main.asp?sLang=E&sLink=http://www.oas.org/juridico/english/cyber.htm.

different localities, states, countries, and regions. These multiscalar ties link them to different types of governmental and nongovernmental entities, all of which play important roles in the transnational regulation of climate change and its impact on the Inuit.

1.2. *Claims*

This section will not attempt to summarize the 167-page petition, but rather examines the ties to place and space within the main conceptual pieces of the petition's argument. The link between U.S. climate change policy and the Inuit's human rights has three main components: (1) the United States contributes a substantial portion of the world's greenhouse gases but is not taking adequate policy steps to reduce those emissions; (2) the resulting phenomenon of global climate change has significant impacts on the Inuit; and (3) these impacts violate rights of the Inuit protected under the Inter-American human rights system. The section addresses the geographic dimensions of each of these claims in turn.

1.2.1. U.S. Approach to Greenhouse Gas Emissions

The first conceptual piece of the petition's argument is that despite the United States' substantial contribution to the world's greenhouse gas emissions, it has failed to develop adequate policies to limit its emissions. The petition focuses on the Bush administration in its argument because it was filed during that period. By that time, President Bush had acknowledged that almost 20% of the world's human-made greenhouse gases originate from within U.S. borders,[37] and that the United States' own projections indicated that its emissions would rise by 42.7% between 2000 and 2020.[38] The United States' rejection of the Kyoto Protocol and its other efforts to obstruct constructive progress on climate change – in the face of its acknowledged substantial contribution – was characterized by the petitioners as a refusal "to take meaningful action to tackle global warming."[39]

This piece of the argument engages a complex mix of places and spaces. As noted in the previous description of the United States, U.S. climate policy is formed through all three branches of government, and a mix of governmental and non-governmental actors operating at multiple domestic scales. Moreover, the treaty regimes interact with domestic policymaking, and so bring a supranational scale into the dynamics. These interactions together represent the geography of "U.S. climate policy."

---

[37] President George W. Bush, Speech Discussing Global Climate Change (June 11, 2001), *available at* http://www.whitehouse.gov/news/releases/2001/06/20010611-2.html.
[38] U.S. Dep't State, United States Climate Action Report 2002, at 73 (2002), *available at* http://unfccc.int/resource/docs/natc/usnc3.pdf.
[39] Inuit Petition, *supra* note 12, at 111.

1.2.2. Impact of Climate Change on the Inuit

The petition's second major conceptual piece is that the supranational phenomenon of global climate change – resulting from these inadequately regulated emissions – has multiple impacts on the Inuit. The 2004 Arctic Climate Impact Assessment documents that temperatures in the Arctic are climbing at twice the average global rate and that the effects of climate change are particularly severe in the region.[40]

These changes have significant implications for the Inuit. Melting permafrost and worsening storms damage their homes. Changes in animal populations threaten their livelihood as hunting becomes more precarious. Ice thaws make it dangerous to use traditional travel routes. The ground is literally shifting under the Inuit's feet, and everything from weather prediction to igloo building is not what it once was.[41]

At first blush, the ties to place and space in this piece of the petition are straightforward. They certainly include the ties of the Inuit petitioners described earlier. However, like the first part of the petition, they involve additional supranational dimensions: namely, the ties of the Arctic to the many places where greenhouse gases are emitted through the complex processes in the oceans and atmosphere that lead to changes in climate. Human-induced climate change links the geographies of U.S. policy and of the harms suffered by the Inuit.

1.2.3. Human Rights Violations

The final conceptual piece of the petition is its linking of the effects of climate change on the Inuit to violations of their rights. The petition claims that the many impacts on the Inuit violate their right to enjoy the benefits of their culture, their right to use and enjoy lands they have traditionally occupied, their right to use and enjoy their personal property, their right to the preservation of health, their rights to life, physical integrity, and security, their right to their own means of subsistence, and their rights to residence and movement and inviolability of the home.[42]

In making this claim, the petition relies upon rights contained in the American Declaration of the Rights and Duties of Man, a regional human rights document.[43] Because the Commission interprets these rights in light of broader international legal developments, the rights claims provide ties not only across the Americas

---

[40] Susan Joy Hassol, Arctic Climate Impact Assessment, Impacts of a Warming Arctic: Arctic Climate Impact Assessment 2004, at 8, *available at* http://www.amap.no/acia/index.html [hereinafter ACIA]. The ACIA resulted from a collaborative project of the international scientific community to document changes in the Arctic. See id.

[41] Inuit Petition, *supra* note 12, at 35–67.

[42] Id. at 75–95.

[43] See id.; American Declaration of the Rights and Duties of Man, O.A.S. Res. XXX, adopted by the Ninth International Conference of American States (1948), *reprinted in* Basic Documents Pertaining to Human Rights in the Inter-American System, OEA/Ser. L.V/I.4 Rev. 9 (2003).

but also to the other places and supranational institutions that together shape international law.⁴⁴

In both its actors and claims, the Inuit petition thus simultaneously engages multiple scales, from the local to international; multiple branches of government, from executive to legislative to judicial; and multiple types of actors, from governmental entities to NGOs to corporations to individuals. Its places and spaces represent a nuanced geography that makes discerning the petition's potential impact difficult.

2. THE INUIT PETITION AS A BRIDGE?

These complex relationships embodied in the Inuit petition raise questions about what its role is and should be. Watt-Cloutier, during that same presentation at the 2005 Conference of Parties, characterized the petition in this way:

> Following more than two years of preparation we have submitted today a petition – this 167-page petition – to the Inter-American Commission on Human Rights based in Washington DC. . . . A declaration from the commission may not [be] enforceable, but it has great moral value. We intend the petition to educate and encourage the United States to join the community of nations in a global effort to combat climate change. . . . I suggested that the Arctic is a bridge between regions of the world. Inuit have the same philosophy. We want to bring people together. Protecting human rights is ground occupied by both reasonable governments and civil society, including Inuit and other Indigenous peoples. This petition is our means of inviting the United States to talk with us and to put this global issue into a broader human and human rights context. Our intent is to encourage and to inform.⁴⁵

Her statement invokes two intersecting geographies underlying the petition. The first is a physical geography, through which the areas of the Arctic where the Inuit live connect the United States, Canada, Greenland, and Russia. The second is a political and cultural geography, in which the petition becomes a dialogue between the United States and indigenous peoples through a shared commitment to human rights protection; the petition thus potentially serves as a bridge between nation-states and civil society.

This part of the chapter builds on Watt-Cloutier's metaphor by considering six dialectical relationships that underlie these two geographies – as well as the other ties to place and space discussed in Section 1 – and the possibilities for moving beyond the limits of these dynamics. In so doing, it owes a debt to and is inspired by the scholarship of Keith Aoki, Richard Ford, Jerry Frug, Madhavi Sunder, and Leti Volpp.⁴⁶ Drawing from their work, this section attempts to engage the complexities

---

⁴⁴ Inuit Petition, *supra* note 12, at 96–102.
⁴⁵ *See* Presentation by Sheila Watt-Cloutier, *supra* note 1.
⁴⁶ For some of the works I particularly use as inspiration for this piece, see Keith Aoki, *Space Invaders: Critical Geography, the "Third World" in International Law and Critical Race Theory*, 45 VILL. L. REV. 913 (2000); Richard Thompson Ford, *The Boundaries of Race: Political Geography in Legal Analysis*,

of perspective and culture raised by the Inuit petition and Watt-Cloutier's framing of it.

The first relationship, environmental protection/human rights, involves the complexities of capturing crosscutting problems through the available substantive categories. The second set of relationships – (1) indigenous peoples/nation-states, (2) local/national/supranational, and (3) private/public – confront the evolving constraints of the international legal system. The third set of relationships – (1) traditional law and culture/international human rights and (2) dialogue/confrontation – explore the distinctions typically made between Western/Northern/Developed and Indigenous legal systems. This section's inquiry into these six dialectics will consider the possibilities for creative bridging.

An exploration of these dialectics, in turn, raises foundational questions about the appropriate role for supranational human rights petitions in the broader context of advocacy involving indigenous peoples. How problematic is it in this context that international human rights protections emerge out of international law and a legal system based on nation-states? What is the value of obtaining these judgments? Can petitions like this one actually bridge these dialectics? Addressing these questions is critical to assessing what might constitute progress in protecting indigenous peoples' rights.

2.1. Dialectics of Substantive Categories: Environmental Protection/Human Rights

The Inuit petition builds on the existing jurisprudence in the Inter-American Commission on Human Rights by presenting an environmental rights harm that is separated in both time and location from the behavior causing it. The previous decisions of the Inter-American Commission and Court demonstrate a receptiveness to the interweaving of environmental harm and human rights violations, especially in the context of indigenous peoples.[47] This case, however, involves a much more diffuse

---

107 HARV. L. REV. 1841 (1994); Richard T. Ford, *Law's Territory (A History of Jurisdiction)*, 97 MICH. L. REV. 843 (1999); Jerry Frug, *The Geography of Community*, 48 STAN. L. REV. 1047 (1996); Gerald E. Frug & David J. Barron, *International Local Government Law*, 38 URB. LAWYER 1 (2006); Madhavi Sunder, *Cultural Dissent*, 45 STAN. L. REV. 495 (2001); Madhavi Sunder, *Piercing the Veil*, 112 YALE L.J. 1399 (2003); Leti Volpp, *The Citizen and the Terrorist*, 49 UCLA L. REV. 1575 (2002); Leti Volpp, *Feminism Versus Multiculturalism*, 101 COLUM. L. REV. 1181 (2001); Leti Volpp, *Migrating Identities: On Labor, Culture, and Law*, 27 N.C. J. INT'L L. & COMM. REG. 507 (2002); Leti Volpp, *"Obnoxious to Their Very Nature": Asian Americans and Constitutional Citizenship*, 8 ASIAN L.J. 71 (2001).

47 The Mayagna (Sumo) Awas Tingni Community v. Nicaragua, Case No. 79, Inter-Am. Ct. H.R., Ser. C (2001); Mary and Carrie Dann v. United States, Case 11.140, Inter-Am C.H.R., 75/02 (2001); Case No. 7615, Inter-Am. C.H.R. 12/85, OAS/Ser.L/V/II.66, doc. 10 rev 1 (1985). For a discussion of this jurisprudence, see Deborah Schaaf & Julie Fishel, *Mary and Carrie Dann v. United States at the Inter-American Commission on Human Rights: Victory for Indian Land Rights and the Environment*, 16 TUL. ENVTL. L.J. 175 (2002); Jennifer A. Amiott, Note, *Environment, Equality, and Indigenous Peoples' Land Rights in the Inter-American Human Rights System*: Mayagna (Sumo) Indigenous Community of Awas Tingni v. Nicaragua, 32 ENVTL. L. 873 (2002); *see also* Osofsky, *Learning from Environmental Justice*, *supra* note 5.

geography in its causal links, a difference that may help to explain the Commission's negative response to the petition. Unlike corporations logging on indigenous peoples' lands, the greenhouse gas emitters are physically separated from the Inuit and the harm is caused through a complex process in the oceans and atmosphere around the globe.

This geography thus requires a further bridging of substantive categories. The environmental rights claims intertwine science and law to represent a complex environmental process in human terms.[48] Although this reliance on science is not novel – both environmental and environmental rights claims routinely rely on scientific analysis[49] – it reinforces the intersectional nature of the petition. In order for the Inuit to ultimately prevail, they must persuade the Commission to crosscut not only different types of law, but also multiple disciplines. Such an approach might not be dialectical synthesis, but rather a recognition of a need to create space for problems that have multiple substantive dimensions.[50]

2.2. *Dialectics of Legal Structure: Civil Society/Westphalia*

The petition does not simply represent a crosscutting, interdisciplinary approach on a substantive level. The previously described geography of the petition implicates further relationships between what Watt-Cloutier calls "civil society" and the nation-state structure on which the formal international legal system rests. This section highlights three of those relationships: indigenous peoples/nation-states, local/national/supranational, and public/private.

2.2.1. *Indigenous Peoples/Nation-States*

Indigenous peoples and nation-states have an uneasy relationship. Conceptions of *terra nullius*, or empty lands, undergirded the colonial project that devastated indigenous peoples.[51] In the United States, despite treaties between the federal government and tribes protecting Native American lands, domestic legal structures and judicial interpretation often continue to perpetuate a process of subordination.[52] More specifically, the Inuit petition comes in the wake of the U.S. government refusing to change its behavior in response to a successful petition by Mary and

---

[48] Inuit Petition, *supra* note 12, at 20–34.
[49] For two recent analyses of these issues, see CARL F. CRANOR, TOXIC TORTS: SCIENCE, LAW & THE POSSIBILITY OF JUSTICE (2006), and Carol M. Rose, *Environmental Law Grows Up (More or Less), and What Science Can Do to Help*, 9 LEWIS & CLARK L. REV. 273 (2005).
[50] See Osofsky, *The Geography of Climate Change Litigation, Part 2*, *supra* note 7.
[51] See Sherene H. Razack, *When Place Becomes Race*, *in* RACE, SPACE, AND THE LAW: UNMAPPING A WHITE SETTLER SOCIETY 1 (Sherene H. Razack ed., 2002).
[52] For an analysis of relevant Supreme Court jurisprudence, see Gloria Valencia-Weber, *The Supreme Court's Indian Law Decisions: Deviations from Constitutional Principles and the Crafting of Judicial Smallpox Blankets*, 5 U. PA. J. CONST. L. 405 (2003). For a comprehensive treatment of federal Indian law, see FELIX COHEN'S HANDBOOK OF FEDERAL INDIAN LAW (2005 ed.).

Carrie Dann – members of the Western Shoshone indigenous peoples – to the Inter-American Commission on Human Rights that challenged the U.S. government's expropriation of their land.[53] Although the United States, as a member of the OAS, has an obligation to abide by the recommendations of the Commission, only the Inter-American Court of Human Rights, which does not have jurisdiction over the United States, can create enforceable decisions;[54] this difficulty would likely have plagued a positive Commission decision regarding the Inuit petition.

In one sense, the Inuit are stepping outside of the constraints of the U.S. legal system and its relationship to Native Americans by petitioning an international body. However, nation-states are deeply imbedded in the structures of the Inter-American human rights system and the Inuit's claims. Nation-states created the Inter-American Commission on Human Rights, and their grant of sovereignty gives the Commission its authority. The petition aims to influence the behavior of a particular nation-state, the United States, on the basis of its OAS membership.[55]

The embeddedness of this attempt to move beyond domestic barriers in the legal authority of the nation-state poses a potential opportunity for a bridge. If international human rights mechanisms are able to provide an effective way to engage subordination by nation-states – which is very much an open question given the barriers to formal enforcement or implementation – they connect nation-state sovereignty with limitations on that sovereign authority. As with the substantive interconnections, this bridge may not involve dialectical dynamics, but rather a hybrid space that allows for the complexities of the problematic historical and current relationship between nation-states and indigenous peoples.

2.2.2. Local/National/Supranational

These dances around nation-state sovereignty invoke a second dialectical relationship of legal structure, that of scale. Law often creates hierarchical divisions based on scale.[56] Debates over federalism in the United States embody this phenomenon in

---

[53] See Observations of the Government of the United States to the Inter-American Commission on Human Rights Report No. 113/01 of Oct. 15, 2001, concerning Case No. 11.140 (Mary and Carrie Dann) (Dec. 17, 2002) [hereinafter, Observations], *available at* http://www.state.gov/s/l/38647.htm ("The United States rejects the Commission's Report No. 113/01 of October 15, 2001, in its entirety."). For a comprehensive analysis of indigenous peoples and international law, see S. JAMES ANAYA, INDIGENOUS PEOPLES IN INTERNATIONAL LAW (2d ed., 2004).

[54] See Statute of the Inter-American Commission on Human Rights arts. 1–2, Oct. 31, 1979, O.A.S. G.A. Res. 447 (IX-0/79), *available at* http://www.iachr.org/Basicos/basic15.htm.

[55] See *supra* notes 33–36 & 53–54 and accompanying text.

[56] The geography literature explores the complex and multiple definitions of the concept of scale, some of which view scale as creating hierarchical relationships. For a summary of different ways of defining scale, see NEIL BRENNER, NEW STATES SPACES: URBAN GOVERNANCE AND THE RESCALING OF STATEHOOD 9 (2004). I have engaged the complexities of scale in more depth in Hari M. Osofsky, *A Law and Geography Perspective on the New Haven School*, 32 YALE J. INT'L L. 421 (2007), and Hari M. Osofsky, *The Intersection of Scale, Science, and Law in* Massachusetts v. EPA, this volume.

their constant navigation between state and federal authority.[57] Furthermore, in an international law context, national authority – in terms of power though not in terms of extent – stands at the top of the traditional Westphalian hierarchy; subnational law has relevance through its interaction with national law, and international law rests on the consent of sovereign and equal nation-states.[58]

But the geography of Inuit petition, described in Section 1, represents multi-scalar regulatory dynamics. International negotiations over climate change, U.S. policy decisions, and state and local decisions to regulate climate change are all interrelated.[59] Moreover, those decisions are causing tangible harm to local communities that are connected to broader communities.

The petition thus challenges advocates and governments to bridge scales in order to engage this problem. Although the petition formally occurs at a supranational level, it interacts with multiple types of actors operating at more than one level of governance. This interplay across levels may be one in which different approaches and dynamics can coexist without synthesis and, in the process, provide the basis for constructive progress.

2.2.3. Private/Public

As the petition engages these issues of sovereignty and scale, it simultaneously pushes at the distinctions the law often makes between public and private.[60] At first blush, the Inuit petition clearly falls within the category of "public." The subject matter,

---

[57] The debate over environmental federalism in the mid-1990s embodies this dynamic. For examples of scholarship arguing for the value of federal environmental regulations, see Kristen H. Engel, *State Environmental Standard-Setting: Is There a "Race" and Is It "to the Bottom"?*, 48 HASTINGS L.J. 271 (1997); Daniel C. Esty, *Revitalizing Environmental Federalism*, 95 MICH. L. REV. 570 (1996); Joshua D. Sarnoff, *The Continuing Imperative (but Only from a National Perspective) for Federal Environmental Protection*, 7 DUKE ENVTL. L. & POL'Y F. 225 (1997); Peter P. Swire, *The Race to Laxity and the Race to Undesirability: Explaining Failures in Competition Among Jurisdictions in Environmental Law*, 14 YALE J. ON REG. 67 (1996). For examples of scholarship arguing for a more limited federal role, see Henry N. Butler & Jonathan R. Macey, *Externalities and the Matching Principle: The Case for Reallocating Environmental Regulatory Authority*, 14 YALE L. & POL'Y REV. & YALE J. ON REG. 23 (1996); Richard L. Revesz, *Rehabilitating Interstate Competition: Rethinking the "Race-to-the-Bottom" Rationale for Federal Environmental Regulation*, 7 N.Y.U. L. REV. 1210 (1992); Richard L. Revesz, *The Race to the Bottom and Federal Environmental Regulation: A Response to Critics*, 82 MINN. L. REV. 535 (1997); Richard B. Stewart, *Environmental Regulation and International Competitiveness*, 102 YALE L.J. 2039 (1993). For an interesting analysis of the problem of spatial mismatch, see William W. Buzbee, *Recognizing the Regulatory Commons: A Theory of Regulatory Gaps*, 89 IOWA L. REV. 1 (2003).

[58] See IAN BROWNLIE, PRINCIPLES OF PUBLIC INTERNATIONAL LAW 287–88 (6th ed. 2003); see also Michael J. Kelly, *Pulling at the Threads of Westphalia: "Involuntary Sovereignty Waiver," Revolutionary International Legal Theory or Return to Rule by the Great Powers*, 10 UCLA J. INT'L L.361 (2005). I have engaged the problems with this model in other pieces. See Hari M. Osofsky, *Climate Change Litigation as Pluralist Legal Dialogue?*, 26A STAN. ENVTL. L.J. 181 & 43A STAN. J. INT'L L. 181 (2007); Osofsky, *The Geography of Climate Change Litigation Part 2, supra* note 7.

[59] See Osofsky, *Climate Change Litigation as Pluralist Legal Dialogue?, supra* note 58.

[60] For a history of the public/private distinction with respect to cities, see Gerald Frug, *A Legal History of Cities*, in THE LEGAL GEOGRAPHIES READER 154 (Nicholas Blomley, David Delany & Richard T. Ford eds., 2001).

international human rights, forms part of public international law. The previously described importance of the nation-state as the provider of authority and as the respondent further suggests the public character of the petition.

This formal, legal portrayal of the petition, however, masks the crucial role of private actors in the problem of greenhouse gas emissions. The dynamics between governmental and private actors in the energy industry, which form a key piece of the regulation of emissions, illustrates the inaccuracy of simply characterizing the petition as public. State sovereignty over natural resources ensures an enmeshing of the extractive industry and governments.[61] Multinational enterprise ties together corporate entities, which are regulated by a multiplicity of governmental actors in different places at multiple scales. The NGOs that advocate for responsible energy production represent a similar web of interconnections.[62]

For the petition to invite meaningful dialogue, therefore, it must reach across the public/private divide. The Inuit's attempt to "encourage and inform" forms part of a transnational regulatory dialogue in which a mix of public and private actors participate.[63] As with the other "dialectics," a full view of the petition involves acknowledging this multiplicity.

2.3. Dialectics of Legal Approaches: Indigenous Legal Systems/"Northern" Legal Systems

These complex issues of how to situate the petition in existing legal structures lead to deeper questions of culture and identity arising in dynamic among indigenous peoples' legal systems and "Northern" legal systems. This choice of language is intentional, as it suggests some of the complexities; the Inuit are certainly farther north than most "Northern" legal systems, and "legal system" means a broad range of things.[64] In particular, this section highlights two sets of relationships that embody these questions. First, it explores the interaction of traditional law and culture with international human rights. Second, it builds from that interaction to consider the dynamic between confrontation and dialogue embodied in the petition.

2.3.1. Traditional Law and Culture/International Human Rights

The national Inuit organization of Canada's statement about cultural origins and history, which is included in part in the petition that was submitted to the

---

[61] See Robert Dufresne, *The Opacity of Oil: Oil Corporations, Internal Violence, and International Law*, 36 N.Y.U. J. INT'L & POL. 331 (2004). For an interesting analysis of corporate responsibility in the context of indigenous peoples' land rights, see Lillian Aponte Miranda, *The Hybrid State-Corporate Enterprise and Violations of Indigenous Land Rights: Theorizing Corporate Responsibility and Accountability under International Law*, 11 LEWIS & CLARK L. REV. 135 (2007).

[62] I have described these dynamics in more depth in previous articles. See Osofsky, *Learning from Environmental Justice*, supra note 5, at 72–76; Osofsky, *The Geography of Climate Change Litigation*, supra note 5, at 1796–97.

[63] For a discussion of judicial dialogue, see *infra* note 72.

[64] Michael Reisman, for example, has argued for the existence of microlaw. *See* W. MICHAEL REISMAN, LAW IN BRIEF ENCOUNTERS (1999).

Inter-American Commission, embodies the complex dance between "traditional" and "Northern" approaches:

> When we speak about the origins and history of our culture, we do so from a perspective that is different from that often used by non-Inuit who have studied our past.... Our past is preserved and explained through the telling of stories and the passing of information from one generation to the next through what is called the oral tradition.... Now the challenge is ours to begin to rebuild an understanding of our past by using all of the information we now have from our legends, our real life stories, our knowledge about the Arctic environment and [its] wildlife and from information now available to us through archeology.[65]

The petition filed by the Inuit with the assistance of major environmental NGOs can be viewed as part of this evolving narrative tradition that attempts to bridge the past and the present in the Inuit's voice. The petition provides a telling of the story of the U.S. responsibility for the devastation climate change has wreaked upon them. And yet this story is told in the language of law, specifically international human rights law rather than traditional law, with outsiders helping in the telling but the Inuit themselves presenting it.[66] As such, the petition embodies the dance between "traditional" and "Northern" legal approaches, a dance that cannot fully be captured through a dialectical perspective.

2.3.2. Dialogue/Confrontation

This categorization of the petition as narrative underscores a related dynamic, the one between dialogue and confrontation. The filing of this petition would generally be regarded as confrontational. Someone with such a perspective likely would characterize it as an effort by the Inuit to force the United States to change its behavior.[67]

But Watt-Cloutier's framing of the petition[68] calls that view into question. If the people filing the petition recognize its limited capacity to formally compel action, can it be viewed as inviting dialogue? Can a "confrontational" mechanism help to foster a much-needed public conversation about these issues? Even if the Commission refuses to reconsider its negative response, has the petition still succeeded in some manner by initiating discourse about climate change and human rights? This ambiguity about how to categorize this petition opens the possibility of bridging dialogue and confrontation.

---

[65] http://www.itk.ca/5000-year-heritage/cultural-origin.php.
[66] Inuit Petition, *supra* note 12.
[67] This notion of confrontation is at the core of the traditional litigation model. For an analysis of the growing possibilities for intermingling between traditional litigation and alternative dispute resolution, see Vance K. Opperman, *The Pros and Cons of ADR, Including ADR/Litigation Hybrids*, 1 SEDONA CONF. J. 79 (2000).
[68] *See* Presentation by Sheila Watt-Cloutier, *supra* note 1.

Together, these two dualisms of legal approaches suggest that a reframing of the petitioning process opens possibilities for progress on indigenous peoples' rights. The petition and Watt-Cloutier's approach to it attempt to transform an international human rights process created by nation-states into a vehicle for articulating and protecting traditional values on an international stage. As this bridging begins to re-narrate[69] what accessing the international legal system means, it provides a potential mechanism for not only addressing particular problems, but also engaging cultural-legal myths that have served to devastate indigenous peoples.[70] The goal of such a dialogue may not be Hegelian synthesis, but rather an opening of multiple pathways for "moving forward."

3. CONCLUDING REFLECTIONS: PETITIONS AND PROGRESS?

The geography of the petition and the dynamics it embodies present a fundamental question: where do these potential bridges leave an analysis of the Inuit petition's capacity to contribute to progress regarding indigenous peoples' rights in general and harms from climate change in particular? This thought piece concludes by exploring potential answers to this question in terms of both legitimacy and value. In so doing, it suggests some of the conceptual and legal creativity needed to move beyond current – often confining – dialectics.

The nation-state system, international institutions, and the individually oriented human rights problems run into serious legitimacy issues in the context of indigenous peoples. Petitions like the one the Inuit filed do not solve those problems, but they engage them constructively. The Inuit themselves, not the outside lawyers, filed the petition. Watt-Cloutier's description of the petition challenges the conventional characterization of it and opens possibilities in the process of doing so. This petition cannot force the United States to change its behavior directly but puts pressure on it to do so, or at the very least to enter a dialogue about its choices.

Conceiving of the petition as part of a narrative and as initiating a dialogue leads to the issue of value. Often, in the United States' conventional confrontational lawyering model, success gets equated with courtroom wins. But a "win" in the context of the Inuit petition is a much more nuanced issue. Is it a turnaround in the Commission's stance that allows for a positive ruling on the petition? Is it the hearing that took place on climate change and human rights? Is it a change in U.S. policy? Is it a new, and much-needed, dialogue about climate change and/or indigenous peoples' rights that is arguably happening as a result of the petition?

[69] Narratives serve as a powerful tool for exposing problems of subordination. *See* Richard Delgado, *Storytelling for Oppositionists and Others: A Plea for Narrative*, 87 MICH. L. REV. 2411 (1988); Stephanie Weinstein & Arthur Wolfson, *Toward a Due Process of Narrative: Before You Lock My Love Away, Please Let Me Testify*, 11 ROGER WILLIAMS U. L. REV. 511 (2006); Matthew L.M. Fletcher, Note, *Listen*, 3 MICH. J. RACE & L. 523 (1998).

[70] For an analysis of those myths, see Razack, *supra* note 51. For a critical analysis of the international legal system, see BALAKRISHNAN RAJAGOPAL, INTERNATIONAL LAW FROM BELOW: DEVELOPMENT, SOCIAL MOVEMENTS AND THIRD WORLD RESISTANCE (2003).

In an ideal scenario, the United States "listens" and changes its behavior dramatically, something that now is happening with respect to federal climate policy through the change in presidential administration. But even short of that, many positive possibilities exist through the bridges that petitions like this one may be able to build. The petition generated publicity that helped to raise awareness about the way in which climate change is impacting the Inuit and about international human rights tribunals as appropriate institutions for addressing crosscutting problems.[71] A statement from the Inter-American Commission on climate change and human rights could be used as persuasive authority in other pending actions addressing climate change and/or environmental rights issues.[72] In its various formal and informal interactions with governments and civil society, the petition becomes a "port of entry"[73] for making progress on these issues.[74]

A final excerpt from the remarks of Watt-Cloutier situates the petition and the importance of reframing our conversation about climate change.

> I have attended three COPs. People rush from meeting to meeting arguing about all sorts of narrow technical points. The bigger picture, the cultural picture, the human picture is being lost. Climate change is not about bureaucrats scurrying around. It is about families, parents, children, and the lives we lead in our communities in

---

[71] A Westlaw search in all news on Nov. 11, 2006, reveals twenty-five news articles in the preceding year that contain the words "Inuit" and "Inter-American."

[72] The use of international and foreign decisions as persuasive authority in Constitutional interpretation has created controversy in the United States, see Agora: *The United States Constitution and International Law*, 98 AM. J. INT'L L. 42 (2004), but judicial dialogue – both formal and informal – only seems to be increasing, Anne-Marie Slaughter, *A Global Community of Courts*, 44 HARV. INT'L L.J. 191 (2003); Anne-Marie Slaughter, *Judicial Globalization*, 40 VA. J. INT'L L. 1103 (2000); Melissa A. Waters, *Mediating Norms and Identity: The Role of Transnational Judicial Dialogue in Creating and Enforcing International Law*, 93 GEO. L.J. 487, 490–97 (2005); see also Rex D. Glensy, *Which Countries Count? Lawrence v. Texas and the Selection of Foreign Persuasive Authority*, 45 VA. J. INT'L L. 357 (2005); Joan L. Larsen, *Importing Constitutional Norms from a "Wider Civilization": Lawrence and the Rehnquist Court's Use of Foreign and International Law in Domestic Constitutional Analysis*, 65 OHIO ST. L.J. 1283 (2004).

[73] Judith Resnik used this term to reference the multiple ways in which "foreign" norms become domesticated. See Resnik, *supra* note 27.

[74] The legal pluralist literature has engaged the implications of multiple normative communities sharing social spaces. See, e.g., Robert M. Cover, *The Supreme Court 1982 Term Foreword: Nomos and Narrative*, 97 HARV. L. REV. 4, 9 (1983) ("A legal tradition is hence part and parcel of a complex normative world.... Law may be viewed as a system of tension or a bridge linking a concept of a reality to an imagined alternative...."); Sally Engle Merry, *Legal Pluralism*, 22 LAW & SOC'Y REV. 869, 870 (1988) ("What is legal pluralism? It is generally defined as a situation in which two or more legal systems coexist in the same social field."). In the international law context, scholars from the New Haven School have described law as "a process of authoritative decision by which members of a community clarify and secure their common interests" and argued that "humankind today lives in a whole hierarchy of interpenetrating communities, from the local to the global." HAROLD D. LASSWELL & MYRES S. MCDOUGAL, JURISPRUDENCE FOR A FREE SOCIETY: STUDIES IN LAW, SCIENCE AND POLICY xxi (1992). For other, more recent pluralist analyses in an international law context, see, for example, Paul Schiff Berman, *Global Legal Pluralism*, 80 S. CAL. L. REV. 1155 (2007); Janet Koven Levit, *A Bottom-Up Approach to International Law Making: The Tale of Three Trade Finance Instruments*, 30 YALE J. INT'L L. 125 (2005); William W. Burke-White, *International Legal Pluralism*, 25 MICH. J. INT'L L. 963 (2004).

the broader environment. We have to regain this perspective if climate change is to be stopped. Inuit understand these connections because we remain a people of the land, ice, and snow. This is why, for us, climate change is an issue of our right to exist as an Indigenous people. How can we stand up for ourselves and help others do the same?[75]

Through its many bridges, the Inuit petition provides an important model for how creative lawyering may help to transform dialogue.

[75] *See* Presentation by Sheila Watt-Cloutier, *supra* note 1.

13

Bringing Climate Change Claims to the Accountability Mechanisms of International Financial Institutions

Jennifer Gleason* and David B. Hunter**

INTRODUCTION

Since the early 1990s, pressure from environmental and human rights groups has pushed international financial institutions (IFIs) to address the sustainable development impacts of the projects they finance. As part of their response, IFIs have increasingly adopted environmental and social policies that require them, among other things, to take environmental and social concerns into account. Some of these policies specifically include impacts associated with climate change. IFIs that have adopted environmental and social policies to guide their lending include multilateral development banks (MDBs) such as the World Bank Group, the Asian Development Bank, and other regional development banks;[1] export credit and insurance agencies, such as the U.S. Export-Import Bank and the United Kingdom's Export Credits Guarantee Department;[2] and private commercial banks, such as Citibank, HSBC, and ABN Amro.[3]

Concern over the implementation of these policies led the same environmental and human rights groups to advocate for the establishment of citizen-driven accountability mechanisms in the IFIs. Beginning with the creation of the World Bank Inspection Panel, nine accountability mechanisms have been established. Although they all differ in terms of their independence and effectiveness, each provides an opportunity for project-affected people to raise concerns about the compliance of

---

* Staff Attorney, Environmental Law Alliance Worldwide (ELAW), http://www.elaw.org, 1877 Garden Ave., Eugene, OR 97403, 541–687-8454, ext. 15, jen@elaw.org.
** Assistant Professor of Law & Director, Program on International and Comparative Environmental Law, American University Washington College of Law, 4801 Massachusetts Ave., NW, Washington, D.C. 20016, 202–274-4415, dhunter@wcl.american.edu.
[1] *See, e.g., infra* text accompanying note 31 (discussing the environmental and social policies of the World Bank).
[2] *See* Organisation for Economic Co-operation and Development, Updated Recommendation on Common Approaches on Environment and Export Credits, TD/ECG(2005)3 (Feb. 22, 2005).
[3] *See* The Equator Principles: A Financial Industry Benchmark for Determining, Assessing and Managing Environmental & Social Risk in Project Financing (headline) (July 2006), *available at* http://www.equator-principles.com (last visited Jan. 28, 2007).

IFI-financed projects.[4] This chapter reviews these mechanisms' potential use in filing claims relating to climate impacts. Unlike the litigation approaches explored by most of the chapters in this book, a pure climate change–related claim has yet to be filed with any of the IFI accountability mechanisms. But the mechanisms have been used to raise concerns about the environmental and social impacts of oil pipelines and other energy-related projects, and these institutions are frequently discussed as potential venues for raising climate change claims.

Section 1 of this chapter reviews the IFIs' impacts on climate change and the importance of raising the profile of contributions to climate change at these institutions. Sections 2 and 3 introduce the accountability mechanisms by focusing on the World Bank Inspection Panel and the International Finance Corporation's (IFC's) Compliance Advisor and Ombudsman (CAO). These two are the most widely used among the IFI accountability mechanisms, and the institutions in which they operate (i.e., the World Bank and the IFC) are the recognized leaders in international development. Section 4 provides a brief overview of accountability mechanisms available at other IFIs.

1. CLIMATE CHANGE AND THE BANKS

Targeting the banks that finance activities resulting in significant greenhouse gas emissions is a potentially effective strategy for several reasons. First, IFIs wield significant influence over what type of projects get funded, particularly in developing countries, providing $30–$40 billion in financing every year.[5] Moreover, some financial institutions play a disproportionate role in climate-changing projects. The World Resources Institute (WRI) reports that "[t]he lending profile of MDBs demonstrates significant concentrations of finance in sectors with substantial greenhouse gas (GHG) emission footprints, including transport, oil and gas, electric power, and mining."[6] The report calculates that 37 percent of the World Bank's lending in 2004 went toward these projects, with an investment of $7.6 billion. The same year, the Inter-American Development Bank invested $730 million (12 percent of its total lending), and in 2003, the European Bank for Reconstruction and Development invested $3.3 billion (27 percent of its total lending), in sectors with substantial impacts on the climate.[7]

---

[4] This chapter discusses the potential for climate claims at the IFI's own accountability mechanisms. Not covered are recent efforts to use domestic courts to compel greater attention to climate impacts in international project finance. See, e.g., Order Denying Defendants' Motion for Summary Judgment, Friends of the Earth, Inc. v. Watson, No. C02–4106 JSW (Aug. 23, 2005) (case brought to require consideration of cumulative climate impacts in the activities financed by the Overseas Public Investment Corporation); Friends of the Earth Germany & GermanWatch, Press Briefing, *German Government Sued over Climate Change* (June 15, 2004).

[5] Bank Information Center website at http://www.bicusa.org/en/Page.About.aspx (last visited Jan. 26, 2007).

[6] Jon Sohn, Smita Nakhooda & Kevin Baumert, *Mainstreaming Climate Change Considerations at the Multilateral Development Banks*, at 1, box 1, World Resources Institute (July 2005).

[7] Id.

Closer review of the World Bank shows the influence of the banks and their connection to climate change. In addition to its direct financing, the World Bank is also an implementing agency of the Global Environment Facility (GEF), which among other roles, acts as the financial mechanism for the United Nations Framework Convention on Climate Change (UNFCCC).[8] Through its Carbon Finance Unit,[9] the Bank supports the global carbon market by financing the purchase of emission credits under the Kyoto Protocol's Clean Development Mechanism, and in 2008, the World Bank launched its $6 billion Climate Investment Fund to support the long-term transition to low-carbon energy systems. The Bank's influence is expanded further by coordinating other donors; mobilizing bilateral, and increasingly, private-sector financing; conducting policy research; and providing technical assistance to borrowing countries.

Beginning in the 1970s, independent observers began to recognize that the World Bank and other IFIs supported some of the most environmentally damaging projects in developing countries. Even assuming good intentions, the size and scale of many of the projects simply dwarfed the environmental legal and policy infrastructure of some of the borrowing countries.[10] In recent years, for example, the World Bank has supported the Baku-Tbilisi-Ceyhan pipeline and the Chad-Cameroon pipeline, which are critical for the future expansion of oil markets and thus likely contributors of significant emissions. Despite the obvious climate implications of such projects, as WRI recently reported, "MDBs do not systematically integrate climate change concerns into their operations. Over 80 percent of World Bank's publicly disclosed lending in the energy sector from 2000 to 2004 did not consider climate change issues in project appraisals and documentation."[11]

A focus on the IFIs is also a sound strategy for environmental activists focused on climate change: those who bring the money to the table often can impose environmental and social improvements on projects, if they deem it necessary. Thus, by greening international finance, environmental advocates can leverage environmental change at the project level. Moreover, many financial institutions are

---

[8] U.N. Framework Convention on Climate Change, Art. 21(3) (1992).

[9] "The World Bank Carbon Finance Unit (CFU) uses money contributed by governments and companies in OECD countries to purchase project-based greenhouse gas emission reductions in developing countries and countries with economies in transition. The emission reductions are purchased through one of the CFU's carbon funds on behalf of the contributor, and within the framework of the Kyoto Protocol's Clean Development Mechanism (CDM) or Joint Implementation (JI)." http://carbonfinance.org/Router.cfm?Page=About&ItemID=1.

[10] See, e.g., Robert Goodland, The Environmental Implications of Major Projects in Third World Development, in MAJOR PROJECTS AND THE ENVIRONMENT 9–16 (P. Morris ed., Oxford 1987). The Bank's poor record in environmental protection, human rights, and poverty alleviation has been well documented. See CATHERINE CAUFIELD, MASTERS OF ILLUSION: THE WORLD BANK AND THE POVERTY OF NATIONS (1997); BRUCE RICH, MORTGAGING THE EARTH: THE WORLD BANK, ENVIRONMENTAL IMPOVERISHMENT, AND THE CRISIS OF DEVELOPMENT (1994); THE WORLD BANK: LENDING ON A GLOBAL SCALE (Jo Marie Greisgraber & Bernhard G. Gunter eds., 1996); INTERNATIONAL BANKS AND THE ENVIRONMENT: FROM GROWTH TO SUSTAINABILITY: AN UNFINISHED AGENDA (Raymond Mikesell & Lawrence Williams eds., 1992).

[11] Jon Sohn, Smita Nakhooda & Kevin Baumert, supra note 6.

concerned about their exposure to climate risk, either in their lending or insurance portfolios, and this can make them receptive to arguments about climate change.

Recognizing the leverage of IFIs to influence projects led activists to push for accountability mechanisms. They hope these mechanisms can raise the profile of local community concerns, bringing them directly to the attention of the IFIs' top decisionmakers. In so doing, they aim to improve decisionmaking with respect to specific project, and more generally sensitize the institution to environmental and social concerns. Accountability mechanisms now exist in six multilateral financial institutions and three bilateral financial institutions.[12]

2. POTENTIAL CLAIMS TO THE WORLD BANK INSPECTION PANEL

The World Bank Group is comprised of five separate institutions: the International Bank for Reconstruction and Development (IBRD), the International Development Association (IDA), the International Finance Corporation (IFC), the Multilateral Investment Guarantee Agency (MIGA), and the International Center for the Settlement of Investment Disputes (ICSID). Together the IBRD and the IDA are most frequently referred to as the "World Bank," a taxonomy we adopt for this chapter.[13]

The World Bank Inspection Panel was created by the Executive Directors of the IBRD and the IDA on September 22, 1993.[14] The three-member Panel is authorized to receive claims about the Bank's failure to comply with its own policies and procedures, including its nonenforcement of loan conditions. According to the Panel's Operating Procedures, the Panel was "established for the purpose of providing people directly and adversely affected by a Bank-financed project with an independent forum through which they can request the Bank to act in accordance with its own policies and procedures."[15]

The Panel has jurisdiction to review any project financed by either the IBRD or IDA. The three-member Panel reports directly to the Board of Executive Directors.[16]

---

[12] These are identified in the chart included *infra* in Section 4.
[13] The primary difference between the IBRD and the IDA is that the IDA provides concessional or low-cost loans to the poorest countries (having per capita annual income below $1025 (in 2005 dollars)). Nearly half of the eighty-one countries that qualify for IDA lending are in sub-Saharan Africa. The IBRD provides loans to other developing countries and countries in economic transition at a near-market rate with longer repayment terms than commercial loans.
[14] IBRD, Resolution No. 93–10; IDA, Resolution No. 93–6 (Sept. 22, 1993).
[15] World Bank Inspection Panel, *The Inspection Panel for the International Bank for Reconstruction and Development and International Development Association: Operating Procedures*, 34 I.L.M. 510, 511 (1995) [hereinafter Inspection Panel Operating Procedures]; *see also* IBRD, Resolution No. 93–10; IDA, Resolution No. 93–6 (Sept. 22, 1993).
[16] The twenty-four-member Board of Executive Directors meets several times a week and among other things has the responsibility to approve every loan proposed by the Bank. A Board of Governors, responsible for broad policy, meets once a year. Voting at the Executive Directors and Board of Governors is based on financial shareholding percentages; the United States has the largest voting share of 17 percent. The G7 comprises approximately 45 percent of the voting shares at the Bank, and all of the donor countries together comprise a solid majority of the vote. Although most countries share an executive director, several, including the United States and China, have their own representative on the Board of Executive Directors. The Board meetings and decisions are not open to the public.

It is thus independent of Bank Management (including the President), which is responsible for promoting or developing Bank projects. To enhance its independence, Panel members cannot have served the Bank in any capacity for the two years preceding their selection and can never work for the Bank again. The Panel's Secretariat has five staff members that support the Panel's investigations and conduct outreach on the Panel's behalf.[17] This section explores how claims might be brought before the World Bank Inspection Panel.

2.1. Filing a Petition

Filing claims to the World Bank Inspection Panel is intended to be simple, with the goal of making the Panel accessible to even the most disadvantaged communities or their representatives. Claims can be filed by any two or more affected parties in the borrower's territory.[18] Affected parties can file through a local representative, but nonlocal representatives are allowed only in "exceptional cases" where "appropriate representation is not locally available."[19] Claims must be in writing and must explain how the affected parties' interests have been, or are likely to be, affected by "a failure of the Bank to follow its operational policies and procedures with respect to the design, appraisal and/or implementation" of a project.[20] The claimants must provide evidence that they gave Bank staff a reasonable opportunity to respond to the allegations.

Upon receiving a complete request for inspection,[21] the Panel registers the claim and forwards a copy to Bank Management, which has twenty-one days to respond.[22] The Panel subsequently has twenty-one days to review Management's response and to make a recommendation to the Executive Directors regarding whether the claim warrants a full investigation.[23]

The Executive Directors have the exclusive authority to authorize or deny a full investigation. Although this initially politicized the Panel process, since changes made in 1999, the Board has supported every Panel recommendation for an investigation.[24] Once an investigation is authorized, the Panel enjoys broad investigatory powers, including access to all Bank staff, documents, and the project site. After the investigation, the Panel issues a report evaluating the Bank's compliance with its

---

[17] For more information, see the Inspection Panel website, *available at* http://www.worldbank.org/inspectionpanel (last visited Aug. 15, 2006).

[18] Inspection Panel Operating Procedures, *supra* note 15.

[19] *Id.* at para 4. The Bank's Board of Executive Directors and, in some circumstances, an individual Executive Director, are also eligible to file a claim.

[20] *Id.* at paras. 4 & 11.

[21] Several types of complaints are explicitly beyond the Panel's jurisdiction, including complaints (i) addressing actions that are the responsibility of parties other than the Bank, (ii) relating to procurement decisions, (iii) filed after a loan's closing date or after 95 percent of the loan has been disbursed, or (iv) regarding matters already heard by the Panel unless justified by new evidence.

[22] Inspection Panel Operating Procedures, *supra* note 15, at 522.

[23] *Id.*

[24] *See World Bank: Conclusions of the Second Review of the World Bank Inspection Panel*, 39 I.L.M. 249, 250 (2000).

policies. Within six weeks, Management must submit to the Executive Directors a report and recommendations in response to the Panel's findings. The Panel's report, Management's recommendations, and the Board's decision are released publicly two weeks after Board consideration.[25]

By the beginning of 2007, the Inspection Panel had received forty-three formal requests for inspection and had registered thirty-eight of them.[26] The Panel has recommended an investigation in twenty-one claims, and the Board has approved investigations in seventeen of those.[27]

Only one claim – relating to the Chad-Cameroon oil pipeline – has explicitly raised climate change impacts. In that claim, local communities affected by the proposed pipeline raised concerns primarily about local impacts on their forests and agricultural lands. Among the alleged insufficiencies in the environmental assessment in that claim, however, was a failure to assess the "impact of the combustion of the oil in the project on climate change."[28] In response, Management asserted that it had adequately assessed the pipeline's direct emissions in accordance with its policies. The Panel ultimately agreed and found further that contribution from the pipeline would be a "tiny 0.15%" of annual global greenhouse emissions.[29] Climate issues were a small part of the overall claim and did not receive significant attention from either the Panel or the civil society advocates monitoring the project. The Chad-Cameroon experience confirms that climate-related concerns can be reviewed by the Panel, but the unsatisfactory results suggest that future claims need to provide more detailed factual and policy-based arguments for raising climate concerns.

Other Inspection Panel claims have raised issues that indirectly have significant ramifications for climate change. For example, in a claim regarding the West African natural gas pipeline (WAGP) in Nigeria,[30] twelve communities challenged the Bank's assessment and approval of the project on several grounds, including the Bank's failure to require that waste gas associated with Nigeria's oil production be captured to supply the pipeline. This waste gas is currently flared in open flames, which contributes significantly to climate change. The communities based their claims on economic development grounds, but, if successful, their claims would have significantly reduced GHG emissions from Nigeria's fossil fuel sector.

The Chad-Cameroon and Nigeria examples suggest that the World Bank Inspection Panel could under certain circumstances raise issues relevant to climate change.

---

[25] Inspection Panel Operating Procedures, *supra* note 15, at 519.
[26] See Inspection Panel, *Requests for Inspection, available at* http://web.worldbank.org/inspectionpanel (last visited Jan. 29, 2007).
[27] See Inspection Panel, *Summary of Inspection Panel Cases* (June 30, 2006), *available at* http://siteresources.worldbank.org/EXTINSPECTIONPANEL/Resources/Table1SummaryofInspectionPanelCases.pdf?&resourceurlname=Table1SummaryofInspectionPanelCases.pdf.
[28] Request for Inspection, World Bank Inspection Panel Request No. RQ 02/2, para. 5 (Sept. 20, 2002).
[29] World Bank Inspection Panel, Investigation Report No. 25734, paras. 87–89 (May 2, 2003).
[30] Request for Inspection by Representatives of the Communities Impacted by the West African Gas Pipeline in Lagos State, Nigeria, World Bank Inspection Panel Request No. RQ 06/03 (Apr. 27, 2006).

The next section will examine existing World Bank policies that could form the basis for such claims.

2.2. Policies Relating to Climate Change

The Inspection Panel is authorized to review compliance with World Bank "operational policies and procedures." Over the past twenty years, the Bank has adopted a set of mandatory environmental and social Operating Procedures (OPs) and associated Bank Procedures (BPs), which are collectively referred to as the Bank's "safeguard" policies. Among other things, these safeguard policies require Bank staff to assess the environmental and social impacts of proposed projects,[31] protect critical natural habitats,[32] compensate people who are involuntarily resettled,[33] and protect the rights of indigenous peoples.[34]

The Bank's policies do not include an explicit and comprehensive approach to climate change, although the Bank clearly recognizes the importance of climate change generally:

> The Bank accepts the IPCC's conclusion that emissions of greenhouse gases from human activities are affecting the global climate. It also believes that the consequences of climate change will disproportionately affect both poor people and poor countries. The World Bank has an important role to play in helping to avert climate change, and it will assist its clients in meeting their obligations under the... UNFCCC. Under the 1997 Kyoto Protocol, some client countries with economies in transition have obligations to reduce emissions of greenhouse gases. Other clients – developing nations – have obligations to measure and monitor GHG emissions within their countries but do not have to reduce emissions yet.[35]

This section explores the Bank's environmental policies relevant to a potential claim based on climate change.

2.2.1. Assessing Climate Impacts

The Environmental Assessment OP[36] (EA Policy) is the most important of the Bank's environmental policies with respect to a potential climate claim.[37] The EA Policy requires that all World Bank–financed projects be screened into one of three categories depending on the extent of environmental impacts associated with the

---

[31] World Bank, Operational Policy 4.01: Environmental Assessment (Jan. 1999) [hereinafter World Bank OP 4.01].
[32] World Bank, Operational Policy 4.04: Natural Habitats (June 2001).
[33] World Bank, Operational Policy 4.12: Involuntary Resettlement (Dec. 2001).
[34] World Bank, Operational Policy 4.10: Indigenous Peoples (July 2005).
[35] WORLD BANK, POLLUTION PREVENTION AND ABATEMENT HANDBOOK 36–37 (1998) (hereinafter *Pollution Handbook*).
[36] World Bank OP 4.01, *supra* note 31.
[37] Other OPs and BPs could be used in a climate claim but are not analyzed in depth here.

project.[38] "Category A" projects, which have "significant adverse environmental impacts that are sensitive, diverse, or unprecedented" must undergo a full environmental assessment with opportunities for consultation by project-affected people.[39] "Category B" projects are those with adverse impacts that are "less adverse," including impacts that are typically site specific, reversible, and capable of being mitigated. The scope of environmental assessment for a Category B project is narrower than that of Category A.[40] "Category C" projects normally do not require any environmental analysis because the project is "likely to have minimal or no adverse environmental impacts."[41] In general, about 10 percent of World Bank projects are classified as Category A, and significantly more (57 percent in 2005) are classified as Category B.

Several provisions of the EA Policy could apply to a climate claim. Some sections specifically mention a project's climate impacts, while others refer more generally to impacts on biodiversity or natural resources, require projects to be environmentally sound,[42] or require compliance with international commitments or domestic law.[43] These provisions could be used to support a petition about the climatic impacts of a project.

Any proposed project that contributes significantly to climate change could arguably be classified as Category A, thus requiring a full environmental assessment and consultation. The Bank's EA policy states, "[A] proposed project is classified as Category A if it is likely to have significant adverse environmental impacts that are sensitive, diverse, or unprecedented. These impacts may affect an area broader than the sites or facilities subject to physical works...."[44] For projects that are likely to have significant regional effects, including presumably those from climate change, the Bank may require a regional environmental assessment with "particular attention to potential cumulative impacts of multiple activities."[45]

Assuming that an EA is required, the EA Policy specifically requires environmental assessments to address the climate impact of a project. The policy states that the assessment must take "into account . . . transboundary and global environmental

---

[38] A fourth category (FI) covers "investment of Bank funds through a financial intermediary, in subprojects that may result in adverse environmental impacts." World Bank OP 4.01(8)(d).
[39] World Bank OP 4.01, *supra* note 31, at para. (8)(a).
[40] *Id.* at para. (8)(b).
[41] *Id.* at para. (8)(c).
[42] *Id.* at para. (1).
[43] The project's EA must take into account the project country's "national legislation . . . and obligations of the country, pertaining to project activities, under relevant international environmental treaties and agreements. The Bank does not finance project activities that would contravene such country obligations, as identified during the EA." OP 4.01(3). *See also* World Bank Operating Procedure 4.00: Piloting the Use of Borrower Systems to Address Environmental and Social Safeguard Issues in Bank-Supported Projects (Mar. 2005) (requiring compliance with domestic law); World Bank Operating Procedure 8.60: Development Policy Lending (Aug. 2004), section 2 (noting that meeting international commitments is a purpose of policy and institutional lending) World Bank OP 8.60, section 2, also requires projects to meet international commitments.
[44] World Bank OP 4.01, *supra* note 31, at para. 8.
[45] World Bank OP 4.01, *supra* note 31, at para. 1(7) Annex A, § 6.

aspects," which specifically include climate change.⁴⁶ An assessment must evaluate adverse impacts on different components of the environment, including biodiversity and ecosystem services, and with respect to certain social issues, including indigenous peoples, resettlement, or cultural property.⁴⁷ The Bank requires that borrowers evaluate the impact of future climate change on the project itself as well. Thus, for example, "projected sea level rise and increased coastal flooding should be considered when evaluating the design of a coastal drainage and wastewater system."⁴⁸

The World Bank's environmental assessment policy is mirrored by policies at many of the other MDBs. These assessment policies also provide the framework for other potential claims, including the need to address alternatives, discussed subsequently.

2.2.2. Evaluating Alternatives

The EA process may be used to push for alternatives or mitigation measures that can reduce the project's climate impact. The Bank's EA Policy is intended to ensure that project sponsors consider measures that have fewer environmental impacts. In this regard, the Bank is supposed to favor "preventive measures over mitigatory or compensatory measures, whenever feasible."⁴⁹

The Environmental Assessment Sourcebook (Sourcebook), referred to in the introductory note to the EA Policy, clarifies the alternatives requirement in the context of climate change:

> [O]ptions to reduce a project's contribution to global change without adversely affecting the cost or success of the project should be evaluated. For example, expansion of domestic coal mining operations is likely to result in methane emissions. Collection of the coalbed methane and use as an energy source would not only reduce the contribution to climate warming, but might also be economic....
>
> When evaluating various alternative projects, not only should potential total gas emissions be considered, but also the particular gases that are released since not all the gases are equally efficient in terms of their greenhouse and ozone depletion capacity. For example, even though natural gas emits approximately 30 percent less CO_2 per unit of energy produced than oil (and over 40 percent less than coal), the production and distribution of natural gas often results in the release of CH_4, which radiatively is a much more effective greenhouse gas than CO_2 (over 20 times more effective, kilogram-for-kilogram, over a 100-year period). Therefore, when considering switching from oil to natural gas in order to reduce CO_2 emissions, the increased potential for CH_4 emissions must also be considered.⁵⁰

---

⁴⁶ World Bank OP 4.01, *supra* note 31, at para. 3 & n. 4.
⁴⁷ The Natural Habitats OP also addresses the impact of a project on biodiversity. It states, "[t]he Bank does not support projects that, in the Bank's opinion, involve the significant conversion or degradation of critical natural habitats." Natural Habitats OP 4.04, *supra* note 32, at para. 4 (footnote omitted); *see also* Indigenous Peoples Policy, *supra* note 34; Resettlement Policy, *supra* note 33.
⁴⁸ The World Bank, *Environmental Assessment Sourcebook*, Chapter 2 at § 28 (1999) (published online at http://www.siteresources.worldbank.org) [hereinafter *Environmental Assessment Sourcebook*].
⁴⁹ World Bank OP 4.01, *supra* note 31, at para. 2.
⁵⁰ *Environmental Assessment Sourcebook*, *supra* note 48, at chap. 2, §§ 26–27 (citation omitted).

In practice, only a few projects have included some assessment of a proposed project's impact on the global climate. Many of these include little more than passing mention of climate impacts and none appear to take seriously the requirement to evaluate alternative approaches in light of climate impacts. For example, the EA for the Power Sector Generation & Reconstruction Project for Albania[51] estimates the likely emissions of carbon dioxide from the proposed facility and estimates that those emissions would represent 0.05 percent of the total carbon dioxide emissions for Albania. The EA notes that Albania had not ratified the Kyoto Protocol at the time the EA was prepared. Similarly, the EA for the WAGP[52] provides information about the project's climate impacts, including some information in the Executive Summary. Interestingly, the EA concludes (with relatively little analysis) that the project will reduce GHG emissions because it will induce "a switch to gas fuel from other fossil fuels (primarily light crude oil) among end-user customers."[53] For the most part, however, World Bank project proponents have not systematically or comprehensively evaluated their climate impacts.

2.2.3. Achieving Minimum Emissions Levels

Potential claimants may be able to raise questions about whether proposed projects meet certain minimum standards or whether alternatives should have been selected. Under the EA Policy, Bank-sponsored projects are normally supposed to comply with the pollution prevention and abatement measures and emissions levels found in the Bank's *Pollution Prevention and Abatement Handbook (Pollution Handbook)*.[54] Taking into account borrower country legislation and local conditions, however, the EA may recommend alternative emission levels and pollution prevention and abatement measures, but only after providing "full and detailed justification for the levels and approaches chosen."[55]

The *Pollution Handbook* includes several provisions relevant to a future climate-based claim. For example, the *Pollution Handbook* states, "The World Bank Group will not invest in a country's energy sector unless that country shows a commitment to improving efficiency, whether by restructuring the sector or reforming its policies."[56] A borrower arguably must demonstrate a commitment to improving

---

[51] World Bank project number P077526.
[52] World Bank project number P082502.
[53] Environmental Assessment for the West Africa Gas Pipeline, Executive Summary at 3.
[54] *Pollution Handbook*, supra note 35.
[55] World Bank OP 4.01 (6). The Inspection Panel only has authority to review compliance with Bank "operational policies and procedures," which specifically do not include "Guidelines and Best Practices and similar documents or statements." See IBRD's and IDA's Resolutions No. 93–10 and No. IDA 93–6, Art. 12 (Sept. 22, 1993). Although the Panel may not have authority to evaluate compliance with the *Pollution Handbook* directly, the Panel cannot evaluate compliance with the EA Policy without ensuring that the project sponsor is either complying with the handbook or has adequately justified any deviation from the handbook's emission levels or pollution prevention and abatement measures.
[56] *Id.* at 33.

energy efficiency before the Bank finances an energy sector project.[57] The *Pollution Handbook* includes sector-specific emissions standards for several sectors that have substantial implications for climate change, including coal-fired power plants and cement manufacturing.[58] Moreover, the *Pollution Handbook* is currently being revised and the applicable standards for climate-relevant sectors are likely to get stricter over time. Projects in these sectors with high greenhouse gas emissions are likely candidates for an Inspection Panel claim.

2.2.4. Consistency with the UNFCCC and Kyoto

A potential claimant may try to tie a specific project to the international obligations of the borrower country under the UNFCCC or the Kyoto Protocol. The Bank is not supposed to finance projects that are inconsistent with the borrowing country's international obligations. In an operational statement that set out the principles for the Bank's operational policies, the Bank committed not to "finance projects that contravene any international environmental agreement to which the member country concerned is a party."[59] The *Pollution Handbook* addresses this issue as well. It states:

> In the case of the UNFCCC and the Kyoto Protocol, the World Bank will ensure that its activities are consistent with these conventions and will actively support its member countries in building capacity and undertaking investments for their implementation. Global environmental externalities can be recognized at the project level and, increasingly, in economic and sector work and in national environmental action plans. Where appropriate, country assistance strategies also include global environmental issues....[60]

Most developing countries do not yet have specific obligations (other than reporting obligations under the UNFCCC or Kyoto Protocol), but this could change in future negotiations. More generally, the lack of policy coherence between the goals of the international climate regime and the World Bank's lending activities could provide the general policy background for a successful Inspection Panel claim.

---

[57] Note that the *Pollution Handbook* also states, "Studies by the World Bank point out the insufficiency of energy efficiency measures alone...." *Pollution Handbook, supra* note 35, at 173.
[58] *Pollution Handbook, supra* note 35, at 275 (cement) and 413 (new thermal power plants).
[59] World Bank, Operational Manual Statement 2.36: Environmental Aspects of Bank Work (1984).
[60] *Pollution Handbook, supra* note 35, at 170. Note, however, that the *Handbook* continues: "It is universally recognized that the energy needs of developing countries are enormous, that increased energy consumption and economic growth will be essential if the living standards are to be raised, that without accelerated development in many countries domestic environmental degradation will worsen, and that the current threat from anthropogenic climate change is caused much more by affluent countries than by the poorer nations. For all these reasons, the Convention and the Protocol make it clear that continued growth of energy and use of fossil fuels in developing countries is consistent with the stipulations of the Convention and the Protocol. But guidance from the parties as to when and how such growth must be moderated in order to maintain this consistency will only evolve over time." *Id.*

2.2.5. Other Potential Policy Violations

In addition to the core issues of environmental assessment, other potential policy concerns may arise in the context of projects affecting climate change. Consultation and access to information, for example, are particularly important to affected peoples. The EA Policy requires consultation with affected groups at least twice during project preparation as well as throughout project implementation.[61] In practice, many of the Bank-sponsored consultation processes are flawed and consultation failures have formed the basis of many claims to the Inspection Panel. Climate-related impacts on indigenous peoples that involve involuntary resettlement or that could potentially harm cultural property or biodiversity might violate related Bank policies.

2.3. *Conclusions Regarding Inspection Panel Claims*

Given its preeminent role in financing and shaping the development paths of its developing country clients, the World Bank will be a critical player in shaping the world's response to climate change. The Inspection Panel offers a significant opportunity for climate advocates to raise the profile of climate-related issues at the World Bank. Climate-based claims can be addressed by the Panel, at least if they are brought by project-affected communities and linked to violations of the Bank's policies. As discussed further in the conclusion, for Panel claims to be successful in influencing Bank policy or mitigating climate impacts from specific projects, claimants will need to complement any Panel-based strategy with media outreach and direct lobbying of the Bank's top management and Executive Directors.

3. THE IFC/MIGA COMPLIANCE ADVISOR AND OMBUDSMAN

The IFC and MIGA are the private sector arms of the World Bank Group.[62] The IFC provides loans to private companies conducting projects in developing countries, and MIGA provides guarantees against civil war, government expropriation, or other political risks. Both IFC lending and MIGA risk guarantees can be critical for leveraging additional private sector capital in developing country projects. The IFC and MIGA are separate, independent organizations within the World Bank Group. Each has its own articles of agreement and bylaws, but they share almost identical Boards of Executive Directors. The president of the World Bank is the top management official for IFC and MIGA as well.

IFC and MIGA are the fastest-growing parts of the World Bank Group and both regularly support environmentally controversial projects.[63] IFC and MIGA support

---

[61] World Bank OP 4.01, *supra* note 31, at para. 14.
[62] The IFC makes loans and equity investments in private-sector projects. The MIGA provides insurance against political risks faced by private sector investments in developing countries (i.e., risks from civil unrest or war).
[63] *See, e.g.*, Pangue Audit Team, *Pangue Hydroelectric Project (Chile): An Independent Review of the International Finance Corporation's Compliance with Applicable World Bank Group Environmental*

for the fossil fuel industry has also grown in recent years with, for example, high-profile support to the Baku-Tbilisi-Ceyhan and Chad-Cameroon pipeline projects. As their role increases, so too does the need to ensure that their policies reflect climate concerns. The IFC and MIGA have developed environmental and social policies in recent years, but neither institution is covered by the World Bank Inspection Panel. Instead, in 1999, then World Bank President James Wolfensohn established a new office of the Compliance Advisor and Ombudsman (CAO) to address environmental and social issues related to IFC- and MIGA-supported projects. The following sections address the CAO and the relevant environmental and social policies of IFC and MIGA.

3.1. Introducing the CAO

The CAO provides people affected by IFC or MIGA projects an opportunity to seek relief, either through dispute resolution or through a compliance review. Unlike the Inspection Panel (which reports directly to the Board of Directors), the CAO reports directly to the Bank President. As set forth in its operational guidelines, the CAO has three distinct roles:

[1] Responding to complaints by persons who are affected by projects and attempting to resolve the issues raised using a flexible, problem solving approach (the Ombudsman role);

[2] Providing a source of independent advice to the President and the management of IFC and MIGA. The CAO will provide advice both in relation to particular projects and in relation to broader environmental and social policies, guidelines, procedures, resources and systems (the Advisory role); and

[3] Overseeing audits of IFC's and MIGA's social and environmental performance, both overall and in relation to sensitive projects, to ensure compliance with policies, guidelines, procedures, and systems (the Compliance role).[64]

At least until recently, the CAO has considered its ombudsman function to be its primary and most important responsibility. The ombudsman function attempts to use a "flexible, problem-solving approach"[65] to resolve the issues raised by affected people. According to the CAO Guidelines, the "aim is to identify problems, recommend practical remedial action and address systemic issues that have contributed to the problems, rather than to find fault." The CAO thus enjoys a proactive and flexible mandate aimed primarily at solving problems.

---

and Social Requirements (World Bank, 1997). For a summary of the project and its role in the development of IFC's environmental policies, see David Hunter, Cristian Opaso & Marcos Orellana, *The Biobio's Legacy: Institutional Reforms and Unfulfilled Promises at the International Finance Corporation*, in DEMANDING ACCOUNTABILITY: CIVIL SOCIETY CLAIMS AND THE WORLD BANK INSPECTION PANEL (Dana Clark, Jonathan Fox & Kay Treakle eds., 2003).

[64] Compliance Advisor and Ombudsman, Operational Guidelines, at 7 (Apr. 2000).
[65] *Id.*

3.2. Filing a Petition

Any party affected or likely to be affected by the social or environmental impacts of an IFC or MIGA project may make a complaint to the ombudsman. Representatives of an affected party may also file a complaint, with appropriate proof of the representation. Complaints must be in writing and must relate to the planning, implementation, or impact of an IFC or MIGA project. The CAO will reject complaints that are "malicious, trivial or which have been generated to gain competitive advantage."[66]

Once the CAO accepts a complaint, the CAO immediately notifies the complainant, registers the complaint, informs the project sponsor, and requests relevant information from the appropriate IFC or MIGA personnel. Management has twenty working days to respond to the request for information. The CAO then undertakes a preliminary assessment to determine how it proposes to handle the complaint, culminating in a specific proposal to the claimant of how the CAO proposes to address their complaint.

The CAO's proposal may include anything from convening informal consultations with IFC, MIGA, or the project sponsor, to organizing a more formal mediation process. Overall, the ombudsman's office seeks to take a proactive and flexible approach where the "aim is to identify problems, recommend practical remedial action and address systemic issues that have contributed to the problems, rather than to find fault."[67]

For cases that raise issues of compliance with IFC or MIGA policies, the CAO will typically include a compliance audit as one potential way forward. Moreover, under the CAO's new proposed rules, if any party (including the claimant) does not wish to continue the problem-solving activities recommended by the ombudsman's office, then the CAO automatically transfers the complaint to the compliance side for a formal compliance audit based on any allegations of noncompliance in the complaint. In carrying out a compliance audit, the CAO has broad investigatory powers, including authority to review IFC or MIGA files; meet with the affected people, IFC or MIGA staff, project sponsors, and host country government officials; conduct project site visits; hold public meetings in the project area; request written submissions from any source; and engage expert consultants to research or address specific issues.[68]

As of 2006, the CAO had received more than fifty-five claims, twenty of which are related to one project: the Baku-Tbilisi-Ceyhan pipeline.[69] Some of these claims have resulted in long and complex involvement by the CAO, for example complaints relating to the Newmont Corporation's Yanacocha gold mine in Peru, and others

---

[66] *Id.* at 17.
[67] *Id.* at 13.
[68] *Id.* at 22–23.
[69] *Retrospective Analysis of CAO Interventions, Trends, Outcomes and Effectiveness* 2006 (summary) available at http://www.cao-ombudsman.org/html-english/aboutretrospectiveanalysis.htm.

have involved relatively limited interventions, for example a case involving one family's inadequate compensation for land lost to the construction of Chile's Pangue Dam. No claim to the CAO has involved climate change, but in a recent challenge to two pulp mills on the border of Uruguay and Argentina, the CAO concluded that the IFC had been deficient in its due diligence with respect to cumulative emissions.[70] Although disputed (and largely ignored by IFC management), representatives of the local community used the CAO findings to raise concerns with other potential financiers of the project, and ultimately one of the two projects was withdrawn.[71] The case illustrates both that the CAO may be used to evaluate issues involving cumulative emissions of greenhouse gases and that the CAO findings may be useful for raising climate-related concerns in other fora.

3.3. IFC Policies Related to Climate Change

In 2006, IFC completely revamped its environmental and social policies, departing substantially from the World Bank's operational policies described previously. The new IFC policy framework includes a "Policy on Social and Environmental Sustainability" that applies to IFC's review and environmental evaluation of a project as a lender and eight "Performance Standards" that apply to the private-sector borrowers who are primarily responsible for implementing the project.

3.3.1. Assessing Climate Impacts

Under the new Performance Standards, all borrowers must "establish and maintain a Social and Environmental Management System appropriate to the nature and scale of the project and commensurate with the level of social and environmental risks and impacts."[72] Among other requirements, the Management System must include a social and environmental assessment that considers "all relevant social and environmental risks and impacts of the project,"[73] including global climate impacts.

The Standards require the risks and impacts to be "analyzed in the context of the project's area of influence,"[74] which apparently includes the project's climate

---

[70] Compliance Advisor/Ombudsman, *CAO Audit of IFC's and MIGA's Due Diligence for Two Pulp Mills in Uruguay: Final Report*, § 6.1 (Feb. 22, 2006).

[71] For a complete description of the legal and other steps taken in opposition to the pulp mills, see the website of Argentina's Center for Human Rights and the Environment, *available at* http://www.cedha.org.ar/en/initiatives/paper_pulp_mills/.

[72] International Finance Corporation, Performance Standards on Social and Environmental Sustainability, Performance Standard No. 1: Social and Environmental Management and Assessment Systems, § 3 (Apr. 30, 2006) [hereinafter Performance Standard No. 1].

[73] *Id.* at § 4.

[74] *Id.* at § 5. While many projects may only impact the specific project site, other projects will have a much broader area of influence. According to the Standard, the area of influence should include "areas potentially impacted by cumulative impacts from further planned development of the project... and... areas potentially affected by impacts from unplanned but predictable developments caused by the project that may occur later or at a different location." Impacts associated with supply chains must also be analyzed under certain specific circumstances. *Id.* at § 6.

impacts. The Standards explicitly state that the assessment will "consider potential transboundary effects... as well as global impacts, such as the emission of greenhouse gasses."[75] The IFC recognizes that:

> While individual project impacts on climate change, ozone layer, biodiversity or similar environmental issues may not be significant, when taken together with impacts created by other human activities, they can become nationally, regionally or globally significant. When a project has the potential for large scale impacts that can contribute toward adverse global environmental impacts, the Assessment should consider these impacts.[76]

In addition to impacts from direct emissions, the IFC appears to require project sponsors to assess the indirect impacts on climate such as the emissions resulting at other locations associated with electricity consumed by the borrower. Moreover, borrowers must also assess significant impacts on ecosystem services, including carbon sequestration and other functions that may influence climate.[77]

3.3.2. Evaluating Alternatives

Under Performance Standard 3, the IFC requires the borrower to

> evaluate technically and financially feasible and cost-effective options to reduce or offset project-related GHG emissions during the design and operation of the project. These options may include, but are not limited to, carbon financing, energy efficiency improvement, the use of renewable energy sources, alterations of project design, emissions offsets, and the adoption of other mitigation measures such as the reduction of fugitive emissions and the reduction of gas flaring.[78]

The requirement to assess alternatives is a core component of the IFC's environmental assessment system,[79] and the explicit reference to such a broad range of options with respect to climate change will likely be critical for future climate claims.

3.3.3. Requirements to Mitigate Climate Impacts

In most projects with environmental impacts, the borrower is required to prepare and implement an action plan to ensure that the project meets the Performance Standards over time. Under the Performance Standards, the borrower must take actions

---

[75] *Id.* at § 6.
[76] International Finance Corporation, Guidance Note 1: Environmental and Social Management Systems, at para. 19 (Apr. 30, 2006).
[77] International Finance Corporation, Guidance Note 6: Biodiversity Conservation and Sustainable Natural Resource Management, at para. G4 (Apr. 30, 2006).
[78] International Finance Corporation, Performance Standards on Social & Environmental Sustainability, Performance Standard 3: Pollution Prevention and Abatement, at para. 11 (April 30, 2006) [hereinafter Performance Standard 3].
[79] Performance Standard 1, *supra* note 72, at para. 9.

to mitigate any environmental and social impacts. Such actions should "favor the avoidance and prevention of impacts over minimization, mitigation, or compensation, wherever technically and financially feasible."[80] Where risks and impacts cannot be avoided or prevented, the projects must nonetheless take measures to ensure compliance with applicable laws and regulations, and meet the requirements of the Performance Standards.[81] Thus, all borrowers are arguably required to identify their project's impact on climate and take steps to minimize such impacts.

More specifically, Performance Standard 3 on Pollution Prevention and Abatement requires clients to "promote the reduction of project-related greenhouse gas emissions in a manner appropriate to the nature and scale of project operations and impacts."[82] Borrowers are also supposed to apply "good international industry practice" in preventing and controlling pollution. This is defined in part by reference to the IFC's Environmental, Health and Safety (EHS Guidelines), the IFC analogue to the World Bank *Pollution Handbook* discussed earlier.[83] The standards found in the EHS Guidelines contain the "performance levels and measures that are normally acceptable."[84] Different sectors may also be subject to specific additional requirements. For example, clients in the oil and gas sector must reduce flaring and venting of gas associated with the extraction of crude oil.[85] Any proposed deviation from these standards must be justified and explained in light of their anticipated environmental and human health impacts.[86] Taken collectively, these Standards should allow potential claimants to use IFC's policies to push for more efficient and climate-friendly technologies.

3.3.4. Measuring and Monitoring GHG Emissions

For projects that emit significant levels of GHGs, the borrower is required to "quantify direct emissions from the facilities owned or controlled within the physical project boundary and indirect emissions associated with the off-site production of power used by the project. Quantification and monitoring of GHG emissions will be conducted annually in accordance with internationally recognized methodologies."[87] The IFC defines "significant" to be more than 100,000 tons of CO_2 equivalent per year for aggregate emissions from direct sources and indirect sources associated with electricity consumption.[88]

---

[80] Performance Standard 1, *supra* note 72, at para. 14.
[81] Performance Standard 1, *supra* note 72, at paras. 13–14.
[82] Performance Standard 3, *supra* note 78, at para. 10.
[83] *See* text accompanying note 54.
[84] Performance Standard 3, *supra* note 78, para. 8.
[85] International Finance Corporation, Guidance Note 3: Pollution Prevention and Abatement, para. G.33 (Apr. 30, 2006).
[86] *Id.* at para. G.24.
[87] Performance Standard 3, *supra* note 78, at para. 10.
[88] *Id.*

3.3.5. Compliance with the UNFCCC and Kyoto Protocol

IFC requires the borrower to "comply with applicable national laws, including those laws implementing host country obligations under international law."[89] Such laws and regulations must also be considered as part of the assessment process.[90] This Standard could be used to raise issues related to a country's domestic and international obligations to measure, report, or reduce greenhouse gas emissions over time. As discussed with respect to the World Bank, most countries eligible for IFC-financed projects do not yet have international obligations to reduce their greenhouse gas emissions, but some may accept such commitments as part of future climate negotiations. Such future commitments could form the basis of CAO petitions.

3.3.6. Other Requirements

Other IFC Performance Standards also might be important for climate-related claims. As in the World Bank context, issues regarding consultation and access to information are frequently the subject of claims by affected peoples.[91] Moreover, climate-related impacts on indigenous peoples, involuntary resettlement, cultural property, and biodiversity would potentially violate Performance Standards relating to each of those issues.[92]

3.4. Conclusions Relating to IFC's Compliance Advisor/Ombudsman

The IFC's private-sector lending is the fastest-growing part of the World Bank, and the IFC is the recognized standard setter for private-sector sustainable finance in developing countries. IFC's new Performance Standards are shaping the environmental approaches at both export credit agencies and private commercial banks. As such, ensuring compliance with IFC's environmental standards will have far-reaching implications for major private-sector projects around the world. The CAO provides an opportunity for local affected communities and their civil society allies to raise not only localized, short-term project impacts but also long-term, cumulative impacts relating to climate change. Although there is little experience thus far in bringing climate-related claims, the new IFC Performance Standards for the first time explicitly require climate-related emissions to be quantified and impacts assessed. In the future, these provisions could form the basis for project-specific claims or for raising more general concerns under the CAO's advisory function.

---

[89] International Finance Corporation, Performance Standards on Social & Environmental Sustainability, Introduction, at para. 3 (Apr. 30, 2006).
[90] Performance Standard 1, *supra* note 72, at para. 4.
[91] *See id.* at paras. 19–22.
[92] *Id.* at paras. 19 & 21–22.

4. MECHANISMS AT OTHER FINANCIAL INSTITUTIONS

In addition to the Inspection Panel and CAO, accountability mechanisms exist in four other MDBs and three bilateral export credit or insurance agencies. The mechanisms and the dates they were created are listed in the table here, as well as whether they are primarily compliance review mechanisms like the Inspection Panel, dispute settlement mechanisms like the CAO's ombudsman function, or both.

Financial institution	Mechanism name (Date)	Compliance review	Dispute resolution
African Development Bank	Independent Review Mechanism (2004)[93]	X	X
Asian Development Bank	Accountability Mechanism (1995/2003)[94]	X	X
European Bank for Reconstruction and Development	Independent Review Mechanism[95] (2003)	X	
InterAmerican Development Bank	Independent Investigation Mechanism[96] (1994)	X	
Export Development Canada	Compliance Officer[97] (2002)	X	X
Japan Bank of International Cooperation	Office of Examiners (2003)[98]	X	
U.S. Overseas Private Investment Corporation (OPIC)	Office of Accountability (2004)[99]	X	X

---

[93] African Development Bank Board of Directors, Resolution B/BD/2004/9 (June 30, 2004).
[94] The Asian Development Bank adopted its original Inspection Function in 1995. In May 2003 it adopted a completely revised Inspection Function that includes a Special Project Facilitator (SPF) and a Compliance Review Panel (CRP). *Review of the Inspection Function: Establishment of a New ADB Accountability Mechanism*, ADB Board Paper, May 8, 2003.
[95] European Bank for Reconstruction and Development (EBRD), Independent Recourse Mechanism (Apr. 29, 2003). The EBRD recently solicited public comments on a significantly improved mechanism, which is expected to be adopted in 2002.
[96] *See* Inter-American Development Bank, *The IDB Independent Investigation Mechanism, Rules and Procedures*, para. 1.1. Washington, D.C. (2000). The IDB has drafted a considerably improved mechanism, but has failed to adopt it for over a year. Information on the IDB Mechanism is available at http://www.iadb.org/cont/poli/investig/brochure.htm.
[97] Press Release, Export Development Canada, Apr. 10, 2002, Information on the EDC compliance officer is available at http://www.edc.ca/corpinfo/csr/compliance_officer/ index_e.htm.
[98] Japan Bank of International Cooperation, Major Rules for Establishment of Examiner for Environmental Guidelines, issued on May 1, 2003, *available at* http://www.jbic.go.jp/english/environ/examiner/index.php (last visited Jan. 26, 2007).
[99] For a description of OPIC's Office of Accountability, see http://www.opic.gov/doingbusiness/accountability/; *see also* Harvey A. Himberg, *The New Accountability and Advisory Mechanism of the Overseas Private Investment Corporation: The Application of International Best Practices of International Financial Institutions* in PROCEEDINGS FROM THE SEVENTH INTERNATIONAL CONFERENCE ON ENVIRONMENTAL COMPLIANCE AND ENFORCEMENT, 307–16 (Durwood Zaelke et al. eds., 2005).

Significant differences do exist among the mechanisms. Some of them, such as Canada's EDC Compliance Officer, can explicitly address human rights. Others, such as the Asian Development Bank's Accountability Mechanism Inspection Function have the explicit authority to monitor implementation of any recommendations. The scope of the environmental and social policies at the financial institutions also can vary considerably, thus affecting the scope of the compliance review process of the mechanisms.

Nonetheless, all of these mechanisms have certain critical features in common: (1) they are designed to serve in compliance review or problem-solving roles; (2) they can be accessed directly by project-affected people or their civil society representatives, typically with a simple letter; (3) they only can review the respective IFI's activities in light of its own environmental and social policies; and (4) their remedies include public reporting and, in some cases, using the IFI's own leverage to force change in the project.

The experience with these mechanisms is far more limited than with the World Bank Inspection Panel and CAO. In part, this reflects a lack of awareness about the mechanisms, but it also reflects concerns among potential claimants that some mechanisms may lack independence and authority to make a difference on the ground. In fact, open questions of effectiveness remain about most of the mechanisms other than the Inspection Panel, and potential claimants should recognize that their claims will often be breaking new ground and in some cases testing the ultimate effectiveness of the mechanisms. Moreover, all of the mechanisms lack traditional enforcement powers (that, for example, one would find in a court). This means, as described further next, that these claims should be integrated into a broader advocacy effort that involves multiple strategies and approaches.

5. CONCLUSION

The IFI accountability mechanisms offer potentially useful opportunities for raising climate-related concerns regarding projects financed by the World Bank or other IFIs. To be sure, these mechanisms do not have the muscular remedies available to courts. They do not have powers to compel actions by the IFIs or the project sponsors. Theirs is mostly a power to make recommendations. The mechanisms also rarely result in projects being enjoined, even while the investigations are under way. And none of the mechanisms have the power to order compensation for injuries, although at times claimants have received some payments as a result of the investigations.

On the other hand, the IFI mechanisms do offer one of the few ways local affected people can raise their concerns at the international level. Generally speaking, about half of the claims filed using IFI mechanisms result in some positive gains for the affected communities.[100] The power of the mechanisms resides mostly in their ability to shine a public spotlight on non-compliance, to bring the IFI or the borrower to the table for meaningful dispute resolution, or to raise project concerns to the political

---

[100] See Clark et al., *supra* note 63.

level of the IFIs. Ultimate decision-making authority, however, remains with the IFI's top management or board of directors. Sometimes, corrective decisions are taken, but neither the claimants nor the accountability mechanisms have any recourse if the IFIs ignore the findings and recommendations.

To be successful in changing a specific project or IFI policy more generally, claims to the accountability mechanisms need to be part of a larger set of strategies. Thus, the ability to run effective, multifaceted campaigns around the submissions to the IFI accountability mechanisms is critical. Such campaigns should include both media outreach and a coordinated political strategy aimed at lobbying the IFI's executive directors or top decisionmakers. In this context, the accountability mechanism's findings can validate civil society concerns and provide legitimacy for claims brought to the political level of the IFIs. By raising the political stakes for the top decisionmakers, such public campaigning can turn Panel recommendations into political action or create the leverage needed to force concessions during mediation.

More specifically with respect to climate change, claims brought through IFI accountability mechanisms when part of a broader campaign could be used to accomplish several goals. First, claims could be successful in getting particular projects to mitigate their anticipated climate impacts. The recent CAO claim regarding the Uruguayan pulp mills provides an example of how the mechanism's findings were used in a variety of forums to stop one of the two projects. In that case, IFC's initial failure to assess cumulative impacts or to conduct proper due diligence over the assessment fueled criticism of the projects.[101] Climate-related claims offer an opportunity to enlist the accountability mechanisms in a closer scrutiny of climate impacts, which in some projects would likely result in improved project efficiency, fuel switching, or other specific steps that could reduce climate impacts.

Beyond the impacts in the particular projects, climate-related claims can also have broader implications. IFIs are leaders in setting policies for project finance in developing countries. By highlighting emissions resulting from specific projects, they can raise the awareness of the climate impacts of the IFI's lending portfolios within a country, regionally, or globally. Pressure at the project level has always been the primary, if not only, way to make broader changes at the IFIs. Past project claims have led to significant policy restructuring at the IFIs. For example, the filing of the Pangue claim to the Inspection Panel led to the first clear adoption of environmental and social policies at the IFC.[102] Similarly, findings from a Panel claim relating to an agricultural development project in China led directly to a restructuring of the

---

[101] See supra notes 72–73 and accompanying text.
[102] The Pangue claim was ruled ineligible by the Inspection Panel (because it related to an IFC project), but the President of the World Bank forwarded the claim to an independent review panel patterned after the Inspection Panel. See Pangue Audit Team, *Pangue Hydroelectric Project (Chile): An Independent Review of the International Finance Corporation's Compliance with Applicable World Bank Group Environmental and Social Requirements* (World Bank, 1997); see also David Hunter, Cristian Opaso & Marcos Orellana, *The Biobio's Legacy: Institutional Reforms and Unfulfilled Promises at the International Finance Corporation*, in Dana Clark et al., supra note 63, at 115–43 (discussing the impact of the panel investigation on IFC's policies).

World Bank's approach to compliance.[103] A focus on climate impacts at the project level could lead to significant improvements in the policies and practices of the IFIs, including stronger policies on assessing climate impacts, on evaluating and financing mitigation steps, and on shifting their portfolios toward increased renewables, energy efficiency, and other low-carbon technologies.

These broad global policy implications have to be balanced with the expected impacts at the local level. Because of the high-profile nature of IFI-financed projects in many countries, the filing of claims to an IFI mechanism can offer significant media opportunities to press for related changes within the national and local context. They can also bring scrutiny and negative, sometimes dangerous, pressures on the local claimants, who will have to face any local fallout from their claim. Those hoping to use the accountability mechanisms to raise climate change impacts must put the needs and goals of the local claimants first. In this way, the IFI mechanisms can serve their function as a validator for local communities who raise often unpopular concerns over potential IFI-financed projects, and, in the process, as a megaphone for those local communities who want to draw attention to the impacts of project-finance on long-term climate change.

[103] See Dana Clark & Kay Treakle, *The China Western Poverty Reduction Project*, in Dana Clark et al., *supra* note 63, at 211, 235–36.

14

Potential Causes of Action for Climate Change Impacts under the United Nations Fish Stocks Agreement

William C. G. Burns*

The seas – all the seas – cry for regulation as a veritable *res communis omnium*.[1]

INTRODUCTION

This chapter examines another potential international forum in which the threat of climate change might be addressed, the Agreement for the Implementation of the Provisions of the U.N. Convention on the Law of the Sea 10 Dec. 1982 Relating to the Conservation and Management of Straddling Fish Stocks and High Migratory Fish Stocks ("UNFSA").[2] Actions under UNFSA could be salutary for several reasons. First, as outlined hereafter, the commercial fisheries sector may be profoundly and adversely affected by climate change.[3] This includes many fish stocks regulated under UNFSA: highly migratory species, which have wide geographic distribution and undertake significant migrations,[4] and straddling stocks, which occur both within and beyond Exclusive Economic Zones (EEZs).[5] Overall, "[m]igratory and

---

* Class of '46 Visiting Professor, Center for Environmental Studies, Williams College, Williamstown, Massachusetts, wburns@williams.edu, 650-281-9126.

[1] Louis Henkin, *Arctic Anti-Pollution: Does Canada Make – or Break – International Law?* 65 AM. J. INT'L L. 131, 136 (1971).

[2] Aug. 4, 1994, U.N. Doc. A/CONF.164/37.

[3] *See* sec. 2, *infra*.

[4] Pacific Fishery Management Council, *Background: Highly Migratory Species*, <http://www.pcouncil.org/hms/hmsback.html>, site visited on Dec. 26, 2006. Highly migratory species include many species of tuna and tuna-like species, oceanic sharks, mackerel, sauries, pomfrets, swordfish, marlin, and sailfish. S.M. Garcia, *World Review of Highly Migratory Species and Straddling Stocks*, UN Food and Agriculture Organization, FAO Fisheries Technical Paper No. 337 (1994), <http://www.fao.org/docrep/003/T3740E/T3740E00.HTM>, site visited on Dec. 26, 2006; NOAA Fisheries, Office of Sustainable Fisheries, *Highly Migratory Species*, <http://www.nmfs.noaa.gov/sfa/hms/>, site visited on Dec. 26, 2006.

[5] Garcia, *supra* note 4, at 4. Overall, about 200 species have been identified as highly migratory species or straddling stocks species. FAO, *The State of World Highly Migratory Straddling and Other High Seas Fishery Resources and Associated Species* 2 (2006), FAO Fisheries Technical Paper No. 495, <ftp://ftp.fao.org/docrep/fao/009/a0653e/A0653E01.pdf>, site visited on Dec. 26, 2006. "Most typically, such stocks frequent the localized edges of wide continental shelves, e.g., the "Flemish

straddling species account for roughly 20% of the total marine catch and include some of the most economically valuable fish populations."[6]

Second, the United States, one of the world's largest emitter of greenhouse gases[7] and a State with an abject record in addressing climate change, was one of the first nations to ratify UNFSA[8] and has played an active leadership role in its implementation.[9] UNFSA thus presents an excellent forum in which to engage the United States, as well as other major greenhouse gas emitters, including the European Union and China, on climate issues. Finally, unlike the other international fora where climate change actions have been pursued to date, UNFSA provides a dispute resolution mechanism with teeth.[10]

A relatively brief chapter necessarily cannot discuss all of the intricate scientific and legal issues that an action of this nature would invoke; rather, it seeks to lay a foundation for further research and discussion. In this pursuit I will (1) outline the potential impacts of climate change on fish species, with an emphasis on the potential impacts of climate change on highly migratory fish species and straddling stocks; (2) provide an overview of UNFSA and potential actions for climate change damages under the Agreement; and (3) briefly discuss potential barriers to such actions.

Cap" in the northwest Atlantic, or the continental slopes..." Jamison E. Colburn, *Turbot Wars, Straddling Stocks, Regime Theory, and a New U.N. Agreement*, 6 J. TRANSNAT'L L. & POL'Y 323, 327 (1997).

[6] W.M. von Zharen, *The Shrinking Sea and Expanding Sovereignty: The Fate of Fisheries*, 15 NAT. RESOURCES & ENV'T 24, 26 (2000).

[7] In 2006, China's greenhouse gas emissions surpassed those of the United States, Netherlands Environmental Assessment Agency, *Chinese CO2 in Perspective*, Press Release, 22 June 2007, <http://www.mnp.nl/en/service/pressreleases/2007/20070622ChineseCO2emissionsinperspective.html>, site visited on June 24, 2007. However, the United States is still responsible for approximately a quarter of the world's cumulative greenhouse gas emissions over the past century. Kevin A. Baumert & Nancy Kete, *Climate Issue Brief*, World Resources Institute (2001), at 1. Additionally, U.S. per capita emissions are approximately 10 times those of China. Id. at 2.

[8] Note, *Fisheries: United States Ratifies Agreement on Highly Migratory and Straddling Stocks*, 1996 COLO. J. INT'L ENVTL. L. & POL'Y 78, 80 (1996).

[9] David A. Balton & Holly R. Koehler, *Reviewing the United Nations Fish Stocks Treaty*, SUSTAINABLE DEV. L. & POL'Y 5, 5–6 (2006), <http://www.wcl.american.edu/org/sustainabledevelopment/2006/06fall.pdf?rd=1>, site visited on Dec. 29, 2006.

[10] See Sec. 3.2, *infra*. By contrast, under the American Convention on the Rights of Man, which is invoked in the Inuit's petition to the Inter-American Commission on Human Rights, the Inter-American Commission's only recourse should it find the United States to have violated the human rights of the Inuit is to issue a report outlining conclusions and nonbinding recommendations. Because the United States is not a member of the Inter-American Court of Human Rights, the Commission cannot refer the case to the Court for a binding decision. Inter-American Commission on Human Rights, *What Is the IACHR?*, <http://www.cidh.org/what.htm>, site visited on Dec. 28, 2006. Similarly, even if the World Heritage Convention were to list World Heritage sites threatened by climate change on its "in danger" list in the future, this would trigger little more than the potential for financial assistance to address the threats under the Convention. World Heritage Convention, Convention Concerning the Protection of the World Cultural and Natural Heritage, Nov. 16, 1972, 27 U.S.T. 37, 1037 U.N.T.S. 151, (entered into force Dec. 17, 1975), at art. 11(4).

1. THE POTENTIAL IMPACTS OF CLIMATE CHANGE ON FISH SPECIES

As Hannesson recently concluded: "The fisheries are even more dependent than agriculture on climatic conditions. While agriculture does up to a point compensate for the shortcomings of nature... the fisheries, which essentially are an advanced form of hunting, are totally dependent on what nature will or will not provide."[11]

Fish species are ectothermic (cold blooded); thus, water temperature is the primary source of environmental impact on fish, including growth and maturity rates, distribution and migration patterns, and incidence of disease.[12] Substantially rising oceanic temperatures throughout this century will likely have negative impacts on highly migratory and straddling stocks species in many regions, especially those near the edge of their temperature tolerance range.[13] For example, the range of colder-water fish species, such as capelin, polar cod, and Greenland halibut, is likely to shrink, resulting in a decline in abundance.[14] A decline in nutrient upwelling as a consequence of increased stratification between warmer surface waters and colder deep water in warming oceans could also result in a decline in bigeye and yellowfin tuna in the central and western Pacific.[15] Tuna species are a particularly important, and dependable, source of revenue for Pacific small island States.[16]

Warming oceans could also radically change the distribution of some straddling stock and high migratory species. For example, rising ocean temperatures could result in a shift of the distribution of herring northward, upsetting a delicate agreement in the Northeast between coastal States who harvest herring within their EEZs and distant water fishing nations (DWFNs)[17] that fish on the high seas.[18] Similarly, shifts in the distribution of cod and haddock in the Barents Sea may necessitate

---

[11] Rögnvaldur Hannesson, *Introduction*, 31, 1, 1 (2007).
[12] William E. Schrank, *The ACIA, Climate Change and Fisheries*, 31 MARINE POL'Y 5, 12 (2007); G.A. Rose, *On Distributional Responses of North Atlantic Fish to Climate Change*, 62 ICES J. MARINE SCI. 1360, 1360 (2005).
[13] *See generally* European Science Foundation, *Impacts of Climate Change on the European Marine and Coastal Environment* (2007), <http://www.vliz.be/docs/Events/JCD/MB_Climate_Change_VLIZ_05031.pdf>, site visited on Apr. 19, 2007.
[14] *Id.* at 12; Robin A. Clark, et al., *North Sea Cod and Climate Change – Modelling the Effects of Temperature on Population Dynamics*, 9 GLOBAL CHANGE BIOLOGY 1669, 1677 (2003).
[15] World Bank, *Cites, Seas and Storms* 27 (2004), <http://siteresources.worldbank.org/INTPACIFICISLANDS/Resources/4-Chapter+5.pdf>, site visited on Dec. 31, 2006.
[16] Emily E. Larocque, *The Convention on the Conservation and Management of Highly Migratory Fish Stocks in the Western and Central Pacific Ocean: Can Tuna Promote Development of Pacific Island Nations?*, 4 ASIAN-PAC. L. & POL'Y J. 82, 87 (2003).
[17] "DWFNs are landlocked states and states that have the fleet capacity to fish distant regions." Julie R. Mack, *International Fisheries Management: How the U.N. Conference on Straddling and Highly Migratory Fish Stocks Changes the Law of Fishing on the High Seas*, 26 CAL. W. INT'L L.J. 313, 316 (1996). "Japan, Russia, South Korea, Spain, Taiwan, and Poland account for almost ninety percent of the world's high seas fish catch." Note, *supra* note 8, at 81.
[18] Elin H. Sissener & Trond Bjørndal, *Climate Change and the Migratory Pattern for Norwegian Spring-Spawning Herring – Implications for Management*, 29 MARINE POL'Y 299, 305 (2005); Francis Neat & David Righton, *Warm Water Occupancy by North Sea Cod*, 274 PROC. ROYAL SOC'Y, BIOLOGY 789, 789 (2007).

renegotiation of existing fisheries agreements between Russia and Norway.[19] Should cooperative management agreements of this nature collapse, it might lead to "strategic overfishing" of stocks that are currently recovering from a historical decline.[20] Warming in the Pacific could similarly result in a redistribution of tuna resources to higher latitudes, such as Japan and the western equatorial Pacific.[21]

Temperature increases will also adversely affect prey species of many straddling stocks and highly migratory species. For example, in the North Atlantic, strong biogeographical shifts in copepod and plankton assemblages associated with warming trends[22] could substantially reduce the abundance of fish in the North Sea and ultimately result in the collapse of the stocks of cod, an important straddling stock species.[23] There are already disturbing portents of this, with warming in the North Sea over the past few decades resulting in key changes in planktonic assemblages, which has resulted in a poor food environment for cod larvae, adversely affecting recruitment success.[24] The decline of stocks has also increased their sensitivity to regional climate warming because of shrinkages in age distribution and geographical range.[25]

There will also be direct biological effects from rising levels of carbon dioxide entering the oceans. Atmospheric carbon dioxide levels increase at a rate of only approximately 50% of human carbon dioxide emissions because of the existence of large ocean and terrestrial sinks.[26] Over the past two centuries, the world's oceans have absorbed 525 billion tons of carbon dioxide, constituting nearly half of carbon emissions over this period.[27] Over the next millennium, it is estimated that the world's oceans will absorb 90% of anthropogenic carbon dioxide currently being released into the atmosphere.[28]

---

[19] European Science Foundation, *supra* note 13, at 23.

[20] *Id.* at 304.

[21] World Bank, *supra* note 15, at 28.

[22] Russell B. Wynn et al., *Climate-Driven Range Expansion of a Critically Endangered Top Predator in Northeast Atlantic Waters*, 3 BIOLOGY LETTERS 529, 530–31 (2007); G. Beaugrand & P.C. Redi, *Long-Term Changes in Phytoplankton, Zooplankton and Salmon Related to Climate*, 9 GLOBAL CHANGE BIOLOGY 801–817 (2003).

[23] Grégory Beaugrand et al., *Reorganization of North Atlantic Marine Copepod Biodiversity and Climate*, 296 SCI. 1692, 1693 (2002). *See also* Anthony J. Richardson & David S. Schoeman, *Climate Impact on Plankton Ecosystems in the Northeast Atlantic*, 305 SCI. 1609–1612 (2004).

[24] Institute for Environment & Sustainability, European Commission Directorate General Joint Research Centre, *Marine and Coastal Dimension of Climate Change in Europe* (2006), at 24, <http://ies.jrc.cec.eu.int/fileadmin/Documentation/Reports/Varie/cc_marine_report_optimized2.pdf>, site visited on Feb. 19, 2007.

[25] *Id.*

[26] Corinnne Le Quéré et al., *Saturation of the Southern Ocean CO_2 Sink Due to Recent Climate Change*, ScienceXpress, May 17, 2007, at 1, <http://www.scienceexpress.org>, site visited on May 27, 2007.

[27] Richard A. Feely, Christopher L. Sabine & Victoria J. Fabry, *Carbon Dioxide and Our Ocean Legacy*, NOAA, Pacific Marine Environmental Laboratory (Apr. 2006), at 1, <http://www.pmel.noaa.gov/pubs/PDF/feel2899/feel2899.pdf>, site visited on Apr. 22, 2007.

[28] J.A. Kleypas et al., *Impacts of Ocean Acidification on Coral Reefs and Other Marine Calcifiers*, Report of a Workshop Sponsored by the NSF/NOAA/UGSG (2006), at 3.

Although chemically neutral in the atmosphere, carbon dioxide in the ocean is chemically active.[29] As carbon dioxide dissolves in seawater, it reacts with water molecules (H_2O) to form a weak acid, carbonic acid (H_2CO_3), the same weak acid found in carbonated beverages. Like all acids, carbonic acid then releases hydrogen ions (H^+) into solution,[30] leaving both bicarbonate ions (HCO_3^{-1}) and, to a lesser extent, carbonate ions (CO_3^{2-}) in the solution as well.[31] The acidity of ocean waters is determined by the concentration of hydrogen ions, which is measured on the pH scale. The higher the level of hydrogen ions in a solution, the lower the pH.[32]

The increase of atmospheric concentrations of carbon dioxide since the advent of the Industrial Revolution has decreased surface pH values by 0.12 units.[33] Although this may not sound like a substantial change, the pH scale is logarithmic.[34] Thus, a 0.1 unit change in pH translates into a 30% increase in hydrogen ions.[35] The pH of the world's oceans now stands at approximately 8.2, with a variation of about ±0.3 units because of local, regional, and seasonal variations.[36] The pH unit change over the past 150 years is probably the greatest seen over the past several million years.[37]

Although increases in ocean acidification have been substantial to date,[38] far more dramatic changes are likely to occur during this century and beyond as a substantial portion of burgeoning levels of anthropogenic carbon dioxide emissions enter the world's oceans. Under a "business as usual" scenario, carbon dioxide emissions are projected to grow at 2% annually during the remainder of this century,[39] although

---

[29] R. Schubert et al., *The Future Oceans – Warming Up, Rising High, Turning Sour*, German Advisory Council on Global Change (2006), <http://www.wbgu.de/wbgu_sn2006_en.pdf>, site visited on Dec. 25, 2007, at 26.

[30] Haruko Kurihara, Shoji Kato & Atsushi Ishimatsu, *Effects of Increased Seawater pCO₂ on Early Development of the Oyster Crassostrea Gigas*, 1 AQUATIC BIOLOGY 91, 91 (2007).

[31] Scott C. Doney, *The Dangers of Ocean Acidification*, SCI. AM., Mar. 2006, at 60.

[32] Royal Society, *Ocean Acidification Due to Increasing Atmospheric Carbon Dioxide*, Policy Doc. 12/05 (2005), at 4, <http://royalsociety.org/displaypagedoc.asp?id=13539>, site visited on Dec. 25, 2007.

[33] Ulf Riebesell, *Effects of CO₂ Enrichment on Marine Phytoplankton*, 60 J. OCEANOGRAPHY 719, 719–20 (2004).

[34] Caspar Henderson, *Paradise Lost*, NEW SCI., Aug. 5, 2006, at 36.

[35] Schubert, et al., *supra* note 29, at 66.

[36] Royal Society, *supra* note 32, at 1.

[37] C. Turley et al., *Reviewing the Impact of Increased Atmospheric CO₂ on Oceanic pH and the Marine Ecosystems*, AVOIDING DANGEROUS CLIMATE CHANGE 67 (Hans Joachim Schellnhuber ed., 2006).

[38] The term "ocean acidification" was coined in 2003 by climate scientists Ken Caldeira and Michael Wickett. Elizabeth Kolbert, *The Darkening Sea*, NEW YORKER, Nov. 20, 2006, at 67, <http://equake.geos.vt.edu/acourses/3114/global_warming/061120nyek-sea.html>, site visited on Dec. 25, 2007. However, it should be emphasized that this term is a bit of a misnomer since seawater is naturally alkaline, and a neutral pH is 7. Thus, it highly unlikely that surface ocean seawater will ever actually become acidic. Y. Shirayama & H. Thornton, *Effect of Increased Atmospheric CO₂ on Shallow Water Marine Benthos*, 110 J. GEOPHYSICAL RES. 1, 1 (2005).

[39] James Hansen et al., *Climate Change and Trace Gases*, PHIL. TRANS. ROYAL SOC'Y A, 1925, 1937 (2007);

emissions have grown far more substantially in the past six years,[40] exceeding even the upper range of the projections of the Intergovernmental Panel on Climate Change (IPCC).[41] The IPCC in its *Special Report on Emissions Scenarios* projected that carbon dioxide emissions could be as high as 37 gigatons of carbon annually by 2100, with the median and mean of all scenarios being 15.5 and 17 GtC, respectively.[42] Atmospheric concentrations of carbon dioxide may reach twice preindustrial levels by as early as 2050[43] and could triple or quadruple by 2100.[44]

The "business as usual" scenario for carbon dioxide emissions during this century, in turn, is projected to result in a tripling of dissolved carbon dioxide in seawater by 2100, producing an additional decline in ocean pH by approximately 0.3 to 0.4 units.[45] Moreover, continued oceanic absorption of carbon dioxide may result in a further decline of pH levels of 0.77 units by 2300, reaching levels not seen for the past 300 million years, with the possible exception of rare, extreme events.[46] These levels will persist for thousands of years even after oceanic concentrations of carbon dioxide begin to decline.[47]

---

[40] Fossil fuel and cement emissions of carbon dioxide increased at a rate of 3.3% annually from 2006–2006, a dramatic acceleration from the rate of 1.3% annually from 1990–1999. Josep G. Canadell, et al., *Contributions to Accelerating Atmospheric CO_2 Growth from Economic Activity, Carbon Intensity, and Efficiency of Natural Sinks*, PROC. NAT'L ACAD. SCI. EARLY EDITION, 10.1073 (2007), at 5. The increasing growth rate of carbon dioxide emissions is attributable to increased economic growth, an increase in carbon dioxide emissions required to produce each additional unit of economic activity, and decreasing efficiency of carbon sinks on land and the oceans. *Id.* at 3.

[41] Michael R. Raupach, et al., *Global and Regional Drivers of Accelerating CO_2 Emissions*, 104(24) PROC. NAT. ACAD. SCI., 10,288, 10,289 (2007).

[42] IPCC, NEBOJSA NAKICENOVIC ET AL., SPECIAL REPORT ON EMISSIONS SCENARIOS (2000), <http://www.ipcc.ch/ipccreports/sres/emission/116.htm>, site visited on Dec. 30, 2007.

[43] James E. Hansen, *Dangerous Human-Made Interference with Climate*, Testimony to the Select Committee on Energy Independence and Global Warming, U.S. House of Representatives, Apr. 26, 2007, <http://www.columbia.edu/~jeh1/testimony_26april2007.pdf>, site visited on Dec. 30, 2007, at 4.

[44] David Talbot, *The Dirty Secret*, TECH. REV. (July/Aug. 2006), <http://www.technologyreview.com/Energy/17054/>, site visited on Dec. 30, 2007; Richard A. Feely et al., *Impact of Anthropogenic CO_2 on the $CaCO_3$ System in the Oceans*, 305 SCI. 362, 362 (2004); Stephen F. Lincoln, *Fossil Fuels in the 21st Century*, 34(8) AMBIO 621, 621 (2005).

[45] G.A. Meehl et al., *Global Climate Projections*, in CLIMATE CHANGE 2007: THE PHYSICAL SCIENCE BASIS. CONTRIBUTION OF WORKING GROUP I TO THE FOURTH ASSESSMENT REPORT OF THE INTERGOVERNMENTAL PANEL ON CLIMATE CHANGE 750 (2007), <http://www.ipcc.ch/pdf/assessment-report/ar4/wg1/ar4-wg1-chapter10.pd>, site visited on Dec. 30, 2007; Björn Rost & Ulf Riebesell, *Coccolithophores and the Biological Pump: Responses to Environmental Changes*, in COCCOLITHOPHORES: FROM MOLECULAR PROCESSES TO GLOBAL IMPACTS 116 (Hans R. Thierstein & Jeremy R. Young eds., 2004), <http://books.google.com/books?id=oIAVyi_GaoAC&pg=PA119&lpg=PA119&dq=coccolithophores+acidification&source=web&ots=4433-fDIg7&sig=LNpJCjhmNa1vgVYEzILe6W-oeiM#PPA99,M1>, site visited on Dec. 30, 2007; Henry Elderfield, *Carbonate Mysteries*, 296 SCI. 1618, 1619 (2002).

[46] J.C. Blackford et al., *Regional Scale Impacts of Distinct CO_2 Additions in the North Sea*, 56 MARINE POLLUTION BULL. 1461, 1466 (2008). *See also* Ben I. McNeil & Richard J. Matear, *Climate Change Feedbacks on Future Oceanic Acidification*, 59(B) TELLUS 191, 191 (2007).

[47] *The Acid Ocean – The Other Problem with CO2 Emission*, RealClimate, July 2, 2005, <http://www.realclimate.org/index.php?p=169>, site visited on Dec. 25, 2007.

Acidification of the oceans will result in a decrease in the concentration of carbonate and related ions that reef-building and other calcifying organisms[48] draw upon to produce calcium carbonate.[49] In recent experiments in which dissolved carbon dioxide was increased to double preindustrial levels, shell- and skeleton-building rates of organisms with carbonate shells and skeletons declined by as much as 50%.[50] Moreover, a recent analysis of the causes of mass extinctions of scleractinian corals and sphinctozoid sponges during the Late Triassic period (with declines of these species of 96.1% and 91.4%, respectively) concluded that substantial declines of ocean pH during that period may have been the primary factor.[51]

Among the species that might be severely affected are a group of thirty-two species of planktonic snail species with calcium carbonate shells, pteropods. Although the species have a global distribution, population densities are highest in polar and subpolar regions, and they are the primary calcifiers in the Southern Ocean.[52]

In the Ross Sea, the subpolar-polar pteropod *Limacina helicina* sometimes replaces krill as the dominant zooplankton species in the ecosystem.[53] A recent study indicates that increased acidification of pteropod habitat in the sea might ultimately result in the disappearance of the species from Antarctic waters, or shift its distribution to lower latitudes.[54] The potential exclusion of the pteropod from other polar and subpolar regions could also have negative impacts on several straddling stock species for which it is a prey species, including North Pacific salmon, mackerel, herring, and cod.[55]

Other potential impacts of reduced pH in the oceans could include disruptions in the carbon cycle and the nutrient ratios, which could adversely affect phytoplankton species critical for many fish species, including straddling stocks and high migratory species,[56] as well as changes in internal acid-base parameters and ion levels in fish species, and reductions in the ability of species to carry oxygen.[57]

---

[48] An example of noncoral reef builders is rudistid bivalves, which secrete calcium carbonate shells or skeletons. Kaustuv Roy & John M. Pandolfi, *Responses of Marine Species and Ecosystems to Past Climate Change*, in CLIMATE CHANGE & BIODIVERSITY 164 (Thomas E. Lovejoy & Lee Hannah eds., 2005).

[49] O. Hoegh-Guldberg et al., *Pacific in Peril*, Greenpeace Rep., Oct. 2000, at 14.

[50] Feely et al., *supra* note 27, at 2.

[51] Michael Hautmann et al., *Catastrophic Ocean Acidification at the Triassic-Jurassic Boundary*, 249(1) N. JB. GEOLOGY PALÄONT. ABH. 119, 122–125 (2008).

[52] James C. Orr et al., *Anthropogenic Ocean Acidification over the Twenty-First Century and Its Impact on Calcifying Organisms*, 437 NATURE 681, 685 (2005).

[53] *Id.*

[54] *Id.*

[55] *Id.*; Feely et al., *supra* note 27, at 3.

[56] P.S. Liss, G. Malin & S.M. Turner, *Production of DMS by Marine Phytoplankton*, in Dimethylsulphide: Oceans, Atmosphere & Climate, European Commission, Proceedings of the International Symposium held in Belgirate, Italy, 13–15 Oct. 1992, at 10, <http://books.google.com/books?id=YMjnw9MBJDMC&pg=PA10&lpg=PA10&dq=coccolithophores+dimethyl+sulphide&source=web&ots=1SrjC-AJLk&sig=mCDJoARMLGabTTxQajLLzrGf68s#PPP1,M1>, site visited on Dec. 31, 2007

[57] Hautmann et al., *supra* note 51, at 122.

Given the severe impacts of that climate change may have on straddling stock and high migratory species, it is germane to next assess the prospects for enhancing their protection through the primary international legal instrument for their management and conservation.

2. UNFSA AND CLIMATE CHANGE

2.1. Overview of UNFSA

The Third United Nations Conference of the Law of Sea convened in 1973 and culminated nine years later in the adoption of the United Nations Convention on the Law of the Sea (UNCLOS).[58] UNCLOS entered into force in 1994 and currently has 148 parties.[59] A major component of UNCLOS is provisions for the regulation of fisheries, with an emphasis on the sovereign rights of coastal States to explore, exploit, conserve, and manage living natural resources, including fish stocks, within their respective 200-mile EEZs.[60] UNCLOS thus extends coastal State jurisdiction over 90% of the world's fish resources, and almost 40% of the world's oceans.[61] The emphasis on coastal State management of fisheries resources was premised on the belief that "entry into fisheries would be controlled, thereby reducing both the potential for overfishing and for overcapitalization of fishing fleets."[62] Moreover, it was hoped that coastal States' authority to enforce regulations against all fishing vessels within their respective EEZs would obviate the problems associated with weak flag-state enforcement of national and international fisheries regulations.[63]

Although many have characterized UNCLOS as "a constitution for the oceans,"[64] it provides only general governing principles for the management of straddling

---

[58] United Nations, Convention on the Law of the Sea, October 10, 1982, 21 I.L.M. 1261 (1982), <http://www.un.org/Depts/los/convention_agreements/texts/unclos/unclos_e.pdf>, site visited on Dec. 26, 2006.

[59] UN Oceans and the Law of the Sea, *Chronological Lists of Ratifications of, Accessions and Successions to the Convention and the Related Agreements as at 01 February 2005*, <http://www.un.org/Depts/los/reference_files/chronological_lists_of_ratifications.htm#The%20United%20Nations%20Convention%20on%20the%20Law%20of%20the%20Sea>, site visited on June 2, 2005.

[60] UNCLOS, *supra* note 58, at art. 58; arts. 61–68.

[61] Derrick M. Kedziora, *Gunboat Diplomacy in the Northwest Atlantic: The 1995 Canada-EU Fishing Dispute and the United Nations Agreement on Straddling and High Migratory Stocks*, 17 N.W. J. INT'L L. & BUS. 1132, 1139 (1996–1997).

[62] Donna R. Christie, *The Conservation and Management of Stocks Located Solely within the Exclusive Economic Zone*, in DEVELOPMENTS IN INTERNATIONAL FISHERIES LAW 396 (Ellen Hey ed., 1999).

[63] Donna R. Christie, *It Don't Come EEZ: The Failure and Future of Coastal States Fisheries Management*, 14 J. TRANSNAT'L L. & POL'Y 1, 2 (2004); Christopher C. Joyner, *Compliance and Enforcement in New International Fisheries Law*, 12 TEMP. INT'L & COMP. L. J. 271, 277–78 (1998).

[64] United Nations, Division for Ocean Affairs and the Law of the Sea, Tommy T.B. Koh, *A Constitution for the Oceans* (1982), <http://www.un.org/Depts/los/convention_agreements/texts/koh_english.pdf>, site visited on Aug. 30, 2005; *Report of the Work of the United Nations Ad Hoc Open-Ended Informal*

stocks and high migratory species. In cases where stocks are found within the EEZs of two or more coastal States, or an EEZ and an area beyond it, UNCLOS merely requires that the pertinent fishing States "seek" to agree upon management measures either directly or through subregional or regional organizations.[65] In the case of highly migratory species, coastal States and other States with nationals fishing in the region are exhorted to cooperate directly or through international organizations "with a view" to ensuring conservation and optimal utilization.[66] A proposal by some coastal States for an arbitration clause was beaten back by DWFNs and subsequently withdrawn.[67] Thus, States may, consistent with the provisions of UNCLOS and in good faith, fail to agree to conservation measures to protect highly migratory and straddling fish stocks.[68]

The lack of binding obligations in UNCLOS for high migratory species and straddling stocks was largely attributable to the fact that fishing in these regions was not considered to be a major issue in the early 1980s.[69] However, as coastal States began to claim their rights within their EEZs, large distant water fishing fleets were increasingly displaced from their traditional fishing grounds, placing rapidly increasing pressures on highly migratory species and straddling stocks.[70] Moreover, technological breakthroughs during this period, including satellite tracking, specially designed nets to compensate for the reduced density of stocks on the high seas, and larger and more efficient vessels, facilitated an ever-expanding scope of fishing operations by distant water fishing nations.[71] Overall, the proportion of catches taken beyond 200-mile EEZs doubled during the 1990s.[72]

These trends quickly took their toll. In 1994, the U.N. Food and Agriculture Organization (FAO) reported that straddling fish stock catches in EEZs and high

---

Working Group to Study Issues Relating to the Conservation and Sustainable Use of Marine Biological Diversity Beyond Areas of National Jurisdiction, at 21, <http://daccessdds.un.org/doc/UNDOC/GEN/N06/277/50/PDF/N0627750.pdf?OpenElement>, site visited on July 8, 2007.

[65] UNCLOS, *supra* note 58, at art. 63.
[66] *Id.* at art. 64.
[67] D.H. Anderson, *The Straddling Stocks Agreement of 1995: An Initial Assessment*, 45(2) INT'L & COMP. L.Q. 463, 465 (1996).
[68] Jon C. Goltz, *The Sea of Okhotsk Peanut Hole: How the United Nations Draft Agreement on Straddling Stocks Might Preserve the Pollack Fishery*, 4 PAC. RIM L. & POL'Y 443, 458 (1995); Mack, *supra* note 17, at 322–23.
[69] FAO, *supra* note 5, at 1; Anderson, *supra* note 67, at 465.
[70] Stuart Kaye, *Implementing High Seas Biodiversity Conservation: Global Geopolitical Considerations*, 28 MARINE POL'Y 221, 222 (2004); Chairman of the Conference at the Opening of the Organizational Session, United Nations Conference on Straddling Fish Stocks and Highly Migratory Fish Stocks, 19 April 1993, at 1, UN. Doc. A/Conf.164/7. Distant water fishing fleets were often subsidized by high-seas fishing nations. Alison Rieser, *International Fisheries Law, Overfishing and Marine Biodiversity*, 9 GEO. INT'L ENVTL. L. REV. 251, 263 (1997).
[71] A. Anna Zumwalt, *Straddling Fish Stock Spawn Fish War on the High Seas*, 3 U.C. DAVIS J. INT'L L. & POL'Y 35, 43 (1997); Rieser, *supra* note 70, at 263.
[72] Note, *Toward a Rational Harvest: The United Nations Agreement on Straddling Fish Stocks and Highly Migratory Species*, 5 MINN. J. GLOBAL TRADE 357, 365 (1999).

seas had been declining since 1989, and that many highly migratory fish stocks, including a majority of tuna species, were depleted, in some cases severely.[73]

In 1992, the participants at the U.N. Conference on Environment and Development called for an intergovernmental conference under the auspices of the United Nations to address to promote effective implementation of UNCLOS provisions related to straddling stocks and highly migratory species.[74] In December, 1992, the U.N. General Assembly, recalling Agenda 21, passed Resolution 47/192, which authorized the convening of the United Nations Conference on Straddling Fish Stocks and Highly Migratory Fish Stocks (UNCSFS).[75]

In 1993, the U.N. General Assembly convened the UNCSFS, culminating in adoption of UNFSA in August of 1995. UNFSA entered into force in December of 2001 and currently has sixty-two Parties,[76] "including most States with significant interests in international fisheries."[77]

The Agreement's overarching objective is to "ensure long-term conservation and sustainable use of straddling fish stocks and highly migratory fish stocks...."[78] The Agreement's primary means of effectuating this is through engendering cooperation between coastal States and States fishing on the high seas, through, inter alia:

- Seeking agreement between coastal States and States on the high seas on necessary measures for conservation of stocks in the high seas areas and straddling stocks through direct agreements and cooperation in Regional Fisheries Management Organizations;[79]
- Collecting and exchanging of critical data with respect to straddling stocks and high migratory species;[80] and
- Expanding the duties of Flag States to ensure enforcement of and compliance with the Convention's provisions, as well as the rights of other States, including port States, to ensure compliance with the Agreement.[81]

---

[73] Giselle Vigneron, *Compliance and International Environmental Agreements: A Case Study of the 1995 United Nations Straddling Fish Stocks Agreement*, 10 GEO. INT'L ENVTL. L. REV. 581, 586 (1998). The status of these stocks remains imperiled a decade later. According to the most recent analysis by the FAO, "about 30 percent of the stocks of highly migratory tuna and tuna-like species, more than 50 percent of the highly migratory oceanic sharks and nearly two-thirds of the straddling stocks and the stocks of other high seas fishery resources are overexploited or depleted." FAO, *supra* note 5, at iv.

[74] United Nations Conference on Environment and Development, Agenda 21, Programme of Action for Sustainable Development, ch. 17, para. 17.49 (1992).

[75] United Nations Conference on Straddling Fish Stocks and Highly Migratory Fish Stocks, United Nations Resolutions and Decisions, 47th Sess., Supp. No. 49, at 145, G.A. Res. 47/192, U.N. Doc. A/47/49 (1992).

[76] U.N. Oceans and Law of the Sea, Status of the Agreement, <http://www.un.org/Depts/los/convention_agreements/convention_overview_fish_stocks.htm%20stocks>, site visited on Dec. 27, 2006.

[77] Balton & Koehler, *supra* note 9, at 7.

[78] UNFSA, *supra* note 2, art. 2.

[79] *Id.* at arts. 7–10.

[80] *Id.* at art. 14.

[81] *Id.* at arts. 19–23.

However, although the focus of UNFSA is on the relationship between coastal States and States fishing in areas beyond EEZs, there are a large number of provisions that could give rise to claims associated with climate change impacts on straddling stocks and highly migratory species.

2.2. UNFSA and Climate Change

UNFSA adopts the well-recognized "no harm rule" of international environmental law, which obliges States to ensure that activities within their jurisdiction or control do not result in injuries to the interests of other States or areas beyond national control.[82] UNFSA provides that "States Parties are liable in accordance with international law for damage or loss attributable to them in regard to this Agreement."[83] Many of the provisions of UNFSA, in turn, could provide the basis for a Party to bring an action against one or more other Parties for climate-related damages to fisheries.

As indicated earlier, the Agreement's primary objective is to ensure the long-term conservation and sustainable use of straddling fish stocks and highly migratory species,[84] mandating that its Parties take conservation and management measures to further this objective. Although the Agreement's primary focus is on the impacts of the harvesting of fish stocks,[85] it clearly contemplates the regulation of other potential activities that could imperil the conservation and sustainable use of such stocks. For example, UNFSA requires the Parties to assess the impacts of "other human activities and environmental factors on target stocks and species belonging to the same ecosystem or associated with or dependent upon the target stocks."[86]

Moreover, the Agreement requires the Parties to "minimize pollution."[87] Although the Agreement doesn't define the term "pollution," Article 4 provides that UNFSA is to be "interpreted and applied in the context of and in a manner consistent with the Convention."[88] Thus, it is germane to look at the definition of pollution provided for in UNCLOS. In pertinent part, UNCLOS defines "pollution of

---

[82] Nuclear Tests (AUSTRALIA v. FRANCE) ICJ Rep. 1973; 2 RESTATEMENT (THIRD) FOREIGN RELATIONS LAW 103 sec. 601, at 103. *See also* RODA VERHEYEN, CLIMATE CHANGE DAMAGE & INTERNATIONAL LAW 146 (2005); Richard S.J. Tol & Roda Verheyen, *State Responsibility and Compensation for Climate Change Damages – A Legal and Economic Assessment*, 32 ENERGY POL'Y 1109, 1110 (2004). As embodied in documents such as Principle 2 of the Rio Declaration, Principle 21 of the Stockholm Declaration, treaties, including the UNFCCC, and the Trail Smelter Arbitration in the 1941, the no-harm rule "has its foundations in the principle of good neighbourliness between States formally equal under international law." *Id.*

[83] UNFSA, *supra* note 2, at art. 35.

[84] *See supra* note 78.

[85] Timothy D. Smith, *United States Practice and the Bering Sea: Is It Consistent with a Norm of Ecosystem Management?*, 1 OCEAN & COASTAL L.J. 141, 150 (1995).

[86] UNFSA, *supra* note 2, at art. 5(d).

[87] *Id.* at art. 5(f).

[88] *Id.* at art. 4.

the marine environment" as: "the introduction by man, directly or indirectly, of substances or energy into the marine environment... which results or is likely to result in such deleterious effects as harm to living resources and marine life... hindrance to marine activities, including fishing...."[89]

Although rising ocean temperatures related to climate change could not reasonably be construed as a "substance" under Article 1.1 of UNCLOS, it would likely be construed by a dispute resolution body as "energy," much as introduction of heat, such as wastewater from production processes, appears to fall under this rubric.[90] Moreover, as developed earlier, the uptake of anthropogenically generated carbon dioxide into the oceans can result in direct deleterious impacts on marine life,[91] which clearly brings carbon dioxide under the definition in Article 1.1 of UNCLOS of an a polluting "substance" introduced into the ocean.

Where necessary, UNFSA also imposes obligations on the Parties to adopt conservation and management measures for "species belonging to the same ecosystem or associated with or dependent upon target species" and to "protect biodiversity of the marine environment."[92] Moreover, the Parties are obligated to ensure adequate implementation and enforcement of such measures "through effective monitoring, control and surveillance."[93] Finally, UNFSA requires the Parties to promote and conduct relevant scientific research. A coherent research agenda is extremely important in the context of climate change to ensure quantification of potential impacts on specific species and to incorporate such impacts into stock assessment processes that are critical for successful long-term management of marine species.[94]

Thus, to the extent that climate change may result in the diminution of certain stocks, or alter their distribution in a way that adversely affects the interests of discrete Parties, a cause of action could arise under the Agreement by which Parties might seek: (1) damages; (2) enforcement of conservation obligations; and (3) a commitment by all Parties to assess the potential impacts of climate change on species regulated under UNFSA.

Rare among international environmental agreements, UNFSA provides for a binding dispute resolution mechanism where efforts to resolve the dispute through nonbinding methods, such as negotiation, inquiry, mediation, or conciliation, have proven to be unavailing. Part VIII of the Agreement applies the dispute resolution mechanism set out in Part XV of UNCLOS to any dispute under the Agreement, even where one or more of the disputants are not Parties to UNCLOS.[95]

---

[89] UNCLOS, *supra* note 58, at art. 1(4).
[90] VERHEYEN, *supra* note 82, at 194–95.
[91] *See* Sec. 2, *infra*.
[92] UNFSA, *supra* note 2, at art. 5(g).
[93] *Id.* at art. 5(l).
[94] Jonathan A. Hare & Kenneth W. Able, *Mechanistic Links Between Climate and Fisheries Along the East Coast of the United States: Explaining Population Outbursts of Atlantic Croaker* (Micropogonias undulates), 16(1) FISHERIES OCEANOGRAPHY 31, 45 (2007).
[95] *Id.* at art. 30(1).

As Jonathan Hafetz observes, UNCLOS "creates a binding system of obligations and dispute resolutions, which confers on a forum international jurisdiction, authority, and implementing powers that exceed those of other international environmental law forums and rival those conferred on the World Trade Organization."[96] Part XV of UNCLOS provides States with four potential fora for settlement of disputes:[97] the International Tribunal for the Law of the Sea (ITLOS);[98] the International Court of Justice; an arbitral panel; or a special arbitral panel.[99] States may choose to declare their choice of forum, but in cases where they have not, or where Parties to a dispute have not accepted the same procedure for dispute settlement, the dispute must be submitted to binding arbitration unless the Parties agree otherwise.[100] To date, the vast majority of Parties to UNCLOS have, de facto, chosen arbitration by their silence on the matter, as have most Parties to UNFSA.[101]

3. POTENTIAL BARRIERS TO CAUSES OF ACTION UNDER UNFSA

A Party to UNFSA pursuing an action based on climate change damages would face some imposing barriers, though none need prove fatal.

3.1. Causation

As Smith and Shearman observe, "establishing legal causation in climate change actions – that is, proving that a defendant's actions caused the harm suffered by a plaintiff – will pose the greatest obstacle for a majority of plaintiffs."[102] Indeed, causation issues have been raised in two international climate cases to date, in the Inuit petition to the Inter-American Commission on Human Rights,[103] and the petitions to the World Heritage Committee to list several sites allegedly threatened by

---

[96] Jonathan L. Hafetz, *Fostering Protection of the Marine Environmental and Economic Development: Article 121(3) of the Third Law of the Sea Convention*, 15 AM. U. INT'L L. REV. 583, 596 (2000).

[97] Under UNCLOS's dispute resolution mechanism "[a]ny decision rendered by a court or tribunal having jurisdiction under this section shall be final and shall be complied with by all the parties to the dispute." UNCLOS, *supra* note 58, at art. 296(1).

[98] *Id.* at Annex VI. The Tribunal is composed of twenty-one judges representing the legal systems of UNCLOS's Parties. *Id.* at arts. 1, 2, 4.

[99] *Id.* at art. 287(1). Special arbitral panels may be convened for disputes involving "(1) fisheries, (2) protection and preservation of the marine environment, (3) marine scientific research, or (4) navigation, including pollution from vessels and by dumping..." *Id.* at Annex VIII, art. 1.

[100] *Id.* at art. 287(3)-(5).

[101] ANDREE KIRCHNER, INTERNATIONAL MARINE ENVIRONMENTAL LAW 22 (2003); U.N. Division for Ocean Affairs and the Law of the Sea, Straddling Stocks Convention, *Declarations*, <http://www.un.org/Depts/los/convention_agreements/fish_stocks_agreement_declarations.htm>, site visited on Dec. 28, 2006. The United States has chosen a special arbitral tribunal for, inter alia, disputes involving fisheries or marine pollution. *Id.* However, since most Parties to UNFSA have chosen either another option for dispute resolution, or none at all, any dispute involving the United States would likely be settled by an arbitration panel.

[102] JOSEPH SMITH & DAVID SHEARMAN, CLIMATE CHANGE LITIGATION 107 (2006).

[103] In the course of the hearing granted by the Commission in March of 2007, Commissioners Abramovitch and Pinheiro pressed the petitioners as to whether the Commission could attribute State responsibility to the United States for the alleged human rights violations to petitioners given

climate change on the List of World Heritage in Danger under the World Heritage Convention.[104] Domestic legal systems, and to some extent international law, draw a distinction between general and specific causation, the former referring to the causal link "between an activity and the general outcome," and the latter to the causal link between a specific activity and specific damage.[105] It is likely that both aspects of causation would be raised in an UNFSA climate action.

3.1.1. General Causation

In many cases, declines of fish stocks or shifts in distribution may be attributable to a number of factors other than, or in conjunction with, climate change, including overfishing,[106] habitat destruction,[107] or diminution of prey species.[108] As a report to the European Commission recently concluded, "it is extremely difficult to separate, in terms of changes in population densities and recruitment, regional climate effects from direct anthropogenic influences."[109] Thus, a Party to UNFSA defending itself against a claim of damages associated with climate change may contend that species decline or distribution shifts cannot be linked solely to climatic factors, and thus the State cannot be held liable under UNFSA. This argument should not prevail. First,

---

that many other States, including States that were not members of the Organization of the American States, were substantial emitters of greenhouse gases. *Response to the Commission's Question on Attribution of Responsibility Submitted by Sheila Watt-Cloutier*, Earthjustice and the Center for International Environmental Law, March 2007.

[104] United States, *Position of the United State [sic] of America on Climate Change with Respect to the World Heritage Convention and World Heritage Sites*, <http://www.elaw.org/assets/word/u.s.climate.US%20position%20paper.doc>, site visited on Sept. 28, 2007. The United States contended, inter alia, that "there is not enough data available to distinguish whether climatic changes at the named World Heritage Sites are the result of human-induced climate change or natural variability." *Id.* at 4. For additional information on the petitions, see Erica J. Thorson, *The World Heritage Convention and Climate Change: The Case for a Climate-Change Mitigation Strategy Beyond the Kyoto Protocol*, this volume.

[105] Richard S.J. Tol & Roda Verheyen, *Liability and Compensation for Climate Change Damages – A Legal and Economic Assessment*, Research Unit Sustainability and Global Change, Hamburg University, FNU-9 (2001), <http://www.fnu.zmaw.de/fileadmin/fnu-files/publication/working-papers/adapcap.pdf>, site visited on Sept. 25, 2007.

[106] Samuel F. Herrick, Jr., et al., *Management Application of an Empirical Model of Sardine-Climate Regime Shifts*, 31 MARINE POL'Y 71, 91 (2007); Gian-Reto Walther et al., *Ecological Responses to Recent Climate Change*, 416 NATURE 389, 393 (2002).

[107] K.I. Matics, *Measures for Enhancing Marine Fisheries Stock in Southeast Asia*, 34(3) OCEAN & COASTAL MGMT. 233–247 (1997).

[108] Michel Potier et al., *Forage Fauna in the Diet of Three Large Pelagic Fishes* (Lancetfish, Swordfish and Yellowfin Tuna) *in the Western Equatorial Indian Ocean*, 83(1) FISHERIES RES. 60–72 (2007); Giovanni Bearzi et al., *Prey Depletion Caused by Overfishing and the Decline of Marine Megafauna in Eastern Ionian Sea Coastal Waters (Central Mediterranean)*, 127 BIOLOGICAL CONSERVATION 373–382 (2006).

[109] Institute for Environment & Sustainability, *supra* note 24, at 21, <http://ies.jrc.cec.eu.int/fileadmin/Documentation/Reports/Varie/cc_marine_report_optimized2.pdf>, site visited on Feb. 19, 2007. *See also* Anna Rindorf & Peter Lewy, *Warm, Windy Winters Drive Cod North and Homing of Spawners Keeps Them There*, 43 J. APPLIED ECOLOGY 445, 445 (2006).

even if other factors may constitute threats to regulated species, climate change is clearly a substantial peril for many of these species. A tribunal or panel could assess the extent of this threat by employing statistical probability analysis to support a finding of liability where a moving party can establish that climate change results in a "material increase in risk."[110] This approach has been embraced by a number of courts in recent years.[111] This would in turn trigger the obligation of major emitters of greenhouse gases that are Parties to UNFSA to adopt measures to reduce these emissions to levels that substantially reduce the threat to high migratory and straddling stock species.[112]

Second, all causation challenges must be considered in light of the regime's precautionary principle provisions. Recognition of the failure of the assimilative capacity paradigm to adequately safeguard the environment led to the formulation of the precautionary principle:

> The precautionary concept advocates a shift away from the primacy of scientific proof and traditional economic analyses that do not account for environmental degradation. Instead, emphasis is placed on: 1) the vulnerability of the environment; 2) the limitations of science to accurately predict threats to the environment, and the measures required to prevent such threats; 3) the availability of alternatives (both methods of production and products) which permit the termination or minimization of inputs into the environment; and 4) the need for long-term, holistic economic considerations, accounting for, among other things, environmental degradation and the costs of waste treatment.[113]

"The precautionary principle can also be viewed as a safeguard against the opportunism of decision-makers in situations of asymmetric information or imperfect monitoring by society."[114] In the context of management and conservation of wildlife species, the principle reflects the recognition that "scientific understanding of

---

[110] See Peter A. Stott, D.A. Stone, & M.R. Allen, *Human Contribution to the European Heatwave of 2003*, 432 NATURE 610 (2004) ("It is an ill-posed question whether the 2003 heatwave was caused, in a simple deterministic sense, by a modification of the external influences on climate – for example, increasing concentrations of greenhouse gases in the atmosphere – because almost any such weather event might have occurred by chance in an unmodified climate. However, it is possible to estimate by how much human activities may have increased the risk of the occurrence of such a heatwave.") Peñalver argues that the "but for" analysis employed by many courts to assess causation, reflecting a "deductive nomological" model of scientific explanation, is inappropriate in causal analysis in toxic tort and climate change cases. He advocates a probabilistic theory of causation that reflects the nature of these phenomena. Eduardo M. Peñalver, *Acts of God or Toxic Torts? Applying Tort Principles to the Problem of Climate Change*, 38 NAT. RESOURCES J. 563, 582–85 (1998). See also S. Greenland & J.M. Robins, *Epidemiology, Justice and the Probability of Causation*, 40 JURIMETRICS 321–40 (2000).

[111] See Fairchild v. Glenhaven, [2002] UKHL 22 (collecting cases from Australia, Canada, and Britain).

[112] UNFSA, *supra* note 2, at art. 5(a).

[113] Ellen Hey, *The Precautionary Concept in Environmental Policy And Law: Institutionalizing Caution*, 4 GEO. INT'L ENVTL. L. REV. 303, 307 (1992).

[114] Ylva Arvidsson, *The Precautionary Principle: Experiences from Implementation into Swedish Law*, IIIEE Reports, 2001:7 (2001), at 11, <http://www.iiee.u.se/information/libary/publicatons/reports/20001/Ylva-Arvidsson.pdf>, site visited on July 8, 2007.

ecosystems is complicated by a host of factors, including complex and cascading effects of human activities and uncertainty introduced by naturally chaotic population dynamics."[115]

UNFSA provides that "States shall apply the precautionary approach widely to conservation, management and exploitation of straddling fish stocks and highly migratory fish stocks in order to protect the living marine resources and preserve the marine environment."[116] Thus, even under scenarios of uncertainty about a given threat, such as climate change impacts, Article 6 of UNFSA provides "[t]he absence of adequate scientific information shall not be used as a reason for postponing or failing to take conservation and management measures."[117] As Colburn observes, "[t]he precautionary approach essentially reverses the process of marine scientific research ("MSR") application in the management of straddling and highly migratory fish stocks, allowing states and RFOs to proceed with conservation measures even in the absence of scientific certainty."[118] Thus, in the context of potential threats posed by climate change to fish species regulated under UNFSA, it can be argued that the Parties have an obligation to take action even in the absence of definitive proof of causation.

3.1.2 Specific Causation

The targeted Party in a climate-related UNFSA action might argue that climate change is caused by a multitude of anthropogenic sources, and thus, any specific harm cannot be attributable to a specific Party, even a large greenhouse emitting State such as the United States or China. This argument should not prevail for two reasons. First, the issue of specific causation would be most germane in cases where a moving Party seeks damages.[119] A Party to UNFSA might not seek monetary damages in pressing a climate change case against another Party. Rather a Party bringing such an action might be exclusively, or in the alternative, seeking a commitment by the targeted Party to fulfill its "duty to cooperate" under the treaty [120] by enacting effective measures to contribute to the goal of "long-term sustainability of straddling fish stocks and highly migratory fish stocks."[121]

Under the terms of UNFSA, as well as customary international law, all treaty obligations must be fulfilled in good faith, the principle of *pacta sunt servanda*.[122]

---

[115] Robert J. Wilder, *Precautionary Principle; Prevention Rather Than Cure*, OCEAN 98, <http://www.wildershares.com/pdf/Ocean98.Nature%20article.Wilder.pdf >, site visited on Sept. 28, 2007.
[116] *Id.* at art. 6(1).
[117] *Id.* at art. 6(2).
[118] Colburn, *supra* note 5, at 347.
[119] VERHEYEN, *supra* note 82, at 248.
[120] UNFSA, *supra* note 2, at art. 5.
[121] *Id.* at art. 5(a).
[122] *Id.* at art. 34; Vienna Convention on the Law of Treaties, May 23, 1969, at art. 26; I.I. Lukashuk, *The Principle of 'Pacta Sunt Servanda' and the Nature of Obligation Under International Law*, 83 AM. J. INT'L L. 513, 513 (1989).

The obligation of good faith, which Henkin has correctly characterized as "the most important principle of international law,"[123] imposes a duty upon treaty Parties to exercise their sovereign rights in a manner that is consistent with their treaty obligations.[124] Moreover, the failure to fulfill treaty obligations in good faith constitutes a breach of treaty obligations and entails international responsibility.[125] Furthermore, a finding of a breach of a treaty obligation would not require the establishment of specific causation:

> It is important to note that injury or material damage is not a prerequisite for the existence of a wrongful act, i.e. for the invocation of State responsibility.... Thus, while a claimant State must, under the [Draft Articles on State Responsibility], show a causal relationship between the activity and the damage caused to be eligible for reparations... the State can, without showing a causal relationship demand, as long as breach of an international obligation has taken place. This is in line with customary law....[126]

Thus, any UNFSA Party failing to make a good faith effort to address its anthropogenic emissions of greenhouse gases, given their potential impact on fish species, could be found to be in violation of the treaty even in the absence of establishment of specific causation. This breach, in turn, would impose an obligation on the breaching Party to cease its wrongful conduct,[127] which in this context would require a Party to reduce its emissions below a threshold that would substantially decrease the risks to interests protected under UNFSA.

Moreover, even in cases where a Party might seek damages under UNFSA, the fact that other States may contribute to climate change need not prove fatal to such an action. As Roda Verheyen notes, "[T]hat a contribution to the legally relevant outcome can be sufficient to establish causation is accepted in many jurisdictions around the world."[128] This includes under the U.S. Restatement of Torts, which provides that "a conduct or event question is a cause in fact of the harm if it is a substantial factor in producing it,"[129] as well as under German law, which provides for holding a person responsible for increases in risk that manifest themselves in damages.[130] Moreover, the International Law Commission has held that a State can

---

[123] LOUIS HENKIN, CONSTITUTIONALISM, DEMOCRACY, AND FOREIGN AFFAIRS 62 (1990).

[124] Declaration on Principles of International Law concerning Friendly Relations and Co-operation among States in Accordance with the Charter of the United Nations, G.A. Res. 2625 (Oct. 24, 1970); Final Act of the Conference on Security and Cooperation in Europe, Aug. 1, 1975, 73 Dep't State Bulletin 323 (1975).

[125] Duncan Currie, *Whales, Sustainability and International Environmental Governance*, 16 RECIEL 45, 53 (2007). *See also* International Law Commission, Draft Articles on Responsibility of States for Internationally Wrongful Acts (2001), at art. 2.

[126] VERHEYEN, *supra* note 82, at 243.

[127] International Law Commission, *supra* note 125, at art. 30.

[128] VERHEYEN, *supra* note 82, at 255.

[129] RESTATEMENT (SECOND) OF TORTS § 431, CMT. A (1965).

[130] VERHEYEN, *supra* note 82, at 255.

be held liable for reparations in cases where it has played a "decisive" role in causing an injury.[131]

3.2. Reluctance of Dispute Resolution Bodies to Address Climate Change

Experience with climate change litigation to date in the United States, at least, has demonstrated some reluctance on the part of members of the judiciary to address climate change issues given their limited scientific expertise. Consider, for example, Justice Scalia's flippant but telling comment in the recent Supreme Court oral arguments in *Massachusetts v. Environmental Protection Agency*:[132]

> JUSTICE SCALIA: . . . your assertion is that after the pollutant leaves the air and goes up into the stratosphere it is contributing to global warming.
> MR. MILKEY: Respectfully, Your Honor, it is not the stratosphere. It's the troposphere.
> JUSTICE SCALIA: Troposphere, whatever. I told you before I'm not a scientist. (Laughter.)
> JUSTICE SCALIA: That's why I don't want to have to deal with global warming, to tell you the truth.[133]

Parties bringing an action before ITLOS or an arbitral panel might experience similar reservations on the part of the dispute resolution body to grapple with the complicated technical issues associated with climate change, especially since the primary area of expertise of tribunal or panel members may be more traditional fisheries issues, such as the impact of harvesting on species. UNFSA provides two mechanisms to help address this concern. First, in cases where "a dispute concerns a matter of a technical nature," the States involved in a dispute may refer the dispute to an "ad hoc expert panel," which will confer with the Parties and seek to resolve the dispute without recourse to binding procedures.[134] A Party seeking to press a climate change claim could certainly seek to engage another Party in such negotiations initially, and should this fail to resolve the dispute, which is likely, seek to introduce the panel's scientific findings in a binding dispute resolution forum.

Additionally, if both Parties agree to it, cases of this nature can be referred to a "special arbitral panel."[135] Under UNCLOS's dispute resolution provisions in this

---

[131] Gaetano Arangio-Ruiz, 2nd Report on State Responsibility, II(1) Y.B. Int'l L. Commission, A/CN.4/426 (1989), at 14.
[132] No. 05–1120, U.S. Supreme Court, Oral Argument, Nov. 29, 2006, <http://www.supremecourtus.gov/oral_arguments/argument_transcripts/05–1120.pdf>, site visited on Dec. 29, 2006.
[133] *Id.* at 23.
[134] UNFSA, *supra* note 2, at art. 29.
[135] *See* note 99, *supra*, and accompanying text.

context, which UNFSA fully incorporates,[136] a panel hearing a climate change–related dispute could be constituted by experts in the fields of fisheries, marine environmental protection, and marine scientific research drawn from the FAO, the United Nations Environment Programme, and the Intergovernmental Oceanographic Commission,[137] all of whom have expertise on the nexus of fisheries and climate change. Of course, as indicated earlier, this provision of UNFSA can only be invoked with the consent of both parties. Thus, there is a very good chance that a party against which a climate action would be brought would refuse, believing that ITLOS or an arbitral panel might be far less likely to grapple with complicated science associated with such a case.

Perhaps an even more imposing barrier to a cause of action under UNFSA may be the perceived threat to the legitimacy of a dispute resolution body should it enter a decision against a hegemonic State that then chose to either ignore the decision or drag its feet. International tribunals carefully marshal their political capital in an effort to preserve and enhance their legitimacy. The primary threat to the legitimacy of a UNFSA dispute resolution body in the context of climate change may be that a powerful State would choose to not comply with the decision given the dramatic policy changes that it might necessitate. As Richard Silk recently noted, States may choose to not to comply with "binding" decisions when they deem it against their interests:

> In international law, even allegedly binding dispute settlement mechanisms such as arbitration may be ignored when a state disagrees with the decision. To illustrate, in the Beagle Channel dispute between Chile and Argentina, Argentina challenged the validity of the arbitrators' decision on dubious grounds and, despite the implausibility of Argentina's repudiation, the decision was never enforced.... Under UNCLOS, there might be strong domestic and international pressures to sign a fishery agreement regardless of the costs of compliance, but when the time for compliance comes, narrower national interests may prevail.[138]

Indeed, the fear that decisions against the United States might be ignored may help to explain the recent rejection of petitions to address climate change by the Inter-American Commission on Human Rights and World Heritage Committee.[139]

CONCLUSION

In a perfect world, the threat of climate change would be effectively addressed through the international institutional responses developed in the 1990s. Unfortunately, the specter of climate change looms larger currently than it did a decade ago,

---

[136] *See* note 95, *supra*, and accompanying text.
[137] UNCLOS, *supra* note 58, at Annex VIII, art. 2(1)(2).
[138] Richard J. Silk, Jr., *Nonbinding Dispute Resolution Processes in Fisheries Conflicts: Fish out of Water?*, 16 Ohio St. J. on Disp. Resol. 791, 800–01 (2001).
[139] *See supra* notes 22–23 and accompanying text.

and the prospects for adequate responses within the UNFCCC framework appear increasingly remote. Now more than ever, those most vulnerable to the impacts of climate change must explore alternatives that may finally galvanize the major greenhouse-emitting States into action. UNFSA is one option that deserves further exploration.

15

Climate Change Litigation: Opening the Door to the International Court of Justice

Andrew Strauss*

INTRODUCTION

In March 2003, I wrote an article for the *Environmental Law Reporter* surveying potential international judicial forums where victims of global warming could bring lawsuits.[1] In the ensuing six years, numerous lawsuits have been brought in the United States and in other countries,[2] and environmentalists can now celebrate

---

* Visiting Professor of Law, Notre Dame Law School, Distinguished Professor of Law, Widener University School of Law. I would like to thank Janet Lindenmuth, Gina Serra, Michael Hubbard, and Warren Rees for their very valuable research help on this chapter.

[1] Andrew L. Strauss, *The Legal Option: Suing the United States in International Forums for Global Warming Emissions*, 33 ENVTL. L. REP. 10, 185 (2003).

[2] *See, e.g.*, Friends of the Earth, Inc. v. Watson, No. C02–4106 JSW, 2005 WL 2035596, at*1 (N.D. Cal. Aug. 23, 2005) (alleging the U.S. Overseas Private Investment Corporation and Export-Import Bank facilitated the financing of projects in developing countries that contributed significantly to global warming without following proper procedures, including the production of Environmental Impact Statements, under the National Environmental Policy Act (NEPA) and the Administrative Procedures Act (APA)); Connecticut v. Am. Elec. Power Co., 406 F. Supp. 2d 265 (S.D.N.Y. 2005) (alleging that five major American power company emitters of carbon dioxide should be held liable under federal and state common law for contributing to the public nuisance of global warming); California v. Gen. Motors, No. 3:06 CV-05755 MJJ (2006) (alleging that the six largest automobile manufacturers should be held liable to the state of California for global warming-related damages under both the federal and state common law of public nuisance); New York v. EPA, No. 06-1148, 2007 U.S. APP. LEXIS 30013 (D.C. Cir. Dec. 26, 2007) (per curium) (challenging the EPA's refusal to add carbon emissions standards to the new power plant source performance standards); Cent. Valley Chrysler-Jeep v. Witherspoon, 2007 WL 135688 E.D.Cal., 2007. January 16, 2007 (challenging the California Air Resources Board's regulations to limit greenhouse gas emissions from motor vehicles on the basis that the regulations are preempted by the Clean Air Act and on other grounds); Ctr. for Biological Diversity v. Norton, N.D. Cal., No. 3:05CV05191 (settled Jan. 2007) (alleging that the Secretary of the Interior had not acted within the statutory period to review the petition of the polar bear as an endangered species due to global warming); Comer v. Murphy Oil, C.A. No. 1:05-cv-00436-LG-RHW (S.D. Miss.) (dismissed) (on appeal to U.S. Court of Appeals for the Fifth Circuit, No. 07–60756) (claiming that, because of their contribution to global warming which warms the waters in the Gulf of Mexico, insurers, chemical companies, oil companies and coal companies are liable for the increasing frequency and severity of Atlantic hurricanes including Hurricane Katrina). For representative cases outside of the United States, see GermanWatch v. Euler Hermes AG, Administrative Court Berlin, Jan. 10, 2006, 10A 215.04, *available at* http://www.climatelaw.org/media/Germany/de.export.decision.pdf; Genesis Power Ltd. v. Franklin Dist. Council, 2005 NZRMA 541 A148/05 (Env't Ct. Auckland); Gbemre v. Shell Petroleum Dev. Co Nigeria Ltd., [2005] F.H.C. FHC/B/CS/153/05 (Nigeria).

their first significant victory. In April 2007, based upon its finding that greenhouse gases are pollutants under Section 202(a)(1) of the U.S. Clean Air Act, the Supreme Court in *Massachusetts v. EPA*[3] held that the U.S. Environmental Protection Agency (EPA) has the authority to regulate greenhouse gases.

Though we are still in the early days of global warming litigation, these lawsuits are having a significant impact on the legal and political climate. In response to a good deal of popular[4] and academic discussion[5] suggesting that those most responsible for the global warming problem be held legally accountable, corporations in the carbon sector are becoming concerned about the extent of their potential legal liability. This concern is one reason they are coming to publicly accept the reality of anthropogenic-caused global warming, and the corresponding need for regulation of greenhouse gas emissions.[6]

Despite the significance of this litigation, however, global warming actions thus far have almost all been brought in domestic rather than international forums. The only exceptions are a petition by the Inuit to the Inter-American Commission on Human Rights,[7] and petitions by environmental groups and others to UNESCO's World Heritage Committee to include various natural sites as world heritage endangered by global warming.[8] While domestic courts are still far and away the primary formal

---

[3] *Massachusetts v. EPA*, 549 U.S. 497 (2007).
[4] *See, e.g.*, Andrew Simms & Andrew Strauss, *America in the Dock: Poor Nations at Risk from Global Warming Are Growing Tired of Talking*, FIN. TIMES, Aug. 22, 2002, at 21. Eoin O'Caroll, *As Earth Warms, Lawsuits Mount*, Christian Sci. Monitor, Feb. 22, 2007, at 12; *Global Warming: Here Come the Lawyers*, BUS. WK., Oct. 30, 2006, David Lynch, *Corporate America Warms to Fight Against Global Warming*, USA TODAY, June 1, 2006, at 1B.
[5] *See, e.g.*, JOSEPH SMITH & DAVID SHEARMAN, CLIMATE CHANGE LITIGATION (2006); JUSTIN R. PIDOT, GLOBAL WARMING IN THE COURTS: AN OVERVIEW OF CURRENT LITIGATION AND COMMON LEGAL ISSUES (2006); RODA VERHEYEN, CLIMATE CHANGE DAMAGE AND INTERNATIONAL LAW: PREVENTION, DUTIES AND STATE RESPONSIBILITY (DEVELOPMENTS IN INTERNATIONAL LAW) (2005); Sara Aminzadeh, *A Moral Imperative: The Human Rights Implications of Climate Change*, 30 HASTINGS INT'L & COMP. L. REV. 231 (2007); Philippe Sands, *International Environmental Litigation and Its Future*, 32 U. RICH. L. REV. 1619 (1999); Rebecca Elizabeth Jones, Comment, *Treading Deep Waters: Substantive Law Issues in Tuvalu's Threat to Sue the United States in the International Court of Justice*, 14 PAC. RIM L. & POL'Y J. 103 (2005); J. Chris Larson, Note, *Racing the Rising Tide: Legal Options for the Marshall Islands*, 21 MICH. J. INT'L L. 495 (2000); Kevin Healy & Jeffrey Tapick, *Climate Change: It's Not Just an Issue for Corporate Counsel – It's a Legal Problem*, 29 COLUM. J. ENVTL. L. 89 (2004).
[6] Jeffrey Ball, *Electricity Group Backs Emissions Caps*, WALL ST. J., Feb. 7, 2007, at A10; Jeffrey Ball, *Conoco Calls for Emissions Cap – Oil Producer Joins Effort to Shape New U.S. Policy on Greenhouse-Gas Limits*, WALL ST. J., Apr. 11, 2007, at A3.
[7] *See* Petition to the Inter-American Commission on Human Rights Seeking Relief from Violations Resulting from Global Warming Caused by Acts and Omissions of the United States (submitted Dec. 7, 2005), at 13–20, *available at* http://www.earthjustice.org/library/reports/ICC_Human_Rights_Petition.pdf; Letter from the Organization of American States to Sheila Watt-Cloutier regarding Petition No. P-1413–05, Nov. 16, 2006, *available at* http://graphics8.nytimes.com/packages/pdf/science/16commissionletter.pdf.
[8] *See* U.N. Educ. Sci. and Cultural Org. [UNESCO], World Heritage Comm., *Decision 29COM 7B.a Threats to World Heritage Properties* (2005), *available at* http://whc.unesco.org/download.cfm?id_document=5941; U.N. Educ. Sci. and Cultural Org. [UNESCO], World Heritage Comm., *Decision 30COM 7.1 The Impacts of Climate Change on World Heritage Properties* (2006), *available at* http://whc.unesco.org/download.cfm?id_document=6728.

institutions of dispute resolution in the world, they are in certain ways ill suited to address the global nature of the climate change problem.[9] For example, in the *Massachusetts* case, the EPA partially based its refusal to regulate carbon emissions on the global dimensions of the climate change problem which raise "important foreign policy issues" that are "the President's prerogative" to address.[10] Also based in part on similar concerns and quoting from that EPA decision, Judge Preska of the Southern District of New York dismissed a claim that the greenhouse gas emissions of the power companies constituted a nuisance.[11] Though both the EPA and Judge Preska address the problem from their vantage point as discrete decision makers within a domestic forum, the implication of their analysis points to the need for global prescriptive and adjudicatory action.

Within the international realm, the one court of general competence is the World Court or the International Court of Justice (ICJ). In terms of status and hold on the public imagination, it is the closest institution we have to a high court of the world. Initiating a global warming case before that body could, therefore, bring significant benefits, but the barriers to initiating such a case are also quite formidable, perhaps fatally so. My intention in this chapter is to contribute to the discussion of global warming litigation with an exploration of both the benefits of and barriers (primarily jurisdictional) to initiating a case. It updates and expands that part of my analysis from the 2003 *Environmental Law Reporter* relating specifically to the ICJ. As with the 2003 article, this chapter is not meant to be the definitive word on possibilities for litigating before the ICJ, but rather a contribution to an evolving exploration of the issue. Because the United Nations Framework Convention on Climate Change (UNFCCC) and the Kyoto Protocol establish the core of the present global warming international legal regime, they both loom large in my analysis. Yet the days of the Kyoto Protocol are numbered, and what will come after is now the subject of intensive negotiations. To the extent (as is likely) that the post-Kyoto regime draws on many of the legal structures and institutional approaches of Kyoto, much of my Kyoto specific analysis will continue to be relevant in the post-Kyoto world.

In Section 1, I continue with a general discussion of the advantages of litigating before the ICJ. In Section 2, I introduce the countries that could be potential applicants and those that could be potential respondents in a global warming suit, and I focus on evaluating the possible jurisdictional basis upon which such a suit could proceed. I conclude this section with a discussion of other procedural and substantive hurdles that would have to be overcome before a case could be decided by the ICJ. In Section 3, I then shift to reviewing briefly the nature of the substantive law that the ICJ would apply. Finally, I conclude by considering the need to view litigation

---

[9] For a view critical of the characterization of the climate change problem as of essentially global dimension, see Hari M. Osofsky, *Is Climate Change "International": Litigation's Diagonal Regulatory Role*, 49 VA. J. INT'L L. 585 (2009).

[10] Control of Emissions from New Highway Vehicles and Engines, 68 Fed. Reg. 52,922, at 52,928 (Sept. 8, 2003).

[11] Connecticut v. Am. Elec. Power Co., F. Supp. 2d 265 (S.D.N.Y. 2005).

before the ICJ in the context of the broader political strategies for responding to the global warming challenge.

1. ADVANTAGES OF LITIGATING BEFORE THE INTERNATIONAL COURT OF JUSTICE

As I will discuss, there are large hurdles to bringing a global warming suit before the ICJ, but the potential benefits of a favorable ruling on treaty negotiations over the future of the climate change regime are significant enough to make a serious exploration of prospective litigation worth the effort. Because the ICJ is the only standing international court whose subject matter competence and membership is not limited, and because of its unique status and visibility, a favorable ruling would contribute to creating a political environment conducive to the furtherance of a post-Kyoto treaty regime that can meaningfully deal with the global warming problem. In a pluralistic world of conflicting opinions, anyone can argue that it is morally wrong for countries not to do what is possible to reduce greenhouse gas emissions. However, a favorable ruling by the ICJ could provide an authoritatively sanctioned reference point around which public opinion can crystallize by imbuing that claim with the official imprimatur of law.[12]

In addition, such a ruling could alter the interstate dynamics of negotiation over the future of the treaty regime. Recalcitrant countries can regard the subjective moral claims of their negotiating partners as deserving of no greater deference in the negotiations than their own contrary claims. But the ability of parties pushing global warming remediation to appeal to neutral determinations of law adds a new dimension to the negotiations. It backs their claims with the venerable weight of respected independent legal authority and gives them the normative high ground in the negotiations.

Furthermore, to the extent that corporations face a credible threat of exposure to climate change litigation, corporate managers are likely to want to reduce that potential by encouraging their governments to join past and/or future international agreements containing clearly identifiable limits to which they can adhere.[13] While

---

[12] For a classic work discussing the role and influence of the International Court of Justice, see Robert Y. Jennings, *The United Nations at Fifty: The International Court of Justice After Fifty Years*, 89 Am. J. Int'l L. 493 (1995). *See also* NAGENDRA SINGH, THE ROLE AND RECORD OF THE INTERNATIONAL COURT OF JUSTICE (1989).

[13] For the classic case study of the interests and role of corporations in promoting regulation, see Gabriel Kolko, THE TRIUMPH OF CONSERVATISM: A REINTERPRETATION OF AMERICAN HISTORY, 1900–1916 (1977). For discussion of the potential legal liability of corporations for climate change, see David A. Grossman, *Warming Up to a Not-So-Radical Idea: Tort-Based Climate Change Litigation*, 28 COLUM. J. ENVTL. L. 1 (2003); J. Kevin Healy and Jeffrey M. Tapick, *Climate Change: It's not Just a Policy Issue for Corporate Counsel – It's a Legal Problem*, 29 COLUM. J. ENVTL. L. 89 (2004). For examinations of how various corporations are adapting to the shifting legal, regulatory, and political environment brought about by climate change, see Miquel Bustillo, *A Shift to Green: Driven by Profit and the Opportunity to Shape Regulations, Major Corporations Are Backing Stronger Measures to Reduce Global Warming*, L.A. TIMES, June 12, 2005, at C1; Jad Mouawad, *Oil Industry Moves to Curb Carbon*

the ICJ would likely rule on State responsibility under international law rather than corporate responsibility under domestic law, its rulings would carry liability implications for corporations. In potential domestic nuisance or negligence cases against corporations for causing harm, it is necessary to establish that the defendant corporation's contribution to the global warming problem contravened some community-wide standard of behavior.[14] A decision by the ICJ could help to establish the existence of such standards and perhaps be a guide as to the limits on corporate greenhouse gas emissions they require.

Finally, as will be discussed in Section 2, the mere fact that countries join a climate change regime does not ensure compliance with that regime. Moreover, even such compliance may not be adequate to meet the whole compliment of their remedial obligations under international law. As Section 3 explains, standards derived from customary international law and general principles of international law as evidenced frequently in judicial and arbitral decisions, solemn declarations, and restatements can also effect the ultimate obligation of states to limit their contribution to the global warming problem. A ruling by the ICJ can help to put pressure on countries to comply with their obligations, and it can help clarify the full extent of these obligations.

2. GETTING INTO THE ICJ

2.1. *Contentious Cases*

2.1.1. Applicants and Respondents

Only countries can bring suits against other countries before the ICJ.[15] Determining which applicant State or States could most effectively bring such a suit would not be simple. Almost all of us today are participants in the carbon economy. We are both contributors to, as well as victims of, global warming. Having said that, some are contributing orders of magnitude more than others to the problem. For example, the average citizen in the United States is responsible for emitting over forty times more greenhouse gases into the environment than the average citizen of Kiribati.[16] And some, in contrast, are bearing the brunt of the effects of global warming and will continue to do so into the foreseeable future. The most obvious applicant

---

Emissions: The Energy Challenge: Big Oil, Small Step, N.Y. Times, June 30, 2006, at C1; Steven Mufson & Juliet Eilperin, *Energy Firms Come to Terms with Climate Change*, Wash. Post, Nov. 25, 2006, at A1; Daniel B. Wood, *On Road to Clean Fuels, Automakers Cover Some Ground*, Christian Sci. Monitor, Dec. 1, 2006, at 01; Claudia H. Deutsch, *Selling Fuel Efficiency the Green Way*, N.Y. Times, Dec. 11, 2006, at C7; John O'Dell, *So Who's the Greenest of Them All?: Well, It Depends on Who You're Talking To: In Any Case, There's Hot Competition Among Car Makers to Lay Claim to the Eco-Friendly Crown*, L.A. Times, Mar. 29, 2000, at 1.

[14] See generally Dan D. Dobbs, The Law of Torts 393–403 (2000).
[15] Statute of the International Court of Justice, effective Oct. 24, 1945, art. 34 (1), 59 Stat. 1031, 1060, T.S. No. 993.
[16] See World Resources Institute Chart of Total Greenhouse Gas Emissions in 2000, *available at* http://cait.wri.org/cait.php?page=yearly (last visited Apr. 30, 2007).

countries, therefore, are those that have contributed little to the problem and are most victimized by it. Low-lying Pacific island countries such as Kiribati whose very existence is imperiled by global warming have been most often mentioned.[17] A few years ago, the small Pacific island nation of Tuvalu, for example, considered trying to bring a claim against the United States before the ICJ.[18]

Another category of applicant countries that has not been considered are developed country parties to the Kyoto Protocol.[19] Specifically identified in Annex 1 to the Protocol, these countries bear almost the entire burden for reducing greenhouse gases.[20] Consistent with the increasingly accepted principle that countries have common but differentiated responsibilities to remediate environmental problems,[21] the Protocol puts the onus on them because of the developed world's disproportionate wealth and historical contribution to the global warming problem. To the extent, therefore, that such developed countries are themselves victims of global warming, a potential claim could be explored against fellow developed countries that are not bearing their share of the responsibility for the global warming problem, either because they do not appear to be on track to meet their emission reduction obligations, including under the Protocol, or they have not acceded to the Protocol and are not otherwise bearing their share of the responsibility for the global warming problem.

Whether either vulnerable developing countries or developed countries that are making a serious effort to deal with the global warming problem could successfully bring a lawsuit before the ICJ presents the threshold question of whether the ICJ would find it had jurisdiction over the dispute. In accordance with the principle of State sovereignty, jurisdiction by the Court must ultimately be based upon State

---

[17] Low-lying coastal states such as Bangladesh are also particularly at risk. For a discussion of the probable effects of global warming on low-lying coastal island states, see Intergovernmental Panel on Climate Change, *Climate Change 2007: Impacts, Adaptation and Vulnerability. Contribution of Working Group II to the Fourth Assessment Report of the Intergovernmental Panel on Climate Change* (2007), at 481. For a discussion of the extent to which international law protects vulnerable island states from harms caused by global warming, see William C.G. Burns, *Potential Implications of Climate Change for the Coastal Resources of Pacific Island Developing Countries and Potential Legal and Policy Responses*, 8(1) HARV. ASIA-PAC. REV. 1–8 (2005). *See also* Nicholas D. Kristof, *Island Nations Fear Sea Could Swamp Them*, N.Y. TIMES, Dec. 1, 1997, at F9.

[18] Koloa Talake, the prime minister who was the driving force behind the lawsuit, lost reelection in August 2002, and the subsequent government did not pursue the litigation. *See* Leslie Allen, *Will Tuvalu Disappear Beneath the Sea? Global Warming Threatens to Swamp a Small Island Nation*, SMITHSONIAN, Aug. 1, 2004, at 44.

[19] Protocol to the United Nations Framework Convention on Climate Change (UNFCCC), art 3, Dec. 11, 1997, 37 I.L.M. 32 (1998), *available at* http://unfccc.int/resources/docs/cpmvp/kpeng.html.

[20] *Id.*

[21] The idea that international agreements should place different burdens on differently situated states predates modern international environmental law. The term first appears explicitly in the United Nations Framework Convention on Climate Change (UNFCCC), *see infra* note 20, but the concept has been integrated into earlier international environmental agreements. For further discussion, see Christopher D. Stone, *Common but Differentiated Responsibilities in International Law*, 98 AM. J. INT'L L. 276 (2004). For an exploration of the idea applied specifically to climate agreements, see Comment, *Rethinking the Equitable Principle of Common but Differentiated Versus Absolute Norms of Compliance and Contribution in the Global Climate Change Context*, 13 COLO. J. INT'L ENVT'L L. & POL'Y 473 (2002).

consent, which can be manifest in three ways. The first way would be for disputing parties to agree to refer a matter to the Court pursuant to Article 36(1) of its Statute.[22] The second way the Court could attain jurisdiction is if under the so-called optional clause of the Article 36(2) of the ICJ Statute, the respondent State has prospectively entered a declaration accepting the compulsory jurisdiction of the Court for the kind of dispute being litigated, and the applicant State has allowed in its own declaration that, in accordance with the *rule of reciprocity*, it would itself be subject to the Court's jurisdiction were it to be sued in a case of a similar nature.[23] Finally, the third way that the Court could gain jurisdiction, also pursuant to Article 36(1), is if the parties have specifically provided for dispute resolution before the Court in a pertinent treaty which is in effect between the parties.[24]

2.1.2. Referral to the ICJ by Mutual Agreement

It is unlikely that a developed country being challenged by either a developing or developed country for a claimed failure to deal sufficiently with its emissions of greenhouse gases would agree to have that claim adjudicated by the ICJ. The ICJ has over time heard many cases under the referral by mutual agreement provision of Article 36(1).[25] However, almost all of them have been in the nature of boundary disputes where the disputing parties both desired an independent and authoritative resolution of a thorny political problem.[26] In the global warming context, it is quite unlikely that a targeted State would see itself as having an interest in exposing itself to a potentially adverse ICJ decision.

2.1.3. Compulsory Jurisdiction under the Optional Clause

The viability, on the other hand, of establishing jurisdiction under Article 36(2) would depend upon the coincident existence of applicant and respondent parties who had accepted the compulsory jurisdiction of the ICJ over such a dispute. Of

---

[22] I.C.J. art. 36(1), Stat. 1031, 1060, T.S. No. 993.
[23] *Id.* at art. 36(2).
[24] *Id.* at art. 36(1).
[25] For some representative cases, see Minquiers and Ecrehos (Fr. v. U.K), 1953 I.C.J. 47 (Nov. 17); Sovereignty over Certain Frontier Land (Belg. v. Neth.), 1959 I.C.J. 209 (June 20); North Sea Continental Shelf (F.R.G. v. Den., F.R.G. v. Neth.), 1969 I.C.J. 3 (Feb. 20); Continental Shelf (Libya v. Malta), 1984 I.C.J. 3 (Mar. 21); Frontier Dispute (Burk. Faso v. Mali), 1986 I.C.J. 554 (Dec. 22); Land, Island and Maritime Frontier Dispute (El Sal. v. Hond., Nicar. intervening), 1992 I.C.J. 351 (Sept. 11); Territorial Dispute (Libya v. Chad), 1994 I.C.J. 6 (Feb. 3); Grabčikovo-Nagymaros Project (Hung. v. Slovk.), 1997 I.C.J. 7 (Sept. 25); Kasikili/Sedudu Island (Bots. v. Namib.), 1999 I.C.J. 1045 (Dec. 13); Pulau Ligitan and Pulua Sipadan (Indon. v. Malay.), 2002 I.C.J. 625 (Dec. 17); Frontier Dispute (Benin v. Niger), 2005 I.C.J. 90 (July 12); Territorial and Maritime Dispute Between Nicaragua and Honduras in the Caribbean Sea, 2007 I.C.J. (Oct. 8).
[26] For reference to the prevalence of land and sea delimitation cases, see TERRY D. GILL, ROSENNE'S THE WORLD COURT: WHAT IT IS AND HOW IT WORKS (2003). *See also* Todd L. Allee & Paul K. Huth, *Legitimizing Dispute Settlement: International Legal Rulings as Domestic Political Cover*, 100 AM. POL. SCI. REV. 219, 220–21, 229–32 (2006) (discussing the domestic political advantages of referring bilateral disputes to the ICJ).

the category of unambiguously developed countries, only two, the United States and Australia, refused timely ratification of the Kyoto Protocol. Australia, however, has now ratified the Protocol, leaving the United States as the sole remaining holdout. And with the Obama administration now in office, the United States is poised to play a meaningful role in negotiating the post-Kyoto regime.[27]

Among the more economically significant countries that are not party to the Kyoto Protocol, Turkey also has neither signed nor ratified the agreement to reduce its greenhouse gas emissions. Other countries that are not party to the Kyoto Protocol include Afghanistan, Andorra, Brunei, Central African Republic, Chad, Comoros, Iraq, Kazakhstan, Saint Kitts and Nevis, San Marino, São Tomé and Principe, Somalia, Tajikistan, Timor-Leste, Tonga, and Zimbabwe. Among the nonmember countries, only Somalia has acceded to the compulsory jurisdiction of the ICJ.[28] Somalia, one of the least developed countries in the world, is in political turmoil and is, in any event, a very low emitter of greenhouse gases. Of the countries that have acceded to the Kyoto Protocol, the most likely targets of an international liability claim would be those whose compliance with that agreement is in question. The primary requirement the Protocol imposes is that the developed country members (Annex 1 Countries) make reductions in their greenhouse gas emissions during the period 2008–2012,[29] and that by 2005 they have made demonstrable progress toward this commitment.[30] In addition, all of the parties to the Kyoto Protocol are also parties to the master agreement, the UNFCCC, which requires more broadly in Article 4.2(a) that the developed countries take measures to mitigate climate change by limiting their anthropogenic emissions of greenhouse gases.[31]

---

[27] *Rudd Takes Australia Inside Kyoto*, BBC, Dec. 3, 2007, *available at* http://news.bbc.co.uk/2/hi/asia-pacific/7124236.stm.

[28] The United States, the only unambiguously developed country not to have now acceded to the Kyoto Protocol, has withdrawn from the compulsory jurisdiction of the ICJ. *See infra* note 45 and accompanying text.

[29] Protocol to the United Nations Framework Convention on Climate Change (UNFCCC), art. 3, Dec. 11, 1997, 27 I.L.M. 32 (1998), *available at* http://unfccc.int/resource/docs/convkp/kpeng.pdf [hereinafter Kyoto Protocol].

[30] *Id.* at art 3.

[31] United Nations Framework Convention on Climate Change, art. 4.2(a), May 9, 1992, 1771 U.N.T.S. Article 4.2(a) in its entirety reads as follows:

> The developed country and other Parties included in Annex I commit themselves specifically as provided for in the following:
>
> (a) Each of these Parties shall adopt national policies and take corresponding measures on the mitigation of climate change, by limiting its anthropogenic emissions of greenhouse gases and protecting and enhancing its greenhouse gas sinks and reservoirs. These policies and measures will demonstrate that developed countries are taking the lead in modifying longer-term trends in anthropogenic emissions consistent with the objective of the Convention, recognizing that the return by the present decade to earlier levels of anthropogenic emissions of carbon dioxide and other greenhouse gases not controlled by the Montreal Protocol would contribute to such modification, and taking into account the differences in these Parties' starting points and approaches, economic structures and resource bases, the need to maintain strong and sustainable economic growth, available technologies and other individual circumstances, as well as the need for equitable and

Among the clearly developed Annex 1 countries that appear most on track to meet their 2008–2012 Kyoto reduction commitments are Britain, Sweden, and Iceland.[32] Of the three, Britain and Sweden have acceded to the ICJ Article 36 optional clause. Austria, Belgium, Denmark, Ireland, Italy, Liechtenstein, Norway, Portugal, Spain,[33] Canada,[34] and New Zealand[35] are among the countries least on track for meeting their 2008–2012 Kyoto reduction commitments and are, therefore, arguably not in compliance with the Kyoto requirements and, more generally, with Article 4.2(a) of the UNFCCC.[36] All of these countries except Ireland and Italy have acceded to the optional clause of the ICJ. Complicating ICJ jurisdiction over them, however, is the fact that the Kyoto Protocol has its own dispute resolution provisions. Article 19 of the Protocol incorporates by reference mutatis mutandis Article 14 of the UNFCCC, which provides first under paragraph 1 that parties can jointly seek settlement of their dispute "though negotiation or any other peaceful means of their own choice."[37] Alternatively, Article 14, Paragraph 2, provides that a complaining party can unilaterally refer a UNFCCC or Protocol dispute to the ICJ or to binding arbitration, providing that each of the parties has entered a prospective declaration accepting the respective forum for the type of dispute in question. If there is no unilateral referral under Paragraph 2, and if the parties are unable to resolve their dispute within twelve months under Paragraph 1, any party to the dispute can submit it to conciliation by a commission established pursuant to the UNFCCC.[38]

---

appropriate contributions by each of these Parties to the global effort regarding that objective. These Parties may implement such policies and measures jointly with other Parties and may assist other Parties in contributing to the achievement of the objective of the Convention and, in particular, that of this subparagraph. *Id.*

[32] *See* EUROPEAN ENVIRONMENT AGENCY, GREENHOUSE GAS EMISSIONS TRENDS AND PROJECTIONS IN EUROPE (2006), *available at* http://reports.eea.europa.eu/eea_report_2006_9/en/eea_report_9_2006.pdf.

[33] European Environment Agency, *E.U. Must Take Immediate Action on Kyoto Targets* (2006), *available at* http://www.eea.europa.eu/pressroom/newsreleases/ghgtrends2006-en.

[34] Ian Austin, *Canada Announces Goals for Reducing Emissions*, N.Y. TIMES, Apr. 27, 2007, at C7.

[35] *NZ Greenhouse Gases Keep Rising*, N.Z. PRESS ASS'N, May 4, 2007.

[36] Complicating a legal action against Austria, Belgium, Denmark, Ireland Italy, Portugal, Spain or any of the fifteen European countries that were members of the European Union at the time the Kyoto Protocol was negotiated is that pursuant to Article 4 of the Protocol those fifteen countries can fulfil their mutual reduction commitments in an aggregate way. The European Environmental Agency maintained that as of late 2008 those countries were on track to meeting their collective commitments. *See* European Environmental Agency, *EU-15 on Target for Kyoto, Despite Mixed Performances* (2008), *available at* http://www.eea.europa.eu/pressroom/newsreleases/eu-15-on-target-for-kyoto-despite-mixed-performances. Because of Australia's late ratification of the Kyoto Protocol at the end of 2007, it only committed to stabilizing greenhouse gases at 108% of 1990 levels by 2012. Even meeting this modified target, however, will be difficult. *See* Rosslyn Beeby, *Push for Quicker Green Target*, CANBERRA TIMES, Feb. 15, 2008, at A15. Japan's compliance is also questionable, but that country is making very significant efforts. *See* Shigeru Sato and Yuji Okada, *Japan Utilities to Buy Carbon Credits: Steel Makers Also Push to Cut Greenhouse Gases in Nation*, INT'L HERALD TRIB., Oct. 12, 2007, at 19. Both Australia and Japan have acceded to the optional clause of the ICJ.

[37] United Nations Framework Convention on Climate Change, art. 14.1, May 9, 1992, 1771 U.N.T.S.

[38] *Id.* at art. 14.6.

To date, no country has opted into binding jurisdiction before the ICJ under Article 14 and neither arbitration nor conciliation procedures called for by the UNFCCC have been established. The failure of countries to enter UNFCCC Article 14 declarations granting the ICJ jurisdiction over matters specifically under the climate change regime should not preclude the Court from adjudicating climate change claims pursuant to those countries' general acceptance of ICJ Article 36 optional clause jurisdiction. States only need consent to the jurisdiction of the Court once, and disputes over treaty interpretation are among the conflicts that the ICJ is empowered to adjudicate under Article 36.[39]

There is, however, a problem. All of the nine countries that have accepted the compulsory jurisdiction of the ICJ under Article 36 – except for Denmark, Liechtenstein, and Norway – have entered reservations to their acceptances excepting disputes which the parties agree to settle by *other means of peaceful settlement*.[40] While the system envisioned in Article 14 would seem to constitute *other means of peaceful settlement*, the fact that no party has opted into Article 14 ICJ jurisdiction, and that neither the procedures for arbitration nor conciliation called for by Article 14 have ever been adopted by the parties, could be interpreted to mean there is, in fact, no final or implementable agreement providing for an *other means of peaceful settlement* under the parties' reservations.[41]

In addition, arguably the fact that the parties to a dispute had previously opted into the optional clause of Article 36 makes settlement by the ICJ an "other peaceful means of [the parties'] own choice" under Paragraph 1 of Article 14, and for parties to have opted into ICJ jurisdiction under Paragraph 2 would have been redundant. It would be harder to make this claim if the mechanisms for arbitration were ever to be established and contesting parties were to have declared their acceptance of arbitration. Of course, the relatively short twelve-month time period envisioned in Paragraph 5 for a party to submit the dispute to conciliation if the parties have not been able to "settle their dispute" would not seem to contemplate the more lengthy process of the ICJ.[42]

---

[39] I.C.J. art. 36(2) (a), Stat. 1031, 1060, T.S. No. 993

[40] For example, the reservation in Austria's declaration reads as follows: "This Declaration does not apply to any dispute in respect of which the parties thereto have agreed or shall agree to have recourse to other means of peaceful settlement for its final and binding decision." Arguably neither negotiation under Article 14.1 nor conciliation under Article 14.6 would constitute "a final and binding decision." The Canadian formulation, on the other hand, reserves from its acceptance of compulsory jurisdiction, "disputes in regard to which the parties have agreed or shall agree to have recourse to some other method of peaceful settlement." For the complete collection of Article 36 declarations accepting the binding jurisdiction of the ICJ, see The International Court of Justice, *Declarations Recognizing the Jurisdiction of the Court as Compulsory*, http://www.icj-cij.org/jurisdiction/index/php?p1=5&p2=1&p3=3.

[41] Supporting such a restrictive reading of a *settlement by other peaceful means* reservation as not divesting the predecessor court to the ICJ of jurisdiction despite a later dispute resolution agreement between the parties, see Electricity Co. of Sofia and Bulgaria, Judgment, 1939 PCIJ (ser.A/B) No. 77, at 62. For further discussion of the meaning of *settlement by other peaceful means* reservations, see Bernard Oxman, *Complementary Agreements and Compulsory Jurisdiction*, 95 Am. J. Int'l L. 277 (2001).

[42] One additional argument a party attempting to use an *other means of peaceful settlement* clause to divest the ICJ of jurisdiction might make is that Article 18 of the Kyoto Protocol constitutes an

My general conclusion is that a persuasive case could be made that the ICJ could assert jurisdiction over disputes under the UNFCCC and the Protocol if they involve countries that have opted into the binding jurisdiction of that Court regardless of whether they have done so subject to an *other means of peaceful settlement* provision. At the end of the day, however, whether the ICJ can assert jurisdiction under the UNFCCC and the Protocol may not be relevant to the larger question of whether it can assert jurisdiction in a climate change case generally. This is because countries attempting to formulate climate change claims so as to achieve maximum impact in an ICJ proceeding would be unlikely to conceptualize them as solely a question of compliance with the UNFCCC and the Kyoto Protocol even if they and their adversaries were party to these agreements.[43] As I discuss in Section 3 of this chapter, other norms of international law may also be relevant as well, and to the extent that a climate change action is framed as a broader question of State responsibility for environmental harm under international law, the dispute resolution provisions of specific treaties would most likely not be directly applicable.[44] After all, the UNFCCC and the Kyoto Protocol do not definitively settle the question of who

---

other means of peaceful settlement. Article 18 directs the parties to "approve appropriate and effective procedures and mechanisms to determine and to address cases of non-compliance with the provisions of [the] protocol." Kyoto Protocol, *supra* note 29, at art. 18. Unlike the dispute resolution provisions of Kyoto Article 19 and UNFCCC Article 14, the parties have taken action to create the compliance mechanisms called for by Article 18. Because Article 18 does not provide for parties to resolve disputes between each other, however, it can more accurately be characterized as a provision dealing with enforcement rather than dispute settlement of the sort envisioned by the declarations.

[43] This is likely as 170 states have now ratified the Kyoto Protocol. See Kyoto Protocol Status of Ratification, *available at* http://unfccc.int/files/kyoto_protocol/background/status_of_ratification/application/pdf/kp_ratification.pdf.

[44] The issues involving the relationship between the Framework Convention on Climate Change and the Kyoto Protocol and other international legal obligations is a complex one involving the relationship between these specific international agreements and more general principles of international law, including customary international law. *See generally* Dinah Shelton, *Normative Hierarchy in International Law*, 100 AM. J. INT'L. L. 291 (2006). For a discussion of the implications of the relationship between treaty law and customary international law in the ICJ's assertion of compulsory jurisdiction in the Nicaragua case, see Monroe Leigh, *Military and Paramilitary Activities in and Against Nicaragua*, 81 AM. J. INT'L L. 206 (1987).

Complicating a comprehensive determination by the ICJ of the extent to which under treaty and customary law a state party to the Kyoto Protocol may be derelict in its obligation to help remedy the global warming problem is that the Kyoto Protocol provides for a variety of financial mechanisms that states can pay into as an alternative to reducing greenhouse gas emissions. Under Article 6 of the Protocol, countries that fail to meet their domestic emissions reduction commitments may contribute financially to the reduction of emissions in other Annex 1 countries, or alternatively, they may buy the right to exceed their own emissions quotas in the form of "emissions reductions units" from other Annex 1 countries who reduce their own emissions by more than their commitments require. Kyoto Protocol, *supra* note 29, at art. 6. Also under the Clean Development Mechanism of Article 12, Annex 1 countries can compensate for exceeding their commitments by funding offsetting projects in developing countries. *Id.* at art. 12. Finally, pursuant to Article 18 of the Protocol, the parties determined that countries that fail to comply with the Kyoto Protocol will be assigned an amount from the second commitment period of a number of tons equal to 1.3 times the amount in tons of excess emissions. *Id.* at art. 18. It is unclear how the ICJ might factor in such a penalty to the overall obligations that a country might have under international law.

should bear the considerable cost of global warming which will persist even if the UNFCCC, and the Kyoto Protocol are fully complied with.

2.1.4. Jurisdiction by Way of Independent Treaty

The ICJ can also take jurisdiction under Article 36 if the parties to the litigation have agreed to an independent treaty with a dispute resolution clause specifying settlement before the ICJ. The difficulty for the purposes of this chapter is to find such a clause in a treaty whose subject matter arguably covers global warming. In my 2003 *Environmental Law Reporter* article, I specifically examined ICJ dispute resolution clauses in independent treaties that might provide for jurisdiction over the United States. Although many countries have entered into treaties with such clauses, the United States makes for the most logical focus of this study as it has rescinded its acceptance of ICJ compulsory jurisdiction.[45] In addition, it continues to be the world's largest per capita emitter of greenhouse gases, and during the Bush administration it refused to ratify the Kyoto Protocol.

In my research, I found that the United States has entered into many Friendship, Commerce, and Navigation (FCN) or other similar treaties. These are general agreements that provide that parties treat each other's citizens as favorably as they treat their own citizens in commercial transactions. Because I thought these agreements might contain generally worded obligations in the nature of good faith between the parties, I looked into FCN treaties and other similar agreements between the United States and coastal and island States[46] that provided for dispute resolution before the ICJ. Typical of the most relevant language to be found in these treaties is the passage from the United States' agreement with Greece: "Each Party shall at all times accord equitable treatment to the persons, property, enterprises and other interests of nationals and companies of the other Party."[47]

Other similarly situated coastal nations with which the United States has such agreements containing roughly the equivalent language and binding dispute resolution before the ICJ are Thailand,[48] the Netherlands,[49] Korea,[50] Denmark,[51]

[45] The United States, in response to the ICJ's determination to assert jurisdiction over it in the Nicaragua case in 1986, withdrew its acceptance of the court's compulsory jurisdiction. Military and Parliamentary Activities in and Against Nicaragua (Nicar. v. U.S.), 1986 I.C.J. 14 (June 27).

[46] As mentioned in Section 2.1.1, island and coastal States are thought to be particularly vulnerable to the ill effects of global warming because of rising sea levels and severe coastal weather. *See supra* note 17 and accompanying text.

[47] Treaty of Friendship, Commerce and Navigation between the United States and the Kingdom of Greece, Aug. 3, 1951, U.S.-Greece, art. I, 5 U.S.T. 1829, 1835.

[48] Treaty of Amity and Economic Relations between the United States of America and the Kingdom of Thailand, May 29, 1966, U.S.-Thail., art. XIII, para. 2, 19 U.S.T. 5843, 5859.

[49] Treaty of Friendship, Commerce and Navigation between the United States of America and the Kingdom of the Netherlands, Mar. 27, 1956, U.S.-Neth., art. XX25, para. 2, 8 U.S.T. 2043, 2083.

[50] Treaty of Friendship, Commerce and Navigation between the United States of America and the Republic of Korea, Nov. 28, 1956, U.S.-S. Korea, art. XXIV, para. 2, 8 U.S.T. 2217, 2227.

[51] Treaty of Friendship, Commerce and Navigation between the United States of America and the Kingdom of Denmark, Oct. 1, 1951, U.S.-Den., art. XXIV, para. 2, 12 U.S.T. 908, 923.

and Ireland.[52] Ethiopia, although no longer a coastal State, in its Treaty of Amity and Economic Relations with the United States has particularly promising language: "There shall be constant peace and firm and lasting friendship between the United States of America and Ethiopia,"[53] and "The two High Contracting Parties reiterate their intent to further the purposes of the United Nations."[54] I could find no such treaties containing provisions providing for binding dispute resolution before the ICJ with small island nations.

The previously mentioned treaties attempt generally to prescribe how each party within its own country should treat the other country's nationals and their property. U.S. greenhouse gas emissions arguably harm foreign nationals and their property within their own countries. It is, of course, possible to argue something along the lines that while the parties may not have specifically contemplated such an application of these treaties, to the extent that they are meant to prescribe against harm to foreign interests inside American jurisdiction, then certainly they cannot have meant to allow a fundamentally more egregious extension of harm by the United States extending outside of its own boundaries.

The ICJ has had the opportunity to rule in a different substantive context on a similar attempt to construe a FCN treaty to provide a basis for jurisdiction in the preliminary phase of *The Case Concerning Oil Platforms (Islamic Republic of Iran v. United States)*.[55] In that case, Iran petitioned the ICJ to accept jurisdiction over a dispute involving the destruction by the U.S. Navy of three Iranian oil complexes during the Iran-Iraq War. The basis for Iran's claim that the Court had jurisdiction was found in the clause allowing for dispute resolution by the ICJ under the United States/Iran FCN treaty, the Treaty of Amity, Economic Relations and Consular Rights.[56] Iran argued that several general treaty provisions of the sort that I have identified were violated by the United States military action. The Court, in finding that it had jurisdiction, accepted the position that the FCN treaty had extraterritorial application. For example, the Court construed the requirement that a Party accord the other Party's nationals fair and equitable treatment as not applying solely within its territory. The decision is, however, somewhat more qualified in its acceptance of the sort of broad interpretation of language that would be helpful in a global warming case.[57] For example, it read the requirement of fair and equitable treatment as not including the protection of a party's nationals from military actions by the other party. The Court, on the other hand, decided that military activities which destroy

---

[52] Treaty of Friendship, Commerce and Navigation between the United States of America and Ireland, Jan. 21, 1950, U.S.-Ir., art. XXIII, 1 U.S.T. 785, 795.
[53] Treaty of Amity and Economic Relations between the United States of America and Ethiopia, Sept. 7, 1951, U.S.-Eth., art VIII, para. 1, 4 U.S.T. 2134, 2141.
[54] Treaty of Amity and Economic Relations between the United States of America and Ethiopia, Sept. 7, 1951, U.S.-Eth., art. I, para. 2, 4 U.S.T. 2134, 2136.
[55] Concerning Oil Platforms (Iran v. U.S.), 1996 I.C.J. 803.
[56] Treaty of Amity, Economic Relations, and Consular Rights, Aug. 15, 1955, U.S.-Iran, 8 U.S.T. 899, 901.
[57] *Concerning Oil Platforms*, 1996 I.C.J. at 814.

or impede the transportation or storage of exports implicate the treaty's requirement that the parties uphold freedom of commerce between their territories.[58] This raises the question of whether such general language could be violated to the extent that a country's contribution to global warming can be shown to affect negatively an FCN treaty partner's ability to engage in commerce (say by indirectly damaging its economy or directly flooding a port city).

Similarly, in the *Nicaragua* case against the United States, referred to earlier, the Court also accepted jurisdiction based in part on a binding ICJ dispute resolution provision in an FCN treaty in force between the parties.[59] In that case, as in the Iran case, military activities arguably more directly impacted upon specific provisions of the treaty than would global warming. Ultimately, then, the jurisdictional question in applying FCN treaties to global warming cases would be whether treaties negotiated in the context of protecting the mutual commercial interests of countries' citizens can be construed to protect them from harm caused by global warming. The *Oil Platforms* and *Nicaragua* cases give reason to believe that such a construction by the ICJ is possible.

2.1.5. Other Procedural and Substantive Issues

In addition to jurisdictional issues, there are other very significant procedural hurdles in contentious (nonadvisory) cases that would have to be overcome before a global warming suit could proceed to the merits of the case. Most significant would be the issue of standing, whether applicants have a sufficiently individualizable interest in litigation as to be able to bring the suit. Alternatively, it could be demonstrated that countries' obligations not to cause serious harm through the emissions of greenhouse gases is an obligation *erga omnes* (i.e., that such obligation is sufficiently important that all States have a legal interest in its enforcement).

Assuming that a tribunal in a global warming lawsuit would accept the scientific consensus that human-created greenhouse gases are a major contributor to global warming, other significant proof problems would remain in bringing such a suit. A connection would need to be drawn between global warming and specific environmental effects.[60] In addition, both assessing prospective damages from global warming and apportioning the extent to which they are attributable to any specific

---

[58] *Concerning Oil Platforms*, 1996 I.C.J. at 819–20.
[59] Military and Paramilitary Activities in and Against Nicaragua (Nicar. v. U.S.), 1986 I.C.J. 14, 116, 136.
[60] As the science of global warming rapidly develops, such connections are becoming easier to establish with reasonable scientific certainty. The highly credible United Nations Intergovernmental Panel on Climate Change, for example, concluded with "very high confidence" in its 4th Assessment Report that there is warming of lakes and rivers in many regions with effects on water quality and that global warming is causing earlier timing of spring events such as leaf-unfolding, bird migration and egg-laying and poleward shifts in ranges on plant and animal species. It additionally concluded with "high confidence" that changes in snow, ice, and frozen ground are increasing ground instability in permafrost regions and rock avalanches in mountain regions and that rising ocean and fresh water temperatures are causing changes in the ice cover, salinity, oxygen levels, and circulation including changes in algal, plankton, and fish abundance in high altitude oceans. *See* Intergovernmental Panel on Climate

country would be challenging and perhaps could render a case infeasible. The law in this area is not unique to global warming,[61] and it is beyond the scope of this chapter to specifically review it. I only note these considerations here as factors to which careful consideration would have to be given in conceiving a contentious global warming case before the International Court of Justice.

2.2. Advisory Opinions

There is another possible avenue that would facilitate an ICJ decision on the legal responsibility of countries to participate meaningfully in the remediation of the global warming problem, but that does not require that the Court have the ability to assert jurisdiction over any specific countries. Pursuant to Article 65 of the ICJ's Statute, the Court is empowered "to give an advisory opinion on any legal question at the request of whatever body may be authorized by or in accordance with the Charter of the United Nations [U.N.] to make such a request."[62] Article 96 of the Charter of the U.N. provides that "[t]he General Assembly or the Security Council may request the [ICJ] to give an advisory opinion on any legal question,"[63] and that "[o]ther organs of the [U.N.] and specialized agencies, which may at any time be so authorized by the General Assembly, may also request advisory opinions of the Court on legal questions arising within the scope of their activities."[64]

Pursuing an advisory opinion was the path followed by the civil society–led initiative to get the ICJ to rule on the legality of nuclear weapons in the 1990s. In that case, both the General Assembly as well as the World Health Organization (WHO) requested an advisory opinion.[65] The Court recognized that the General Assembly

---

Change (IPCC), Working Group II, Climate Change 2007: IMPACTS, ADAPTATION AND VULNERABILITY, SUMMARY FOR POLICYMAKERS 1–3 (2007), *available at* http://www.ipc.ch (last visited Feb. 27, 2008).

In addition, courts themselves seem increasingly receptive of such conclusions. In determining that the state of Massachusetts claimed sufficient injury for standing to bring suit in *Massachusetts v. EPA*, the U.S. Supreme Court observed that:

> The harms associated with climate change are serious and well recognized. Indeed, the [National Research Council Report] itself – which EPA regards as an "objective and independent assessment of the relevant science," identifies a number of environmental changes that have already inflicted significant harms, including "the global retreat of mountain glaciers, reduction in snow-cover extent, the earlier spring melting of rivers and lakes [and] the accelerated rate of rise of sea levels during the 20th century relative to the past few thousand years ... "
> Massachusetts v. EPA, 549 U.S. 497, 521 (Apr. 2, 2007) (citations omitted).

[61] *See, e.g.*, Summers v. Tice, 199 P.2d 1 (Cal. 1948) (where two hunters negligently fired their shotguns in the direction of the plaintiff on a hunting trip, the burden of proof is on the defendant to absolve herself of liability); Sindell v. Abbott Labs., 607 P.2d 924 (Cal. 1980) (where almost 200 manufacturers produced DES, a toxic compound that caused the plaintiffs' cancer, the court held each defendant liable for the proportion of the judgment represented by its share of the market).

[62] I.C.J., art. 65, 59 Stat. 1031, 1063, T.S. No. 993.

[63] Charter of the United Nations, effective Oct. 24, 1945, art. 96(1), 59 Stat. 1031, 1052, T.S. No. 993.

[64] *Id.* at art. 96(2).

[65] The WHO was authorized by the General Assembly to request advisory opinion from the ICJ pursuant to the agreement governing its relationship to the United Nations. *See* Agreement Between the United

could request an advisory judgment in the matter, but it ruled against the WHO.[66] It explained that the WHO was authorized to "deal with the effects on health of the use of Nuclear Weapons, or of any hazardous activity, and to take preventative measures aimed at protecting the health of populations in the event of such weapons being used or such activities engaged in."[67] The Court concluded, however, that, "[w]hatever those effects might be, the competence of the WHO to deal with them is not dependent on the legality of the acts that caused them."[68]

The Court is not technically bound by prior decisions,[69] but as a practical matter, it does tend to follow them, and the global warming case would seem to be very similar. Perhaps it could be distinguished because of the WHO's need to be involved in ongoing strategies for adapting to global warming as it relates to public health. Given global warming's likely effect on agriculture, the other potential candidate to request an advisory opinion would be the Food and Agricultural Organization (FAO) in Rome,[70] but it would likely face the same problem as the WHO.

The Security Council, especially given the ability of any one of its permanent five members to cast a veto, would not be likely to authorize a request for an advisory opinion. The General Assembly would seem to be more promising. Pursuant to Article 18 of the U.N. Charter, "important" questions require a two-third's majority of the General Assembly.[71] The ICJ, however, agreed to render an opinion in the nuclear weapons case with only a majority (of less than two thirds) voting in favor. Even this lower threshold could, however, be difficult to achieve. Unlike the nuclear weapons case where only a handful of countries actually had nuclear weapons, many countries are significant emitters of greenhouse gases. Depending on how narrowly the question presented to the ICJ could be framed, these countries might well be reluctant to charge the ICJ with coming to a determination that could implicate the legality of their own emissions.

One disadvantage of the advisory approach is that in terms of publicity value (which is helpful for achieving the benefits I refer to in Section 1 of this chapter) identifiable applicants and respondents in contentious cases might better capture the public imagination than would a simple statement of the law in an advisory case. Recommending the advisory approach, however, is its simplicity. It requires no imaginative theories of jurisdiction, and it avoids singling out countries simply

---

Nations and the World Health Organization, adopted by the First World Health Assembly, 10 Jul. 1948, art. X.

[66] Legality on the Threat or Use of Nuclear Weapons (United Nations), 1996 I.C.J. 226, 235.
[67] Id.
[68] Id.
[69] I.C.J., art. 59, 59 Stat. 1031, 1063, T.S. No. 993.
[70] The FAO has also been authorized by the General Assembly to request advisory opinions from the ICJ pursuant to the FAO's agreement governing its relationship to the United Nations. See Agreement between the United Nations and the Food and Agricultural Organization of the United Nations. Feb. 1947, art. IX. Para. 2.
[71] Charter of the United Nations, art. 18, para. 3, provides that "[d]ecisions on other questions ... shall be made by a majority of the members present and voting." Id.

because they are subject to the jurisdiction of the Court. Ultimately, it has the advantage of articulating a clear legal standard equally applicable to all states.

3. A BRIEF LOOK AT THE LAW THE ICJ WOULD APPLY

With the exception of the UNFCCC and its Kyoto Protocol, the international community has not developed specific treaties to deal explicitly with the normative dimensions of the global warming problem. Asked to decide comprehensively upon the responsibility of States to ameliorate global warming, the Court would also look to other international treaties of a more general nature, customary norms of international law, and general principles of international law.[72] To help ascertain the content of the relevant principles of customary international law and general principles of law, the ICJ would refer to such secondary materials as general restatements and codifications of the law as well as nonbinding judicial precedents from various tribunals. It would also look to multilateral declarations of States.[73] It is beyond the scope of this chapter to review specific conceptions of how these sources and materials interact to create a coherent body of international law or to construct a theory of state responsibility for global warming emissions. What follows, rather, is an overview of the basic building blocks for the construction of such a theory.

3.1. *General Restatements and Codifications of the Law*

Because much of international law is derived from customary international law and general principles of law, the norms as they develop in the messy world of politics and statecraft often lack the clear precision of treaties or domestic statutes. For this reason, those working within the international system rely relatively heavily on various restatements and codifications of the law that attempt to give clarity to areas where international law is amorphous. Of particular relevance to ascertaining the responsibility of States for global warming is the law on State responsibility for transboundary harm and transboundary pollution in general. Arguably global warming, which is caused by gases released mostly within the various countries causing the whole of the planetary climate system to warm, is not exactly the same as pollutants released in one country causing direct transboundary harm in another. The central legal principles that are pertinent to State responsibility for causing environmental harm outside their own borders are relevant, however, to considering the problem of global warming.

Ultimately, the principle behind holding countries liable for transboundary pollution is drawn from one of the most basic precepts of all legal systems that legal actors should be responsible for the harm that they do to others. Several expert bodies,[74]

---

[72] I.C.J., art. 38, 59 Stat. 1031, 1063, T.S. No. 993.
[73] Id.
[74] The views of these bodies on international law generally tend to be fairly subjective, and the relative weight which a court should accord the opinions of these bodies when they differ is not well defined.

official and unofficial, have proclaimed their own international environmental law variations on this precept.

One relevant pronouncement comes from the American Law Institute (ALI) in its Restatement (Third) of the Foreign Relations Law of the United States. The ALI is composed of eminent lawyers, judges, and law professors in the United States, and its restatements are considered by courts and legal professionals within the United States to be the most authoritative unofficial reporters of the applicable law in areas where clear statutory guidance tends to be lacking. The relevant provisions from Section 601, State Obligations with Respect to Environment of Other States and the Common Environment, are potentially helpful in the context of climate change. They assert that:

(1) A state is obligated to take such measures as may be necessary, to the extent practicable under the circumstances, to ensure that activities within its jurisdiction or control
  (a) conform to generally accepted international rules and standards for the prevention, reduction and control of injury to the environment of another state or of areas beyond the limits of national jurisdiction; and
  (b) are conducted so as not to cause significant injury to the environment of another state or of areas beyond the limits of national jurisdiction.[75]

A frequently cited similar, although arguably slightly stronger, statement of the law can be found in Article 3 of the International Law Association's Rules on International Law Applicable to Transfrontier Pollution.[76] The International Law Association is a private expert body.

The most authoritative international body of expert reporters is the U.N.'s International Law Commission. Established by the General Assembly pursuant to the U.N. Charter, the members of the Commission, international lawyers who serve in their individual capacities, attempt to both codify and "progressively develop" international law. Some of the International Law Commission's works are adopted by the General Assembly as declarations and some eventually become treaties. Over many years, the International Law Commission has been heavily involved in attempting to define the law of State responsibility. Probably most relevant is its work on recently adopted International Liability for Injurious Consequences Arising out of Acts Not Prohibited by International Law (Prevention of Transboundary Damage from Hazardous Activities), which according to its terms applies "to activities not prohibited by international law which involve a risk of causing significant transboundary harm through their physical consequences."[77] Its language requires States to "take appropriate measures to prevent significant transboundary harm or at any event to

---

[75] RESTATEMENT (THIRD) OF FOREIGN RELATIONS LAW § 601(1) (1986).
[76] Montreal Rules of International Law Applicable to Transfrontier Pollution, art. 3(1), Int'l Law Assn., Rep. 60th Conf., at 1–3 (1982).
[77] Draft Articles on the Prevention of Transboundary Damage from Hazardous Activities, International Law Commission, 53rd Sess., Supp. No. 10, ch. V.E. 1, art. I, U.N. Doc. A/56/10 (2001).

minimize the risk thereof" and to "cooperate in good faith and, as necessary, seek the assistance of one or more competent international organizations in preventing significant transboundary harm or at any event in minimizing the risk thereof."[78] Other works by the Commission may also be relevant.

3.2. Precedent

The *Trail Smelter* arbitration[79] decision is generally considered to be the lead case in the area of State liability for transboundary pollution. The dispute resulted from injuries caused in the U.S. state of Washington from sulfur dioxide discharged by a smelter plant in British Columbia, Canada, in the 1930s. Following diplomatic protests by the United States, the two countries agreed to submit the matter to arbitration. In its decision, the arbitrator proclaimed a general principle of international law that would be very helpful to establishing State liability for greenhouse gas emissions. Citing a well-known treatise of the day,[80] the arbitrator stated that "[a] State owes at all times a duty to protect other States against injurious acts by individuals from within its jurisdiction,"[81] and later in the decision he went on to add that

> [n]o state has the right to use or permit the use of its territory in such a manner as to cause injury by fumes in or to the territory of another or the properties or persons therein, when the case is of serious consequences and the injury is established by clear and convincing evidence.[82]

State actions in more recent and well-known cases would not be as helpful in demonstrating the pervasive present-day acceptance of a principle of State liability for transboundary pollution. Most important is the Chernobyl nuclear accident, where the Ukraine refused to acknowledge liability and, in fact, the international community paid for the costs of decommissioning the reactors.[83] Also unhelpful is the *Sandoz Chemical Fire* case, which involved a fire at a Sandoz corporation warehouse in Switzerland. The fire resulted in thousands of cubic meters of chemically contaminated water seeping into the Rhine and constituted one of the worst environmental disasters ever in Western Europe. None of the States affected brought claims against Switzerland.[84] Finally, in the 1997–1998 *Asian Haze* case, a thick smoky haze caused by fires used to clear forests in the Indonesian provinces of Kalimantan and

---

[78] *Id.*
[79] Trail Smelter (U.S. v. Can.), 3 R.I.A.A. 1938, 1965 (Mar. 11, 1941).
[80] Clyde Eagleton, RESPONSIBILITY OF STATES IN INTERNATIONAL LAW (1928).
[81] Trail Smelter, *supra* note 79 at 79.
[82] *Id.* at 90.
[83] *See* Margaret Cocker, *Chernobyl's No. 4 Reactor Remains Crumbling Threat, Mismanagement Snarls the Multibillion-Dollar Cleanup Effort in Ukraine*, ATLANTA J. & CONST., Apr. 23, 2000 (discussing the Ukraine's use of the disaster as leverage to get increased foreign aid); *see also* A JOINT REPORT OF THE OECD NUCLEAR ENERGY AGENCY AND THE INTERNATIONAL ATOMIC ENERGY AGENCY, INTERNATIONAL NUCLEAR LAW IN THE POST-CHERNOBYL PERIOD (2006).
[84] *See Sandoz to Pay Rhine Pollution Claims, Swiss Chemical Company to Reimburse Claimants*, FIN. TIMES UK, Nov. 14, 1986.

Sumatra spread across Southeast Asia. Despite the costly disruption of air travel and other business activities and significant adverse health and environmental effects, neighboring Southeast Asian countries did not make official diplomatic claims to the effect that Indonesia should be held legally responsible for the costs of the problem.[85]

All of these cases may be distinguished from global warming by their unique facts. The Ukraine, for example, was poor and unable to well afford the cost of decommissioning the reactor on its own.[86] Sandoz privately provided compensation for individual victims of the disaster.[87] Finally, Southeast Asian governments, in accordance with ASEAN diplomatic protocol, used diplomacy, rather than formal legal claims, to encourage Indonesia to take action to avoid recurrence.[88] The international environmental precedent relevant to a global warming case is, therefore, inconclusive.

3.3. *Treaties and Soft Law Declarations*

Treaties are usually considered to be the most authoritative source of international law. The UNFCCC treaty standards prescribing state action related to global warming are likely to be the most generally applicable in a global warming suit because of States' almost universal participation in it, including by the United States. As discussed in Subsection 2.1.3, Article 4.2(a) of the UNFCCC specifically commits developed countries to limit their anthropogenic emissions of greenhouse gases.[89] Other "principles" of the convention specified in Article 2 are likely to be important as well in interpreting this commitment. For example, Article 3(1) provides:

> The Parties should protect the climate system for the benefit of present and future generations of humankind, on the basis of equity and in accordance with their common but differentiated responsibilities and respective capabilities. Accordingly, the developed country Parties should take the lead in combating climate change and the adverse effects thereof.[90]

---

[85] Instead, beginning in 1997, there has been joint ASEAN efforts at haze prevention pursuant to the Regional Haze Action Plan. In 2003, the ASEAN Agreement on Transboundary Haze Pollution entered in force. *See* ASEAN Agreement on Transboundary Haze Pollution, June 10, 2002, *available at* http://www.aseansec.org/pdf/agr_haze.pdf. The treaty provides for the use of zero burning and controlled-burning practices and for the deployment of a Panel of ASEAN Experts on Fire and Haze Assessment and Coordination. The problem, however, continues to persist. *See Indonesia Downbeat on Stopping Fires Causing Haze*, ASIAN ECON. NEWS, Dec. 11, 2006; *see also* Haze Online, Main Page, http://www.haze-online.or.id/ (last visited Feb. 27, 2008).
[86] *See* sources *supra* note 83.
[87] *See Sandoz to Pay Rhine Pollution Claims*, *supra* note 84.
[88] *See* sources *supra* note 85.
[89] *See supra* note 31 and accompanying text.
[90] United Nations Framework Convention on Climate Change, art 3.1, May 9, 1992, 1771 U.N.T.S. Also helpful in supporting a climate change law suit would be Article 3.3, which provides:

> The Parties should take precautionary measures to anticipate, prevent or minimize the causes of climate change and mitigate its adverse effects. Where there are threats of serious or irreversible damage, lack of full scientific certainty should not be used as a reason for postponing such

The extensive state adherence to the UNFCCC is the result of the general perception that the articles that I have referenced place no precisely definable legal limitations on states. Given that treaty's obligatory language regarding remediation of the global warming problem, particularly by developed countries, it is quite possible, however, that the ICJ would decide this not to be the case.

Also discussed in Subsection 2.1.3, the Kyoto Protocol places obligations on developed countries to meet specific targets for reducing their contribution to global warming between 2008 and 2012.[91] Because of the different ways in which the Kyoto obligations can be met,[92] as well as that Protocol's more limited membership, its contribution to the theory of a global warming case is likely to be much more complex. Other treaties also could possibly be relevant to constructing an international global warming suit. Among them is the Straddling Fish Stocks Agreement examined by Wil Burns in this book, as well as the Convention on Long-Range Transboundary Air Pollution[93] and certain of its protocols. This latter treaty regime regulates some pollutants which affect global warming, and contains general language possibly helpful in a global warming suit.

The two primary declarations relevant to liability for emissions of greenhouse gases are the Stockholm Declaration and the Rio Declaration. The Stockholm Declaration came out of the 1972 Stockholm Conference on the Human Environment, often considered the progenitor of the modern environmental movement. It was adopted by a vote of 103 to 0 with 12 abstentions. Principle 21 of the Declaration is most apposite. It provides that:

> States have, in accordance with the Charter of the United Nations and the principles of international law, the sovereign right to exploit their own resources pursuant to their own environmental policies and the responsibility to ensure that activities within their jurisdiction or control do not cause damage to the environment of other States or of areas beyond the limits of national jurisdiction.[94]

In 1992, twenty years after Stockholm, the second major global environmental conference, and one of the largest diplomatic gatherings in history, took place in Rio de Janeiro. It was the Earth Summit, officially called the United Nations Conference on the Environment and Development. One of the principal outcomes of this

---

    measures, taking into account that policies and measures to deal with climate change should be cost-effective so as to ensure global benefits at the lowest possible cost. To achieve this, such policies and measures should take into account different socio-economic contexts, be comprehensive, cover all relevant sources, sinks and reservoirs or greenhouse gases and adaptation, and comprise all economic sectors. Efforts to address climate change may be carried out by interested Parties. *Id.* at art. 3.3.

[91] *See supra* note 29 and accompanying text.
[92] *See supra* note 45.
[93] Convention on Long-Range Transboundary Air Pollution, Nov. 13, 1979, TIAS No. 10,541, *reprinted in* 18 I.L.M. 1442 (1979).
[94] Stockholm Declaration on the Human Environment, princ. 21, Report of the United Nations Conference on the Human Environment, Stockholm, June 5–16, 1972, U.N. Doc. A/CONF.48/14/Rev.1, U.N. Sales No. E.73.II.A.14, pt. 1, ch. 1 (1973), *reprinted in* 11 I.L.M. 1416 (1972).

conference was the Rio Declaration which was adopted by consensus. Principle 2 of that declaration is identical to Principle 21 of the Stockholm Declaration, except that the words "and developmental" are inserted between "environmental" and "policies."[95] Because the legal authority of declarations, and the relationship of treaties to each other and to other sources of international law, are not well settled within the international system, there are varied conceptual possibilities for how these legal instruments can be tailored into a coherent theory of a global warming case.

4. CONCLUSION

These are hopeful times in the short history of our efforts to remediate the global warming problem. For the first time, the issue seems to have penetrated deeply into the global mass political consciousness. Foundation money is flowing into climate change initiatives. It has become fashionable for celebrities and public personalities to associate themselves with the cause. Former Vice President Gore and the Intergovernmental Panel on Climate Change won the 2007 Noble Peace Prize for their work on global warming. Venture capital and other forms of financing are flowing into researching and developing alternatives to greenhouse gas–emitting technologies. The Obama administration's commitment to climate and energy issues appears to be ushering in a new era of U.S. efforts.

Yet there is reason to be sober in our assessment. Most climate scientists agree that greenhouse gas reduction targets currently being proposed are not sufficient to avert potentially cataclysmic effects. What's more, viewing the present concern from an historical perspective gives another reason for pause. We have seen before a pattern of great environmental awakening only to be followed by mass political denial. Building upon the publication of Rachel Carson's *Silent Spring* in 1962, the modern environmental movement was born of an emerging consciousness that we share one small finite planet. After a sustained period of growing awareness and action, however, environmental matters largely went out of fashion in the 1980s. Then, heralded by *Time* magazine's choice of "endangered earth" as its "Planet of the Year" for 1989, and fueled by the end of the Cold War in the 1990s, concern for the environment again resurfaced in the popular consciousness. But this was once more followed by a decline in interest, especially after the terror attacks of September 11, 2001.[96]

---

[95] Rio Declaration on Environment and Development, Aug. 12, 1992, U.N. Doc. A/CONF.151/5/REV.1 (1992), 31 I.L.M. 876.

[96] For a discussion of changing environmental attitudes in the United States specifically and the methodology of measuring them, see Chapter 3, Stability: Have Environmental Attitudes Changed over Time? *in* DEBORAH LYNN GUBER, THE GRASSROOTS OF A GREEN REVOLUTION (2003); *see also* TOM W. SMITH, TRENDS IN NATIONAL SPENDING PRIORITIES, 1973–2006, 23 (2007) (documenting results of U.S. public opinion polls demonstrating that support for environmental spending rose at the immediate end of the cold war and fell after the terror attacks of 2001). For a discussion of attitudes in the United States regarding global warming specifically, see Matthew C. Nisbet & Teresa Myers, *Twenty Years of Public*

Whatever political vagaries influence attempts to counteract global warming, there is likely a constructive role for litigation in general and perhaps for the ICJ in particular. But any such role needs to be seen as complementary to a broader political strategy. For example, the trust necessary for parties to succeed in good faith negotiations over global warming could well be undermined by certain parties initiating legal actions against others. On the other hand, as a spur to recalcitrant parties, litigation could have the benefits described in Section 1 of this chapter.

We are still in the early stages of the global warming phenomenon. There likely will be different generations of lawsuits, probably evolving over time to deal less with the raising of political consciousness and more with the allocation of losses and adaptation costs. Litigation is poised to play a role, and the ICJ with its unique status and visibility could make an important contribution. My hope in this chapter has been to further a discussion of how the door to that forum might be opened.

Opinion About Global Warming, 71 PUB. OPINION Q. 13 (Fall 2007) (reporting on Gallup Poll results showing that between 1989 and 1991 about one-third of respondents worried "a great deal" about global warming with results fluctuating in the 1990s, falling after the 2001 terror attacks and now rebounding).

16

The Implications of Climate Change Litigation: Litigation for International Environmental Law-Making

David B. Hunter*

INTRODUCTION

Everyone is talking about climate change. Climate change has been on the cover of almost every U.S. magazine in the past couple of years, including *Vanity Fair, Time, Newsweek,* the *Economist,* and even *Sports Illustrated,* on such television shows as *Oprah* and *The Tonight Show,* and in the movie theaters with Al Gore's *An Inconvenient Truth* and *Who Killed the Electric Car?* To be sure, this media attention is driven first by the increasingly clear scientific connection between greenhouse gas concentrations, climate change, and real impacts affecting real people. But the growing public awareness of climate change is also being driven by the actions of lawyers and other climate advocates who are increasingly litigating climate change in the world's courts, commissions, and congresses. Climate change even made an appearance before the U.S. Supreme Court.[1] Win or lose (and some will surely win, as they did in the U.S. Supreme Court), these litigation strategies are significantly changing and enhancing the public dialogue around climate change.

This chapter discusses the awareness-building impacts of climate litigation as well as related impacts such strategies may have on the development of climate law and policy – even if many of the individual cases lose.[2] The chapter does not discuss the significant implications if a tort action in the United States or the Inuit human rights claims, for example, were ultimately to prevail. Such precedents, which would obviously be far reaching, are discussed in the various chapters of this book addressing difficult litigation strategies. The primary focus here is on the

---

* Assistant Professor of Law & Director, Program on International and Comparative Environmental Law, American University Washington College of Law, 4801 Massachusetts Ave., NW, Washington, D.C. 20016, 202-274-4415, dhunter@wcl.american.edu.
[1] Massachusetts v. EPA, 549 U.S. 497 (2007).
[2] *See also* Stephanie Stern, *State Action as Political Voice in Global Climate Change Policy: A Case Study of the Minnesota Environmental Cost Valuation Regulation,* this volume (discussing how climate change actions by states can strengthen their political influence in the climate debate); JOSEPH SMITH & DAVID SHEARMAN, CLIMATE CHANGE LITIGATION: ANALYSING THE LAW, SCIENTIFIC EVIDENCE & IMPACTS ON THE ENVIRONMENT, HEALTH & PROPERTY 12 (2006) (noting public awareness-building impact and motivation of some of the climate litigation).

implications of climate litigation simply by virtue of cases having been filed. In fact, the debate over whether specific theories will prevail or what remedies can be fashioned in a specific case misses much of the significance of these litigation strategies. Just the acts of preparing, announcing, filing, advocating, and forcing a response have significant impacts.

Climate advocates are necessarily pushing the development of the law in new directions. The world's legal systems – both international and national – have never seen a challenge quite like climate change. The science involves complexities of global ecology that are of a scale new to the courts. Nearly all of our activities, whether as individuals, corporations, or governments, contribute to the problem and almost everyone is affected. The entire world is at once simultaneously both a potential plaintiff and defendant. Climate change presents significant geographic complexities, with significant implications for jurisdiction and the shaping of remedies.[3] Climate change also presents difficult temporal problems, with emissions today mixing with emissions from yesterday to cause impacts in the future. This geographic and temporal distance between the wrongs (e.g., the emissions) and the injuries presents new challenges for law.

The unique aspects of climate change have forced climate advocates to innovate and to develop creative new strategies internationally and domestically. They have had to push for the progressive development of the law and related institutions, emphasizing not only the differences but the similarities of climate change with more familiar issues. Viewed in this light, climate change is just another, albeit distinctly modern, common law nuisance, threat to cultural property, or human rights violation. In this respect, the climate change advocates are right: climate change may be global, it may be complex, but climate change is also strikingly familiar. Real people, typically those already marginalized with few resources, will suffer real harm because of the activities of others. Isn't this precisely what the law is meant to address?

1. THE FOCUS ON VICTIMS

Climate advocates' focus on specific injuries in specific situations has far-reaching implications for climate policy more generally. In the Kyoto negotiations or in previous national climate policy debates, the focus has primarily been on climate change's global impacts: average temperature increases, average sea level rise, average changes in precipitation. With the rise of climate litigation strategies, however, the focus necessarily shifts to the specific injuries being asserted by the plaintiffs or

---

[3] *See, e.g.*, Hari M. Osofsky, *The Geography of Climate Change Litigation: Implications for Transnational Regulatory Governance*, 83 WASH. U. L.Q. 1789 (2005); Hari M. Osofsky, *The Inuit Petition as a Bridge? Dialectics of Climate Change and Indigenous Peoples' Rights*, this volume; *see also* Kirsten Engel, *Harmonizing Regulatory and Litigation Approaches to Climate Change Mitigation: Incorporating Tradable Emissions Offsets into Common Law Remedies*, 155 U. PA. L. REV. 1563 (2007).

claimants: the impacts on New England's ski industry,[4] California's coastline,[5] the life and culture of the Inuit,[6] the survival of polar bears or penguins,[7] or the grandeur of Mount Everest or Glacier National Park.[8]

Advocates have had to compile and present detailed assessments of climate impacts in ways that highlight the many regional and local impacts of climate change. In *Connecticut v. American Electric Power Co.*, for example, the New England states documented impacts that included declining snowpack and ice; increased loss of life and public health threats from heat-related illnesses and smog; impacts on the San Francisco Bay, Jamaica Bay National Wildlife Refuge, and other coastal resources from storm surges and permanent sea level rise; declining water levels in the Great Lakes; increases in temperatures in the upper surfaces of the Great Lakes; and rapid declines in forest resources, including New York's Adirondack State Park, among other regionally specific allegations.[9] Similarly, California, in *California v. General Motors*, detailed impacts of global warming that are already occurring in California and related costs the state is incurring in response. These impacts include, for example, a decline in snowpack in the Sierra Nevada range due to an increase in average winter temperatures; the costs of rebuilding levees to prevent seawater infiltration and other impacts of sea level rise on the Sacramento Bay and Delta; increased floods from earlier spring runoffs; and beach preservation efforts to reverse increased beach erosion from sea level rise.[10]

---

[4] Complaint, Connecticut v. Am. Elec. Power Co., 406 F. Supp. 2d 265 (S.D.N.Y. 2005) (No. 04 Civ. 5669(LAP)) [hereinafter Connecticut v. AEP Complaint].

[5] Complaint, California v. Gen. Motors Corp., No. C06–05755 (N.D. Cal., Sept. 20, 2006) [hereinafter California v. Gen. Motors Complaint].

[6] *See* Center for International Environmental Law, An Inuit Petition to the Inter-American Commission on Human Rights for Dangerous Impacts of Climate Change at 35–69 (2004), *available at* http://www.ciel.org/Publications/COP10_Handout_EJCIEL.pdf [hereinafter Inuit Petition] (describing impacts on "every aspect of Inuit life and culture").

[7] Center for Biological Diversity, Petition to List the Polar Bear (Ursus maritimus) as a Threatened Species under the Endangered Species Act before the Secretary of the Interior (Feb. 16, 2005), *available at* http://www.biologicaldiversity.org/swcbd/SPECIES/polarbear/petition.pdf [hereinafter Polar Bear Petition]; Center for Biological Diversity, Petition to List 12 Penguin Species under the Endangered Species Act before the Secretary of the Interior (Nov. 28, 2006), *available at* http://www.biologicaldiversity.org/swcbd/SPECIES/penguins/PenguinPetition.pdf [hereinafter Penguin Petition].

[8] *See, e.g.*, Petition to the World Heritage Committee for Inclusion of the Waterton-Glacier International Peace Park on the List of World Heritage in Danger and for Protective Measures and Actions (Feb. 16, 2006), *available at* http://law.lclark.edu/org/ielp/objects/Waterton-GlacierPetition2.15.06.pdf [hereinafter Waterton-Glacier UNESCO Petition]. Other petitions were filed to list the Mesoamerican Barrier Reef in Belize, Huarascán National Park in Peru, Sagarmatha National Park in Nepal, and the Great Barrier Reef in Australia. *See* Climate Justice Programme, UNESCO Danger-Listing Petitions Presented (Nov. 17, 2004), *available at* http://www.climatelaw.org/media/UNESCO.petitions.release [hereinafter UNESCO Petitions]. *See generally* UNESCO, World Heritage Centre, Predicting and Managing the Effects of Climate Change on World Heritage, WHC-06/30.COM/7.1, Annex 4 (June 26, 2006) [hereinafter World Heritage Climate Report].

[9] Connecticut v. AEP Complaint, *supra* note 4, at paras. 112–17, 121–27, 132–35.

[10] California v. Gen. Motors Complaint, *supra* note 5, at paras. 46–56.

This focus on specific injuries is critical for building political support; such cases link climate change with the lives of ordinary people. Reports of a global increase in temperature of 1°F or even 5°F have little meaning to most people. The impact is much more understandable when an Inuit expresses implications of climate change for their lives, when the glaciers of Nepal are melting, or when descriptions of drowning or cannibalistic polar bears are reported on the news. The Inuit human rights petition, for example, provides thirty-five pages on impacts of climate change on their life and culture. The petition details changes in Arctic ice conditions and the resulting dangers for Inuit travel, the reduction in materials (thick ice) for building traditional igloos, and the deterioration of wildlife harvests because of declining populations of caribou, seals, polar bears, and other animals.[11] In short, the petition tells a story about the impacts of climate change in human terms far removed from the antiseptic discussion of greenhouse gas concentrations or global mean temperatures that have traditionally predominated international climate negotiations.

The storytelling quality of "cases" thus makes climate change more tangible and more immediate, which significantly changes the tone of the climate debate.[12] If real victims – such as islanders or the Inuit – are in a room pressing their stories, it is harder for others to bluster about how climate change is a hoax or is unimportant because some regions may benefit from warming or will be able to adapt relatively easily. At the very least, addressing climate change takes on a renewed urgency when one moves from the abstraction of sea level rise, for example, to questions of how to treat climate refugees from South Pacific islands or how to shore up the eroding California coastline. A focus on victims increases the saliency of questions about compensation and adaptation to climate change, and the urgency of mitigating climate change to avoid even worse impacts in the future.[13] This builds momentum at both the national and international levels for stronger climate policymaking.

2. IMPLICATIONS FOR CLIMATE POLICY

2.1. *Implications for Climate Science*

Climate litigation's focus on victims and on specific impacts has implications for how we use climate science and on what climate science is conducted. Every litigation strategy requires the collection, synthesis, and presentation of climate science in support of its claims. This process highlights and makes more accessible to a wider audience the expanding research and analysis on specific local and regional climate impacts.

---

[11] Inuit Petition, *supra* note 6, at 35–69.
[12] The story-telling or narrative quality of cases has spawned significant scholarship. *See, e.g.*, Daniel A. Farber & Suzanna Sherry, *Telling Stories out of School: An Essay on Legal Narratives*, 45 STAN. L. REV. 807 (1993); PETER BROOKS & PAUL GEWIRTZ, LAW'S STORIES: NARRATIVE AND RHETORIC IN THE LAW (1996).
[13] *See infra* Section 2.2 (discussing impacts of litigation strategies on the development of international climate policy).

This has proved particularly true of the reports issued by the Intergovernmental Panel on Climate Change (IPCC),[14] which have been cited as the scientific basis by most of the climate plaintiffs or petitioners.[15] The IPCC reports attract particular attention because they compile and summarize the international consensus on climate science at a specific point in time. Moreover, the IPCC's practice of explicitly bounding its views of the likelihood of certain scientific conclusions in terms of numeric probabilities not only assists international policymakers at the U.N. Framework Convention on Climate Change (UNFCCC) but also offers lawyers scientific conclusions that are useful in explaining and meeting the standards for causation. This reliance on the IPCC's reports presents a two-way validation: the IPCC's prestige and international status provides a convenient and effective affirmation of the claimant's factual allegations (at least with respect to global climate trends) and, at the same time, use of the IPCC (and particularly its acceptance, if it ensues, by other institutions as authoritative) adds legitimacy and prestige to the IPCC and its reports. This has been evidenced by the enormous, mostly positive media attention the IPCC's Fourth Assessment has received since the beginning of its release in 2007, and the dominant role it now plays in public discourse over climate science. One can also expect that the Fourth Assessment will be central to the next generation of climate cases and claims. Although some may argue that the IPCC's reports are not meant to be used for direct advocacy in specific cases, the IPCC's screening and presentation of the emerging science provides an important service in allowing litigants and adjudicators alike to ground advocacy strategies and opinions in the current scientific consensus.

The IPCC reports are not the only scientific studies to play a significant role in climate litigation. The Inuit Petition, for example, relied heavily on the Arctic Climate Impact Assessment, a comprehensive regional report released by the Arctic Council and International Arctic Science Committee.[16] That 2004 report concluded that the Arctic was "experiencing some of the most rapid and severe climate change on Earth."[17] California's complaint against the automobile industry also highlighted the Assessment.[18] Such use of the Assessment has helped to raise awareness of its findings in ways that would have been unlikely without it forming part of controversial and novel litigation strategies.

---

[14] *See* INTERGOVERNMENTAL PANEL ON CLIMATE CHANGE [IPCC], WORKING GROUP I, CLIMATE CHANGE 2007: THE PHYSICAL SCIENCE BASIS (2007), *available at* http://ipcc-wg1.ucar.edu/wg1/wg1-report.html [hereinafter IPCC, 2007 PHYSICAL SCIENCE BASIS]; IPCC, WORKING GROUP II, CLIMATE CHANGE 2007: IMPACTS, ADAPTATION AND VULNERABILITY 2 (2007), *available at* http://www.ipcc.ch/spm13apr07.pdf.

[15] *See, e.g.*, Massachusetts v. EPA, 549 U.S. 497, 508–509 (2007); California v. Gen. Motors Complaint, *supra* note 5, at paras. 24, 26, 31; Connecticut v. AEP Complaint, *supra* note 4, at paras. 80, 88, 92–93.

[16] *See, e.g.*, Inuit Petition, *supra* note 6, at 35; *see also* INT'L ARCTIC SCIENCE COMM. & THE ARCTIC COUNCIL, ARCTIC CLIMATE IMPACT ASSESSMENT (Nov. 2004) [hereinafter Arctic Climate Assessment].

[17] California v. Gen. Motors Complaint, *supra* note 5, at para. 37 (quoting Arctic Climate Assessment, *supra* note 16).

[18] California v. Gen. Motors Complaint, *supra* note 5, at paras. 37–38.

Climate litigation strategies not only rely on emerging science but also will influence the development of climate science both directly and indirectly. Some domestic climate cases in several countries have been filed with the goal of improving the assessment of climate impacts and the use of climate science. In *Massachusetts v. EPA*, the U.S. Supreme Court required the government to make a reasoned judgment on whether emissions of carbon dioxide are endangering public health and welfare as an initial step in determining whether to regulate carbon dioxide as an air pollutant under the Clean Air Act.[19] At the project level, cases in the United States,[20] Germany,[21] and Australia[22] have sought (sometimes successfully) to require under national law the consideration of climate impacts in project finance or permitting. In Australia, for example, greenhouse gas emissions and resulting climate impacts must be assessed in coal mining and power plant operations, which will increase the scientific basis for decision making in those sectors.[23] A recent lawsuit in the United States is aimed at compelling the United States to complete a National Assessment of climate impacts, which was required by Congress to be completed by 2004.[24] Other U.S. cases seek to force the assessment of climate change impacts[25] or the consideration of such impacts in permitting decisions.[26]

[19] *Massachusetts*, 549 U.S. at 532–35; *see also* Coke Oven Envtl. Task Force v. EPA, No. 06–1131 (D.C. Cir. filed Apr. 7, 2006) (pending challenge to EPA's refusal to regulate carbon dioxide emissions in setting new source performance standards under the Clean Air Act).

[20] Friends of the Earth v. Mosbacher, No. C02–4106 JSW, 2007 WL 962955 (N.D. Cal. Mar. 30, 2007) (order denying plaintiffs' motion for summary judgment and granting in part and denying in part defendants' motion for summary judgment); Friends of the Earth v. Watson, No. C02–4106 JSW, 2005 WL 2035596 (N.D. Cal. Aug. 23, 2005) (order denying defendants' motion for summary judgment). These cases settled in February 2009. *See* Joint Motion for Dismissal with Prejudice, Friends of the Earth v. Spinelli, No. C02–4106 JSW (Feb. 6, 2009).

[21] *See* Press Release, GermanWatch & BUND, German Government Sued over Climate Change (June 15, 2004), *available at* http://www.climatelaw.org/media/german.suit/press.release.pdf (announcing lawsuit against the German Federal Ministry of Economics and Labour to compel disclosure of the climate change contribution made by those projects financed by the German export credit agency, Euler Hermes AG); Bund & Germanwatch v. German Fed. Ministry of Econ. and Labour [BMWA], Beschluss, Verwaltungsgericht [VG Berlin] [Local Administrative Court], Jan. 10, 2006, VG 10 A 215.04 (2006), *translated at* http://www.climatelaw.org/media/Germany/de.export.decision.eng.doc (order entering settlement with legal opinion).

[22] Australian Conservation Found. v. Minister for Planning, Administrative Decision, (2004) VCAT 2029 (holding that the Australian Planning and Environment Act requires consideration of greenhouse gas emissions and resulting climate impact in licensing coal mining and power plant operations); Wildlife Preservation Soc. of Queensland Proserpine/Whitsunday Branch v. Ministry for the Env't & Heritage (2006) FCA 736 (upholding decisions by the Australian environment ministry to license two coal mines, despite their failure to consider climate impacts on natural heritage sites). For information on climate-related cases brought in Australia, see the website of the Australian Climate Justice Program, *available at* http://www.cana.net.au/ACJP/cases.php?case_table=cases_aust (last visited May 28, 2007).

[23] Australian Conservation Found.v. Minister for Planning, (2004) VCAT 2029; *see also* Smith & Shearman, *supra* note 2 (discussing Australian Conservation Foundation).

[24] Complaint, Ctr. for Biological Diversity v. Brennan, No. C06–7061 (N.D. Cal. Nov. 14, 2006).

[25] *See, e.g.*, Watson, 2005 WL 2035596.

[26] Nw. Envtl. Def. Ctr. v. Owens Corning Corp., 434 F. Supp. 2d 957 (D. Or. 2006) (opinion and order) (holding that plaintiff environmental organization had standing to challenge a permit application that would have permitted significant releases of a potent greenhouse gas (HCFC-142b)).

In other cases, expanding climate science may be an indirect or secondary outcome of the litigation effort.[27] The petitions to the World Heritage Committee, for example, triggered a series of activities and reports that are aimed in part at reviewing the nature and scale of the risks posed to World Heritage properties arising specifically from climate change.[28] More generally, climate litigation efforts may provide an incentive to some scientists to prioritize certain questions that they might otherwise ignore. Questions of attribution, for example, become particularly relevant for litigation strategies aimed at securing compensation for those affected or for driving corrective action by identifying those responsible.[29] The science of attribution is gaining ground; one recent study, for example, found that the human contribution to the 2003 European heat wave increased the potential of risk of such weather from four to ten times.[30] Approximately 22,000 to 35,000 people died from heat-related deaths, 75 percent of whom would have been likely to survive for more than a year without such heat.[31] Such studies will be critical in shaping future climate litigation strategies.

Finally, climate litigation is shaping the tone of the debate over climate science. In journalistic or political approaches to climate, the views of climate skeptics were previously given equal weight to the broad consensus views regarding science. In climate litigation forums, however, such skeptics may be asked to submit affidavits or even face cross-examination of their views. This ground-truthing of climate science may screen out and discredit those fringe scientists whose positions may not be able to withstand the scrutiny that comes from adversarial proceedings, particularly in domestic courts. To be sure, some opinions questioning the adequacy of climate science for judicial review have and will occur,[32] but recent cases, including the

---

[27] See, e.g., Inuit Petition supra note 6, at 118 (seeking as one remedy that the "U.S. take into account the impacts of U.S. greenhouse gas emissions on the Arctic and affected Inuit in evaluating and before approving all major government actions"); see also Bund & Germanwatch v. German Fed. Ministry of Econ. and Labour [BMWA], Beschluss, Verwaltungsgericht [VG Berlin] [Local Administrative Court], Jan. 10, 2006, VG 10 A 215.04, translated at http://www.climatelaw.org/media/Germany/de.export.decision.eng.doc (entering order requiring Hermes to assess impacts of its financial decisions on climate change).

[28] See UNESCO, Announcement of World Heritage, Climate Change and World Heritage: Expert Meeting, March 16–17, 2006, available at http://whc.unesco.org/en/events/301 (last visited Dec. 16, 2006).

[29] See, e.g., Myles Allen et al., Scientific Challenges in the Attribution of Harm to Human Influence on Climate, 155 U. PA. L. REV. 1353 (2007).

[30] Id.; see also Myles Allen, Liability for Climate Change, 421 NATURE, 891–92 (Feb. 27, 2003); Peter A. Stott et al., Human Contribution to the European Heatwave of 2003, 432 NATURE, at 610 (Dec. 2, 2004); Simone Bastianoni, Federico M. Pulselli & Enzo Tiezzi, The Problem of Assigning Responsibility for Greenhouse Gas Emissions, 49 ECOLOGICAL ECON. 253 (2004) (discussing difficulties in assigning responsibility for greenhouse gas emissions).

[31] See Allen, supra note 29.

[32] Massachusetts v. EPA, 127 S.Ct. 1438, 1463–71 (Roberts, C.J., dissenting); Re Xstrata Coal Queensland Pty Ltd & Ors, [2007] QLRT 33 (holding that plaintiffs had not proven a causal link between climate change and carbon emissions); Korsinsky v. EPA 5 No. 05 Civ. 859 (NRB), 2005 WL 2414744 (S.D.N.Y. Sept. 29, 2005).

U.S. Supreme Court decision in *Massachusetts v. EPA*, are tending to support and recognize the general scientific consensus regarding climate change.[33] When courts and other highly credible institutions validate the basic science of climate change, the general public's perception of the climate debate shifts from *whether* climate change is occurring to what the appropriate remedies should be. For the public, judicial decisions can move the debate from an esoteric one among scientists to an issue *decided* by impartial judges whose job it is to resolve such matters.

2.2. Implications for the Climate Negotiations

Climate change litigation strategies have been at least partly a response to the perceived weakness of the international climate regime. Initially, many of the litigation strategies were designed as an indirect response to the decisions by Australia and the United States to withdraw from the Kyoto Protocol.[34] More recently, a Canadian environmental group filed a lawsuit asking the courts to declare Canada in noncompliance (or imminent noncompliance) with the UNFCCC and Kyoto Protocol.[35] The application for judicial review alleges that Canada's Ministries of Environment and Health are in violation of section 166 of the Canadian Environmental Protection Act, which requires them to act "if the Ministers [of the Environment and Health] have reason to believe that a substance released from a source in Canada into the air creates, or may reasonably be anticipated to contribute to (a) air pollution in a country other than Canada; or (b) air pollution that violates, or is likely to violate, an international agreement binding on Canada in relation to the prevention, control or correction of pollution."[36] According to the application, the Government of Canada's own reports estimate that its actual emissions will be nearly 40 percent higher than that which is allowed under the Kyoto Protocol.[37] Although this is the first lawsuit in the world aimed specifically at enhancing compliance with the international climate regime, many of the other climate litigation strategies have also been designed at least in part to increase the political will for stronger international climate change policy.[38]

---

[33] *Massachusetts*, 127 S.Ct. at 1455–58; *see also, e.g., In re* Quantification of Envtl. Costs, 578 N.W.2d 794, 799 (Minn. Ct. App. 1998) (upholding Commission finding that carbon dioxide negatively affects the environment).

[34] *See* William C. G. Burns & Hari M. Osofsky, *Overview: The Exigencies That Drive Potential Causes of Action for Climate Change*, this volume.

[35] Application, Friends of the Earth v. Her Majesty the Queen, Minister of the Env't & Minister of Health, No. T-914-07 (Federal Court Ottawa, May 28, 2007), *available at* http://www.sierralegal.org/reports/notice_of_application07_05_29.pdf (application for judicial review of the Canadian government's actions, emitting greenhouse gases, in violation of section 166 of the *Canadian Environmental Protection Act*, the UNFCCC, and the Kyoto Protocol). The case was subsequently dismissed as being unjusticiable, and is being appealed. *See* Friends of the Earth v. Canada, 2008 FC 1183 (CanLII).

[36] *Id.*

[37] *Id.*

[38] *See, e.g.,* Stern, *supra* note 2 (noting that a Minnesota climate regulation was a "statement of political opposition to ineffective national and global climate change policies").

The litigation efforts thus should not be seen in isolation from the negotiations under the UNFCCC and the Kyoto Protocol. The Conference of the Parties (CoP) to the UNFCCC and the Meetings of the Parties (MoP) to the Protocol are now enormous events that bring together a broad range of nontraditional parties to discuss a wide range of responses to climate change. Many of the principal players in climate litigation are also active in international negotiating and policymaking processes. In the "epistemic community"[39] that has emerged around climate negotiations, climate advocates find both a ready audience for spreading the news of litigation and for seeking the same goals that they are seeking through the litigation. The CoP/MoP community is thus a critical venue for developing strategies, identifying partners, reaching out to the press, building legitimacy and credibility for the litigation, and developing factual experts that can support the litigation.

For climate advocates, the CoP/MoP presents additional opportunities for pursuing their specific goals and they actively seek to influence discussions at the negotiations. The Inuit, for example, held "side-events" at three UNFCCC CoPs before filing their petition,[40] and they chose the CoP as the place for formally announcing their intent to file the petition. This brought attention to their claims and their concerns, both for the filing of the petition but also in the negotiations as well. So, too, the civil society coalition that submitted petitions to the World Heritage Committee, as well as the Secretary General of UNESCO, have held events at the UNFCCC CoP to highlight the impacts of climate change on World Heritage sites.[41]

High-profile climate litigation strategies in the United States have also helped to undermine the U.S. opposition to the Kyoto Protocol, particularly its efforts to derail the launch of negotiations for the second reporting period under Kyoto. At the 2005 CoP/MoP in Montreal, the United States sought to enlist Australia, China, and India in a united front against the European push for negotiations of future commitments under the Kyoto Protocol. The U.S. strategy failed in part because of the multiplicity of U.S. voices at the negotiations (including local government officials, former President Bill Clinton, and several Senators) that argued action was occurring in the United States, that the Bush administration was isolated, and that the United States would likely engage in future international negotiations after the next president took office.[42] The presence of high-profile alternative U.S. voices and

---

[39] *See generally* Peter Haas, Introduction: Epistemic Communities and International Policy Coordination, 46 INT'L ORG. 1 (1992) (defining epistemic communities as "networks of professionals with recognized expertise and competence in a particular domain and an authoritative claim to policy-relevant knowledge within that domain or issue-area."); Robert Keohane & Joseph Nye, Transgovernmental Relations and International Organizations, 27 WORLD POL. 39 (1974). According to these and other authors in international relations, the recurrent meetings of these epistemic communities at, for example, annual meetings of multilateral environmental regimes link government and nongovernment officials in a more effective and dynamic, long-term policymaking process.

[40] Inuit Petition, *supra* note 6, at 117.

[41] Statement of Koichiro Matsuura, UNESCO Director General, to the 12th Conference of the Parties to the UNFCCC (Nov. 2006), *available at* http://whc.unesco.org/en/activities/396/.

[42] *See, e.g.*, Pew Center on Global Climate Change, COP 11 and COP/MOP 1 Montreal, *available at* http://www.pewclimate.org/what_s_being_done/in_the_world/cop11/ (last visited Apr. 19, 2007);

actions thus emboldened negotiators to set out a future negotiation schedule, more confident that the United States would eventually come back to the table.

Harder to judge is the impact climate litigation strategies will have on the climate change regime if some of these cases prove successful. On one hand, taking climate change issues to other forums may seem to undermine the monopoly the climate secretariat might like to have on the issue. On the other hand, by focusing other institutions on climate impacts, the actions may help petitioners to be more active and productive players in the climate negotiations and create mechanisms for the integration of the climate regime with other institutions (e.g., human rights tribunals, financial institutions, or other treaty regimes). By forcing other institutions to take climate into account, climate litigation will create opportunities for policy coherence across international governance, even if through ad hoc cases. Claims to the World Bank Inspection Panel or the International Finance Corporation's Compliance Advisor/Ombudsman, for example, could seek to force those financial institutions to implement UNFCCC-approved methodologies for measuring, evaluating, or reducing greenhouse gas emissions.[43]

The focus on remedies that is inherent to climate litigation may influence future debates at the UNFCCC over adaptation. Certainly, the portrayal of specific harm to victims *today*, as opposed to general impacts tomorrow, is likely to force climate negotiators and the UNFCCC secretariat to focus on adaptation and compensation sooner than it otherwise would. This could increase funding available under the regime to respond to the needs of victims. In the most extreme scenarios, the threat of civil liability could conceivably lead industry and others to promote a liability regime under the UNFCCC that would both clarify the rules of liability and essentially cap private-sector liability – much as has been done with environmental damage from nuclear facilities[44] and oil spills.[45]

Andrew Buncombe & Geoffrey Lean, *Climate campaigners claim greatest ever success at Montreal*, INDEPENDENT, Dec. 11, 2005, *available at* http://environment.independent.co.uk/article332384.ece; Planktos Inc., The 2005 Montreal COP/MOP in Review, Dec. 15, 2005, *available at* http://www.planktos.com/Newsroom/The2005MontrealCOPMOPinReview.html; *see also* Int'l Inst. for Sustainable Dev., Special Report on Selected Side Events at COP 11 & Kyoto Protocol COP/MOP 1: Events Convened on Monday, 5 Dec. 2005, *available at* http://www.iisd.ca/climate/cop11/enbots/enbots1707e.html (last visited on Apr. 23, 2007) (summarizing a panel on subnational initiatives including a presentation by the N.Y. State Attorney General's office regarding recent climate change cases brought in the United States).

[43] See Jennifer Gleason & David B. Hunter, *Bringing Climate Change Claims to the Accountability Mechanisms of the International Financial Institutions*, this volume.

[44] Paris Convention on Third Party Liability in the Field of Nuclear Energy, July 29, 1960, 956 U.N.T.S. 251; Int'l Atomic Energy Agency [IAEA], Vienna Convention on Civil Liability for Nuclear Damage, IAEA Doc. INFCIRC/500 (May 21, 1963); Brussels Convention Relating to Civil Liability in the Field of Maritime Carriage of Nuclear Material (Dec. 17, 1971).

[45] International Convention on Civil Liability for Oil Pollution, Nov. 29, 1969, 973 U.N.T.S. 3, 9 I.L.M. 45; Protocol of 1992 to the International Convention on the Establishment of an International Fund for Compensation for Oil Pollution Damage, 1971, Nov. 27, 1992, 1953 U.N.T.S. 373 (1996); Convention on Civil Liability for Oil Pollution Damage resulting from Exploration for and Exploitation of Seabed Mineral Resources, Nov. 1977, 16 I.L.M. 1450.

The relationship between remedies in climate litigation and in the climate regime goes both ways. Steps identified and supported by the UNFCCC may help shape remedies in climate litigation, which could remove a major obstacle for successful climate advocacy. Some analysts, for example, have already proposed that remedies in climate litigation should include the requirement to buy carbon offsets endorsed in the climate regime.[46] The climate regime may also be the appropriate forum for a broader remedial response for those who are victims of climate change. If the number of climate refugees increases, for example from sea level rise, a more comprehensive U.N. remedial response may be necessary and would likely come under the auspices of the UNFCCC. Viewed in this light, the climate change litigation strategies are clearly supportive of and a potential catalyst for a stronger and more comprehensive UNFCCC regime.

3. IMPLICATIONS FOR INTERNATIONAL LAW GENERALLY

3.1. Promoting the Progressive Development of International Law

Whether international law will evolve to address climate change impacts effectively is still an open question, but just the act of filing climate-based petitions or complaints advances innovative arguments and pushes international law in new directions. The Inuit Petition to the Inter-American Commission on Human Rights, for example, requires the interpretation and application of rights to the use and enjoyment of traditional lands, to the benefits of culture, to property, to the preservation of health, life, physical integrity, security, and a means of subsistence, and to residence, movement, and inviolability of the home.[47] The petition invites the Commission to continue its recent jurisprudence extending the Inter-American system's human rights protections to the intersection of human rights and the environment.[48] The Inuit petition also presents important and well-supported arguments for the progressive development of international environmental law, including specific reference to U.S. obligations under the UNFCCC and the Kyoto Protocol and to emerging principles of law, including the principle not to cause transboundary environmental harm, the principle of sustainable development, and the principle of precaution.[49] Even if the Commission (as now seems likely) will not pursue the petition directly, both the petition and the ensuing dialogue at the Commission will further the

---

[46] See Engel, *supra* note 3; *see also* Mandatory CO_2 Credit Purchases Eyed as Remedy in Climate Change Suits, INSIDE EPA.COM (Nov. 24, 2006), *available at* http://www.law.arizona.edu/news/Press/Engel112706-2.pdf (quoting proposal from Kirsten Engel).

[47] Inuit Petition, *supra* note 6, at 74–95; *see also* Osofsky, The Inuit Petition, *supra* note 3 (discussing the human rights and environment linkages in the Inuit claim).

[48] *See, e.g.*, Case of the Mayagna (Sumo) Indigenous Community of Awas Tingni, Judgment, 2001 Inter-Am. Ct. H.R. (ser. C) No. 79 (Aug. 31, 2001); *see also* Additional Protocol to the American Convention on Human Rights in the Area of Economic, Social and Cultural Rights, art. 11(1), Nov. 14, 1988, 28 I.L.M. 161 (1989).

[49] Inuit Petition, *supra* note 6, at 97–101.

potential future interpretation of the links between international environmental and human rights law.[50]

These initial efforts to use new areas of the law, such as the law relating to human rights or cultural heritage, may spawn other innovative efforts to build policy coherence between different fields of international law and climate change. On April 17, 2007, for example, the U.N. Security Council held its first briefing on the security implications of climate change. That brought significant attention to the important linkages between climate change and national security.[51] The links between climate change and other fields of international law have triggered substantial scholarship as well as potentially innovative litigation strategies, including links between climate change and international trade law,[52] the law of the sea and fisheries conservation,[53] international finance,[54] coporate social responsibility,[55] and the international protection of wetlands.[56] Taken collectively, these efforts not only explore new aspects of their respective fields but contribute substantially to building policy and legal coherence between the fields of international law – an outcome that is important for sustainable development generally and for international responses to climate change more specifically.

3.2. Strengthening International Institutions

One of the most important outcomes of the current climate litigation strategies is that they may strengthen certain international institutions simply by using them. The

---

[50] The Commission held a broader hearing on the connection between climate change and human rights. See Letter from Ariel E. Dulitzky, Assistant Executive Sec'y, Organization of American States, to Sheila Watt-Cloutier, Chair, Inuit Circumpolar Conference, et al. (Feb. 1, 2007) (on file with author).

[51] See, e.g., U.N. Council Hits Impasse over Debate on Warming, N.Y. TIMES, Apr. 18, 2007, available at http://www.nytimes.com/2007/04/18/world/18nations.html; Andrew C. Revkin & Timothy Williams, Global Warming Called Security Threat, N.Y. TIMES, Apr. 15, 2007, at 25, col. 4.

[52] See, e.g., Andrew L. Strauss, The Legal Option: Suing the United States in International Forums for Global Warming Emissions, 33 ENVTL. L. REP. (ENVTL. L. INST.) 10185 (2003); Andrew L. Strauss, The Case for Utilizing the World Trade Organization as a Forum for Global Environmental Regulation, 3 WIDENER L. SYMP. J. 309 (1998).

[53] See William C. G. Burns, Potential Causes of Action for Climate Change Impacts under the United Nations Fish Stocks Agreement, SUSTAINABLE DEV. L. & POL'Y 34–38 (Winter 2007); William C. G. Burns, Potential Causes of Action for Climate Change Damages in International Fora: The Law of the Sea Convention, 1(2) INT'L J. SUSTAINABLE DEV. L. & POL'Y 27–51 (2006).

[54] See Gleason & Hunter, supra note 43.

[55] See Cornelia Heydenreich, GermanWatch Raises Complaint against Volkswagen: Climate Damaging Business Strategy Violates OECD Guidelines for Multinational Enterprises, GermanWatch Briefing Paper, May 2007, available at http://www.germanwatch.org/corp/vw-hg07e.pdf; Beschwerde gegen die Volkswagen AG unter den OECD-Leitsatzen fur Multinationale Unternehmen, May 7, 2007, available at http://www.germanwatch.org/corp/vw-besch.pdf (petition filed in Germany challenging on climate change grounds Volkswagen's operations as violating the OECD guidelines on multinational enterprises).

[56] Delmar Blasco, Secretary General of the Convention on Wetlands, Statement to the 6th Conference of the Parties to the U.N. Framework Convention on Climate Change, Den Haag, The Netherlands (Nov. 20, 2000).

question of whether or how existing international institutions can address what may be the most important environmental question of our time speaks to the relevance of the institutions themselves. If an institution with an environmental mandate, or at least some relationship to sustainable development, cannot be called into service to address an issue of the magnitude of climate change, what is its relevance more generally?

Appealing to the World Heritage Convention, for example, shines the spotlight on that Convention and enables UNESCO and the World Heritage Committee to raise the importance of protecting World Heritage sites from climate threats. Such petitions force the governments to address the impacts of climate change on cultural and natural heritage. They also provide an opportunity for the Committee to demonstrate its relevance and that of the World Heritage Convention to modern threats, like climate change, that arise indirectly from the processes of globalization and industrialization as opposed to direct, deliberate choices by individual host governments or corporations. Even if the ultimate decision of the Commission (to reject the petitions and adopt a more general strategy for addressing climate change threats to cultural heritage) was likely a politically motivated compromise, it may nonetheless provide the Committee with a long-term platform to highlight links between climate change and cultural heritage. By showing some well-reasoned restraint in expanding its scope to embrace climate change, it may strengthen the long-term credibility and trust the Committee has with member governments, while still garnering support from the petitioners and civil society organizations.[57]

The same can be said for the petition to the Inter-American Commission on Human Rights. The petition helps further the Commission's reach to situations other than traditional civil and political rights. Although the Commission has initially rejected the petition for providing insufficient information to demonstrate a violation of the American Convention,[58] the petition did prompt the Commission to hold, and invite the petitioners to, an unprecedented hearing on the "relationship between human rights and global warming."[59] Like the World Heritage Committee's approach described previously, this response appears to be a compromise that keeps the door open for the Commission to continue to explore climate change in the context of the InterAmerican commitments to human rights. The Commission's reach is thus extended to embrace climate change, albeit not yet through a formal, expansive interpretation of the underlying legal instruments.

To some extent these cross-over petitions – that is, those that make international institutions address an issue (climate change) that is normally outside of their respective mandates – position the institutions to be more relevant for the complexities

---

[57] *See, e.g.,* UNESCO Adopts Climate Change Strategy for World Heritage Sites, ENV'T NEWS SERV., July 11, 2006 (quoting several petitioners supportive of the Committee).
[58] *See* Letter from Ariel E. Dulitzky, Assistant Executive Sec'y, Organization of American States, to Paul Crowley, Legal Rep. (Nov. 16, 2006), *available at* http://graphics8.nytimes.com/packages/pdf/science/16commissionletter.pdf.
[59] *See* Letter from Ariel E. Dulitzky, *supra* note 50.

of sustainable development more generally. Thus, invitations to address the intersection of human rights and climate at the Inter-American Commission, trade and climate at the World Trade Organization, or finance and climate in the case of the international financial institution accountability mechanisms, are invitations for these institutions to show that they can address the complex and integrated aspects of contemporary sustainable development issues.

4. STRENGTHENING THE DEMOCRATIZATION OF GLOBAL ENVIRONMENTAL GOVERNANCE

Climate litigation at all levels is democratizing global environmental law and policymaking. Although the scale, scope, and methods of participation by civil society in the formal climate negotiations have been substantial, at the end of the day everything from the agenda to the final outcome of international treaty negotiations – and the climate change regime is no exception – is appropriately monopolized by governments. Civil society can observe, propose, pressure, prod, and even parody, but ultimately its role in international negotiations is limited.

Not so in terms of litigation. Climate change litigation empowers civil society to shape the agenda in ways not allowed in formal negotiations. It was civil society, for example, that put climate change on the agenda of the World Heritage Committee and the Inter-American Human Rights Commission. Approval to file the petitions was not solicited nor needed, from either the governments or the relevant international institutions. Civil society's exercise of this agenda-creating authority contributes to the ongoing changes seen in who participates and influences international policy. Climate litigation at the national level also helps to democratize climate policy. Clearly, this is the case in the United States, where subnational government units (e.g., the states of Massachusetts, Connecticut, and California, as well as municipalities, such as, Oakland, California, and Boulder, Colorado), frustrated with the lack of federal action under the Bush administration, have taken strong action on climate change – thus expressing their keen interest in participating and shaping climate policy.[60] Similarly, Australian civil society claimants have put climate change on the agenda of otherwise reluctant government agencies.[61] Although *legal* actions, these were also *political* statements intended to pressure the respective governments on climate change and to show the world that at the subnational level, at least, many in the United States and Australia support stronger actions on climate change.

5. TRANSNATIONAL CLIMATE ADVOCACY NETWORKS

Climate litigation efforts are also changing the nature and scope of transnational advocacy networks focused on climate change. The existence of such networks is now

---

[60] *See* Connecticut v. AEP Complaint, *supra* note 4 (plaintiffs include nine states); Friends of the Earth v. Watson, 2005 WL 2035596 (2005) (plaintiffs include Boulder, Colorado, Arcata, California, and Oakland, California); *see also* Hari M. Osofsky, *Climate Change Litigation as Pluralist Legal Dialogue?*, 26 STANFORD ENVTL. L.J. 182 and 43 STANFORD J. INT'L L. (2007).
[61] *Australian Conservation Foundation*, (2004) VCAT 2029.

widely recognized as having significant influence on environmental governance.[62] Climate change policy, generally, benefits from what is among the most well-networked and cooperative of all transnational environmental advocacy movements. Climate change has been a global policymaking priority for more than fifteen years now, and the depth, sophistication, and trust that has built up in transnational climate advocacy networks is unprecedented in international environmental governance. Climate negotiations are host to literally thousands of civil society representatives. The Climate Action Network (CAN), a major network for organizing and coordinating civil society input into the climate negotiations boasts 365 nongovernmental organizations as members and seven regional offices around the world;[63] it is well organized and very visible at the negotiations.

For the most part, CAN and its affiliated organizations and networks have focused their work on influencing the international negotiations, but the advent of the climate litigation strategies outlined in this book reveal a subtle, but important, shift in the strategies and scope of the climate advocacy networks. This shift entails a greater focus on advocating for specific remedies for particular harms, an extension to multiple forums beyond the UNFCCC Conferences of the Parties, and the inclusion of new advocacy organizations with a clearer focus on legal strategies. The climate litigation network is now its own transnational network, albeit arguably a subset of the broader climate networks exemplified by CAN. An advocacy statement calling for the national and international enforcement of climate-related laws, for example, was explicitly endorsed by nearly seventy-five advocates from twenty-six countries; this reflects both global support and cooperation in the strategy of bringing climate litigation claims.[64]

Although it may be too soon to predict, the cooperation in sharing information, strategies, and expertise that is evident in the emerging climate litigation strategies – seen perhaps most readily in the coordinated efforts to file claims under the World Heritage Convention – may herald a new era of transnational cooperation that is designed less for influencing broad international policy and more for using domestic and national forums to bring coordinated impact litigation. This collaborative advocacy will both strengthen the individual cases and serve to highlight the need for a global response. Such a coordinated and integrated litigation strategy, which is emerging in climate change, could also appear in the future with other global environmental issues such as ozone depletion, mercury pollution, or fisheries losses.

CONCLUSION

It is hard to judge how much, if at all, the pressure from climate change litigation will contribute to broader changes in climate policy, but it certainly is influencing the

---

[62] *See, e.g.*, MARGARET KECK & KATHRYN SIKKINK, ACTIVISTS BEYOND BORDERS: ADVOCACY NETWORKS IN INTERNATIONAL POLITICS (1998).
[63] *See* http://www.climatenetwork.org.
[64] *See* http://www.climatelaw.org.

debate. Many of the climate advocates that have brought actions thus far have been motivated substantially (if not primarily) by the goal of raising the profile of climate change in the hopes of building political will to force more ambitious efforts to address the issue. Certainly, the state attorneys general who brought climate-related claims in the United States did so at least partly to pressure for national or statewide climate policies. In California, for example, the litigation was one piece of a multi-part effort to move forward on climate change, which has included setting ambitious emission reduction targets, issuing new fuel efficiency standards, and establishing the framework for a cap-and-trade program for greenhouse gases.[65]

Much of the litigation is directly aimed at forcing political action. The Inuit petition to the Inter-American Commission on Human Rights was aimed at using the moral and political persuasion of a formal human rights finding to isolate the United States and build both international and domestic pressure on the government to take stronger action. Domestic actions in the United States, Germany, Australia, and other countries have also sought to compel government actions relating to climate change.[66] These actions range from requirements to assess climate impacts at the project level,[67] to incorporate climate change into public financing decisions,[68] or to compel government agencies to regulate carbon dioxide and other greenhouse gases as injurious pollutants.[69] Even when domestic actions fail, they may indirectly build pressure for legislative and policy action. In the United States, for example, dismissal of the *Connecticut v. AEP* complaint on political question grounds put the spotlight on the political branches of government for a solution.

Climate litigation also ripples through the private sector, receiving the attention of industries that have potential exposure to climate liability. Plaintiff-side tort lawyers are talented, resourceful, patient, and well financed, and many of them believe climate change either now or in the future will present very real opportunities for successful litigation.[70] In response, corporations and their attorneys now speak openly about the emerging "litigation risk" from climate change.[71] Major U.S. law firms now

---

[65] *See, e.g.,* Global Warming Solutions Act of 2006, Cal. State Code, Div. 1, Sec. 38500 (2006).

[66] *See generally* Smith & Shearman, *supra* note 2.

[67] *Australian Conservation Foundation*, (2004) VCAT 2029.

[68] *Friends of the Earth v. Mosbacher*, No. C02–4106 JSW, 2007 WL 962955 (N.D. Cal. Mar. 30, 2007); *Friends of the Earth v. Watson*, No. C02–4106 JSW, 2005 WL 2035596 (N.D. Cal. Aug. 23, 2005).

79 *Massachusetts v. EPA*, 127 S.Ct. 1438, 1438 (2007).

[70] *See, e.g.,* David A. Grossman, *Warming Up to a Not-So-Radical Idea: Tort-Based Climate Change Litigation*, 28 COLUM. J. ENVTL. L. 1, 9–33 (2003); Matthew F. Pawa & Benjamin A. Krass, *Global Warming as a Public Nuisance: Connecticut v. American Electric Power*, 16 FORDHAM ENVTL. L. REV. 407 (2005).

[71] *See, e.g.,* Vincent S. Oleszkiewicz & Douglas B. Sanders, *The Advent of Climate Change Litigation Against Corporate Defendants*, 35 ENV'T REP. (BNA) 2365 (Nov. 12, 2004) ("Despite the uncertainties, it may not be too early to prepare for the possibility of litigation. Next steps for potential defendants may include a preliminary risk assessment of their exposure to litigation and potential defenses...."); *Global Warming: Here Come the Lawyers*, BUS. WK. ONLINE, Oct. 30, 2006 (quoting Kevin Healy, a partner with the law firm of Bryan Cave, that in the wake of recently filed lawsuits he now advises corporate clients that they need to take "reasonable" steps to pare back emissions to reduce their legal exposure); Kristin Choo, *Feeling the Heat: The Growing Debate over Climate Change Takes on Legal*

routinely market their abilities and successes in climate litigation,[72] and litigation (and the related regulatory) risk are important factors in motivating companies to take proactive steps to reduce their greenhouse gas emissions and related climate impacts.[73]

Thus, the turn to climate litigation and related litigation is reshaping how we think and respond to the climate change challenge – regardless of whether individual cases prevail. But, of course, climate change advocates hope to win. They seek specific and far-reaching remedies. The Inuit Petition, for example, seeks to have a plan established and implemented to protect Inuit culture and resources, including, inter alia, the land, water, snow, ice, and plant and animal species used or occupied by the Inuit.[74] The state attorneys general in *Connecticut v. AEP* seek to have the courts impose a cap on greenhouse gas emissions from the five largest emitting utilities in the United States.[75] The State of California seeks compensation for costs it is already incurring from climate change.[76] These are substantial remedies that would not only improve the plight of the specific plaintiffs, but would also make important contributions to the climate policy debate. Obviously, a court's use of its injunctive powers could lead to direct emissions reductions in the United States, but so too would a monetary damage judgment, which would reverberate throughout the private industry sector, forcing corporations to take proactive steps to reduce their exposure to climate liability.

Nor are victories in climate litigation a chimera. The recent U.S. Supreme Court decision in *Massachusetts v. EPA*, which is forcing the EPA to revisit whether to regulate carbon under the Clean Air Act, a reassessment embraced by the Obama administration, is the most well known climate victory. The Supreme Court found that the risk of rising sea levels alleged by the plaintiffs was sufficiently "real" to afford Massachusetts standing to raise its climate change–based claim.[77] Other courts in the United States and Australia, for example, have extended standing to private

---

Overtones, A.B.A. J., 29, 30, July 2006 (quoting Professor John Dernbach: "The prospect of liability is a serious matter for people who understand climate change and take it seriously."); Christina Ross, Evan Mills & Sean Hecht, *Limiting Liability in the Greenhouse: Insurance Risk-Management Strategies in the Context of Global Climate Change*, 26 STANFORD ENVTL. L.J. 251, 274 and 43 STANFORD J. INT'L L. (2007).

[72] *See, e.g.*, Sidley Austin, LLP, Climate Change Advisory Nov. 21, 2006, *available at* http://www.sidley.com/db30/cgi-bin/pubs/ClimateChangeUpdate11.21.06.pdf.

[73] *See The Climate Group, Carbon Down, Profits Up* (2d ed., 2005) (compiling an extensive list of voluntary emissions targets accepted by corporations).

[74] Inuit Petition, *supra* note 3, at 118.

[75] *Australian Conservation Foundation*, (2004) VCAT 2029.

[76] Friends of the Earth v. Mosbacher, No. C02–4106 JSW, 2007 WL 962955 (N.D. Cal. Mar. 30, 2007) (rejecting summary judgment motion in a case arguing that the U.S. Overseas Private Insurance Company must conduct an assessment of the climate impacts of the projects they finance); Friends of the Earth v. Watson, No. C02–4106 JSW, 2005 WL 2035596 (N.D. Cal. Aug. 23, 2005) (same). These cases were settled in February 2009. *See* Joint Motion for Dismissal with Prejudice, Friends of the Earth v. Spinelli, No. C02–4106 JSW (Feb. 6, 2009).

[77] Massachusetts v. EPA, 127 S.Ct. 1438, 1438, 1455–56 (2007).

parties pressing climate change claims.[78] Significant substantive victories have also required the assessment of climate impacts in the permitting of greenhouse gas–emitting activities,[79] in decisions to provide financing,[80] and in requirements to reduce gas flaring associated with oil refineries.[81] These victories are likely just the tip of the litigation iceberg, but win or lose, climate litigation strategies have harkened in a new era of climate politics.

---

[78] *Owens Corning Corp.*, 434 F. Supp. 2d 957 (holding that plaintiff environmental organization had standing to challenge a permit application that would have permitted significant releases of a potent greenhouse gas (HCFC-142b)); *Watson*, 2005 WL 2035596 (upholding standing of environmental organization to bring a case seeking that a U.S. government agency include climate change in their environmental assessments). *But see* Korsinsky v. EPA No. 05 Civ. 859 (NRB), 2005 WL 2414744 (S.D.N.Y. Sept. 29, 2005) (rejecting standing of an individual in a climate change tort action).

[79] *See, e.g., Australian Conservation Foundation*, (2004) VCAT 2029.

[80] *Mosbacher*, 2007 WL 962955 (rejecting summary judgment motion in a case arguing that the U.S. Overseas Private Insurance Company must conduct an assessment of the climate impacts of the projects it finances); *Watson*, 2005 WL 2035596; Bund & Germanwatch v. German Federal Ministry of Econ. & Labour [BMWA], Beschluss, Verwaltungsgericht [VG Berlin] [Local Administrative Court] Jan. 10, 2006, VG 10 A 215.04 (2006), *translated at* http://www.climatelaw.org/media/Germany/de.export.decision.eng.doc.

[81] Climate Justice Programme, Court Orders Nigerian Gas Flaring to Stop (Nov. 14, 2005), *available at* http://www.snm.nl/pdf/0500_2.7_court_orders_nigarian_flaring_to_stop__background_paper.pdf.

17

Conclusion: Adjudicating Climate Change across Scales

Hari M. Osofsky*

INTRODUCTION

This book explores climate change litigation in its many existing and potential variations. As this volume was being written, the number of relevant cases and their impact increased dramatically. Most notably, the U.S. Supreme Court's decision in *Massachusetts v. EPA*[1] – together with a cultural shift symbolized by Al Gore and the Intergovernmental Panel on Climate Change (IPCC) winning the Nobel Peace Prize[2] – transformed the policy and litigation landscape. The election of President Obama changed things further, with his commitment to active U.S. participation in international climate treaty negotiations and to a robust federal regulatory approach to greenhouse gas emissions.[3]

The concluding chapter grapples with how this evolution impacts the way in which we should view the role of litigation as part of transnational regulation of climate change. Although we are still early in this story that Peter Roderick's foreword began,[4] some of the impacts of climate change litigation are already clear. David Hunter's chapter, *The Implications of Climate Change Litigation: Litigation for International Environmental Law-Making*, does an excellent job of analyzing these cases as vehicles for promoting greater public awareness.[5] This chapter

---

* Associate Professor, Washington and Lee University School of Law; B.A., J.D., Yale University. The author can be reached at osofskyh@wlu.edu. I would like to thank Wil Burns and Stefanie Herrington for their excellent editorial input, and Joshua, Oz, and the newly arrived Scarlet Gitelson for their loving support.

[1] Massachusetts v. EPA, 549 U.S. 497 (2007).

[2] The Nobel Peace Prize 2007, *available at* http://nobelprize.org/nobel_prizes/peace/laureates/2007/ (visited Nov 17, 2007).

[3] *See* President Barack Obama, Address to Joint Session of Congress (Feb. 24, 2009), *available at* http://www.whitehouse.gov/the_press_office/Remarks-of-President-Barack-Obama-Address-to-Joint-Session-of-Congress/; Remarks by the President on Jobs, Energy Independence, and Climate Change, East Room of the White House, (Jan. 26, 2009), *available at* http://www.whitehouse.gov/blog _post/Fromperiltoprogress/; Obama for America, Barack Obama and Joe Biden: New Energy for America, *at* http://www.barackobama.com/pdf/factsheet_energy_speech_080308.pdf (last visited Dec. 22, 2008).

[4] *See* Peter Roderick, *Foreword*, in this volume.

[5] *See* David B. Hunter, *The Implications of Climate Change Litigation: Litigation for International Law-Making*, in this volume.

builds upon his analysis by focusing back on the issues of scale raised in the Introduction.

As a scientific and legal matter, climate change is multiscalar. In other words, emissions, impacts, and the legal structures that interact with them are simultaneously individual, local, state, national, international, and every level in between.[6] The book's chapters demonstrate the complex scalar dynamics both with respect to the problem and with respect to litigation over it. At the end of their analyses, however, two foundational issues remain: (1) At what scales should the various aspects of the problem of climate change be regulated? (2) What role does and should litigation play in establishing appropriate regulatory scale?

To answer these questions, one must have some idea of what "scale" means. At the surface, such a definition seems attainable, especially when conversing in the language of law. Each suit was brought in a particular tribunal, which operates at a designated level of governance. This book's organization reflects such an understanding that tribunals are constituted at particular scales; the chapters focus on cases that we have labeled as subnational, national, or supranational.

A glance at any one of these chapters, however, reveals the danger of such a simple characterization. For example, my chapter on the U.S. Supreme Court case *Massachusetts v. EPA*[7] is situated in the "national" part of the book.[8] This categorization of the lawsuit is clearly rational based on the adjudicating tribunal and the case's focus on federal law. But it fails to capture other aspects of the dispute's scale; the parties on both sides of the case included local and state governments, focusing primarily on potential threats to local and state-based interests, as well as nongovernmental and corporate entities with ties at multiple scales. And, as discussed in that chapter, the "federal law" dispute in the case centered around questions of the appropriate scale at which to regulate climate change.[9]

This chapter interweaves themes of regulation and rescaling to reflect upon the significance of these lawsuits. It argues that the problem of climate change and adjudication over it are simultaneously multiscalar and scale-dependent.[10] In other words,

---

[6] For an in-depth exploration of climate change as a multiscalar regulatory problem, see Hari M. Osofsky, *Is Climate Change "International"?: Litigation's Diagonal Regulatory Role* 49 VA. J. INT'L L. 585 (2009) (draft manuscript on file with author).

[7] See Massachusetts v. EPA, 127 S. Ct. 1438 (2007).

[8] See Osofsky, *The Intersection of Scale, Science, and Law in* Massachusetts v. EPA, in this volume.

[9] See id.

[10] This chapter uses the term "multiscalar" to mean connected to more than one scale. It uses the term "scale-dependent" to mean tied to a particular scale. As discussed in more depth in Section 2, the concept of scale is a highly contested one in the literature of many disciplines. For discussion of that contestation in the geography literature, see NEIL BRENNER, NEW STATE SPACES: URBAN GOVERNANCE AND THE RESCALING OF STATEHOOD 9 (2004); Neil Brenner, *The Limits to Scale? Methodological Reflections on Scalar Structuration*, 25 PROGRESS HUM. GEOGRAPHY 591 (2001); Sallie A. Marston, *The Social Construction of Scale*, 24 PROGRESS HUM. GEOGRAPHY 219 (2000); and Sallie A. Marston & Neil Smith, *States, Scales and Households: Limits to Scale Thinking? A Response to Brenner*, 25 PROGRESS HUM. GEOGRAPHY 615 (2001). For interdisciplinary analyses of these issues more directly tied to environmental regulatory problems, see Michael Mason, *Transnational Environmental Obligations: Locating New Spaces of Accountability in a Post-Westphalian Global Order*, 26 TRANSACTIONS INST.

as the previous chapters illustrate, individuals, localities, states, nations, regional supranational bodies, international entities, and other actors at many intermediate levels make emissions decisions, suffer the impacts of climate change, and bring, defend against, and adjudicate these suits. However, because lawsuits and petitions, as a formal matter, are adjudicated at particular levels of governance over time, an engagement of specific decision-making scales also must inform an exploration of these actions' significance.

Section 1 of this chapter explores the role that climate litigation plays in regulatory rescaling,[11] as illustrated through the contestation described in the book's chapters. Section 2 then argues that climate change litigation provides a valuable complement to treaty, legislative, and executive action because it fosters needed interaction across levels of government. The chapter concludes by considering next steps for climate change litigation as part of the regulatory discourse about this problem.

1. RESCALING THROUGH CLIMATE CHANGE LITIGATION

This section focuses on the rescaling role that climate change litigation plays by considering what regulatory scale is and how these cases interact with it. As noted, the definitional issue with which this chapter begins has been explored extensively in the geography literature on scale. Although human geographers increasingly agree that scale is socially constructed, foundational debates have raged over the past decade about what scale is and how it might be relevant to analyses of regulation.[12] An issue from these dialogues with particular salience for climate change litigation is the question of scale's fixity and fluidity.[13] Namely, to what extent are the categories that this book organizes itself around – subnational, national, supranational – terms that

---

BRITISH GEOGRAPHERS 407 (2001); Hari M. Osofsky, *The Intersection of Scale, Science, and Law in Massachusetts v. EPA*, 9 OR. REV. INT'L L. 233 (2007) (republished in this volume), and Nathan F. Sayre, *Ecological and Geographical Scale: Parallels and Potential for Integration*, 29 PROGRESS HUM. GEOGRAPHY 276, 281 (2005).

[11] Rescaling processes are ones in which individuals or entities attempt to jump levels. See Sayre, *supra* note 10, at 285. In the context of climate change litigation, I have described attempts at rescaling in *Massachusetts v. EPA*. See Osofsky, *The Intersection of Scale, Science, and Law in* Massachusetts v. EPA, *supra* note 10.

[12] See *supra* note 10

[13] For discussion of issues of fixity and fluidity, see Kevin R. Cox, *Spaces of Dependence, Spaces of Engagement and the Politics of Scale, Or: Looking for Local Politics*, 17 POL. GEOGRAPHY 1, 20–21 (1998); David Delaney & Helga Leitner. *The Political Construction of Scale*, 16 POL. GEOGRAPHY 93, 93 (1997); Andrew Herod, *Scale: The Local and the Global*, in KEY CONCEPTS IN GEOGRAPHY 229, 234, 242 (Sarah L. Holloway, Stephen P. Rice & Gill Valentine eds., 2003); Deborah G. Martin, *Transcending the Fixity of Jurisdictional Scale*, 17 POL. GEOGRAPHY 33, 35 (1998); Anssi Paasi, *Place and Region: Looking Through the Prism of Scale*, 28 PROGRESS HUM. GEOGRAPHY 536, 542–43 (2004); Neil Brenner, *Between Fixity and Motion: Accumulation, Territorial Organization and the Historical Geography of Spatial Scales*, 16 ENVT. & PLANNING D: SOC'Y & SPACE 459, 461 (1998); Erik Swyngedouw, *Excluding the Other: The Production of Scale and Scaled Politics*, in GEOGRAPHIES OF ECONOMIES 167, 169 (Roger Lee & Jane Wills eds., 1997); Erik Swyngedouw, *Neither Global nor Local: "Glocalization" and the Politics of Scale*, in SPACES OF GLOBALIZATION: REASSERTING THE POWER OF THE LOCAL 137, 141 (Kevin R. Cox ed., 1997).

actually have meaning? As government and civil society interact with the problem of climate change, are these scales steady organizational groupings or constantly shifting spaces of engagement? How does scale shape regulatory decision making and vice versa?

And yet for all this literature's engagement with scale, it rarely separates out law from other sociopolitical ordering.[14] "Regulation" lumps together formal and informal social ordering. Although such an approach avoids the formalism that too often dominates the legal literature, and that at times prevents an analysis of all relevant stakeholders and decision makers,[15] it can undervalue the role of formal legal institutions in shaping and being shaped by scale.

Law and legal institutions are structured at specific levels of governance, which, despite all of the shifts wrought by globalization, stay relatively stable most of the time. The tribunals adjudicating climate change litigation and the laws that they are relying upon generally are constituted at specific, fixed scales. For example, the Inter-American Commission on Human Rights is a supranational, regional body established through the Organization of American States.[16] At the other end of the spectrum, the Minnesota Court of Appeals is a judicial body created by the state of Minnesota to interpret its laws.[17]

The fluidity in the scales of this litigation comes not from the tribunals themselves, then, but rather from the multiscalar nature of the problem of climate change and regulatory efforts to address it. These "fixed" entities, in their stability, provide a framework in which contestation across scales can take place. The aim of this litigation is not to shift the scales of the tribunals and what law they can consider, but rather to rescale aspects of regulating greenhouse gas emissions and impacts.

The cases thus debate the appropriateness and necessity of regulatory entities at different scales taking particular steps to address global climate change. For example, should the World Heritage Commission be addressing the impacts of climate change

---

[14] *But see* Dennis R. Judd, *The Case of the Missing Scales: A Commentary on Cox*, 17 POL. GEOGRAPHY 29, 30–31 (1998) (analyzing the effect of federalism in the United States).

[15] The legal pluralist literature, for example, engages the importance of addressing the multiple normative communities – formal and informal – that share social spaces. Robert M. Cover, *The Supreme Court 1982 Term Foreword: Nomos and Narrative*, 97 HARV. L. REV. 4 (1983); Sally Engle Merry, *Legal Pluralism*, 22 LAW & SOC'Y REV. 869 (1988); Emmanuel Melissaris, *The More the Merrier? A New Take on Legal Pluralism*, 13 SOC. & LEGAL STUD. 57 (2004); Ambreena Manji, *'Like a Mask Dancing': Law and Colonialism in Chinua Achebe's Arrow of God*, 27 J. LAW & SOC'Y 626 (2000); Dalia Tsuk, *The New Deal Origins of American Legal Pluralism*, 29 FLA. ST. U. L. REV. 189 (2001); Paul Schiff Berman, *Global Legal Pluralism*, 80 S. CAL. L. REV. 1155 (2007). Similarly, the New Haven school of international law views law as "a process of authoritative decision by which members of a community clarify and secure their common interests" and argues that "humankind today lives in a whole hierarchy of interpenetrating communities, from the local to the global." HAROLD D. LASSWELL & MYRES S. MCDOUGAL, JURISPRUDENCE FOR A FREE SOCIETY: STUDIES IN LAW, SCIENCE AND POLICY xxi (1992).

[16] For a discussion of the petition to the Inter-American Commission on Human Rights, see Hari M. Osofsky, *The Inuit Petition as a Bridge? Beyond Dialectics of Climate Change and Indigenous Peoples' Rights*, in this volume.

[17] For a discussion of the Minnesota case, see Stephanie Stern, *State Action as Political Voice in Climate Change Policy: A Case Study of the Minnesota Environmental Cost Valuation Regulation*, in this volume.

on protected world heritage sites? If so, what should its role be?[18] If the impacts of climate change threaten species, at what point should they be listed as endangered under the U.S. Endangered Species Act (ESA)? What specific obligations should such listing entail?[19] When localities in Victoria, Australia, engage in environmental assessments of planned projects, should that process include an examination of climate impacts? If so, which climate impacts?[20]

Moreover, these regulatory questions are made even more complex by the mixed public-private nature of the decisions involved. When can U.S. states sue federal agencies to compel them to regulate greenhouse gas emissions? When can they sue the major corporate emitters directly? Should governmental victims of climate change be treated differently than private property owners? Than indigenous communities? Because climate change results from emissions by individuals, governments, and corporations at multiple scales, the litigation embodies dynamic interactions among relevant public and private parties.

Litigation's mix of scalar fluidity and fixity thus makes it a particularly helpful tool for rescaling. Although individual regulators and corporations operate at specific, even if multiple, scales within a legal framework, the tribunals have the power to rule upon what is appropriate at a given scale. In so doing, they help to shape the scale at which regulation occurs. *Massachusetts v. EPA*, for example, involved a dispute over whether climate change was a "state," "federal," or "international" problem. Procedurally, the case hinged on whether the harms of climate change were small scale enough, in terms of both time and space, for states to have standing to sue. Substantively, the court had to determine whether the Clean Air Act, a federal law, created obligations for the U.S. Environmental Protection Agency to regulate motor vehicles' greenhouse gas emissions; in the process, questions abounded over whether climate change was "too big" to regulate at that level or at the state level. The U.S. Supreme Court's answer – to which the Obama administration, unlike the Bush administration that preceded it, is rapidly responding[21] – established as appropriate interest in the problem at a state scale and regulation at a federal scale.[22]

---

[18] For a discussion of the World Heritage Commission petitions, see Erica J. Thorson, *The World Heritage Convention and Climate Change: The Case for a Climate-Change Mitigation Strategy Beyond the Kyoto Protocol*, in this volume.

[19] For a discussion of the U.S. Endangered Species Act actions, see Brendan R. Cummings & Kassie R. Siegel, *Biodiversity, Global Warming, and the U.S. Endangered Species Act: The Role of Domestic Wildlife Law in Addressing Greenhouse Gas Emissions*, in this volume.

[20] For a discussion of the pending cases in Australia over coal mining, see Lesley K. McAllister, *Litigating Climate Change at the Coal Mine*, in this volume.

[21] For the Bush administration's response, see Advance Notice of Proposed Rulemaking: Regulating Greenhouse Gas Emissions under the Clean Air Act, EPA-HQ-OAR-2008-0318 (July 11, 2008), available at http://www.epa.gov/climatechange/anpr.html. For the Obama administration's response thus far, see Proposed Endangerment and Cause or Contribute Findings for Greenhouse Gases Under Section 202(a) of the Clean Air Act; Proposed Rule 74 Fed. Reg. 18885 (proposed Apr. 24, 2009) (to be codified at 40 C.F.R. ch. 1); John M. Broder, *E.P.A. Clears Way for Greenhouse Gas Rules*, N.Y. TIMES, Apr. 17, 2009, at A15, available at http://www.nytimes.com/2009/04/18/science/earth/18endanger.html.

[22] For an analysis of these issues, see Osofsky, *The Intersection of Scale, Science, and Law in* Massachusetts v. EPA, *supra* note 10.

At the supranational level, the Inter-American Commission on Human Rights, in response to the Inuit petition, had to decide whether it should attempt to push the United States to regulate climate change more aggressively. Or in scalar terms, should a regional supranational body pressure a nation-state to address more effectively this multiscalar problem that is producing harms in indigenous communities in two nation-states within its region? Ultimately, the Commission decided to hold a more general hearing rather than address the petition directly. In holding such a hearing, it opened the question of whether climate change was a regional human rights issue.[23]

When viewed as a whole, rather than in individual snapshots, then, the litigation serves as a lever in regulatory contestation over how to address this looming problem. In a variety of fixed fora, petitioners attempt to reshape the regulatory map. This role raises an important practical and normative question with which Section 2 will grapple; namely, do these regulatory rescaling efforts serve a constructive role in creating appropriate responses to climate change?

2. THE VALUE OF REGULATORY RESCALING

This section considers the significance of these lawsuits connected to multiple levels of government in which debates over regulatory scale take place. However, its analysis of the value of climate change litigation's rescaling role faces two major complexities. First, the question of these cases' "constructive" role has both descriptive and normative dimensions. Descriptively, the dynamics play out in the particular ways highlighted in Section 1 and, as a result, impact the regulatory environment. However, an inquiry into the role that this litigation plays cannot stop with mere description; fully engaging issues of "constructiveness" requires a normative judgment about how litigation should fit into broader regulatory strategy. One's underlying values will impact such a judgment significantly.

Second, and at least as important, this litigation has both formal and informal impacts. As the chapters of this book reflect, these cases have evolved over the past few years from creative advocacy to some courtroom and administrative victories. Those particular decisions, which represent a small portion of the overall adjudication associated with climate change, create binding obligations for regulatory entities and, as such, serve as a formal part of the regulatory process. But the expressive – that is, social norm creating – role of this litigation arguably has been more significant than the gains from implementing those particular judgments.[24] Both formally successful suits and those with little hope of achieving binding results have together helped to change the regulatory landscape by putting pressure on a wide range of

---

[23] For further discussion of this case, see Osofsky, *The Inuit Petition as a Bridge*, *supra* note 16.
[24] For an exploration of expressivism in the War on Terror context, see Mark A. Drumbl, *The Expressive Value of Prosecuting and Punishing Terrorists: Hamdan, the Geneva Conventions, and International Criminal Law*, 75 GEO. WASH. L. REV. 1165, 1169 (2007).

individuals and entities to act.[25] For instance, the Hazelwood Mines dispute in Australia, discussed in Lesley McAllister's chapter, not only mandated that greenhouse gas emissions be included as part of environmental assessment but also resulted in the first-ever Greenhouse Gas Reduction Deed between the Victorian government and International Power Hazelwood.[26] The listing of the polar bear as "threatened" under the U.S. ESA, as described in the chapter by Brendan Cummings and Kassie Siegel, was accompanied by extensive media analysis of climate change.[27] At times, those involved in filing the petitions have even highlighted their expressive role explicitly. For example, Sheila Watt Cloutier, then chair of the Inuit Circumpolar Conference, acknowledged the Inuits' human rights petition as fundamentally about opening a dialogue with the United States regarding climate change and human rights.[28]

Together, these complexities mean that a full engagement of litigation's regulatory role requires a difficult socio-legal analysis. Namely, this rescaling takes place in a broader policy and cultural context that interacts both with the formal litigation impacts and with how people view their value. For some, the litigation only has value currently because of gaps in the treaty regime and national policies. If those were corrected – which appears more likely in the United States since the election of President Obama – they would want to minimize the opportunity for much of this advocacy.[29]

For others, such as the editors of this volume, the litigation has value as a regulatory mechanism whether or not policy steps are taken, and should not be preempted with the implementation of stronger policies.[30] Beyond their gap-filling role, these suits allow concerned individuals and entities to highlight impacts and inequities, as

[25] Numerous legal theories explore these dynamics and their impact on international lawmaking. For example, transnational legal process analyzes the interpretation, internalization, and enforcement of norms through interactions like those described in these cases. See Harold Hongju Koh, *Why Transnational Law Matters*, 24 PENN ST. INT'L L. REV. 745(2006); Harold Hongju Koh, *Jefferson Memorial Lecture: Transnational Legal Process after September 11th*, 22 BERKELEY J. INT'L L. 337, 339 (2004); Harold Hongju Koh, *Transnational Legal Process*, 75 NEB. L. REV. 181 (1996); Harold Hongju Koh, *Why Do Nations Obey International Law*, 106 YALE L.J. 2599 (1997). Global legal pluralists and the New Haven school that often roots their analysis explore the multiple normative communities that shape the lawmaking process. See LASSWELL & MCDOUGAL, supra note 15; Berman, supra note 15.
[26] See Lesley K. McAllister, *Litigation Climate Change at the Coal Mine*, supra note 20.
[27] See Cummings and Siegel, supra note 19.
[28] See Presentation by Sheila Watt-Cloutier, Chair, Inuit Circumpolar Conference Eleventh Conference of Parties to the UN Framework Convention on Climate Change Montreal, Dec. 7, 2005, http://www.inuitcircumpolar.com/index.php?ID=318&Lang=En; see also Osofsky, supra note 16.
[29] See, e.g., Eric A. Posner, *Climate Change and International Human Rights Litigation: A Critical Appraisal*, 155 U. PA. L. REV. 1925, 1925 (2007) ("Litigation seems attractive to many people mainly because the more conventional means for addressing global warming – the development of treaties and other international conventions, such as the Kyoto Accord – have been resisted by governments.").
[30] See, e.g., Hari M. Osofsky, *Climate Change Legislation in Context*, 102 NW. U. L. REV. COLLOQUY 245, 249 (2008) ("[L]itigation plays a crucial role in the regulation of climate change and the legislation should not attempt to preempt access to courts too broadly. Rather, the statutory scheme should provide a clear basis for concerned individuals and organizations to address inadequate regulation by government and failures by major emitters to reduce their production of greenhouse gases.").

well as to maintain pressure on governments to address additional aspects of the problem. Moreover, to the extent that both climate treaties and national legislation contemplate implementation of their mandates over many decades, the specter of climate change litigation may help to steel the resolve of the policymakers who succeed the drafters of such mandates.

Furthermore, how one regards the appropriateness of rescaling through litigation likely depends on how one views the optimal scale of climate regulation. Those who argue for "scaling up" – that is, view the problem as one only appropriate for larger scale management – over the course of litigation express concern about opinions that "scale down" but think more positively of ones that prevent smaller-scale regulation. Their opponents, who value smaller-scale regulation, generally have the opposite perspective.[31]

Regardless of one's stance on litigation's role as part of a well-developed climate regulation regime, however, the current situation is one of regulatory insufficiency. As the Introduction makes clear, the Kyoto Protocol has struggled to reach its goals because of nonparticipation of major emitters like the United States and many Parties' difficulties in meeting their commitments. Moreover, it does not go far enough to achieve the reductions that scientists say are necessary. Although recognition of the problem has increased dramatically in recent years, and efforts are under way to create a post-2012 regime that will be more effective than the Kyoto Protocol, it appears unlikely that the treaty regime alone will be able to "solve" the problem, even with the United States participating more constructively under President Obama.[32]

With the increasing recognition of the problem, national and subnational regulatory efforts are also developing rapidly, but they probably will not go far enough fast enough. At a national level, major emitters are under pressure to address their emissions, and they likely will regulate more effectively than they have in the past. When proposed regulations are compared with the pace of emissions and atmospheric change, however, countries may not have sufficient political will to make drastic enough reductions.[33] An ever-increasing number of state and local-level governments are committing to incorporating climate mitigation and adaptation policies into their laws and planning efforts, even governments that might have appeared improbable a few years ago, such as Tulsa, Oklahoma.[34] Although those smaller-scale efforts represent significant emissions – the international network of localities

---

[31] For analysis of scalar battles in climate change litigation, see Osofsky, *supra* note 6; Osofsky, *supra* note 10.

[32] See William C. G. Burns & Hari M. Osofsky, *Overview: The Exigencies That Drive Potential Causes of Action for Climate Change*, in this volume.

[33] See id.

[34] See Kevin McCarty, Bloomberg, Palmer Lead USA and World Mayors on Climate Protection: US Mayors Climate Agreement Hits 500 Milestone, U.S. MAYOR NEWSPAPER (May 21, 2007), available at http://www.usmayors.org/USCM/us_mayor_newspaper/documents/05_21_07/pg1_NYC_climate.asp (last visited Sept. 10, 2008); see also Hari M. Osofsky & Janet Koven Levit, *The Scale of Networks? Local Climate Coalitions*, 8 CHI. J. INT'L L. 409 (2008).

working on climate change, for example, represents roughly 15% of the global total[35] – their efforts often are not well integrated with larger-scale ones.[36]

Despite the multiscalar nature of climate change and the way in which it interacts with a wide substantive range of regulation at different levels, regulatory collaboration that involves multiple levels of government and a wide range of entities within each level is relatively rare. At most, these efforts tend to be predominantly vertical, with a top-down or bottom-up scheme involving different levels of government, or predominantly horizontal, with different governmental entities at the same level working together.[37] Although many policymakers acknowledge the importance of crosscutting efforts and innovative initiatives exist, much more progress needs to be made in thinking through what I have elsewhere termed "diagonal regulation," which involves approaches that interconnect efforts both vertically and horizontally.[38]

The current regulatory environment and its limitations reinforce the importance of the socio-legal role that climate change litigation plays. The adjudication provides a mechanism for dialogue and awareness, in addition to a more formal forcing or limiting role, in a regulatory environment in which policies have not caught up to the problem.[39] At least as important, it creates diagonal interactions through which different levels and branches of regulators interact and grapple with what is needed. These cases help to bring attention to regulatory options and debates, and push policymakers to address more nuances of the problem in the process.[40]

Thoughtful people may continue to disagree as to the normative implications of litigation's role. However, until executive and legislative branches are able to construct effective multiscalar regulatory mechanisms – which poses a difficult challenge even for politicians committed to addressing the problem – litigation's formal and informal interactions likely will continue to play an essential role in the overall regulatory framework. And as discussed previously, even with a more effective policy regime, these cases may continue to provide an important mechanism for expressing grievances and keeping pressure on decision makers.[41] The combination of discontent with existing efforts and a wide range of legal mechanisms applicable to this crosscutting problem make courtrooms and other quasi-judicial fora important loci for dialogue among disparate actors across levels of governance about how to address climate change most appropriately.

---

[35] See ICLEI Global, *About CCP*, http://www.iclei.org/index.php?id=811 (last visited Sept. 10, 2008).
[36] For example, Mayor Gavin Newsom discussed this lack of integration as a problem, as well as some of the city's efforts to collaborate with entities at multiple scales, in answer to a question I asked following his keynote address at the conference *Surviving Climate Change: Adaptation and Innovation*, University of California, Hastings, College of the Law, Monterey Institute of International Studies, Apr. 4, 2008.
[37] See Osofsky, *supra* note 6.
[38] See id.
[39] See Hunter, *supra* note 5.
[40] See Osofsky, *supra* note 6.
[41] See *supra* Section 2.

3. THE FUTURE OF CLIMATE CHANGE LITIGATION

Thus far, this concluding chapter has focused on spatial scale, and the role that climate change litigation plays in helping regulators grapple with the many levels of governance with which the problem intersects. However, both the problem itself and attempts to address it must also engage complexities of time. As a scientific matter, climate change happens over long periods of time. Current changes result from past emissions and present emissions will cause future changes. These multiple time scales of the problem mean that regulatory efforts always must interlink the past and the future to decide what is appropriate now in terms of ascription of responsibility for climate change related damages and emissions reduction mandates.[42]

Moreover, the cases described in this book themselves span more than one time. Some of them concluded in the past,[43] others are ongoing,[44] and still others only exist as a possible future.[45] The disputes discussed in each chapter reflect the many time scales on which climate regulation takes place. Is there enough scientific certainty about how past emissions have resulted in present change, or current emissions will result in future change? What will be accomplished by acting now and what are the risks of failing to act? Which people and entities have the obligation and/or right to act when?

Ultimately, then, the future of this litigation, in terms of both what will and what should happen, takes place in a spatially and temporally multiscalar context. How quickly climate change creates impacts people care about and how different levels of governance around the world respond will determine what suits people are motivated to bring and their likelihood of success. The less effective we are in addressing the issue of climate change, the more salient these actions will become.[46]

Although significant uncertainty continues to surround that interaction, the current regulatory gaps suggest that the impacts of climate change will likely become more significant before our regulatory efforts catch up to the problem, assuming optimistically that they eventually will.[47] In the near term, then, these suits probably

---

[42] For analyses of some of the intergenerational complexities of addressing climate change, see EDWARD A. PAGE, CLIMATE CHANGE, JUSTICE AND FUTURE GENERATIONS (2006); Burns H. Weston, *Climate Change and Intergenerational Justice: Foundational Reflections*, 9 VT J. ENVTL. L. 375 (2008).

[43] For example, Stern's chapter focuses on an action that took place in the mid-1990s but that has ongoing implications. *See* Stern, *supra* note 17.

[44] Some of the nuisance suits described in the chapters by David Grossman and Jeffrey Stempel, for instance, are currently pending. *See* David A. Grossman, *Tort-Based Climate Litigation*, in this volume; Jeffrey W. Stempel, *Insurance and Climate Change Litigation*, in this volume.

[45] For example, the potential actions described in chapters by William Burns; Jennifer Gleason and David Hunter; Andrew Strauss; and Mary Wood have yet to be filed. *See* William C. G. Burns, *Potential Causes of Action for Climate Change Impacts under the United Nations Fish Stocks Agreement*, *supra*; Jennifer Gleason & David B. Hunter, *Bringing Climate Change Claims to the Accountability Mechanisms of International Financial Institutions*, in this volume; Andrew Strauss, *Climate Change Litigation: Opening the Door to the International Court of Justice*, in this volume; Mary Christina Wood, *Atmospheric Trust Litigation*, in this volume.

[46] Even those more skeptical of the value of this litigation acknowledge that regulatory failures are creating a context for such actions. *See* Posner, *supra* note 29, at 1925.

[47] *See* Burns & Osofsky, *supra* note 32.

will continue the explosive growth – accompanied by some formal successes – that we have seen over the past several years. Moreover, as courts continue to set precedents and the problem evolves, particular legal strategies will become more or less effective.

Even if regulatory efforts improve, as appears more likely in the United States since the 2008 elections, these suits may still remain an important lever within transnational regulation of climate change. Their ability to rescale and to connect people across scales, both spatial and temporal, makes them an important piece in an ongoing regulatory dialogue. These lawsuits provide unique opportunities for people to raise concerns and serve as an impetus for action. Unless widespread agreement exists on an appropriate crosscutting regulatory solution to this problem, courts and other fora likely will remain a key space in which people contest and create climate regulation.

Index

AB32, 31, 39
Access to Environmental Information Act, 24
ACF. *See* Australian Conservation Foundation
ACIA. *See* Arctic Climate Impact Assessment
acidity, of oceans, 318–320
ad hoc expert panel, 331
Ad Hoc Working Group on Long-term Cooperative Action, 19
adder approach, 33
Administrative Procedure Act (APA), 164, 334
African Charter on Human and People's Rights, 179, 183
African Development Bank, 310–311
Agenda for Global Growth and Stability, 13
AGIP, 179
Alaska Center for the Environment v. Browner, 122
Albania, 301
albedo, 88
Alexander v. Chattahoochee Valley Community College, 248
algae, 5
ALI. *See* American Law Institute
Alien Tort Claims Act, 178, 190
American Convention on Human Rights, 186
American Declaration of the Rights and Duties of Man, 25, 281
American Law Institute (ALI), 351
Antarctic, 7
Anvil Hill, 59
Aoki, Keith, 282
APA. *See* Administrative Procedure Act
Arctic Climate Impact Assessment (ACIA), 154, 281, 361
Argentina, 306, 332
Arizona Center for Law in the Public Interest v. Hassell, 104
asbestos, 118, 225, 248–249
Asian Development Bank, 292, 310, 311
Asian Haze, 352

Asia-Pacific Partnership on Clean Development and Climate, 12
assumption of risk, 205, 206
Aswan Dam, 258
atmospheric trust litigation, 99–125
   carbon accounting, 115–118
   carbon fiduciary obligation, 109–113
   carbon orphan shares, 113–114
   collateral benefits of, 122–124
   co-tenancy, 106–107
   declaratory relief, 115
   enforcement and, 114–124
   injunctive relief, 121–122
   nested jurisdictions, 118–119
   public trust assets, 108–109
   public trust law, 101–104
   res of the trust, 104–106
   in United States, 112
Australia, 362
   actions in, 23
   Anvil Hill, 59
   brown coal in, 49–50
   coal in, 48–50
   Commonwealth v. Tasmania, 261
   EPBC Act, 53
   greenhouse gases in, 48–49
   Hazelwood Power Station, 50–55
   Isaac Plains Mine, 55–59
   Kyoto Protocol and, 50, 256, 340–341
   Sonoma Mine, 55–59
   standing to sue in, 63–64
Australian Conservation Foundation (ACF), 23, 52, 362
Australian Conservation Foundation v. Latrobe City Council, 23
Australian Conservation Foundation v. Minister for Planning, 362
Austria, 341–342
awareness building, 122, 357
Axelrod, Robert, 95

Baird, Douglas G., 95
Baku-Tbilisi-Ceyhan pipeline, 294, 303–304
Bali Action Plan, 13, 19
Balling, Robert, 36
Bangladesh, 339
Bank Procedures (BPs), of World Bank, 298
Barron, David, 87
Belgium, 341–342
Berlusconi, Silvio, 16
best available science, 158, 165
bicarbonate ions (HCO_3^{-1}), 318
Bill of Rights, 182
binding resolution
   ICJ and, 342–343
   UNFSA and, 325
biodiversity, 145–172
biological effects, 6
Blair, Tony, 94
Blank, Yishai, 85
Bowel Coal, 56
Boxer, Barbara, 277
BPs. *See* Bank Procedures
breach of duty, 203–206
Britain, 341–342
brown coal
   in Australia, 49–50
   CO_2 from, 49
   Hazelwood Power Station and, 50–51
BUND, 24
burden of proof, 131
Burns, Wil, 354
Bush, George W., 11, 90, 276
   greenhouse gases and, 13
Buss v. Superior Court, 246

CAA. *See* Clean Air Act
CAFE. *See* corporate average fuel economy
California, 42, 141–142
   AB32 in, 31, 39
   CO_2 and, 15
   public nuisance and, 196
   San Bernardino County in, 74
California Global Warming Solutions Act of 2006, 76
California v. General Motors Corp., 196, 204, 334, 359
   justiciability in, 214
   liability insurance and, 231
   pollution exclusion and, 244–245
   relief in, 223
CAN. *See* Climate Action Network
Canada, 16
   actions in, 22
   Inuit Circumpolar Conference and, 276
   Kyoto Protocol and, 341–342, 364
   radioactive waste in, 186
   UNFCCC and, 364
Canadian Environmental Protection Act, 22, 364
CAO. *See* Compliance Advisor and Ombudsman
cap-and-trade program, 46
carbon accounting, 118
carbon budget, 119–121
carbon dioxide (CO_2), 3
   from brown coal, 49
   California and, 15
   cement and, 319
   from coal, 4
   from crude oil, 4
   electricity and, 32
   environmental cost value for, 36
   fossil fuels and, 3–4, 319
   Hazelwood Power Station and, 51
   IFC and, 308
   natural gas and, 300
   oceans and, 8, 317
   from transportation, 80
   United States and, 80
   urban politics and, 74, 78
carbon fiduciary obligation, 109–113
carbon orphan shares, 113–114
carbon trading, 18
carbonate ions (CO_3^{2-}), 318
carbonic acid (H_2CO_3), 318
Carson, Rachel, 355
The Case Concerning Oil Platforms, 346
causation
   generic, 216–219
   proximate, 219–222
   specific, 216–219
   in tort-based litigation, 216–222
   UNFSA and, 326–331
CDM. *See* Clean Development Mechanism
cement, 3–4
   CO_2 and, 319
Central Valley Chrysler-Jeep v. Witherspoon, 334
Centennial Coal, 59
Center for Biological Diversity, 22, 152, 162
Center for Biological Diversity v. Brennan, 162
Center for Clean Air Policy, 81
Center for International Environmental Law, 273
CEQ. *See* Council on Environmental Quality
CH_4. *See* methane
Chad-Cameroon pipeline, 294, 297
   IFC and, 303–304
   MIGA and, 303–304
Chernobyl, 352
Chevron, 179
Chile, 306, 332
China, 18
   coal-fired plants in, 6, 62

energy demands in, 6
fossil fuels and, 17–18
greenhouse gases and, 315
UNFSA and, 315
Ciborowski, Peter, 37
City of Bloomington, Ind. v. Westinghouse Electric Corp., 219
City of Milwaukee v. State, 105, 212
class action, 118
Clean Air Act (CAA), 20–21, 101, 212
  Massachusetts v. EPA and, 134
  public nuisance and, 213
Clean Water Act (CWA), 212
Clear Skies Initiative, 11
Climate Action Network (CAN), 371
Climate Change Initiative, 95
Clinton, Bill, 94, 365
Clinton Foundation, 95
CO_2. *See* carbon dioxide
$CO_3{}^{2-}$. *See* carbonate ions
coal. *See also* brown coal
  in Australia, 48–50
  CO_2 from, 4
  electricity and, 12
  mines, 48–71
coal-fired plants
  in China, 62
  greenhouse gases and, 62
  in India, 62
  Kyoto Protocol and, 62
  steady power flow from, 44
  in United States, 6, 62
coastal erosion, 6
coastal flooding, 9
cod, 316
Colombia, 186
Columbia River, 120
Comer v. Murphy Oil, 334
Comer v. Nationwide Mutual Insurance, 21–22
command-and-control functions, 87–89
common law
  federal, 210–213
  preemption and, 210–214
  state, 213–214
common pool resources, 73, 80
Commonwealth v. Tasmania, 261
compact development, 92
comparative fault, 205
Compliance Advisor and Ombudsman (CAO), 293, 303–309
Compliance Review Panel (CRP), 310
Conference of the Parties (CoP), 365
Connecticut v. American Electric Power Company, 171, 195, 204, 334, 359
  justiciability in, 214

pollution exclusion and, 245
relief in, 223
constructivism, 93–95, 380
consultation process, in ESA, 167–170
consumer expectation test, 202
contributory negligence, 205
Convention on Long-Range Transboundary Air Pollution, 354
CoP. *See* Conference of the Parties
copepod, 317
coral reefs, 7–8
  ESA and, 149–154
  fish and, 8
corporate average fuel economy (CAFE), 168
corporations, multinational, 188, 190
corpus. *See* res of the trust
co-tenancy, 106–107
Council on Environmental Quality (CEQ), 67
County of Oneida v. Oneida Indian Nation of New York, 212
critical habitat, 148, 166
CRP. *See* Compliance Review Panel
crude oil, 4
Center for Biological Diversity v. Norton, 334
Cummings, Brendan, 26, 381
CWA. *See* Clean Water Act

damages
  apportionment of, 227–229
  liability for, 227–229
  restrictions on, 226–227
  significant, 234
  standards for, 223–224
  types of, 224–226
danger list, 255
deaths, 5
Declaration of the Rights of Man, 182
declaratory relief, 115
defenses, for tort-based litigation, 203–206
deforestation. *See* forests
Democratic Party, 91
democratization, 370
Denmark, 345–346
  Kyoto Protocol and, 341–342
Department of Energy, 117
Department of Environmental Protection v. Jersey Central Power & Light Co., 105
design defects
  products liability and, 201–203
  risk-benefit test for, 202
diagonal regulation, 383
dialectical regulation, 274, 283–286
Directors and Officers liability insurance (D&O), 244
disease, 5, 9

distal water fishing nations (DWFNs), 316
D&O. *See* Directors and Officers liability insurance
Doremus, Holly, 130, 131
downstream greenhouse gases, 48, 63
duty to cooperate, 329
duty to defend, 235, 236–237
duty to warn, 204
DWFNs. *See* distal water fishing nations
Dwyer, John, 40

EA Policy. *See* Environmental Assessment OP
Earth Summit. *See* United Nations Conference on Environment and Development
Earthjustice, 273
ECHR. *See* European Court of Human Rights
ecological processes, 52–53
Ecuador, 186
Edelman, Murray, 39
EE Act. *See* Environmental Effects Act
EES. *See* Environmental Effects Statement
EEZs. *See* Exclusive Economic Zones
EHS. *See* Environmental Health and Safety
EIS. *See* environmental impact statement
electricity
  CO_2 and, 32
  coal and, 12
  in United States, 12
elkhorn coral, 149–154
Endangered Species Act (ESA), 22, 26, 101
  consultation process in, 167–170
  coral reefs and, 149–154
  greenhouse gases and, 145–172
  jeopardy and, 167–170
  NGOs and, 149
  polar bears and, 154–162, 381
  take prohibition in, 170–171
energy demands, 6, 302
Energy Information Agency, 18
enforcement
  atmospheric trust litigation and, 114–124
  of carbon budget, 119–121
Engel, Kirsten, 40
Environment Court, 72–98
Environment Protection and Biodiversity Conservation Act (EPBC Act), 53, 56
Environmental Assessment OP (EA Policy), 298
Environmental Assessment Sourcebook, 300
environmental cost valuation
  for CO_2, 36
  for lead, 36
  in Minnesota, 31–47
  for nitrogen oxide, 36
  for sulfur dioxide, 36
Environmental Defence Society (Inc.) v. Auckland Regional Council, 79
Environmental Effects Act (EE Act), 51–52
Environmental Effects Statement (EES), 51
Environmental Health and Safety (EHS), 308
environmental impact, 61–70
Environmental Impact Assessment Act, 187
environmental impact statement (EIS), 65
Environmental Planning & Assessment Act (EP&A), 59
  indirect greenhouse gases and, 61
Environmental Protection Agency (EPA), 20–21. *See also Massachusetts v. EPA*
  New York v. EPA, 334
environmental risk, 230
environmental services, 117
EPA. *See* Environmental Protection Agency
EP&A. *See* Environmental Planning & Assessment Act
EPBC Act. *See* Environment Protection and Biodiversity Conservation Act
epistemic community, 365
Equal Protection Clause, 184
ESA. *See* Endangered Species Act
Ethiopia, 345–346
EU. *See* European Union
European Bank for Reconstruction and Development, 293, 310–311
European Convention on Human Rights and Fundamental Freedoms, 188
European Court of Human Rights (ECHR), 187, 188
European Union (EU), 13
  Kyoto Protocol and, 16
  UNFSA and, 315
Exclusive Economic Zones (EEZs), 314
  UNCLOS and, 322
expected injury, 241–242
experts, 110
  ad hoc panel of, 331
Export Credits Guarantee Department, 292
Export Development Canada, 310–311
Export-Import Bank, 21, 65, 292
Exxon Mobil, 184

fair-share principle, 116
  Kyoto Protocol and, 111
FCN. *See* Friendship, Commerce and Navigation
Federal Ministry for Environment and Heritage, 55–59
First Assessment Report, 38
Fischel, William, 83
fish, 5. *See also* United Nations Fish Stocks Agreement

coral reefs and, 8
technology and, 322
Fish and Wildlife Service (FWS), 145
flaring. *See* natural gas flaring
flood insurance, 231
Food and Agriculture Organization (FAO), 322, 331, 349
food production, 9–10
Ford, Richard, 282
foreseeable future, 164
foreseeable risk, 202
forests, 4, 8
Fort Mojave Indian Tribe v. United States, 105
fortuity, 241–242
fossil fuels, 3. *See also specific fuels*
    China and, 17–18
    CO_2 and, 3–4, 319
    IFC and, 303–304
    MIGA and, 303–304
France, 182
free riders, 83
freedom of information, 187
Friends of the Earth Canada, 21, 22, 65, 206, 334, 362, 372
Friends of the Earth v. Laidlaw Environmental Services, 207
Friends of the Earth v. Mosbacher, 362, 372
Friends of the Earth v. Watson, 21, 65, 334
Friendship, Commerce and Navigation (FCN), 345
Frug, Gerald, 87, 282
FWS. *See* Fish and Wildlife Service

G8, 13, 14
G77, 19
Garber v. Whittaker, 116
Gbemre, Jonah, 179
GCI. *See* Global Climate Initiative
Geer v. Connecticut, 102, 108
GEF. *See* Global Environment Facility
general liability insurance (CGL). *See* liability insurance
generic causation, 216–219
Genesis Power Ltd. and the Energy Efficiency and Conservation Authority v. Franklin District Council, 23, 72–98, 334
    background of, 76–80
    United States and, 74–75
    wind turbines and, 76
Georgia v. Tennessee Copper Co., 108
GermanWatch, 23–24, 334
GermanWatch v. Euler Hermes AG, 334
Germany, 362
    Access to Environmental Information Act of, 24
    actions in, 23–24

Glacier National Park, 269
glaciers, 6, 269
Gleason, Jennifer, 27
Global Change Research Act of 1990, 162
Global Climate Initiative (GCI), 11
Global Environment Facility (GEF), 294
    UNFCCC and, 294
Gore, Al, 355, 357, 375
Gray, Peter, 60
Gray v. Minister for Planning, 23
Great Barrier Reef World Heritage Area, 53, 57, 255
Greece, 345
Greenhouse Gas Reduction Deed, 54, 381
greenhouse gases, 2, 3. *See also* indirect greenhouse gases
    in Australia, 48–49
    Bush and, 13
    cap-and-trade program for, 46
    China and, 315
    coal-fired plants and, 62
    downstream, 48
    ESA and, 145–172
    Hazelwood Power Station and, 51
    IFC and, 308
    indirect, 48
    intensity, 11
    OPEC and, 10
    as planning issue, 52
    quantification of, 117
    Republic Party and, 91–92
    UNFCCC and, 353
    United States and, 10, 13, 280, 315
    upstream, 63
Greenland, 7, 276
Greenpeace New Zealand v. Northland Regional Council and Mighty River Power Limited, 22–23
Grossman, David, 26
Guerra v. Italy, 187

H_2CO_3. *See* carbonic acid
H_2O. *See* water vapor
haddock, 316
Hafetz, Jonathan, 326
Hall, Dale, 161
Hamilton, Bruce, 83
handguns, 199, 204
Hardin, Garrett, 73, 78
Hazelwood Power Station, 50–55
    brown coal and, 50–51
    CO_2 and, 51
    greenhouse gases and, 51
HCO_3^{-1}. *See* bicarbonate ions
Hensler, Deborah, 248

Her Majesty v. City of Detroit, 109
herring, 316
Hirschman, Albert, 41
honesty-in-pleading requirements, 249
Huascarán National Park, 255
human rights, 173–192. *See also specific human rights organizations*
  in Colombia, 186
  in India, 186
  Inuit Petition and, 281–282
  tradition of, 181–185
Hunter, David, 27, 122, 375
Hurricane Katrina, 196, 230

IBRD. *See* International Bank for Reconstruction and Development
Iceland, 341–342
ICJ. *See* International Court of Justice
ICLEI. *See* International Council on Local Environmental Initiatives
ICSID. *See* International Center for the Settlement of Investment Disputes
IDA. *See* International Development Association
Idaho Forest Industry v. Hayden Lake Watershed Improvement District, 105
IEA. *See* International Energy Agency
IFC. *See* International Finance Corporation
IFIs. *See* international financial institutions
Illinois Central Railroad Co. v. Illinois, 103
Illinois v. City of Milwaukee, 211
In the Matter of Quantification of Environmental Costs, 32, 36
An Inconvenient Truth (Gore), 357
independent treaty, 345–347
India, 18
  coal-fired plants in, 6, 62
  energy demands in, 6
  human rights in, 186
indigenous peoples, 284–285. *See also* Inuit
indirect greenhouse gases, 48, 68–70
  EP&A and, 61
  Nathan Dam and, 59
Industrial Revolution, 3
industry practice, 205
initial policy determination, 215
injunctive relief, 120, 121–122
  standards for, 223–224
institutionalism, 95–97
  Mayor's Climate Change Protection Agreement and, 95
insurance, 230–251
intentional injury, 241–242
Inter-American Commission on Human Rights, 25, 174, 186, 279, 369, 378
  Inuit Petition and, 272

Inter-American Development Bank, 293, 310–311
Intergovernmental Oceanographic Commission, 332
Intergovernmental Panel on Climate Change (IPCC), 2, 265, 361
  criticism of, 36
  First Assessment Report of, 38
  Minnesota and, 43
  Nobel Peace Prize to, 355, 375
  Special Report on Emissions Scenarios of, 319
  United Nations Environment Programme and, 36
  USGS and, 160
  World Meteorological Organization and, 36
International Bank for Reconstruction and Development (IBRD), 295
International Center for the Settlement of Investment Disputes (ICSID), 295
International Council for Local Environmental Initiatives Climate Protection Campaign, 279
International Council on Local Environmental Initiatives (ICLEI), 95
International Court of Justice (ICJ), 27, 334–356
  advisory opinions, 348–350
  binding resolution and, 342–343
  jurisdiction of, 345–347
  Kyoto Protocol and, 339
  standing to sue and, 347
  United States and, 341, 345
International Covenant on Civil and Political Rights, 183, 186
International Development Association (IDA), 295
International Energy Agency (IEA), 6
international environmental lawmaking, 357–374
International Finance Corporation (IFC), 293, 295
  Baku-Tbilisi-Ceyhan pipeline and, 303–304
  CAO of, 303–309
  Chad-Cameroon pipeline and, 303–304
  CO_2 and, 308
  EHS of, 308
  fossil fuels and, 303–304
  greenhouse gases and, 308
  Performance Standards of, 306
  Policy on Social and Environmental Sustainability of, 306
  soft law declarations, 353–355
  treaties, 353–355
international financial institutions (IFIs), 292–313
International Law Commission, 351–352
International Liability for Injurious Consequences Arising from Hazardous Activities, 351

international local government law, 87
International Paper Co. v. Ouelette, 213
International Power, 50–51
international relations theory, 85
  constructivism in, 93–95
  structural realism in, 86
International Tribunal for the Law of the Sea (ITLOS), 326, 331
International Union for Conservation of Nature (IUCN), 163
Inuit Circumpolar Conference, 174, 276, 381
Inuit Petition, 25, 272–290, 367
  geography of, 275–282
  human rights and, 281–282
  Inter-American Commission on Human Rights and, 272
IPCC. *See* Intergovernmental Panel on Climate Change
Ireland, 341–342, 345–346
Isaac Plains Mine, 55–59
Islamic Republic of Iran v. United States, 346
Italy, 341–342
ITLOS. *See* International Tribunal for the Law of the Sea
IUCN. *See* International Union for Conservation of Nature
Izaak Walton League, 35

Japan
  as DWFN, 316
  Kyoto Protocol and, 16
Japan Bank of International Cooperation, 310–311
jeopardy, 147
  ESA and, 167–170
JI. *See* Joint Implementation
Jicarilla Apache Tribe v. Supron Energy Corp., 104
joint action theory, 190
Joint Implementation (JI), 294
jurisdiction
  of ICJ, 345–347
  by independent treaty, 345–347
  justiciability and, 214–216
  nested, 118–119
  preemption and, 210–214
  tort-based litigation and, 206–216
justiciability, 214–216

Kantor, Paul, 87
Keck, Margaret, 93
Kempthorne, Dirk, 161
Kittlitz's murrelet, 165
Klein, Allan, 32, 35
Kyoto Protocol, 11, 341–342
  Annex I Parties and, 17, 268, 341–342, 344
  Australia and, 50, 256, 340–341
  Canada and, 364
  Clean Development Mechanism (CDM), 294
  coal-fired plants and, 62
  EU and, 16
  fair-share principle and, 111
  ICJ and, 339
  Japan and, 16
  MoP to, 365
  Turkey and, 341
  UNFCCC and, 341
  United States and, 11, 18, 256, 340–341
  WHC and, 265
  World Bank and, 302

Laborers Local 17 Health & Benefit Fund v. Philip Morris, Inc., 219
Lake Mich. Federation v. U.S. Army Corps of Engineers, 100
land use
  in United States, 82
  urban politics and, 72–98
Large Cities Climate Change Leadership Group, 95
Latrobe Planning Scheme, 51
lead, 36
Leonard v. Nationwide Ins. Co., 231
liability insurance, 231–235
  California v. General Motors Corp. and, 231
  defense cost recoupment and, 245–246
  duty to defend, 236–237
  economic implications of, 246–251
  expected injury and, 241–242
  fortuity and, 241–242
  history of, 235–236
  intentional injury and, 241–242
  occurrences and, 241–242
  political implications of, 246–251
  pollution exclusion in, 245
  structure of, 235–236
  triggering of, 239–241
liberalism, 90–93
Lieberman-Warner Climate Security Act, 15
Liechtenstein, 341–342
Lindzen, Richard, 36
litigation insurance, 235
Livingstone, Ken, 94
local governments. *See* urban politics
Lopez-Ostra v. Spain, 187
Luntz, Frank, 132

Maheu, René, 258
major federal actions, 62
malnutrition, 5
marine areas, 56
marine scientific research (MSR), 329

market share theory of liability, 247
Massachusetts v. EPA, 20–21, 26, 65, 127–144, 331, 336, 362, 364, 379
  actors in, 134–135
  CAA and, 134
  claims of, 135–140
  domestic vs. international, 140–141
  international implications of, 140–143
  local vs. state vs. federal, 141–142
  public vs. private, 142–143
  scale vs. science in, 134
  sea levels and, 137
  standing to sue in, 135–138, 207
  substantive claims in, 138–140
Mattoon v. City of Pittsfield, 213
Mayagna (Sumo) Awas Tingni Community v. Nicaragua, 283
Mayor's Climate Change Protection Agreement, 84, 94
  institutionalism and, 95
McAllister, Lesley, 26, 381
MDBs. See multilateral development banks
Meetings of the Parties (MoP), 365
methane (CH_4), 3, 5
  natural gas and, 300
Michaels, Pat, 36
Michie v. Great Lakes Steel Division, 227
MIGA. See Multilateral Investment Guarantee Agency
migratory species, 56
Millennium Ecosystem Assessment Series, 259
Minnesota
  environmental cost value regulation in, 31–47
  IPCC and, 43
  Minnesota Court of Appeals, 38–39
  Minnesota Pollution Control Agency (MPCA), 35, 37
  Minnesota Public Utilities Commission, 32, 35
  Minnesota Supreme Court, 43
Minnesotans for an Energy Efficient Environment, 35
monk seals, 146
MoP. See Meetings of the Parties
MPCA. See Minnesota Pollution Control Agency
MSR. See marine scientific research
multilateral development banks (MDBs), 292, 294
Multilateral Investment Guarantee Agency (MIGA), 295, 303–309
  Baku-Tbilisi-Ceyhan pipeline and, 303–304
  Chad-Cameroon pipeline and, 303–304
  fossil fuels and, 303–304
multinational corporations, 188, 190
multiscalar, 376
municipal cost recovery rule, 227
musical suburbs, 83

NAACP v. Township of Mount Laurel, 120
NAAQS. See National Ambient Air Quality Standards
NARUC. See National Association of Regulatory Commissioners
Nathan Dam, 58
  indirect greenhouse gases and, 59
National Ambient Air Quality Standards (NAAQS), 138
National Association of Regulatory Commissioners (NARUC), 43
National Environmental Policy Act (NEPA), 21, 62, 101, 334
  CEQ and, 67
national heritage places, 56
National Marine Fisheries Service (NMFS), 145
natural capital, 117
natural gas
  CH_4 and, 300
  CO_2 and, 300
natural gas flaring, 24
  in Nigeria, 173–192
  in United States, 176
natural greenhouse effect, 3
negligence, 203–206
  contributory, 205
  state of the art and, 205
NEPA. See National Environmental Policy Act
Nepal, 255
nested jurisdictions, 118–119
Netherlands, 345–346
Netherlands Environmental Assessment Agency, 17
New South Wales Land and Environment Court, 59
New Urbanists, 81
  compact development and, 92
New York v. EPA, 334
New Zealand, 22–23
  Environment Court of, 72–98
  Genesis Power Ltd. and the Energy Efficiency and Conservation Authority v. Franklin District Council, 72–98
  Kyoto Protocol and, 341–342
  Resource Management Act of, 73, 75
Newman, Peter, 87
Newmont Corporation, 305
Newsom, Gavin, 94
NGOs. See nongovernmental organizations
Nicaragua, 283, 345
Nickels, Greg, 119
Niger Delta, 175–178
  oil spills in, 176
Nigeria
  actions in, 24

Environmental Impact Assessment Act of, 187
  natural gas flaring in, 173–192
  OPEC and, 176
  WAGP in, 297
Nigerian National Petroleum Corporation
    (NNPC), 175, 179
nitrogen oxide, 36
nitrous oxide, 4–5
NMFS. *See* National Marine Fisheries Service
NNPC. *See* Nigerian National Petroleum
    Corporation
nonderivative harms, 226
nongovernmental organizations (NGOs), 18. *See
    also specific organizations*
  ESA and, 149
  World Heritage Committee and, 255
Nonintercourse Act of 1793, 212
Northern legal systems, 287–289
Norway, 341–342
nuclear actions, 56
nuisance. *See* public nuisance
Nw. Envtl. Def. Ctr. v. Owens Corning Corp., 362

O_3. *See* ozone
OAS. *See* Organization of American States
Obama, Barack, 13, 14, 19, 31, 75–76, 130, 144, 172,
    211, 273, 276, 278, 341, 355, 373, 375, 379, 381,
    382
obligation *erga omnes*, 347
occurrences, liability insurance and, 241–242
oceans, 5. *See also* sea levels
  acidity of, 318–320
  CO_2 and, 8, 317
OECD Guidelines for Multinational Enterprises,
    23–24
oil spills, 176
O'Neill, Tip, 92
Opasa v. Factoran, 101
OPEC
  greenhouse gas and, 10
  Nigeria and, 176
Operating Procedures (OPs), 298
OPIC. *See* Overseas Private Investment
    Corporation
OPs. *See* Operating Procedures
Organization of American States (OAS), 279, 378
Organization of Economic Cooperation and
    Development, 11
orphan share. *See* carbon orphan shares
Osofsky, Hari, 26, 27, 383
Ostrom, Elinor, 80
other means of peaceful settlement, 343
OTL. *See* owner's, landlord's, and tenant's
    insurance
outstanding universal values, 258, 265, 269

Overseas Private Investment Corporation (OPIC),
    21, 65, 310–311
owner's, landlord's, and tenant's insurance
    (OTL), 235
ozone (O_3), 3

*Pacific Coast Federation of Fishermen's
    Associations v. U.S. Bureau of Reclamation*,
    110
Pachauri, Rajendra, 265
Pacific island developing countries (PIDCs), 7–8
pacta sunt servanda, 262, 329
Pangue Dam, 306
PE Act. *See* Planning and Environmental Act of
    1987
Pelosi, Nancy, 277
Performance Standards, 306
permafrost, 5, 281
Peru
  Huascarán National Park in, 255
  Yanacocha gold mine in, 305
Peterson, Paul, 86
pH, 318–320
Philippines, 101
Phoenix, Arizona, 119
PIDCs. *See* Pacific island developing countries
PL. *See* public liability insurance
plankton, 5, 317
Planning and Environmental Act of 1987
    (PE Act), 50
Poland, 316
polar bears, 145
  ESA and, 154–162, 381
  as threatened species, 162
Policy on Social and Environmental
    Sustainability, 306
political question doctrine, 214
pollution exclusion
  Connecticut v. American Electric Power Co.
    and, 245
  in liability insurance, 245
  State of California v. General Motors Corp.
    and, 244–245
Pollution Prevention and Abatement Handbook
    (World Bank), 301
popsicle test, 82
Porter, Michael, 86
Portland, Oregon, 84
Portugal, 341–342
Power Sector Generation & Reconstruction, 301
precautionary principle, 328–329
preemption
  common law and, 210–214
  in tort-based litigation, 210–214

Prevention of Transboundary Damage from
    Hazardous Activities, 351
prisoner's dilemma, 95–97
private city, 87
products liability, 199–206
    design defects and, 201–203
    warning defects and, 200–201
property taxes, 82
proximate causation, 219, 222
    public nuisance and, 220
pteropods, 320
public goods, 41, 102
    urban politics and, 82
public liability insurance (PL), 235
public nuisance, 195–199
    CAA and, 213
    California and, 196
    proximate causation and, 220
public trust assets, 108–109
public trust law, 101–104
public vs. private, 142–143
    Inuit Petition and, 286–287

QCoal, 56
quasi-realism, 86–90
Queensland, 55–59

Rabe, Barry, 42
race to the top, 42
radioactive waste, 186
Ratner, Steve, 190
reciprocity, rule of, 340
recovery plan. *See* carbon budget
relief
    declaratory, 115
    injunctive, 120, 121–122, 223–224
    in tort-based litigation, 222–229
*Report on the State of Conservation of Waterton
    Glacier International Peace Park,*
    269
Republican Party, 91–92
res communes, 108
res of the trust, 104–106
Resnik, Judith, 41
Resource Management Act, 73, 75
Restatement of Torts, 111, 196, 219
    UNFSA and, 330
Rio Declaration, 268, 354
risk
    assumption of, 205, 206
    environmental, 230
    foreseeable, 202
risk-benefit test, for design defects, 202
Roderick, Peter, 375
Ross Sea, 320

Royal Dutch/Shell Group, 173–192
    profits of, 184
rule of reciprocity, 340
Russia, 18
    Inuit Circumpolar Conference and, 276

Sagarmatha National Park, 255
Saleska, Scott, 40
salmon, 120, 147
San Bernardino County, 74
Sandoz Chemical Fire, 352
Sassen, Saskia, 87
Savitch, H.V., 87
Sayre, Nathan, 130, 131, 132
scale, 130, 285
    debates over, 132–133
    as lens on science and law, 133
    vs. science, 134
    scientific uncertainty and, 138
scale-dependent, 376
scientific uncertainty, 131–132
    scale and, 138
sea levels, 7, 359
    coastal flooding and, 9
    Massachusetts v. EPA and, 137
Seattle, Washington, 118, 119–120
Seminole Nation v. United States, 104
shared interests. *See* co-tenancy
Shell. *See* Royal Dutch/Shell Group
Shell Petroleum Development Company, 190
Siegel, Kassie, 26, 381
Sierra Club, 84
significant damage, 234
Sikkink, Kathryn, 93
Silent Spring (Carson), 355
Silk, Richard, 332
Sindell v. Abbott Labs, 348
Sinden, Amy, 26
smart growth, 91
snail darter, 146
soft law declarations, 353–355
solar radiation, 2
Somalia, 341
Sonoma Mine, 55–59
South Korea, 345–346
    as DWFN, 316
Spain
    as DWFN, 316
    Kyoto Protocol and, 341–342
Special Project Facilitator (SPF), 310
Special Report on Emissions Scenarios (IPCC),
    319
special solicitude, 207, 209
species. *See also* Endangered Species Act
    extinction, 8–9

migratory, 56
threatened, 56, 162
specific causation, 216–219
SPF. *See* Special Project Facilitator
staghorn coral, 149–154
standing to sue, 63–66
  in Australia, 63–64
  ICJ and, 347
  in *Massachusetts v. EPA*, 135–138, 207
  in tort-based litigation, 206–210
  in United States, 64–65
state action
  by Minnesota, 31–47
  problem with, 188–191
  variation in, 46
State Electricity Commission of Victoria, 51
State of California v. General Motors Corp., 21
State of Connecticut v. American Electric Power Co., 21
state of the art, 206
  negligence and, 205
State Parties, 258–259, 261–264
  UNFSA and, 324
  United States and, 268
State v. City of Bowling Green, 105
Stempel, Jeff, 26
Stern Review on the Economics of Climate Change, 10
Stern, Stephanie, 26
Stockholm Declaration, 354
Straddling Fish Stocks Agreement, 354
Strauss, Andrew, 27, 332
structural realism
  in international relations theory, 86
  urban politics and, 86
substantiality, 219–222
substantive claims, 138–140
sulfur dioxide, 36
Summer v. Tice, 348
Sunder, Madhavi, 282
surface temperatures, 2
Sweden, 341–342
symbolic regulation, 31, 39–42

Taiwan, 316
take prohibition, 170–171
Talaka, Koloa, 339
TANs. *See* transnational advocacy networks
Target for U.S. Emissions Reductions, 110
  carbon accounting and, 118
Te Iwi O Ngati Te Ata, 76
Tennessee Valley Authority v. Hill, 146
terra nullius (empty lands), 284

Thailand, 345–346
Thornley, Andy, 87
Thorson, Erica Jayne, 27
threatened species, 56
  polar bears as, 162
Thucydides, 85
Tiebout, Charles, 82
Time (magazine), 355
tort-based litigation, 193–230
  Alien Tort Claims Act, 178, 190
  breach of duty, 203–206
  causation in, 216–222
  defenses, 203–206
  jurisdiction and, 206–216
  negligence, 203–206
  preemption in, 210–214
  products liability, 199–206
  public nuisance, 195–199
  relief in, 222–229
  Restatement of Torts, 111, 196, 219, 330
  standing to sue in, 206–210
  substantiality in, 219–222
trading up, 42
tragedy of the commons, 73, 78
Trail Smelter arbitration, 352
transnational advocacy networks (TANs), 93, 94, 370–371
transnational legal process, 381
transportation
  CO_2 from, 80
  in United States, 80
travaux préparatoires, 263
treaties, 353–355
Treaty of Amity, Economic Relations, and Consular Rights, 346
triggering, of liability insurance, 239–241
Trisolini, Katherine, 26
trusts. *See* atmospheric trust litigation
tuna, 316
Turkey, 341
Tuvalu, 339

UNCLOS. *See* United Nations Convention on the Law of the Sea
UNCSFS. *See* United Nations Conference on Straddling Fish Stocks and Highly Migratory Fish Stocks
under color of law test, 189
UNESCO. *See* United Nations Educational, Scientific and Cultural Organization
UNFCCC. *See* United Nations Framework Convention on Climate Change
UNFSA. *See* United Nations Fish Stocks Agreement
Union of Concerned Scientists, 110

United Nations
  FAO, 322
  Rio Declaration of, 268, 354
  WHO and, 348–349
United Nations Conference on Environment and Development, 323, 354
United Nations Conference on Straddling Fish Stocks and Highly Migratory Fish Stocks (UNCSFS), 323
United Nations Convention on the Law of the Sea (UNCLOS), 314, 321–323
  EEZs and, 322
United Nations Educational, Scientific and Cultural Organization (UNESCO), 257
United Nations Environment Program, 2
  FAO, 332
  IPCC and, 36
  urban politics and, 96
United Nations Fish Stocks Agreement (UNFSA), 27, 314–333
  binding resolution and, 325
  causation and, 326–331
  China and, 315
  European Union and, 315
  Restatement of Torts and, 330
  State Parties and, 324
  United States and, 315
United Nations Framework Convention on Climate Change (UNFCCC), 10, 109, 361
  Annex I Parties, 17, 268, 341–342, 344
  Canada and, 364
  Carbon Finance Unit (CFU), 294
  CoP to, 365
  GEF and, 294
  greenhouse gas and, 353
  Kyoto Protocol and, 341
  World Heritage Commission and, 266
  World Bank and, 302
United Nations Human Rights Committee, 186
United States, 362
  actions in, 20–22
  atmospheric trust litigation in, 112
  Bill of Rights of, 182
  carbon trading and, 18
  CO_2 and, 80
  coal-fired plants in, 6, 62
  Department of Energy of, 117
  electricity in, 12
  Energy Information Agency of, 18
  Genesis Power Ltd. and the Energy Efficiency and Conservation Authority v. Franklin District Council and, 74–75
  greenhouse gas and, 10, 13, 280, 315
  ICJ and, 341, 345
  Inuit Circumpolar Conference and, 276
  Kyoto Protocol and, 11, 18, 256, 340–341
  land use in, 82
  Mayor's Climate Change Protection Agreement in, 84, 94
  natural gas flaring in, 176
  NEPA in, 62
  Nicaragua and, 345
  standing to sue in, 64–65
  State Parties and, 268
  transportation in, 80
  UNFSA and, 315
  WHC and, 267
United States v. 1.58 Acres of Land, 107
United States v. Gouveia, 184
United States v. Metro. Dist. Comm'n, 122
United States v. White Mountain Apache Tribe, 105
unreasonable injury, 195
upstream greenhouse gases, 63
uranium mining, 56
urban politics, 72–98
  CO_2 and, 74, 78
  land use and, 72–98
  liberalism and, 90–93
  public goods and, 82
  quasi-realism and, 86–90
  structural realism and, 86
  United Nations Environment Program and, 96
  VMT and, 83
Uruguay, 306
U.S. Geological Survey (USGS), 160
  IPCC and, 160
USGS. *See* U.S. Geological Survey

vehicle miles traveled (VMT), 81
  urban politics and, 83
Verheyen, Roda, 330
Victoria
  EE Act of, 51–52
  PE Act of, 50
Victorian Civil and Administrative Tribunal, 50
Vienna Convention, 262, 263
Villaraigosa, Antonio, 94
VMT. *See* vehicle miles traveled
Volkswagen, 23–24
Volpp, Leti, 282

WAGP. *See* West African natural gas pipeline
warning defects, 200–201
water vapor (H_2O), 3
Waterton-Glacier Peace Park, 256, 269–270
Watt-Cloutier, Sheila, 272, 381

West African natural gas pipeline (WAGP), 297
Western Fuels Association, 38
Westphalian model, 84–85
Wet Tropics World Heritage Area, 57
wetlands, 56
WHC. *See* World Heritage Convention
White House Council on Environmental Quality, 14
WHO. *See* World Health Organization
Who Killed the Electric Car? (film), 357
Wildlife Preservation Soc. of Queensland Proserpine/Whitsunday Branch v. Ministry for the Env't & Heritage, 362
Wildlife Preservation Society of Queensland (WPS), 55–59
Winberger v. Romero-Barcelo, 115
wind turbines, 76
Wiwa, Ken Saro, 178, 190
Wolfensohn, James, 304
Wood, Mary, 26
World Bank, 292, 294
    BPs of, 298
    CFU of, 294
    Environmental Assessment OP of, 298
    Kyoto Protocol and, 302
    OPs of, 298
    pipelines and, 294
    Pollution Prevention and Abatement Handbook of, 301
    UNFCCC and, 302
World Bank Inspection Panel, 292
    claims with, 295–303
World Energy Outlook (IEA), 6
World Health Organization (WHO), 5, 348
    United Nations and, 348–349
World Heritage Committee, 25, 363
    NGOs and, 255
World Heritage Convention (WHC), 255–271, 369
    Kyoto Protocol and, 265
    mitigation strategy of, 265–268
    UNFCCC and, 266
    United States and, 267
World Heritage List, 258
World Heritage properties, 56
World Meteorological Organization, 2, 36
World Resources Institute (WRI), 293
WPS. *See* Wildlife Preservation Society of Queensland
WRI. *See* World Resources Institute

Yanacocha gold mine, 305
Young v. Bryco Arms, 219

Zasloff, Jonathan, 26